Microorganisms: Genetic Diversification

Microorganisms: Genetic Diversification

Edited by **Nigel Hogan**

R CALLISTO REFERENCE

New York

Published by Callisto Reference,
106 Park Avenue, Suite 200,
New York, NY 10016, USA
www.callistoreference.com

Microorganisms: Genetic Diversification
Edited by Nigel Hogan

International Standard Book Number: 978-1-63239-459-0 (Hardback)

Contents

Preface

It is often said that books are a boon to mankind. They document every progress and pass on the knowledge from one generation to the other. They play a crucial role in our lives. Thus I was both excited and nervous while editing this book. I was pleased by the thought of being able to make a mark but I was also nervous to do it right because the future of students depends upon it. Hence, I took a few months to research further into the discipline, revise my knowledge and also explore some more aspects. Post this process, I begun with the editing of this book.

Microorganisms are a group of organisms that are characterized with the quality of being visible only with the help of a microscope. This book unveils new topics revealing the scale of genetic diversity of microorganisms existing in various environmental circumstances. The complexity and variety of microbial populations is by far the uppermost among all living organisms. The variety of microbial communities and their ecologic roles are being constantly analyzed in soil, water, plants and animals, and in tremendous environments such as the arctic deep-sea vents or high saline lakes. The growing accessibility of PCR-based molecular developers allows thorough researches and valuation of genetic range in microorganisms. The motive of the book is to offer a glance into the dynamic procedures of genetic diversity of microorganisms by providing the views of experts who are involved in creation of bright ideas and methods employed for the assessment of genetic diversity, often from different perspectives. This book should be helpful to amateurs and experts in the field of molecular biology.

I thank my publisher with all my heart for considering me worthy of this unparalleled opportunity and for showing unwavering faith in my skills. I would also like to thank the editorial team who worked closely with me at every step and contributed immensely towards the successful completion of this book. Last but not the least, I wish to thank my friends and colleagues for their support.

Editor

Part 1

Microbial Genetic Diversity

Diversity of Heterolobosea

Tomáš Pánek and Ivan Čepička
Charles University in Prague,
Czech Republic

1. Introduction

Heterolobosea is a small group of amoebae, amoeboflagellates and flagellates (ca. 140 described species). Since heterolobosean amoebae are highly reminiscent of naked lobose amoebae of Amoebozoa, they were for a long time treated as members of Rhizopoda (Levine, 1980). The class Heterolobosea was established in 1985 by Page and Blanton (Page & Blanton, 1985) by uniting unicellular Schizopyrenida with Acrasida that form multicellular bodies. Later, it was suggested that Heterolobosea might be related to Euglenozoa (e.g., *Trypanosoma*, *Euglena*) instead of other amoebae (Cavalier-Smith, 1998; Patterson, 1988). This assumption based on the cell structure was supported also by early multigene phylogenetic analyses (Baldauf et al., 2000). Currently, the Heterolobosea is nested together with Euglenozoa, Jakobida, Parabasalia, Fornicata, Preaxostyla, *Malawimonas*, and *Tsukubamonas* within the eukaryotic supergroup Excavata (Hampl et al., 2009; Rodríguez-Ezpeleta et al., 2007; Simpson, 2003; Yabuki et al., 2011). The excavate organisms were originally defined on the basis of the structure of flagellar system and ventral feeding groove (Simpson & Patterson, 1999). However, Heterolobosea have lost some of these structures (Simpson, 2003).

The most important heterolobosean taxon is the genus *Naegleria* as *N. fowleri* is a deadly parasite of humans (Visvesvara et al., 2007) and *N. gruberi* is a model organism in the research of assembly of the flagellar apparatus (Lee, 2010). Both the species have been studied in detail for decades and genome sequence of *N. gruberi* was recently published (Fritz-Laylin et al. 2010). On the other hand, the other heteroloboseans are considerably understudied and undescribed despite their enormous ecological and morphological diversity. Many heteroloboseans have adapted to various extreme environments; halophilic, acidophilic, thermophilic, and anaerobic representatives have been described. Few heteroloboseans are facultative endobionts of both vertebrates and invertebrates. *Naegleria fowleri* and *Paravahlkampfia francinae* are even able to parasitize humans (Visvesvara et al., 2007, 2009). The genus *Stephanopogon*, whose members are multiflagellate, was once considered to be a primitive ciliate and was affiliated with Heterolobosea only on the basis of cell structure and phylogenetic position. Finally, acrasids have developed a simple form of aggregative multicellularity and represent the only known multicellular excavates.

2. Morphological diversity

Most heteroloboseans are unicellular and uninucleate, though several species are multinucleate at least in part of their life cycle, e.g., *Stephanopogon* spp., *Gruberella flavescens*,

Pseudovahlkampfia emersoni, Fumarolamoeba ceborucoi, Willaertia magna, and *Psalteriomonas lanterna* (Broers et al., 1990; De Jonckheere et al., 2011b; Page, 1983; Sawyer, 1980; Yubuki & Leander, 2008). All heteroloboseans lack a typical, "stacked" Golgi apparatus. Mitochondria of Heterolobosea are oval, elongated or cup-shaped and possess flattened, often discoidal cristae. Few species are anaerobic and their mitochondria lack cristae. The mitochondrion of Heterolobosea is often closely associated with rough endoplasmic reticulum.

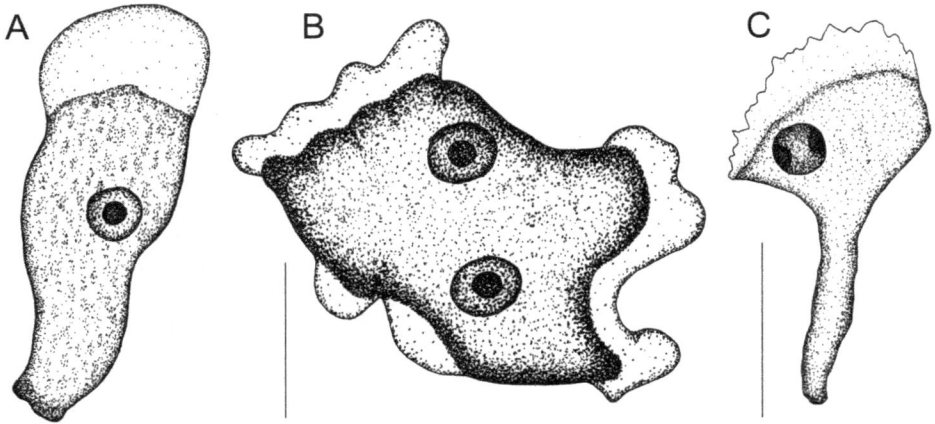

Fig. 1. Amoebae of Heterolobosea. **A**, *Acrasis rosea*; **B**, *Fumarolamoeba ceborucoi*; **C**, flabellate form of *Stachyamoeba lipophora*. Scale bars = 10 μm. After De Jonckheere et al., 2011b; Page, 1988; Olive & Stoianovitch, 1960).

Typical life-cycle of Heterolobosea consists of amoeboid, flagellate and resting stage (a cyst). However, one or two stages are unknown and presumably have been reduced in many taxa. Heterolobosean amoebae bear no flagella. Interestingly, the flagellar apparatus including basal bodies is assembled *de novo* during the transformation to the flagellate (see below). The amoebae are relatively uniform in shape and size. The locomotive forms are usually of the "limax" type (i.e. cylindrical monopodial amoebae, see fig. 1A) and move rapidly with eruptive lobopodia. Amoebae of some species, e.g., *Fumarolamoeba ceborucoi*, form subpseudopodia in all directions (De Jonckheere et al., 2011b; see fig. 1B). The locomotive form of *Stachyamoeba lipophora* is usually flattened ("flabellate") and its single pseudopodium bears many short subpseudopodia (Page, 1987; see fig. 1C). Many heterolobosean amoebae form a posterior uroid, sometimes with long uroidal filaments. *Vahlkampfia anaerobica* was reported to form a floating form (Smirnov & Fenchel, 1996). The heterolobosean amoebae do not possess any cytoskeleton-underlain cytostomes. However, the amoeba of *Naegleria fowleri* forms so-called amoebastomes, sucker-like surface structures that aid in phagocytosis (Sohn et al., 2010). The amoeboid stage is unknown (and possibly completely lost) in genera *Lyromonas*, *Pharyngomonas, Pleurostomum, Percolomonas*, and *Stephanopogon*.

Most heterolobosean flagellates have a groove-like cytostome that rises subapically (see fig. 2A-D). On the other hand, cytostomes of genera *Tetramitus, Heteramoeba*, and *Trimastigamoeba* open anteriorly (fig. 2E, G). Several heterolobosean flagellates, e.g., *Tetramitus* spp. and *Heteramoeba clara*, have a distinct collar or rim that circumscribes the anterior end of the cell body (see fig. 2E). In *Tetramitus rostratus* and *Pleurostomum flabellatum* the collar is drawn out into a short rostrum (fig. 2F). The cytostome of *T. rostratus* and *P. flabellatum* has a broad opening and curves into the cell to form a microtubule-supported tubular feeding apparatus (fig. 2F). The cytostome of *Trimastigamoeba phillippinensis* is a gullet-like tube with flagella rising from its bottom (fig. 2G). *Pharyngomonas kirbyi*, the basal-most lineage of Heterolobosea, has a subtle ventral groove and sub-anterior curved cytopharynx (fig. 2A). The cytostome of genera *Naegleria*, *Willaertia* and *Euplaesiobystra* has been reduced (fig. 2H).

Heterolobosean flagellates typically possess two (Heteramoeba, Euplaesiobystra, Pleurostomum, Pocheina, most Naegleria and some Tetramitus species) or four (Lyromonas, Willaertia, Percolomonas, Pharyngomonas, Tetramastigamoeba, few Naegleria and most Tetramitus species) flagella which arise at the anterior end of the feeding apparatus. Only few heterolobosean species have a different number of flagella. However, the number of flagella may vary among individuals of a single species. For example, most Tetramitus jugosus and Oramoeba fumarolia flagellates are biflagellate, but cells with more flagella (up to 10 in O. fumarolia) were found as well (Darbyshire et al., 1976; De Jonckheere et al., 2011a). Psalteriomonas lanterna has four nuclei, four mastigonts, each with four flagella, and four ventral grooves (fig. 2D). Representatives of genus Stephanopogon have over one hundred flagella (fig. 2I).

The flagella are usually equal in length. Alternatively, some flagella may be longer than the other ones (*Percolomonas* spp., *Pharyngomonas kirbyi*). All four flagella of *Percolomonas descissus* beat synchronously and drive water with food particles into the cytostome. *P. cosmopolitus* often attaches to the substrate by the tip of the longest flagellum. Most unattached cells move with a skipping motion across the substrate, as the trailing flagellum repeatedly makes and breaks contact with the surface (Fenchel & Patterson, 1986). Two flagella of *Pharyngomonas kirbyi* are directed anteriorly and actively beat during swimming. The cells can attach to the substrate using these flagella. The remaining two flagella are directed posteriorly and beat slowly. They are used during feeding to drive the water into the cytostome (Park & Simpson, 2011).

In quadriflagellate heteroloboseans, the basal bodies of flagella are arranged into two linked similar dikinetids rather than a single tetrakinetid. Such an unusual organization of the mastigont is called "double bikont". The arrangement of the pairs between each other can be orthogonal (e.g., *Tetramitus rostratus*), in tandem (e.g., *Percolomonas descissus*) or side-by-side (*Pharyngomonas kirbyi, Percolomonas sulcatus*). The arrangement of basal bodies in a pair can be orthogonal (*Pharyngomonas kirbyi*), parallel or near parallel (other heteroloboseans) (Brugerolle & Simpson, 2004; Park & Simpson, 2011). The flagellar apparatus of most heterolobosean flagellates possesses only two structures characteristic for Excavata as defined by Simpson (2003). In contrast, the mastigont of *Pharyngomonas kirbyi*, the deepest-branching heterolobosean, is more plesiomorphic and displays additional two or three excavate features (for details see Park & Simpson, 2011). The arrangement of basal bodies within a pair of flagella also seems to be more plesiomorphic in *Pharyngomonas* than that of the other heteroloboseans. In addition, the flagellar apparatus of *Percolomonas sulcatus* seems to be more plesiomorphic as well and is the most obvious example of the double bikont organization (Brugerolle & Simpson, 2004; Park & Simpson, 2011). On the other hand, it lacks the additional excavate features observed in *Ph. kirbyi*.

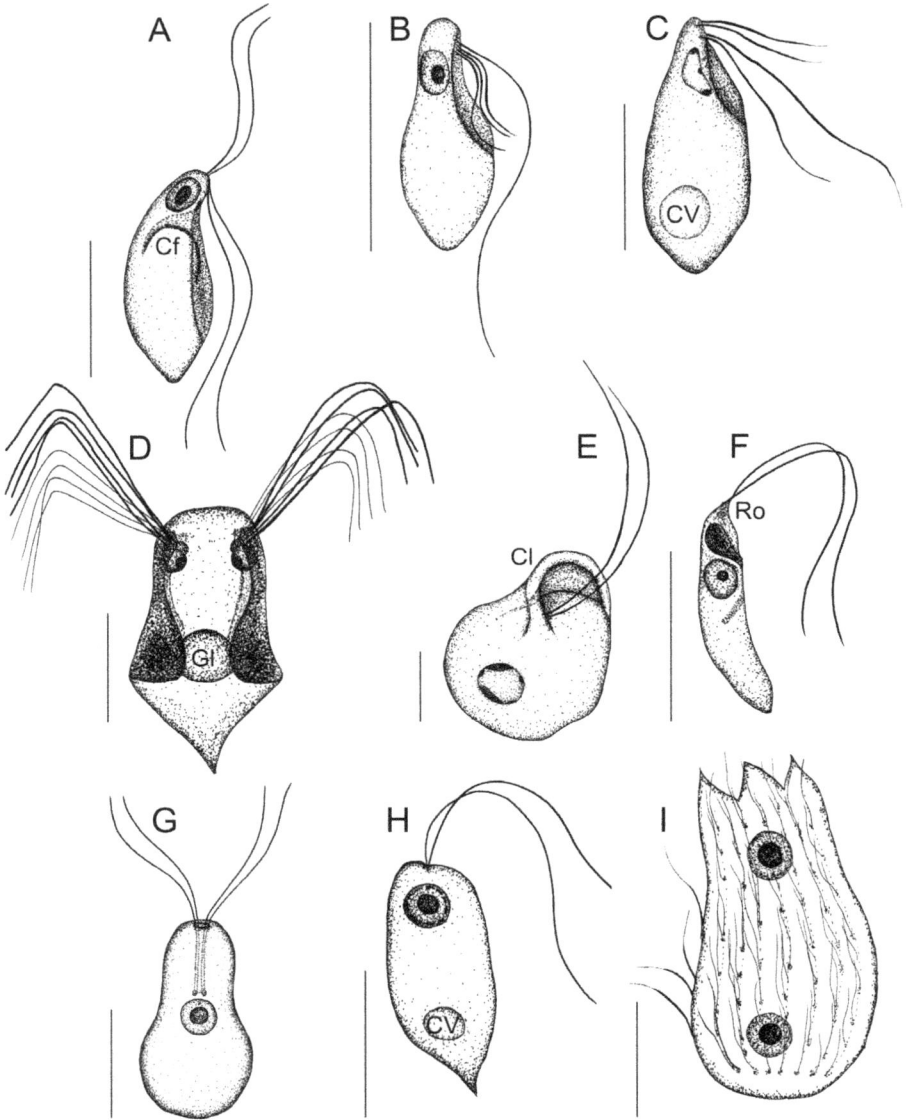

Fig. 2. Heterolobosean flagellates. **A**,*Pharyngomonas kirbyi*; **B**, *Percolomonas cosmopolitus*; **C**, *Percolomonas descissus*; **D**, *Psalteriomonas lanterna*; **E**, *Heteramoeba clara*; **F**, *Pleurostomum flabellatum*; **G**, *Trimastigamoeba philippinensis*; **H**, *Naegleria gruberi* **I**,. *Stephanopogon minuta*. Cf – cytopharynx; Cl – collar; CV – contractile vacuole; Gl – globule of hydrogenosomes; Ro – rostrum. Scale bars = 10 μm. After Broers et al., 1990; Bovee, 1959; Brugerolle & Simpson, 2004; Droop, 1962; Fenchel & Patterson, 1986; Page, 1967, 1988; Park et al., 2007; Park & Simpson, 2011; Yubuki & Leander, 2008.

Members of the eukaryovorous genus *Stephanopogon* are strikingly different from the other heteroloboseans. Their vase-shaped and curved cell bodies possess several longitudinal rows of flagella and two isomorphic nuclei. The cytostome is slit-shaped, dorsally supported by a lip, and accompanied by ventral barbs in most species. The ventral side of the cell bears more than 100 flagella, while only ca. 13 flagella arise from the dorsal side (Yubuki & Leander, 2008).

The cyst is the third heteroloboscan life stage. The cyst wall usually consists of two layers, ectocyst and endocyst. They are either closely associated to each other or can be separated and thus easily recognized by light microscope. The surface of the cyst is wrinkled, rough or smooth, and can be sticky (e.g., in *Paravahlkampfia*). Most heterolobosean cysts have no pores and presumably excyst by a wall rupture as in representatives of *Paravahlkampfia* (Visvesvara et al., 2009). The cyst of genera *Tulamoeba* and *Monopylocystis* has a single pore that penetrates the wall and is sealed with a mucoid plug (Park et al., 2009). Members of genera *Willaertia*, *Naegleria*, *Marinamoeba*, *Pernina*, and *Euplaesiobystra* have pores in the cyst wall. The cyst pores of *Naegleria*, *Willaertia* and *Pernina* are similar to each other in that the pores penetrate both the endocyst and ectocyst. In contrast, the pores of *Euplaesiobystra hypersalinica* do not penetrate the endocyst wall (Park et al., 2009). The cyst morphology of the genus *Tetramitus*, including the presence and number of pores, is highly variable. In many heteroloboseans the cyst stage is unknown (*Pleurostomum*, *Neovahlkampfia*, *Sawyeria*, *Psalteriomonas*, *Lyromonas* etc.).

Members of the Acrasidae have developed an additional stage in their life cycle, a simple multicellular fruiting body (sorocarp) formed by an aggregation of amoebae. The Acrasidae is the only known multicellular lineage of Excavata. The cells are in the mature sorocarp differentiated into two types: basal stalk cells and distal spore cells. Unlike in *Dictyostelium* (Amoebozoa: Dictyosteliida), where the stalk-forming cells undergo programmed cell death, the stalk-forming cells of the Acrasidae do not lose their viability. The most studied species of multicellular heteroloboseans is *Acrasis rosea*. Its sorocarps are complex with many branches ("arborescent"; fig. 4). Fruiting bodies of the recently described species *A. helenhemmesae* are simpler, uniseriate, and with only two or three bottle-shaped stalk cells (Brown M.W. et al., 2010). In contrast, the sorocarps of the putative acrasid *Pocheina flagellata* are globular (Olive et al., 1983).

3. Life cycles of Heterolobosea

Present knowledge on the life cycle of most heteroloboseans is fragmentary and it has been studied in detail only in *Naegleria gruberi*. The main active stage of *N. gruberi* is the amoeba which relies on actin-based cytoskeleton (Walsh, 2007) and has no flagella, basal bodies or cytoplasmic microtubules. It normally feeds, moves, and divides. Under certain conditions the amoeba rapidly transforms to the flagellate or the cyst stage (fig. 3). The transformation to the flagellate stage is triggered by various stressors, such as changes in temperature, osmolarity or availability of nutrients. The flagellate of *N. gruberi* is a temporary stage persisting only for several hours. It does not divide and feed, and has no cytostome. This is a typical feature of most *Naegleria* species. In *Naegleria minor* and *Naegleria robinsoni*, however, the juvenile flagellates possess four flagella and divide once to form biflagellate cells similar to flagellates of the other *Naegleria* species (De Jonckheere, 2002). It was reported that the flagellate of *N. gruberi* plays a crucial role in shuttling from the benthos to the water surface

(Preston & King, 2003). Interestingly, the whole microtubule cytoskeleton of the flagellate, including flagella and their basal bodies, is formed *de novo* during the transformation from the amoeba (Fulton, 1977; Fulton & Dingle, 1971; Lee, 2010). The transformation is incredibly fast being completed within ca. two hours.

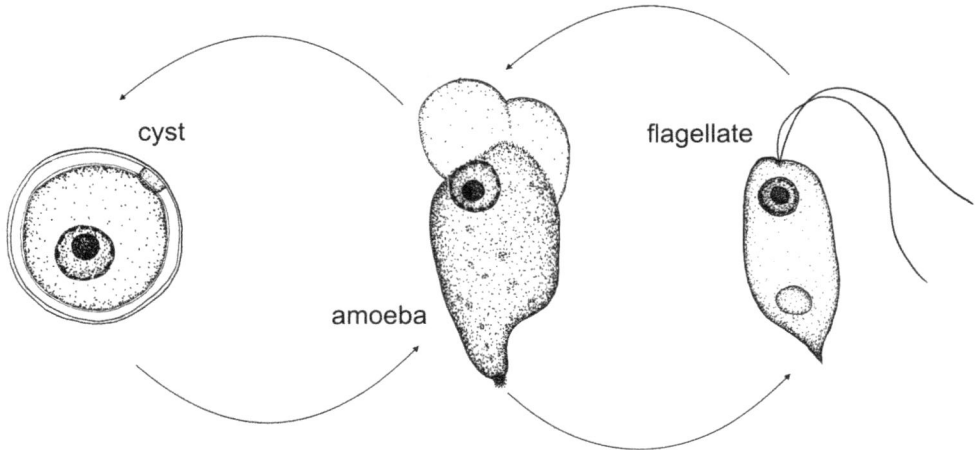

Fig. 3. The life cycle of *Naegleria gruberi*. After Page, 1967.

In contrast to *Naegleria*, flagellates of many other heteroloboseans are able to feed and divide, sometimes for long periods without reverting to the amoeba. In some heteroloboseans the flagellate is the only trophic stage and the ability to form amoebae has been presumably lost. On the other hand, the flagellate is unknown from even more heteroloboseans (see above). The ability to encyst is usually connected with the amoeba stage. The exceptions are *Percolomonas cosmopolitus* and *Stephanopogon* spp. which lack the amoeboid stage and encyst as flagellates (Fenchel & Patterson, 1986; Lwoff, 1936; Raikov 1969).

The amoeba-to-flagellate transformation of many heterolobosean species may be more or less successfully induced *in vitro* (see Page, 1988). In other species, the attempts were unsuccessful. The ability to form flagellates *in vitro* may vary also within the genus. For example, although most *Naegleria* and many *Tetramitus* species are known to produce flagellates, it seems that some of them have lost the ability (e.g., De Jonckheere, 2007; De Jonckheere et al., 2001). However, as culture requirements of almost no heteroloboseans have been studied in detail, some of them may be unable to transform *in vitro* and the observed inability to form certain life stages can be thus artificial. Indeed, strains of particular species known to produce flagellates (*Psalteriomonas lanterna, Heteramoeba clara, Willaertia*, some *Tetramitus* species) or cysts (*Percolomonas cosmopolitus*) were observed to lose the ability after a prolonged cultivation (Broers et al., 1990; Droop, 1962; Fenchel & Patterson, 1986; Page, 1988). In addition, De Jonckheere et al. (2011a) showed that the number of transformed flagellates of *Oramoeba fumarolia* depends on the type of bacterial prey.

The life cycle of acrasids is different from those described above and contains an additional multicellular stage, the sorocarp (see above, fig. 4). Individual amoebae of acrasids normally feed on bacteria. When starving, the amoebae aggregate to form sorocarps (Bonner, 2003).

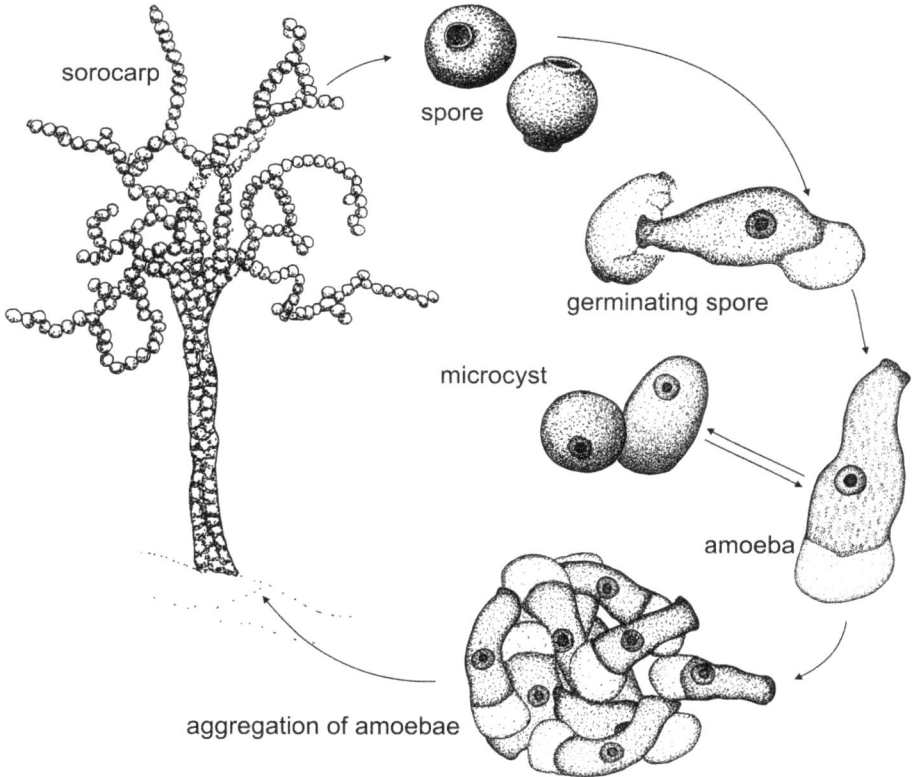

Fig. 4. Life cycle of *Acrasis rosea*. After Olive & Stoianovitch, 1960.

Both slightly different cell types of the fruiting body (basal stalk cells and distal spore cells) are capable of germination through excystment (Bonner, 2003; Brown M.W. et al., 2010). Both cell types of *Pocheina flagellata* are able to produce both amoebae and flagellates (Olive et al., 1983). The formation of multicellular bodies by the aggregation of individual cells is not unique feature of acrasids and has been well documented in several unrelated eukaryotic groups (see Brown M.W. et al., 2011) including the metazoan *Buddenbrockia plumatellae* (Morris & Adams, 2007). Unlike in *Dictyostelium*, the most-studied organism with the aggregative multicellularity, no motile ("slug") stage is formed during the ontogenesis of sorocarps of acrasids and their stalk-forming cells do not undergo the programmed cell death.

The question of the sexuality of Heterolobosea has not been elucidated yet. Although the sexual reproduction of several heterolobosean species has been discussed (e.g., Bunting, 1926; Droop, 1962; Fritz-Laylin et al. 2010; Olive et al., 1961; Olive, 1963; Pernin et al., 1992), no direct evidence has been found and the nature of the putative sexual processes remains unclear. Some authors hypothesize that the amoeba is the diploid stage whereas the flagellate is haploid and represents the gamete (Droop, 1962; Fulton, 1993). In fact, majority of the experiments and observations have to be revised and repeated using modern techniques.

The strongest evidence for sexuality of Heterolobosea has been brought by studies on *Naegleria*. Pernin *et al*. (1992) investigated the genetic structure of a natural population of *N. lovaniensis* by an isoenzyme analysis of 71 strains isolated in France. Analysis of single locus variation revealed that most strains were close to Hardy-Weinberg equilibrium. It indicated segregation and recombination between alleles. Recovery of relatively high number of distinct genotypic associations and the absence of linkage disequilibrium between genotypes at the different loci also supported the existence of recombination. In addition, the level of heterozygosyty in *N. gruberi* genome was reported as typical for sexual organism and most of meiotic genes defined by Ramesh (2005) were discovered (Fritz-Laylin et al., 2010). On the other hand, wild populations of *Naegleria gruberi* and *N. australiensis* showed large departures from Hardy–Weinberg equilibrium, low levels of heterozygosity, and strong linkage disequilibrium (Pernin & Cariou, 1997). These findings led to the conclusion that *N. gruberi* and *N. australiensis* have a predominantly clonal genetic structure in the wild. Different species of *Naegleria* thus could have different reproductive strategies.

4. Ecology of Heterolobosea

Heteroloboseans are heterotrophic prostists that inhabit a wide range of different habitats worldwide. Most heteroloboseans are bacteriovores, although cannibalism was reported in some species. Members of genus *Stephanopogon* are able to feed on diatoms and other eukaryotes. Most species of Heterolobosea live in soil and freshwater sediments. The number of marine species (30 – 50‰ salinity) is relatively low (e.g., *Neovahlkampfia damariscottae*, *Stephanopogon* spp., *Monopylocystis visvesvarai*, *Pseudovahlkampfia emersoni*). On the other hand, adaptations to various non-canonical enviroments occurred repeatedly in several heterolobosean lineages.

Heteroloboseans play a very important role in hypersaline habitats. About one third of species of heterotrophic protists recorded from this environment belong to Heterolobosea (e.g., *Pleurostomum* spp., *Pharyngomonas kirbyi*, *Euplaesiobystra hypersalinica*, and *Tulamoeba peronaphora*). However, the halophilic species do not form monophyletic group and differ in the response to various salinity levels. *Tulamoeba peronaphora* grows in the culture at 75‰ – 250‰ salinity, *Pharyngomonas kirbyi* up to 250‰, *Euplaesiobystra hypersalinica* and *Pleurostomum flabellatum* flourish in more than 300‰ salinity. The latter two species are true extremophiles, because they live in nearly salt-saturated solutions (Park et al. 2007, 2009; Park & Simpson, 2011). Some other heteroloboseans have adapted to extremely acidic habitats with pH < 3. Sheehan et al. (2003) detected DNA of *Naegleria* sp. from a thermal stream with pH 2.7. Amaral Zettler et al. (2002) discovered DNA of uncultured *Paravahlkampfia* sp. from the River of Fire (pH of 2.0). Another heterolobosean DNA sequences were reported in a recent study of the River of Fire (Amaral-Zettler et al., 2011). The only cultured acidophilic heterolobosean is *Tetramitus thermacidophilus* isolated from an acidic hot spring. This species flourishes at pH from 1.2 to 6 with the optimal pH of 3.0 (Baumgartner et al., 2009). Particular heterolobosean species differ in the range of temperature at which they are able to grow. Many of them are thermophilic. For example, *T. thermacidophilus* and *Oramoeba fumarolia* grow in temperature up to 54 ºC (Baumgartner et al., 2009; De Jonckheere et al., 2011a). *Marinamoeba thermophila*, *Fumarolamoeba ceborucoi* and *Euplaesiobystra hypersalinica* grow up to 50 ºC (De Jonckheere et al., 2009, 2011b; Park et al., 2009). Several heteroloboseans, importantly including pathogenic *Naegleria* strains, survive and divide in temperatures around 40 – 45 ºC (De Jonckheere 2007; Guzmán-Fierros et al.,

2008; Park et al., 2007, 2009). In contrast to thermophilic heteroloboseans, there are also few reports on psychrophilic species adapted to cold environments. The growth optimum of *Vahlkampfia signyensis* is 10 °C and the cells die when the temperature exceeds 20 °C (Garstecki et al., 2005). *Tetramitus vestfoldii* isolated from microbial mat of a brackish Antarctic lake grows at 5 °C (Murtagh et al., 2002).

Representatives of at least two heterolobosean lineages have adapted to the life in anoxic/microoxic habitats (i.e. habitats without oxygen/with low concentration of oxygen). Mitochondria of most of them do not possess cristae. The first lineage is represented by the extreme halophile *Pleurostomum flabellatum*, the second one is more diversified and comprises *Psalteriomonas lanterna*, *Sawyeria marylandensis*, *Monopylocystis visvesvarai*, and most probably also *Percolomonas descissus*, *Lyromonas vulgaris* and *Vahlkampfia anaerobica* (Broers et al., 1990, 1993; O'Kelly et al., 2003; Smirnov & Fenchel, 1996). Mitochondrial derivates of *Psalteriomonas lanterna* and *Sawyeria marylandensis* were studied in detail and it was shown that they have been transformed to hydrogenosomes (Barberà et al., 2010; de Graaf et al., 2009). Interestingly, presumably aerobic *Naegleria gruberi* recently appeared to be a facultatively anaerobic protist. Its mitochondria possess cristae and a genome, and are probably equipped to function in both aerobic and anaerobic conditions (Fritz-Laylin et al., 2010; Ginger et al., 2010; Opperdoes et al., 2011).

Some heteroloboseans were reported to be endobionts or even pathogens of both invertebrates and vertebrates including humans. *Naegleria fowleri* causes primary amoebic meningoencephalitis (PAM, PAME, see Visvesvara et al., 2007), rare (235 reported cases worldwide, see De Jonckheere, 2011), but rapidly fatal disease of humans and other mammals. The total number of cases has been probably underestimated because *N. fowleri* lives in warm waters and it could be expected that most cases occur in tropical regions where the possibility of diagnosis is limited (De Jonckheere, 2011). Humans are typically infected while recreating in warm fresh water. In contrast to other CNS-infecting amoebae, *N. fowleri* infects primarily healthy individuals. The amoebae entry the central nervous system through the olfactory neuroepithelium and destroy host cells. Without prompt diagnosis and intervention, the patients die usually within two weeks of exposure; about 97 % of patients do not survive the infection. In addition to *N. fowleri*, pathogenicity was suggested also for *N. australiensis* and *N. italica* on the basis of tests on mice (De Jonckheere, 2002). There is a single report on PAM-like disease caused by *Paravahlkampfia francinae* (Visvesvara et al., 2009). In contrast to PAM caused by *N. fowleri*, the affected patient recovered within a few days. Several strains of *Vahlkampfia* sp., *Tetramitus ovis* and *Paravahlkampfia* sp. were isolated from keratitis patients (Aitken et al., 1996; Alexandrakis et al., 1998; De Jonckheere & Brown S., 2005a; Dua et al. 1998; Kinnear, 2003; Ozkoc et al., 2008; Walochnik et al., 2000). However, their importance in pathogenesis is unclear and no direct evidence of their pathogenity was indicated. Heteroloboseans were found also in the gut of animals (e.g., *Tetramitus* spp., *Paravahlkampfia ustiana*, *Percolomonas sulcatus*) and gills, skin, and internal organs of fish (*Naegleria* spp.) (e.g., Brugerolle & Simpson, 2004; Dyková et al., 2001, 2006; Schuster et al., 2003).

5. Taxonomy of Heterolobosea

The taxon Heterolobosea was created by Page & Blanton (1985) as a class unifying orders Schizopyrenida (limax-type amoebae, often with the flagellate stage) and Acrasida (aggregative amoebae forming multicellular sorocarps) on the basis of the common presence of limax amoeba with eruptive lobopodia, discoidal mitochondrial cristae, and the absence of a stacked Golgi apparatus. However, it was later shown that several organisms with different morphology are closely related to the Heterolobosea (*Pharyngomonas*) or even form its internal branches (*Percolomonas, Stephanopogon, Lyromonas, Psalteriomonas, Pleurostomum*). Currently, two concepts of Heterolobosea, here called Heterolobosea *sensu lato* and Heterolobosea *sensu stricto*, respectively, exist. The concept of Heterolobosea *sensu lato* emphasizes monophyletic taxa and includes all aforementioned genera in Heterolobosea. It means, in fact, that Heterolobosea *sensu lato* is a group containing all descendants of the last common ancestor of *Pharyngomonas* and *Naegleria*. This concept is currently favored by most authors and we follow it as well.

In contrast, some authors emphasize the original definition of Heterolobosea *sensu* Page & Blanton (1985). The absence of mitochondrial cristae and microbodies in *Lyromonas, Sawyeria,* and *Monopylocystis,* the presumed absence of the amoeboid stage in *Percolomonas, Stephanopogon* and *Pharyngomonas* (but not in *Pleurostomum*), and the different arrangement of flagella of *Pharyngomonas* are considered so important that these genera cannot be members of Heterolobosea. Instead, they are classified in separate classes closely related to the Heterolobosea *sensu stricto*. The taxon corresponding to the Heterolobosea *sensu lato* was named Percolozoa (see Cavalier-Smith, 1991, 1993, 2003). The most recent version of this concept is represented by Cavalier-Smith & Nikolaev (2008). They divide the phylum Percolozoa into four classes: Pharyngomonadea (*Pharyngomonas*), Percolatea (*Percolomonas, Stephanopogon*), Lyromonadea (*Lyromonas, Psalteriomonas, Sawyeria, Monopylocystis*), and Heterolobosea (the rest of genera). The latter three classes are united within the subphylum Tetramitia whereas *Pharyngomonas* is the only member of the subphylum Pharyngomonada. Although we do not follow this concept because Heterolobosea *sensu stricto* is highly paraphyletic, we accept the division of Heterolobosea (*sensu lato*) into Pharyngomonada and Tetramitia as it is supported by phylogenetic analyses (see below).

The Pharyngomonada comprises a single family, Pharyngomonadidae, with two species of the genus, *Pharyngomonas*. Pharyngomonads are flagellates with four flagella; amoebae and cysts are unknown. In contrast to Tetramitia, basal bodies within a pair are arranged orthogonally. A large pharynx opens into the anterior end of the longitudinal ventral groove. The mastigont system of *Pharyngomonas* is more plesiomorphic than that of Tetramitia and shows more features of typical excavates (Park & Simpson, 2011).

Synapomoprhies of Tetramitia include parallel or nearly parallel basal bodies in a pair and a specific 17-1 helix in the secondary structure of SSU rRNA molecule (Cavalier-Smith & Nikolaev, 2008; Nikolaev et al., 2004). Tetramitia are currently classified into seven families though some authors recognize only three families of Heterolobosea (e.g., Patterson et al., 2002; Smirnov & Brown, 2004). In addition, several heterolobosean genera have not been assigned to any family (e.g., *Oramoeba, Fumarolamoeba, Pernina, Tulamoeba, Euplaesiobystra*). The paraphyletic family Vahlkampfiidae contains most heterolobosean genera (e.g., *Allovahlkampfia, Fumarolamoeba, Marinamoeba, Naegleria, Neovahlkampfia, Paravahlkampfia, Pseudovahlkampfia, Solumitrus, Tetramastigamoeba, Tetramitus, Vahlkampfia,* and *Willaertia*). It was defined by the presence of amoebae of the limax type and persistence of the nucleolus during mitosis. In addition, the genus *Pleurostomum*, whose amoeboid stage is unknown or has been lost, was placed within Vahlkampfiidae on the basis of its phylogenetic position (Park et al., 2007). In contrast to Vahlkampfiidae, the nucleolus of members of the family

Gruberellidae (genera *Gruberella* and *Stachymoeba*) disintegrates during mitosis. Members of two monotypic families, Lyromonadidae and Psalteriomonadidae, are anaerobic and possess acristate mitochondria. Park et al. (2007) suggested that Lyromonadidae might be a synonym of Psalteriomonadidae. *Lyromonas vulgaris* forms only flagellates with a single kinetid and longitudinal ventral groove, while *Psalteriomonas lanterna* is able to form also amoebae and its flagellates possess four mastigonts, nuclei and longitudinal ventral grooves. The anaerobic genera *Monopylocystis* and *Sawyeria* which were not placed into any family in the original description (O'Kelly et al., 2003) are sometimes placed into family Lyromonadidae or Psalteriomonadidae (Cavalier-Smith, 2003; Cavalier-Smith & Nikolaev, 2008). However, the authors did not specify in which family they classify the genera. The genus *Percolomonas* (family Percolomonadidae) comprises flagellates with four flagella (three in one species) and a longitudinal ventral groove. The amoeba stage is unknown or has been lost. The genus *Percolomonas* is most probably polyphyletic (see later). The peculiar eukaryovorous flagellates of the genus *Stephanopogon* with two nuclei and multiplicated flagella comprise the family Stephanopogonidae. Finally, heteroloboseans forming multicellular sorocarps have been accommodated within the family Acrasidae. Cells of their sorocarps are differentiated into morphologically distinct spores and stalk cells.

In addition to the aforementioned families, the family Guttulinopsidae comprising genera *Guttulinopsis* and *Rosculus* was sometimes affiliated with Heterolobosea (e.g., Smirnov & Brown, 2004), though many authors do not consider them as heteroloboseans (Page & Blanton, 1985).

The concept of the genus level of Heterolobosea was historically based on the cyst morphology and the presence/absence of the flagellate stage. All vahlkampfiids which do not form flagellates and whose cysts do not have pores were classified within the genus *Vahlkampfia* (see Brown S. & De Jonckheere, 1999). However, results of phylogenetic analyses showed that *Vahlkampfia* is polyphyletic. Consequently, genera *Paravahlkampfia*, *Neovahlkampfia*, *Allovahlkampfia*, *Fumarolamoeba*, and *Solumitrus* were created to accommodate species of *Vahlkampfia*-like morphology (Anderson et al., 2011; Brown S. & De Jonckheere, 1999; De Jonckheere et al., 2011b; Walochnik & Mulec, 2009). In addition, several former *Vahlkampfia* species were transferred to the genus *Tetramitus* (Brown & De Jonckheere, 1999). Moreover, *Vahlkampfia anaerobica* described by Smirnov & Fenchel (1996) is morphologically almost identical with *Monopylocystis visvesvarai* and is likely congeneric or conspecific with it. Phylogenetic position of several described *Vahlkampfia* species remains unknown. This applies also on the type species, *V. vahlkampfi*. Therefore, Brown & De Jonckheere (1999) nominated *V. avara* as a new type species of *Vahlkampfia*. However, such a change is not possible according to the International Code of Zoological Nomenclature and the current status of the genus *Vahlkampfia* is chaotic.

Although we agree with the classification of organisms of *Vahlkampfia*-like morphology into several independent genera based solely on molecular-phylogenetic analysis, we do not recognize the genus *Solumitrus* as it is closely related to *Allovahlkampfia* with identical morphology. The genetic distance between *S. palustris* and *A. spelaea* SSU rDNA is 93.7 % (Anderson et al., 2011) which is comparable to genetic distances within other heterolobosean genera. The situation is more complicated by the fact that *Allovahlkampfia* itself has been defined solely on the basis its phylogenetic position (Walochnik & Mulec, 2009). The authors did not include the genus *Acrasis* into their analysis and it was showed later that it is the closest relative of *Allovahlkampfia* (Brown M.W. et al., 2010; present study). The morphology of *Acrasis* and *Allovahlkampfia* amoebae has not been compared and it cannot be ruled out that *Allovahlkampfia spelaea* and *Solumitrus palustris* are, in fact, members of the genus *Acrasis* with unknown or reduced multicellular stage.

The genus *Percolomonas* is even more problematic than *Vahlkampfia*. It was created by Fenchel & Patterson (1986) for the species *P. cosmopolitus* (originally described as *Tetramitus cosmopolitus*) and was later broadened to accommodate flagellates with one long and three shorter flagella inserting at the anterior end of a longitudinal feeding groove (Larsen & Patterson, 1990). In addition, the triflagellate *P. denhami* was described later (Tong, 1997). However, it was soon realized that most *Percolomonas* species were totally unrelated to the type species *P. cosmopolitus*. They were removed from *Percolomonas* and accommodated in genera *Trimastix* (Preaxostyla), *Carpediemonas* (Fornicata), which is, in fact, biflagellate, *Chilomastix* (Fornicata), and *Pharyngomonas* (Heterolobosea) (Bernard et al., 1997, 2000; Cavalier-Smith & Nikolaev, 2008; Ekebom et al., 1996). Moreover, the mastigont structure of *P. sulcatus* and *P. descissus* is so different from that of *P. cosmopolitus* (Brugerolle & Simpson, 2004) that they should be removed from the genus *Percolomonas* as well. The phylogenetic position of the remaining *Percolomonas* species (*P. denhami*, *P. similis* and *P. spinosus*) is uncertain and no sequence data are currently available. Finally, *P. cosmopolitus*, the only certain member of the genus *Percolomonas*, is possibly paraphyletic (see below) and its true identity is unknown.

6. Species concept of Heterolobosea

Ca. 140 species of Heterolobosea have been described so far. The species were originally distinguished on the basis of light-microscopic morphology. The cyst was the most important life stage as its morphological variability was sufficient to recognize most then-known species (see Page, 1988). Unlike the cysts, amoebae of particular heterolobosean species are usually indistinguishable and have been rarely used for the species identification. Finally, although Page (1988) stressed the importance of heterolobosean flagellates in taxonomy as their morphology is quite diverse (see above), their descriptions were often insufficient and cannot be considered in taxonomical studies. Moreover, the flagellates are unknown and presumably absent in many heteroloboseans.

Since it is often impossible to differentiate between pathogenic and non-pathogenic *Naegleria* strains solely on the basis of their morphology, various biochemical and immunological methods have been applied (see Page, 1988). Later on, several new species of *Naegleria* were described on the basis of molecular markers (see De Jonckheere, 2002, 2011). This continuous process reached a peak in the explicit formulation of a molecular species concept of the genus *Naegleria* based on the ITS region (De Jonckheere, 2004). The concept was soon expanded to cover the whole Vahlkampfiidae (De Jonckheere & Brown S., 2005b). It was shown that vahlkampfiid ITS region is extremely variable. Interestingly, even strains with identical SSU rDNA sequences may slightly differ in the ITS region (Baumgartner et al., 2009; De Jonckheere & Brown S., 2005b). According to De Jonckheere & Brown S. (2005b), almost any two vahlkampfiid strains differing in ITS1, 5.8S rDNA or ITS2 sequences should be classified as different species even when their morphology and ecology are identical. The difference in the ITS region was often used as an accessory criterion in addition to morphological identification. However, several vahlkampfiid species were described solely on the basis of minute differences in the ITS region without morphology being effectively involved. This was stated e.g. in the diagnosis of *Naegleria canariensis* (De Jonckheere, 2006): "Because of the morphological similarity of the cysts with those of other *Naegleria* spp. molecular identification is required. The species can be identified from the ITS2 sequence, which differs by 2 bp substitutions from that *N. gallica*. The ITS1 and 5.8S rDNA sequence is identical to that of *N. gallica*. ". Similarly, *Tetramitus anasazii* and *T. hohokami* were distinguished solely on the basis of few differences in the ITS region (De Jonckheere, 2007).

Although we agree that the ITS region represents a quite effective DNA barcode of the Vahlkampfiidae and can be used for rapid determination of strains belonging to already described species, we are convinced that the concept is misleading when used for new species description. The biggest problem is that it assumes zero level of intraspecific polymorphism within the Vahlkampfiidae. However, the true variability of a vast majority of vahlkampfiid species is virtually unknown. It was convincingly shown that certain closely related *Naegleria* species differ in the ITS region sequence (De Jonckheere, 2004). There is, however, no reason to believe that it is true for the whole genus *Naegleria* or even for the whole Vahlkampfiidae, and, on the other hand, that the ITS region displays no intraspecific polymorphism. Indeed, it is a well-documented fact that there is some degree of ITS polymorphism within certain Vahlkampfiidae species. ITS regions of different strains of *Tetramitus jugosus* differ in a single nucleotide (De Jonckheere et al., 2005). Despite strains of *Naegleria fowleri* are even more variable, they are considered to be conspecific instead of belonging to several different species (De Jonckheere, 2004, 2011). Dyková et al. (2006) even showed that there exists an intragenomic polymorphism in *Naegleria clarki*. All these examples show that the ITS region should be no longer used for new species descriptions until the problem of species concept and intraspecific polymorphism within Heterolobosea is settled.

7. Phylogeny of Heterolobosea

Although many heterolobosean species have been successfully transferred into culture and many strains have been deposited into culture collections, the phylogeny of Heterolobosea has not yet been satisfactorily elucidated. Virtually all molecular-phylogenetic analyses with reasonable taxon sampling are based on a single locus, SSU rDNA (e.g., Brown M.W. et al., 2010; Cavalier-Smith & Nikolaev, 2008; De Jonckheere et al., 2011a, 2011b; Nikolaev et al., 2004; Park & Simpson, 2011; Park et al., 2007, 2009). To evaluate the resolving power of SSU rDNA, we performed a phylogenetic analysis which included members of all heterolobosean genera whose sequences are available, and important environmental sequences affiliated with Heterolobosea. Results of our analysis (fig. 5) were in agreement with previous studies.

Heterolobosea robustly split into two lineages, Pharyngomonada and Tetramitia. We divide here the Tetramitia into six clades whose interrelationships and internal phylogeny remain unresolved. Unfortunately, all the clades but E are currently indefinable on the basis of morphology. The current heterolobosean taxonomy is not consistent with delineation of these six monophyletic groups and has to be revised in the future. The family Vahlkampfiidae is highly paraphyletic.

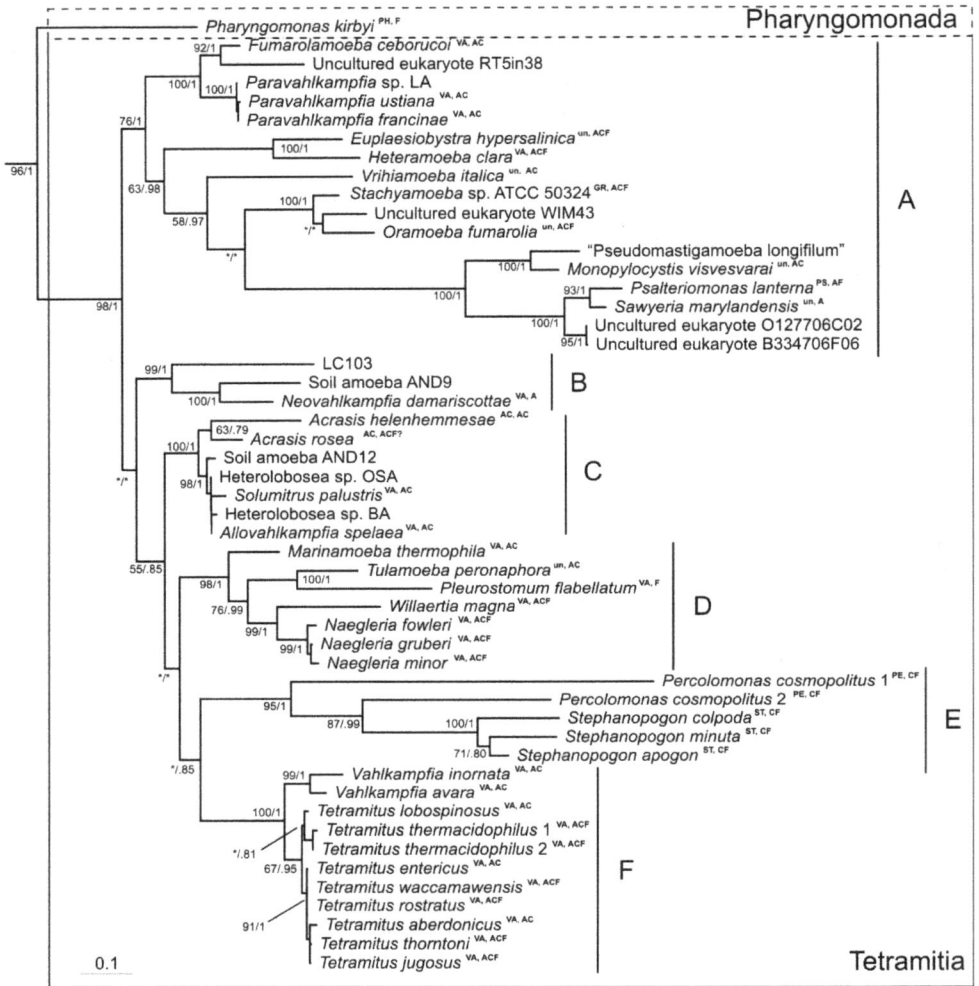

Fig. 5. Phylogenetic tree of Heterolobosea based on SSU rDNA sequences. The tree topology was constructed by the maximum likelihood method (ML) in RAxML 7.2.6 under the GTRGAMMAI model, and by the Bayesian method in MrBayes 3.1.2. under the GTR + Γ + I + covarion model. RAxML 7.2.6 was used for bootstrapping (1000 replicates). The tree was rooted with representatives of other Excavata lineages (outgroups were removed from the tree). The values at the nodes represent statistical support (ML bootstrap values/Bayesian posterior probabilities). Support values below 50%/.50 are represented by asterisks. Heterolobosea are divided into two branches (Pharyngomonada and Tetramitia). Six main clades of Tetramitia are labeled. The classification is indicated: AC – Acrasidae, GR – Gruberellidae, PE – Percolomonadidae, PH – Pharyngomonadidae, PS – Psalteriomonadidae, ST – Stephanopogonidae, un – unassigned, VA – Vahlkampfiidae, The occurrence of particular life stages is indicated: A – amoeba, C – cyst, F – flagellate.

Clade A consists of genera *Paravahlkampfia, Fumarolamoeba, Euplaesiobystra, Heteramoeba, Stachyamoeba, Vrihiamoeba, Oramoeba, Psalteriomonas, Sawyeria, Monopylocystis,* and several undetermined heteroloboseans (strains RT5in38, WIM43, and 'Pseudomastigamoeba longifillum', O127706C02, B334706F06). Moreover, morphological data suggest that *Vahlkampfia anaerobica, Lyromonas vulgaris,* and *Percolomonas descissus* belong to this clade as well (see above). Many members of the clade A display a unique morphology of the nucleolus. Typical heteroloboseans have nucleus with a single central nucleolus. In contrast, *Heteramoeba, Sawyeria, Monopylocystis, Stachyamoeba, Percolomonas descissus, Vahlkampfia anaerobica,* and at least some strains of *Psalteriomonas* possess parietal nucleoli or a thin ring of nucleolar material near the nuclear membrane. *Neovahlkampfia damariscottae* and undetermined heteroloboseans AND9 and LC103 constitute the clade B. Although it is quite robust (bootstrap support 99), its position within Tetramitia is uncertain. In some previous analyses the clade formed the basal branch of Tetramitia (Brown M.W. et al., 2010; Park & Simpson, 2011; Park et al. 2007, 2009) while it branched more terminally in the others (Cavalier-Smith & Nikolaev; 2008; De Jonckheere et al., 2011a, 2011b; Nikolaev et al., 2004; this study). *Acrasis, Allovahlkampfia, Solumitrus* and undetermined heteroloboseans BA, OSA, and AND12 formed tetramitian clade C. All representatives of the clade inhabit freshwater sediments and soil (however, the data are unavailable for strains BA and OSA). Members of the clade D (*Marinamoeba, Tulamoeba, Pleurostomum, Willaertia,* and *Naegleria*) live in wide range of habitats. Interestingly, at least some members of all the genera are able to grow at higher temperatures (40 – 50 °C). The Clade E comprises *Percolomonas cosmopolitus* and the genus *Stephanopogon* (i.e. Percolatea *sensu* Cavalier-Smith & Nikolaev, 2008). Both *Stephanopogon* and *P. cosmopolitus* form long branches in the phylogenetic trees and it cannot be ruled out that their grouping is, in fact, a result of long-branch attraction. However, Yubuki & Leander (2008) identified three morphological features shared by *Stephanopogon* and *P. cosmopolitus* suggesting that the clade E might be monophyletic. In addition, both *Stephanopogon* spp. and *P. cosmopolitus* have lost the amoeba stage and their ability to encyst as flagellates is unique among heteroloboseans. Tetramitian clade F is formed by the remaining genera *Vahlkampfia* and *Tetramitus*. Morphology and ecology of the genus *Tetramitus* is extremely diverse including characteristics used for generic determination (presence and number of cyst pores, number/presence of flagella, marine or freshwater lifestyle, etc.).

Our view on the evolution of Heterolobosea has completely changed after the application of methods of molecular phylogenetics. The analyses of sequence data are currently the only efficient tool for pinpointing relationships between species and genera although it is unable to resolve interrelationships between particular tetramitian clades. The analyses suggested that some morphologically well-defined genera (e.g., *Vahlkampfia* and *Percolomonas*) were polyphyletic. On the other hand, *Tetramitus* spp. is so diverse that it was impossible to group them into a single genus solely on the morphological base. Although it is currently impossible to define morphological synapomorphies of particular tetramitian clades, Heterolobosea itself and both its subphyla, Tetramitia and Pharyngomonada, seem to be well defined on both molecular and morphological level.

8. Conclusion

During the last decade, Heterolobosea have attracted considerable interest because of their extraordinary morphological, ecological and physiological diversity. Members of the most studied genus *Naegleria* are medicinally important or became model organisms in cell biology. The insight obtained from the genome sequence of *N. gruberi* considerably improved understanding of the early eukaryotic evolution. However, our knowledge about Heterolobosea as a whole is still seriously limited. Since it is currently unclear whether Heterolobosea are sexual or asexual organisms, the biological species concept is not applicable. On the other hand, the current species concept of Heterolobosea based on the ITS region is misleading and most probably considerably overestimates the real number of extant species. Heterolobosean phylogeny is unclear as well. Although the monophyly of Heterolobosea and its split into Pharyngomonada and Tetramitia is strongly supported by both cell structure and molecular-phylogenetic analyses, the internal phylogeny of Tetramitia has not yet been satisfactorily elucidated. Since 18S rDNA has not sufficient resolving power, it is necessary to perform multigene phylogenetic analyses in order to improve the hetrolobosean phylogeny. In addition, it is important to obtain sequence data from so-far uncharacterized, potentially important taxa, such as *Percolomonas sulcatus* and *Gruberella flavescens*. There is also a strong possibility that some already-known enigmatic eukaryotes will be shown to belong to Heterolobosea as well. This has already happened in the case of the ciliate-resembling genus *Stephanopogon*. Finally, it is a well-known fact that the current taxonomy of Heterolobosea, particularly the family level, does not reflect the phylogeny and should be changed. This cannot be, however, achieved before the heterolobosean phylogeny is resolved.

9. Acknowledgment

This work was supported by grants from the Czech Ministry of Education, Youth and Sport of the Czech Republic (project MSM0021620828), the Czech Science Foundation (project P506/11/1317) and the Grant Agency of Charles University (project 21610). We would like to thank Pavla Slámová for preparing the line drawings.

10. References

Aitken, D., Hay, J., Kinnear, F.B., Kirkness, C.M., Lee, W.R. & Seal, D.V. (1996). Amebic keratitis in a wearer of disposable contact lenses due to a mixed *Vahlkampfia* and *Hartmanella* infection. *Ophthalmology*, Vol.103, No.3, (March 1996), pp. 485-494, ISSN 0161-6420

Alexandrakis, G., Miller, D. & Huang, A.J.W. (1998). Amebic keratitis due to *Vahlkampfia* infection following corneal trauma. *Archives of Ophthalmology*, Vol.117, No.7, (July 1998), pp. 950-951, ISSN 0003-9950

Amaral Zettler, L.A., Gómez, F., Zettler, E., Keenan, B.G., Amils, R. & Sogin, M.L. (2002). Eukaryotic diversity in Spain's River of Fire. *Nature*, Vol.417, No.6885, (May 2002), p. 137, ISSN 0028-0836

Amaral-Zettler, L.A., Zettler, E.R., Theroux, S.M., Palacios, C., Aguilera, A. & Amils, R. (2011). Microbial community structure across the tree of life in the extreme Río Tinto. *ISME Journal*, Vol.5, No.1, (January 2011), pp. 42-50, ISSN 1751-7362

Anderson, O.R., Wang, W., Faucher, S.P., Bi, K. & Shuman, K.A. (2011). A new heterolobosean amoeba *Solumitrus palustris* n. g., n. sp. isolated from freshwater marsh soil. *Journal of Eukaryotic Microbiology*, Vol.58, No.1, (January-February 2011), pp. 60-67, ISSN 1550-7408

Baldauf, S.L., Roger, A.J., Wenk-Siefert, I. & Doolittle, W.F. (2000). A kingdom-level phylogeny of eukaryotes based on combined protein data. *Science*, Col.290, No.5493, (November 2000), pp. 972-977, ISSN 0036-8075

Barberà, M.J., Ruiz-Trillo, I., Tufts, J.Y., Bery, A., Silberman, J.D. & Roger, A.J. (2010). *Sawyeria marylandensis* (Heterolobosea) has a hydrogenosome with novel metabolic properties. *Eukaryotic Cell*, Vol.9, No.12, (December 2010), pp. 1913-1924, ISSN 1535-9778

Baumgartner, M., Eberhardt, S., De Jonckheere, J.F. & Stetter, K.O. (2009). *Tetramitus thermacidophilus* n. sp., an amoeboflagellate from acidic hot springs. *Journal of Eukaryotic Microbiology*, Vol.56, No.2, (March-April 2009), pp. 201-206, ISSN 1550-7408

Bernard, C., Simpson, A.G.B. & Patterson, D.J. (1997). An ultrastructural study of a free-living retortamonády, *Chilomastix cuspidata* (Larsen & Patterson, 1990) n. comb. (Retortamonadida, Protista). *European Journal of Protistology*, Vol.33, No.3, (August 1997), pp. 254-265, ISSN 0932-4739

Bernard, C., Simpson, A.G.B. & Patterson, D.J. (2000). Some free-living flagellates (Protista) from anoxic habitats. *Ophelia*. Vol.52, No.2, (May 2000), pp. 113-142, ISSN 0078-5326

Bonner, J.T. (2003). Evolution of development in the cellular slime molds. *Evolution & Development*, Vol.5, No.3, (May-June 2003), pp. 305-313, ISSN 1520-541X

Bovee, E.C. (1959). Studies on amoeboflagellates. 1. The general morphology and mastigonts of *Trimastigamoeba philippinensis* Whitmore 1911. *Journal of Protozoology*, Vol.6, No.1 (February 1959), pp. 69-75, ISSN 0022-3921

Broers, C.A.M., Stumm, C.K., Vogels, G.D. & Brugerolle, G. (1990). *Psalteriomonas lanterna* gen. nov., sp. nov., a free-living ameboflagellate isolated from fresh-water anaerobic sediments. *European Journal of Protistology*, Vol.25, No.4, (June 1990), pp. 369-380, ISSN 0932-4739

Broers, C.A.M., Meijers, H.H.M., Symens, J.C., Stumm, C.K., Vogels, G.D. & Brugerolle, G. (1993). Symbiotic association of *Psalteriomonas vulgaris* n. spec. with *Methanobacterium formicicum*. *European Journal of Protistology*, Vol.29, No.1, (February 1993), pp. 98-105, ISSN 0932-4739

Brown, S. & De Jonckheere, J.F. (1999). A reevaluation of the amoeba genus *Vahlkampfia* based on SSU rDNA sequences. *European Journal of Protistology*, Vol.35, No.1, (February 1999), pp. 49-54, ISSN 0932-4739

Brown, M.W., Silberman, J.D. & Spiegel, F.W. (2010). A morphologically simple species of *Acrasis* (Heterolobosea, Excavata), *Acrasis helenhemmesae* n. sp. (2010). *Journal of Eukaryotic Microbiology*, Vol.57, No.4, (July-August 2010), pp. 346-353, ISSN 1550-7408

Brown, M.W., Silberman, J.D. & Spiegel, F.W. (2011). "Slime molds" among the Tubulinea (Amoebozoa): molecular systematics and taxonomy of *Copromyxa*. *Protist*, Vol.162, No.2, (April 2011), pp. 277-287, ISSN 1434-4610

Brugerolle, G. & Simpson, A.G.B. (2004). The flagellar apparatus of heteroloboseans. *Journal of Eukaryotic Microbiology*, Vol.51, No.1, (January-February 2004), pp. 96-107, ISSN 1550-7408

Bunting, M. (1926). Studies of the life-cycle of *Tetramitus rostratus* Perty. *Journal o Morphology and Physiology*, Vol.42, No.1, (June 1926), pp. 23-81, ISSN 0095-9626

Cavalier-Smith, T. (1991). Cell diversification in heterotrophic flagellates. In: *The biology of free-living heterotrophic flagellates*, D.J. Patterson & J. Larsen, (Eds.), 113-131, Clarendon Press, ISBN 978-0-19-857747-8, Oxford, UK

Cavalier-Smith, T. (1993). Kingdom Protozoa and its 18 phyla. *Microbiological Reviews*, Vol.57, No.4, (December 1993), pp. 953-994, ISSN 0146-0749

Cavalier-Smith, T. (1998). A revised six-kingdom system of life. *Biological Reviews*, Vol.73, No.3, (August 1998), pp. 203-266, ISSN 1464-7931

Cavalier-Smith, T. (2003). Protist phylogeny and the high-level classification of Protozoa. *European Journal of Protistology*, Vol.39, No.4, (September 2003), pp. 338-348, ISSN 0932-4739

Cavalier-Smith, T. (2010). Kingdoms Protozoa and Chromista and the eozoan root of the eukaryotic tree. *Biology Letters*, Vol.6, No.3, (June 2010), pp. 342-345, ISSN 1744-9561

Cavalier-Smith, T. & Nikolaev, S. (2008). The zooflagellates *Stephanopogon* and *Percolomonas* are a clade (class Percolatea: phylum Percolozoa). *Journal of Eukaryotic Microbiology*, Vol.55, No.6, (November-December 2008), pp. 501-509, ISSN 1550-7408

Darbyshire, J.F., Page, F.C. & Goodfellow, L.P. (1976). *Paratetramitus jugosus*, an amoeboflagellate of soils and fresh water, type-species of *Paratetramitus* nov. gen. *Protistologica*, Vol.12, No.3, pp. 375-387, ISSN 0033-1821

de Graaf, R.M., Duarte, I., van Alen, T.A., Kuiper, J.W.P., Schotanus, K., Rosenberg, J., Huynen, M.A. & Hackstein, J.H.P. (2009). The hydrogenosomes of *Psalteriomonas lanterna*. *BMC Evolutionary Biology*, Vol.9, pp. 287, ISSN 1471-2148

De Jonckheere, J.F. (2002). A century of research on the amoeboflagellate genus *Naegleria*. *Acta Protozoologica*, Vol.41, No.4, (November 2002), pp. 309-342, ISSN 0065-1583

De Jonckheere, J.F. (2004). Molecular definition and the ubiquity of species in the genus *Naegleria*. *Protist*, Vol.155, No.1, (March 2004), pp. 89-103, ISSN 1434-4610

De Jonckheere, J.F. (2006). Isolation and molecular identification of vahlkampfiid amoebae from an island (Tenerife, Spain). *Acta Protozoologica*, Vol.45, No.1, (February 2006), pp. 91-96, ISSN 0065-1583

De Jonckheere, J.F. (2007). Molecular identification of free-living amoebae of the Vahlkampfiidae and Acanthamoebidae isolated from Arizona (USA). *European Journal of Protistology*, Vol.43, No.1, (January 2007), pp. 9-15, ISSN 0932-4739

De Jonckheere, J.F. (2011). Origin and evolution of the worldwide distributed pathogenic amoeboflagellate *Naegleria fowleri*. *Infection, Genetics* and *Evolution*, in press

De Jonckheere, J.F., Brown, S., Dobson, P.J., Robinson, B.S. & Pernin, P. (2001). The amoeba-to-flagellate transformation test is not reliable for the diagnosis of the genus *Naegleria*. Description of three new *Naegleria* spp. *Protist*, Vol.152, No.2, (July 2001), pp. 115-121, ISSN 1434-4610

De Jonckheere, J.F. & Brown, S. (2005a). Isolation of a vahlkampfiid amoeba from a contact lens: *Tetramitus ovis* (Schmidt, 1913), comb. nov. *European Journal of Protistology*, Vol.41, No.2, (April 2005), pp. 93-97, ISSN 0932-4739

De Jonckheere, J.F. & Brown, S. (2005b). The identification of vahlkampfiid amoebae by ITS sequencing. *Protist*, Vol.156, No.1, (June 2005), pp. 89-96, ISSN 1434-4610

De Jonckheere, J.F., Brown, S., Walochnik, J., Aspock, H. & Michel, R. (2005). Morphological investigation of three *Tetramitus* spp. which are phylogenetically very closely related: *Tetramitus horticolus*, *Tetramitus russelli* n. comb. and *Tetramitus pararusselli* n. sp. *European Journal of Protistology*, Vol.41, No.2, (April 2005), pp. 139-150, ISSN 0932-4739

De Jonckheere, J.F., Baumgartner, M., Opperdoes, F.R. & Stetter, K.O. (2009). *Marinamoeba thermophila*, a new marine heterolobosean amoeba growing at 50 degrees C. *European Journal of Protistology*, Vol.45, No.3, (August 2009), pp. 16-23, ISSN 0932-4739

De Jonckheere, J.F., Baumgartner, M., Eberhardt, S., Opperdoes, F.R. & Stetter, K.O. (2011a). *Oramoeba fumarolia* gen. nov., sp. nov., a new marine heterolobosean amoeboflagellate growing at 54 °C. *European Journal of Protistology*, Vol.47, No.1, (January 2011), pp. 16-23, ISSN 0932-4739

De Jonckheere, J.F., Murase, J. & Opperdoes, F.R. (2011b). A new thermophilic heterolobosean amoeba, *Fumarolamoeba ceborucoi*, gen. nov., sp. nov., isolated near a fumarole at a volcano in Mexico. *Acta Protozoologica*, Vol.50, No.1, pp. 41-48, ISSN 0065-1583

Droop, M.R. (1962). *Heteramoeba clara* n. gen., n. sp., a sexual biphasic amoeba. *Archiv für Mikrobiologie*, Vol.42, No.3, pp. 254-266, ISSN 0003-9276

Dua, H.S., Azuara-Blanco, A., Hossain, M. & Lloyd, J. (1998). Non-*Acathamoeba* amebic keratitis. *Cornea*, Vol.17, No.6, (November 1998), pp. 675-677, ISSN 0277-3740

Dyková, I., Kyselová, I., Pecková, H., Oborník, M. & Lukeš, J. (2001). Identity of *Naegleria* strains isolated from organs of freshwater fishes. *Diseases of Aquatic Organisms*, Vol.46, No.2, (September 2001), pp. 115-121, ISSN 0177-5103

Dyková, I., Pecková, H., Fiala, I. & Dvořáková, H. (2006). Fish-isolated *Naegleria* strains and their phylogeny inferred from ITS and SSU rDNA sequences. *Folia Parasitologica*, Vol.53, No.3, (September 2006), pp. 172-180, ISSN 0015-5683

Ekebom, J., Patterson, D.J. & Vørs, N. (1996). Heterotrophic flagellates from coral reef sediments (Great Barrier Reef, Australia). *Archiv für Protistenkunde*, Vol.146, No.3-4, (February 1996), pp. 251-272, ISSN 0003-9365

Fenchel, T. & Patterson, D.J. (1986). *Percolomonas cosmopolitus* (Ruinen) n. gen., a new type of filter feeding flagellate from marine plankton. *Journal of the Marine Biological Association of the United Kingdom*, Vol.66, No.2, (May 1986), pp. 465-482, ISSN 0025-3154

Fritz-Laylin, L.K., Prochnik, S.E., Ginger, M.L., Dacks, J.B., Carpenter, M.L., Field, M.C., Kuo, A., Paredez, A., Chapman, J., Pham, J., Shu, S.Q., Neupane, R., Cipriano, M., Mancuso, J., Tu, H., Salamov, A., Lindquist, E., Shapiro, H., Lucas, S., Grigoriev, I.V., Cande, W.Z., Fulton, C., Rohksar, D.S. & Dawson, S.C. (2010). The genome of *Naegleria gruberi* illuminates early eukaryotic versatility. *Cell*, Vol.140, No.5, (March 2010), pp. 631-642, ISSN 0092-8674

Fulton, C. (1977). Cell differentiation of *Naegleria gruberi*. *Annual Reviews of Microbiology*, Vol.31, (October 1977), pp. 597-627, ISSN 0066-4227

Fulton, C. (1993). *Naegleria*: A research partner for cell and developmental biology. *Journal of Eukaryotic Microbiology*, Vol.40, No.4, (July-August 1993), pp. 520-532, ISSN 1066-5234

Fulton, C. & Dingle, A.D. (1971). Basal bodies, but not centrioles, in *Naegleria*. *Journal of Cell Biology*, Vol.51, No.3, (December 1971), pp. 826-836, ISSN 0021-9525

Garstecki, T., Brown, S. & De Jonckheere, J.F. (2005). Description of *Vahlkampfia signyensis* n. sp. (Heterolobosea), based on morphological, ultrastructural and molecular characteristics. *European Journal of Protistology*, Vol.41, No.2, (April 2005), pp. 119-127, ISSN 0932-4739

Ginger, M.L., Fritz-Laylin, L.K., Fulton, C., Cande, W.Z. & Dawson, S.C. (2010). Intermediary metabolism in protists: a sequence-based view of facultative anaerobic metabolism in evolutionarily diverse eukaryotes. *Protist*, Vol.161, No.5, (December 2010), pp. 642-671, ISSN 1434-4610

Gray, M.W., Lang, B.F., Burger, G. (2004). Mitochondria of protists. *Annual Reviews of Genetics*, Vol.38, (December 2004), pp. 477-524, ISSN 0066-4197

Guzmán-Fierros, E., De Jonckheere, J.F. & Lares-Villa, F. (2008). Identification of *Naegleria* species in recreational areas in Hornos, Sonora. *Revista Mexicana de Biodiversidad*, Vol.79, No.1, (June 2008), pp. 1-5, ISSN 1870-3453

Hampl, V., Hug, L., Leigh, J.W., Dacks, J.B., Lang, B.F., Simpson, A.G.B. & Roger, A.J. (2009). Phylogenomic analyses support the monophyly of Excavata and resolve relationships among eukaryotic "supergroups". *Proceedings of the National Academy of Sciences of the United States of America*, Vol.106, No.10, (March 2009), pp. 3859-3864, ISSN 0027-8424

Kinnear, F.B. (2003). Cytopathogenicity of *Acanthamoeba*, *Vahlkampfia* and *Hartmanella*: Quantitative & qualitative *in vitro* studies on keratocytes. *Journal of Infection*, Vol.46, No.4, (May 2003), pp. 228-237, ISSN 0163-4453

Larsen, J. & Patterson, D.J. (1990). Some flagellates (Protista) from tropical sediments. *Journal of Natural History*, Vol.24, No.4, (July-August 1990), pp. 801-93, ISSN 0022-2933

Lee, J. (2010). *De novo* formation of basal bodies during cellular differentiation of *Naegleria gruberi*: Progress and hypotheses. *Seminars in Cell & Developmental Biology*, Vol.21, No.2, (April 2010), pp. 156-162, ISSN 1084-9521

Levine, N.D., Corliss, J.O., Cox, F.E.G., Deroux, G., Grain, J., Honigberg, B.M., Leedale, G.F., Loeblich, A.R., Lom, J., Lynn, D., Merinfeld, E.G., Page, F.C., Poljansky, G., Sprague, V. Vavra, J. & Wallace, F.G. (1980). A newly revised classification of the Protozoa. *Journal of Protozoology*, Vol.27, No.1, (Ferbuary 1980), pp. 37-58, ISSN 0022-3921

Lwoff, A. (1936). Le cycle nucleaire de *Stephanopogon mesnili* Lw. (Cilié Homocaryote). *Archives de Zoologie Expérimentale et Générale*, Vol.78, pp. 117-132, ISSN 0003-9667

Morris, D.J. & Adams, A. (2007). Sacculogenesis of *Buddenbrockia plumatellae* (Myxozoa) within the invertebrate host *Plumatella repens* (Bryozoa) with comments on the evolutionary relationships of the Myxozoa. *International Journal for Parasitology*, Vol.37, No.10, (August 2007), pp. 1163-1171, ISSN 0020-7519

Murase, J., Kawasaki, M., De Jonckheere, J.F. (2010). Isolation of a new heterolobosean amoeba from a rice field soil: *Vrihiamoeba italica* gen. nov., sp. nov. *European Journal of Protistology*, Vol.46, No.3, (August 2010), pp. 164-170, ISSN 0932-4739

Murtagh, G.J., Dyer, P.S., Rogerson, A., Nash, G.V. & Laybourn-Parry, J. (2002). A new species of *Tetramitus* in the benthos of a saline antarctic lake. *European Journal of Protistology*, Vol.37, No.4, (February 2002), pp. 437-443, ISSN 0932-4739

Nikolaev, S.I., Mylnikov, A.P., Berney, C., Fahrni, J., Pawlowski, J., Aleshin, V.V. & Petrov, N.B. (2004). Molecular phylogenetic analysis places *Percolomonas cosmopolitus* within Heterolobosea: evolutionary implications. *Journal of Eukaryotic Microbiology*, Vol.51, No.5, (September-October 2004), pp. 501-509, ISSN 1550-7408

O'Kelly C.J., Silberman, J.D., Amaral Zettler, L.A., Nerad, T.A. & Sogin, M.L. (2003). *Monopylocystis visvesvarai* n. gen., n. sp. and *Sawyeria marylandensis* n. gen., n. sp.: The new amitochondrial heterolobosean amoebae from anoxic environments. *Protist*, Vol.154, No.2, (July 2003), pp. 281-290, ISSN 1434-4610

Olive, L.S. (1963). The question of sexuality in cellular slime molds. *Bulletin of the Torrey Botanical Club*, Vol.90, No.2, (March 1963), pp. 144-153, ISSN 0040-9618

Olive, L.S. & Stoianovitch, C. (1960). Two new members of Acrasiales. *Bulletin of the Torrey Botanical Club*, Vol.87, No.1, (January 1960), pp. 1-20, ISSN 0040-9618

Olive, L.S., Dutta, S.K. & Stoianovitch, C. (1961). Variation in the cellular slime mold *Acrasis rosea*. *Journal of Protozoology*, Vol.8, No.4, (November 1961), pp 467-472, ISSN 0022-3921

Olive, L.S., Stoianovitch, C. & Bennett, W.E. (1983). Descriptions of acrasid cellular slime molds: *Pocheina rosea* and a new species, *Pocheina flagellata*. *Mycologia*, Vol.75, No.6, (November-December 1983), pp. 1019-1029, ISSN 0027-5514

Opperdoes, F.R., De Jonckheere, J.F. & Tielens, A.G.M. (2011). *Naegleria gruberi* metabolism. *International Journal for Parasitology*, Vol.41, No.9, (August 2011), pp. 915-924, ISSN 0020-7519

Ozkoc, C., Tuncay, S., Delibas, S.B., Akisu, C., Ozbek, Z., Durak, I. & Walochnik, J. (2008). Identification of *Acanthamoeba* genotype T4 and *Paravahlkampfia* sp. from two clinical samples. *Journal of Medical Microbiology*, Vol.57, No.3, (March 2008), pp. 392-396, ISSN 0022-2615

Page, F.C. (1967). Taxonomic criteria for limax amoebae, with descriptions of 3 new species of *Hartmanella* and 3 of *Vahlkampfia*. *Journal of Protozoology*, Vol.14, No.3, (August 1967), pp. 499-521, ISSN 0022-3921

Page, F.C. (1983) *Marine gymnamoebae*, Institute of Terrestrial Ecology, Culture Centre of Algae and Protozoa, ISBN 0-904282-75-9, Cambridge, UK

Page, F.C. (1987). Transfer of *Stachyamoeba lipophora* to the class Heterolobosea. *Archiv für Protistenkunde*, Vol.133, No.3-4, pp. 191-197, ISSN 0003-9365

Page, F.C. (1988). *A new key to freshwater and soil gymnamoebae*, Freshwater Biological Association, ISBN 1-871105-02-1, Ambleside, UK

Page, F.C. & Blanton, R.L. (1985). The Heterolobosea (Sarcodina, Rhizopoda), a new class uniting the Schizopyrenida and the Acrasidae (Acrasida). *Protistologica*, Vol.21, No.1, pp. 121-132, ISSN 0033-1821

Park, J.S., Simpson, A.G.B., Lee, W.J. & Cho, B.C. (2007). Ultrastructure and phylogenetic placement within Heterolobosea of the previously unclassified, extremely halophilic heterotrophic flagellate *Pleurostomum flabellatum* (Ruinen 1938). *Protist*, Vol.158, No.3, (July 2007), pp. 397-413, ISSN 1434-4610

Park, J.S., Simpson, A.G.B., Brown, S. & Cho, B.C. (2009). Ultrastructure and molecular phylogeny of two heterolobosean amoebae, *Euplaesiobystra hypersalinica* gen. et sp. nov. and *Tulamoeba peronaphora* gen. et sp. nov., isolated from an extremely hypersaline habitat. *Protist*, Vol.160, No.2, (May 2009), pp. 265-283, ISSN 1434-4610

Park, J.S. & Simpson, A.G.B. (2011). Characterization of *Pharyngomonas kirbyi* (= *"Macropharyngomonas halophila"* nomen nudum), a very deep-branching, obligately halophilic heterolobosean flagellate. *Protist*, Vol.55, No.6, (November-December 2011), pp. 501-509, ISSN 1434-4610

Patterson, D.J. (1988). The evolution of Protozoa. *Memórias do Instituto Oswaldo Cruz*, Vol.83, Suppl.1. (November 1988), pp. 580-600, ISSN 0074-0276

Patterson, D.J., Rogerson, A. & Vørs, N. (2002). Class Heterolobosea, In: *The Illustrated Guide to the Protozoa*, J.J. Lee, G.F. Leedale & P. Bradbury, (Eds.), 1104-1111, Society of Protozoologists, ISBN 1-891276-23-9, Lawrence, KS

Pernin, P., Ataya, A. & Cariou, M.L. (1992). Genetic structure of natural populations of the free-living amoeba, *Naegleria lovaniensis*. Evidence for sexual reproduction. *Heredity*, Vol.68, No.2, (February 1992), pp. 173-181, ISSN 0018-067X

Pernin, P. & Cariou, M.L. (1997). Evidence for clonal structure of natural populations of free-living amoebae of the genus *Naegleria*. *Genetical Research*, Vol.69, No.3, (June 1997), pp. 173-181, ISSN 0016-6723

Preston, T.M. & King, C.A. (2003). Locomotion and phenotypic transformation of the amoeboflagellate *Naegleria gruberi* at the water-air interface. *Journal of Eukaryotic Microbiology*, Vol.50, No.4, (July-August 2003), pp. 245-251, ISSN 1550-7408

Raikov, I.B. (1969). The macronucleus of ciliates, In: *Research in Protozoology*, Vol.3, T.T. Chen (Ed.), 1-128, Pergamon Press, ISBN 0080032362, London, UK

Ramesh, M.A., Malik, S.B. & Logsdon, J.M. (2005). A phylogenomic inventory of meiotic genes: Evidence for sex in *Giardia* and an early eukaryotic origin of meiosis. *Current Biology*, Vol.15, No.2, (January 2005), pp. 185-191, ISSN 0960-9822

Rodríguez-Ezpeleta, N., Brinkmann, H., Burger, G., Roger, A.J., Gray, M.W., Philippe, H. & Lang, B.F. (2007). Toward resolving the eukaryotic tree: The phylogenetic positions of jakobids and cercozoans. *Current Biology*, Vol.17, No.16, (August 2007), pp. 1420-1425, ISSN 0960-9822

Sawyer, T.K. (1980). Marine amoebae from clean and stressed bottom sediments of the Atlantic Ocean and Gulf of Mexico. *Journal of Protozoology*, Vol.27, No.1, (February 1680), pp. 13-32, ISSN 0022-3921

Schuster, F.L., De Jonckheere, J.F., Moura, H., Sriram, R., Garner, M.M. & Visvesvara, G.S. (2003). Isolation of a thermotolerant *Paravahlkampfia* sp. from lizard intestine: Biology and molecular identification. *Journal of Eukaryotic Microbiology*, Vol.50, No.5, (September-October 2003), pp. 373-378, ISSN 1066-5234

Sheehan, K.B., Ferris, M.J. & Henson, J.M. (2003). Detection of *Naegleria* sp. in a thermal, acidic stream in Yellowstone National Park. *Journal of Eukaryotic Microbiology*, Vol.50, No.4, (July-August 2003), pp. 263-265, ISSN 1066-5234

Simpson, A.G.B. (2003). Cytoskeletal organization, phylogenetic affinities and the systematics in the contentious taxon Excavata. *International Journal of Systematic and Evolutionary Microbiology*, Vol.53, No.6, (November 2003), pp. 1759-1777, ISSN 1466-5026

Simpson, A.G.B. & Patterson, D.J. (1999). The ultrastructure of *Carpediemonas membranifera* (Eukaryota) with reference to the "Excavate hypothesis". *European Journal of Protistology*, Vol.35, No.4, (December 1999), pp. 353-370, ISSN 0932-4739

Smirnov, A.V. & Brown, S. (2004). Guide to the methods of study and identification of soil gymnamoebae. *Protistologia*, Vol.3, No.3, pp. 148-190, ISSN 0033-1821

Smirnov, A.V. & Fenchel, T. (1996). *Vahlkampfia anaerobica* n. sp. and *Vannella peregrinia* n. sp. (Rhizopoda) – anaerobic amoebae from a marine sediment. *Archiv für Protistenkunde*, Vol.147, No.2, (September 1996), pp. 189-198, ISSN 0003-9365

Sohn, H.J., Kim, J.H., Shin, M.H., Song, K.J. & Shin, H.J. (2010). The Nf-actin gene is an important factor for food-cup formation and cytotoxicity of pathogenic *Naegleria fowleri*. *Parasitology Research*, Vol.104, No.4, (March 2010), pp. 917-924, ISSN 0932-0113

Tong, S.M. (1997). Heterotrophic flagellates from the water column in Shark Bay, western Australia. *Marine Biology*, Vol.128, No.3, (June 1997), pp. 517-536, ISSN 0025-3162

Visvesvara, G.S., Moura, H. & Schuster, F.L. (2007). Pathogenic and opportunistic free-living amoebae: *Acanthamoeba* spp., *Balamuthia mandrillaris*, *Naegleria fowleri*, and *Sappinia diploidea*. *FEMS Immunology and Medical Microbiology*, Vol.50, No.1, (June 2007), pp. 1-26, ISSN 0928-8244

Visvesvara, G.S., Sriram, R., Qvarnstrom, Y., Bandyopadhyay, K., Da Silva, A.J., Pieniazek, N.J. & Cabral, G.A. (2009). *Paravahlkampfia francinae* n. sp. masquerading as an agent of primary amoebic meningoencephalitis. *Journal of Eukaryotic Microbiology*, Vol.56, No.4, (July-August 2009), pp. 357-366, ISSN 1550-7408

Walochnik, J., Haller-Schober, E.M., Kölli, H., Picher, O., Obwaller, A. & Aspöck, H. (2000). Discrimination between clinically relevant and nonrelevant *Acanthamoeba* strains isolated from contact lens-wearing keratitis patients in Austria. *Journal of Clinical Microbiology*, Vol.38, No.11, (November 2000), pp. 3932-3936, ISSN 0095-1137

Walochnik, J. & Mulec, J. (2009). Free-living amoebae in carbonate precipitating microhabitats of karst caves and a new vahlkampfiid amoeba, *Allovahlkampfia spelaea* gen. nov., sp. nov. *Acta Protozoologica*, Vol.48, No.1, pp. 25-33, ISSN 0065-1583

Walsh, C.J. (2007). The role of actin, actomyosin and microtubules in defining cell shape during the differentiation of *Naegleria* amoebae into flagellates. *European Journal of Cell Biology*, Vol.86, No.2, (February 2007), pp. 85-98, ISSN 0171-9335

Yabuki, A., Nakayama, T., Yubuki, N., Hashimoto, T., Ishida, K.I. & Inagaki, Y. (2011). *Tsukubamons globosa* n. gen., n. sp., a novel excavate flagellate possibly holding a key for the early evolution in "Discoba". *Journal of Eukaryotic Microbiology*, Vol.58, No.4, (July-August 2011), pp. 319-331, ISSN 1550-7408

Yubuki, N. & Leander, B.S. (2008). Ultrastructure and molecular phylogeny of *Stephanopogon minuta*: An enigmatic microeukaryote from marine interstitial environments. *European Journal of Protistology*, Vol.44, No.4, (November 2008), pp. 241-253, ISSN 0932-4739

2

Archaeal Diversity and Their Biotechnological Potential

Birgül Özcan
Mustafa Kemal University,
Faculty of Sciences and Letters,
Biology Department,
Turkey

1. Introduction

The curiosity of identifying, grouping, and naming organisms according to their established natural relationship has been the subject of a great interest since ancient times. Today, instead of the traditional rank-based biological classification, phylogenetic systematics, which aims at postulating phylogenetic trees rather than focusing on what taxa to delimit, has been used commonly. Carl Woese is the one who first realized that the ribosome, the ubiquitous molecular structure that conducts protein synthesis, offers a way to investigate systematically the relationships between all forms of life. Woese's approach was to determine the sequences of the RNAs that make up the ribosome, particularly the small subunit of ribosomal RNA (rRNA) (Woese et al., 1990). Comparisons of nucleotide sequences of ribosomal genes from different organisms allowed understanding of the evolutionary relationships between the organisms: the higher the similarity or difference between the rRNA sequences, the more or less closely related the organisms are. Instead of the commonly accepted subdivision of living organisms into the five kingdoms: *Monera, Protista, Fungi, Animalia,* and *Plantae* (Whittaker, 1969), Woese and his colleagues proposed subdivision into three higher taxa: *Archaea, Bacteria,* and *Eukarya,* first, they called them primary kingdoms and then domains (Woese et al., 1990). The sequencing of rRNA genes became one of the main tools for the construction of phylogenetic backbone of microbial classification and today each new description of *Bacteria* and *Archaea* must be accompanied by the complete 16S rRNA sequence of the type strain (Ludwig & Klenk, 2001; Yarza et al. 2010). Although the 16S rDNA gene has been tremendously useful for establishing the molecular phylogeny of prokaryotes over the last three decades, it suffers from the same limits as any other single-gene phylogenetic approach does. The identification of microbial isolates by whole-cell mass spectrometry (WC-MS) is being recognized as one of the latest tools bringing a revolution in microbial diagnostics, with the potential of bringing to an end many of the time-consuming and man-power-intensive identification procedures that have been used for decades. Apart from applications of WC-MS in clinical diagnostics, other fields of microbiology also have adopted the technology with success. MALDI-TOF MS shows particular potential usefulness for applications in environmental microbiology, e.g., to rapidly reveal cryptic species in large batches of related isolates (Clermont et al., 2009; Welkera & Moore, 2011).

Majority of the living beings thrive in environments having physically and geochemically temperate conditions. The extreme environments found on the planet are generally inhabited by microorganisms, which belong to the archaeal and bacterial domains of life. Extreme environments comprise the sites including physical variables as -20°C to +113°C (like stratosphere and hydrothermal vents), ≤120 Mpa (for hydrostatic pressures in the deep sea), aw ≈ 0.6 (for the activity of water in salt lakes) and ≈ 0.5< pH < 11 (for acidic and alkaline biotopes) (Woese et al., 1990). Archaeal ecology is generally accepted as synonymous with extreme environments in the point of the human being view. Representatives of *Archaea*, however, occur everywhere: in samples from ocean water, ocean sediments, freshwater lakes, soil, solid gas hydrates, tidal flat sediments, plant roots, peatlands, petroleum-contaminated aquifers, human subgingival area, skin and gastrointestinal tract and as a simbiyont within the sponge (Cavicchioli et al., 2003; Mills et al., 2005; DeLong, 2005; Knittel et al. 2005; Fierer et al., 2007; Brochier-Armanet et al. 2008; Oxley et al. 2010; Kong, 2011).

The majority of extremophiles belongs to the *Archaea*, the third domain of the living organisms together with *Eukarya* and *Bacteria* as explained before (Woese et al., 1990). The *Archaea* are a prokaryotic domain known to be often associated with habitats of extreme temperature, salinity and pH, and their presence in constantly cold marine waters is also well documented (Karr et al., 2006). Archaeal 16S rRNA community analysis has demonstrated that novel groups of *Archaea* are also abundant in the open ocean, soil and freshwater ecosystems as well (Buckley et al., 1998; Falz et al., 1999). *Archaea* exist in a broad range of habitats, and as a major part of global ecosystems, may contribute up to 20% of Earth's biomass (DeLong & Pace, 2001). *Archaea*, the most recently recognized domain, contains cultivated members that span a fairly limited range of phenotypes, represented by extreme halophiles, Sulfate-reducers and sulfur-metabolizing thermophiles, and methanogens. The first-discovered *Archaea* were extremophiles, which can be divided into four main physiological groups. These are the halophiles, thermophiles, alkaliphiles, and acidophiles.

Organisms from the domain *Archaea* differ fundamentally from *Eukarya* and *Bacteria* in several genetic, biochemical, and structural properties. Archaeal species have been classified as an early-branching evolutionary offshoot of the domain *Bacteria* and have long been considered to represent a primitive form of life that thrives only in extreme environments such as hot springs, salt lakes, or submarine volcanic habitats. However, recent researches have shown that *Archaea* are more physiologically diverse and ecologically widespread than was previously thought. Like *Bacteria*, *Archaea* are commonly mesophilic, and some members are known to be closely associated with eukaryotic hosts, including humans. For instance, high numbers of methane-producing *Archaea* (methanogens) have been detected in the gastrointestinal tract, vagina, and oral cavity (Belay et al., 1990; Vianna et al., 2006) and recently non-methanogenic *Archaea* including members of the *Crenarchaeota, Thermococcales, Thermoplasmatales* and *Halobacteriaceae*, have been detected in human faeces (Oxley et al., 2010). They are now recognized as a component of human microbiota and it is subjects of debate wheather archaea are cause of any disease in human. Although it was shown that members of the domain *Archaea* are found in greater abundance in dental plaque from sites with periodontal disease than in plaque from non-diseased sites (Lepp et al., 2004), it is generally assumed that archaea are not a cause of human disease (Vianna et al., 2006). The isolation of archaeal strains from root-canal samples by amplification of 16S rDNA were at first unsuccessful, however, a recent study has confirmed that members of the methanogenic archaea, similar to *Methanobrevibacter oralis*, can be detected in root canal samples (Vianna et al., 2006). Finally, the presence of archaea in root canals has been confirmed and this provides new insights into the polymicrobial communities in endodontic infections associated with clinical symptoms.

Archaea, one of the three domains of life on Earth, is predominantly composed of two major phyla: the *Euryarchaeota*, the *Crenarchaeota*. In addition to these two major phyla, the *Korarchaeota* (Elkins, 2008), the *Nanoarchaeota* (Huber et al., 2002), the *Thaumarchaeota* (Brochier-Armanet et al., 2008) and the *Aigarchaeota* (Nunoura et al., 2011) have been proposed to be potential phylum-level taxonomic groups within the *Archaea*; however, the establishment of these phyla is still controversial (Takai & Nakamura, 2011). This division, based on small-subunit rRNA phylogeny, is also strongly strengthened by comparative genomics and phenotypic characteristics. The first archaeal genome was sequenced in 1996 and so far 52 genomes of *Archaea* have been sequenced. The cultured *Crenarchaeota* are composed of four orders: *Caldisphaerales*, *Desulfurococcales*, *Sulfolobales* and *Thermoproteales* (Chaban et al., 2006). *Euryarchaeota* are composed of nine orders *Methanobacteriales*, *Methanococcales*, *Methanomicrobiales* (Balch et al., 1979), *Methanosarcinales* (Boone et al., 2001), *Halobacteriales* (Grant & Larsen 1989), *Thermoplasmatales* (Reysenbach, 2001), *Thermococcales* (Zillig et al., 1987), *Archaeoglobales* and *Methanopyrales* (Huber & Stetter, 2001).

The metabolic diversity of archaea is quite similar to bacteria in many aspects. Except for methanogenesis, all metabolic pathways discovered in archaea also exist among bacteria. Archaeal species can be either heterotrophs or autotrophs and use a large variety of electron donors and acceptors. Photosynthesis based on chlorophyll has not been found in *Archaea*, whereas photosynthesis based on bacteriorhodopsin, once believed unique to halophilic archaea, has been recently found in planktonic bacteria as well (Forterre, 2002). Beside their specific rRNA, archaea can be distinguished from bacteria by the nature of their membrane glycerolipids that are ethers of glycerol and isoprenol, whereas bacterial and eukaryal lipids are characterized as esters of glycerol and fatty acids (Kates, 1993). Archaeal glycerolipids are also 'reverse lipids', since the enantiomeric configuration of their glycerophosphate backbone is the mirror image of the configuration found in bacterial and eukaryal lipids. Another difference between *Archaea* and *Bacteria* is the absence of murein in archaea, whereas this compound is present in the cell wall of most bacteria. *Archaea* exhibits a great diversity of cell envelopes (Kandler & Konig, 1998), most archaea have a simple S-layer of glycoproteins covering the cytoplasmic membrane, whereas a few of them (*Thermoplasmatales*) only have a cytoplasmic membrane containing glycoproteins. The difference between *Archaea* and *Bacteria* at the molecular level is exemplified by the resistance of archaea to most antibiotics active on bacteria. Early studies on the molecular biology of archaea have shown that this resistance was due indeed to critical differences in the antibiotic targets (Zillig, 1991).

Ammonia oxidation carried out by microorganisms has global importance in nitrogen cycling and is often thought to be driven only by ammonia-oxidizing bacteria; however, the recent finding of new ammonia-oxidizing organisms belonging to the archaeal domain challenges this notion. Two major microbial groups are now considered to be involved in ammonia oxidation. These are chemolithoautotrophic ammonia-oxidizing bacteria and ammonia-oxidizing archaea. The first isolated ammonia-oxidizing archaeon , *Nitrosopumilus maritimus*, from a tropical marine aquarium tank (Konneke et al., 2005), was reported to include putative genes for all three subunits (*amoA*, *amoB*, and *amoC*) of ammonia monooxygenase that is the key enzyme responsible for ammonia oxidation. Ammonia-oxidizing archaea are determined to thrive in various habitats including marine and fresh waters, hot/thermal springs, soils, and wastewater treatment systems (Youa et al., 2009).

The fascinating ability of members of the *Archaea* to thrive in extremes of temperature, salt and pH as well as in other seemingly hostile niches has generated substantial interest in the molecular mechanisms responsible for mediating survival in the face of such environmental challenges. Despite the obvious potential of extremophilic archaea to yield many commercially appealing enzymes, thermostable DNA polymerases remain the only major class of molecule to have been effectively exploited in a wide range of PCR protocols.

2. Halophilic Archaea

Halophilic archaea (also called halobacteria or haloarchaea) are present in high abundance and are often the dominant prokaryotes in hypersaline environments on the Earth such as solar salterns, hypersaline lakes, the Dead Sea, hypersaline microbial mats and underground salt deposits (Oren, 2002). Most known halophiles are relatively easy to grow, and genera such as *Halobacterium*, *Haloferax*, and *Haloarcula* have become well known models for studies of the archaeal domain because they are much simpler to handle than methanogenic and hyperthermophilic archaea (Ma et al., 2010).

The haloarchaea are a monophyletic group including all known aerobic, obligate halophilic archaea. All are chemoorganotrophs; most utilize carbohydrates or amino acids and grow optimally between 3.4-4.5 M NaCl and generally require a minimum of 1.5 M NaCl for growth, some even grow well in saturated (>5 M) NaCl (Grant et al., 2001). In this respect, many haloarchaea appear ecologically equivalent in terms of resource and physiological requirements and therefore exhibit considerable overlap in their fundamental niches, defined as the combination of conditions and resources which allow a species to maintain a viable population (Begon et al., 1986). However, there are notable exceptions to this observation. For example, while most halophilic archaea exhibit optimal growth at near neutral pH, many are alkaliphilic and require at least pH 8.5 for growth (Grant et al., 2001). Recent molecular studies have revealed the presence of halophilic archaea in several low salinity environments (Elshahed et al., 2004). In addition, Purdy and colleagues (2004) isolated haloarchaea from a coastal salt marsh that exhibited optimal growth at 10% (1.7 M) NaCl but could grow slowly at 2.5% (0.43 M) NaCl. Microbial life has adapted to environments that combine high salt concentrations with extremely high pH values. Alkaline soda lakes in Africa, India, China, and elsewhere with pH values of 11 and higher and salt concentrations exceeding 300 g/l (5.1 M) are teeming with life (Oren, 2002). Hypersaline environments are found in a wide variety of aquatic and terrestrial ecosystems. These environments are inhabited by halotolerant microorganisms but also halophilic microorganisms ranging from moderate halophiles with higher growth rates in media containing between 0.5 M and 2.5 M NaCl to extreme halophiles with higher growth rates in media containing over 2.5 M NaCl (Ventosa et al., 1998). Aerobic, anaerobic, and facultative anaerobic microbes belonging to domains *Archaea* and *Bacteria* have been recovered from these extreme ecosystems, where they participate in overall organic matter oxidation (Oren, 2002; Moune et al., 2003; Hedi et al., 2009).

Hypersaline habitats are a kind of extreme environment dominated by halophilic archaea, which require a minimum of 9% (w/v) (1.5 M) NaCl for growth (Grant et al., 2001). In general, haloarchaeal strains require high salt concentration for growth and cell integrity. They, with some exceptions, lyse or lose viability in low salt concentrations or distilled water, and water sensitivity or lysis-resistence has been a key differentiation criterion between halococci and other haloarchaea (Grant et al., 2001). The most well known haloarchaea *Halobacterium salinarum*, for example, requires at least 2.5 M NaCl for growth and cells lose their morphological integrity instantaneously at less than 1 M NaCl (Kushner, 1964). Another representative, *Halococcus morrhuae* does not lyse in distilled water (Grant et al., 2001).

Over the last decade, the diversity of halophilic archaea in various hypersaline environments has been examined and more fully characterized. Included among these studies are naturally occurring salt lakes, hypersaline microbial mats and man-made solar salterns (Benlloch et al., 2001; Maturrano et al., 2006). In another study the microbiota in colonic mucosal biopsies from patients with inflammatory bowel disease, 16S rDNA sequences representing a phylogenetically rich diversity of halophilic archaea from the *Halobacteriaceae* were determined. The study revealed a multitude of undefined bacterial taxa and a low diversity of methanogenic archaea (Oxley et al., 2010). Representatives of *Archaea*, the third domain of life, were generally thought to be limited to environmental extremes of the earth. However, the discovery of archaeal 16S rRNA gene sequences in water, sediment and soil samples has called into question the idea of archaea as obligate extremophiles (Purdy et al., 2004).

As mentioned above based on the relationship with NaCl, the most salt requiring archaea are found in the order *Halobacteriales* and *Methanosarcinales*. They belong to the phylum *Euryarchaeota*; no halophilic representatives have yet been identified within the *Crenarchaeota* (Oren, 2008). Within the small subunit rRNA gene sequence-based tree of life there are three groups of prokaryotes that are both phylogenetically and physiologically coherent and consist entirely or almost entirely of halophiles. Within the *Euryarchaeota* the order *Halobacteriales* occurs with a single family, the *Halobacteriaceae* (Oren, 2002). The numbers of genera is reached to 35 (November 2011- Euzebylist) and *Salarchaeaum* is the last added extremely halophilic genus (Shimane et al., 2011). Most species of *Halobacteriaceae* are true extreme halophiles according to Kushner's definition (Kushner, 1978), however, *Halobacteriaceae* contains some species which can grow in low salinity for instance, *Haloferax sulfurifontis* (Elshahed et al., 2004), *Haladaptatus paucihalophilus* "the spesific epithet refers to low salt loving", (Savage et al., 2007) and *Halosarcina pallida* (Savage et al., 2008).

Diverse 16S rRNA gene sequences related to haloarchaea were recovered from tidal marine and salt marsh sediments, suggesting the existence of haloarchaea capable of growth at lower salt levels. In a recent study it was reported that two of three newly isolated genotypes had lower requirements for salt than previously cultured haloarchaea and were capable of slow growth at sea-water salinity (2.5% w/v NaCl). They reported the existence of archaea that could grow in non-extreme conditions and of a diverse community of haloarchaea existing in coastal salt marsh sediments and they concluded that the ecological range of these physiologically versatile prokaryotes is much wider than previously supposed (Purdy et al., 2004). Halophilic adaptation of organisms has been the subject of the great interest. Halophilic microorganisms have developed various biochemical strategies to adapt to high saline conditions, such as compatible solute synthesis to maintain cell structure and function (Tehei, 2002). These solutes are clearly of industrial interest. Besides these metabolical and physiological features, halophilic microorganisms are known to play important roles in fermenting fish sauces and in transforming and degrading waste and organic pollutants in saline waste waters (Hedi et al., 2009). All halophilic archaea studied balance the high osmolarity of their environment by having an at least equimolar intracellular salt concentration, KCl instead of NaCl in well-energized cells (Sopa, 2006). It has been shown that typical haloarchaeal proteins differ from mesohalic proteins by having a high fraction of acidic residues and a reduced fraction of basic residues. The genome sequences have corroborated that result and shown that a theoretical 2D gel of a haloarchaeon differs remarkably from that of other organisms (Tebbe et al., 2005). The cytoplasmic membranes of halophilic archaea of the family *Halobacteriaceae* contain interesting ether lipids and often have retinal proteins (bacteriorhodopsin, halorhodopsin, and sensory rhodopsins). It is known that unsaturated ether lipids are far more common in the halophilic archaea than generally assumed. Such unsaturated diether lipids were earlier reported from the psychrotolerant haloarchaeon *Halorubrum lacusprofundi* (Gibson et al., 2005).

Many alkaliphiles are at the same time halophilic, and many useful enzymes applied in the detergent industry (washing powders), the textile industry, and other processes were obtained from bacteria growing in saline alkaline lakes. Halophilic enzymes (typical for halophilic archaea and bacteria) are characterized by an excess of acidic amino acids and subsequent negative surface charge. This feature allows effective competition for hydration water and enables function in solutions of low water activity, including organic solvent/water mixtures. The immediate advantages for enzyme technology are as follows: higher salt and heat tolerance, a catalytic environment which enables use of less polar educts, and potential reversal of hydrolytic reactions, all of which make them powerful candidates for industrial biocatalysts (Ma et al., 2010). The increase of salinity and nitrate/nitrite concentrations in soils and ground waters in the last few decades has focused much attention on the physiological and molecular mechanisms involved in salt-stress tolerance and nitrate metabolism by microorganisms. Physiological studies carried out with *Haloferax mediterranei* have revealed that it is resistant to very high nitrate (up to 2 M) and nitrite (up to 50 mM) concentrations (Bonete et al., 2008). Microorganisms are in general sensitive to low nitrate and nitrite concentrations. The inhibitory effect of these nitrogen compounds is due to the extreme toxicity of nitrite and nitric oxide produced upon nitrate reduction (Bonete et al., 2008).

3. Alkaliphilic Archaea

Alkaliphile are microorganisms that grow very well at pH values between 9 and 12 or grow only slowly at the near neutral pH value of 6.5 (Horikoshi, 1999). The best examples of naturally occurring alkaline environments are soda deserts and soda lakes. Extremely alkaline lakes, for example, Lake Magadi in Kenya and the Wadi Natrun in Egypt, are probably the most stable highly alkaline environments on Earth, with a consistent pH of 10.5 to 12.0 depending on the site (Horikoshi, 1999). Alkaliphilic haloarchaea are a specialized group of obligate extreme halophiles that require high salinity as well as high pH (8.5-11) and low Mg^{2+} for growth (Kamekura et al., 1997; Xu et al., 2001). Alkaliphilic haloarchaea are commonly isolated from soda lakes but they have also been isolated and detected from solar salterns and other hypersaline environments with acidic, neutral and alkaline pHs (Gareeb & Setati, 2009).

Alkaliphilic halophilic archaea are classified in the genera *Natronobacterium, Natronococcus* and *Natronomonas* (Kamekura et al., 1997), and some recent isolates are accommodated in *Halalkalicoccus* (Xue et al., 2005) and *Natronolimnobius* (Itoh et al., 2005). These five genera are so far composed exclusively of alkaliphilic strains. On the other hand, the genera *Haloarcula, Natronorubrum, Halobiforma, Natrialba, Haloterrigena, Halorubrum* and *Halostagnicola* consist of both neutrophilic and alkaliphilic species. *Halorubrum* and *Haloarcula* spp often form the majority of isolates is inhabited neutral brines (Grant et al., 2001). In a recent study isolates belonging to the genera *Natrialba, Natronococcus* and *Natronorubrum* were recovered from brine samples at evaporator ponds in Botswana (Gareeb & Setati, 2009).

Hypersaline soda lakes are mostly inhabited by alkaliphilic representatives of halophilic archaea that could be in numbers of 10^7 to 10^8 /ml in soda lake brines (Horikoshi, 1999). A novel haloalkaliphilic archaeon from Lake Magadi was isolated and characterized (McGenity & Grant, 1993). It was revealed that cells of this isolate contained large gas vacuoles in the stationary phase of growth, and colonies produced by these archaea were bright pink. Xu et al. (1999) isolated two haloalkaliphilic archaea from a soda lake in Tibet. The strains were gram negative, pleomorphic, flat, nonmotile, and strictly aerobic. Their growth required at least 12% (2 M) NaCl and occurred between pH 8.0 and 11 with an optimum at pH 9.0 to 9.5. DNA-DNA hybridization results suggested that the two strains belonged to different species of the same genus. Recently, A novel haloalkaliphilic archaeon from commercial rock salt imported to Japan from China was isolated and characterized (Nagaoka et al., 2011). The isolation of a haloalkaliphilic archaea, grown optimally at pH 10.0, from a saline–alkaline soil was reported (Wang et al., 2010).

The isolates displayed typical haloalkaliphilic growth characteristics with optimal growth at pH 9–10. Halophilic methanogens were isolated from various neutral saline areas and natural hypersaline environments (Boone et al., 2001). These strains showed the optimal growth at temperatures near 40°C and, in medium containing 0.5 to 2.5 M NaCl, at pH values near 7. Zhilina and Zavarzin (1994) described bacterial communities which inhabited in alkaline lakes, and in particular the diversity of anaerobic bacteria developing at pH 10 was exhibited. A new obligate alkaliphilic, methylotrophic methanogen was isolated from Lake Magadi (Kevbrin et al., 1997). Based on its phenotypic and genotypic properties, the isolate found to be belonged to *Methanohalophilus zhilinaeae*. It was an obligate alkaliphile and grew optimally within pH 9.2.

The alkaliphiles are unique microorganisms, with great potential for microbiology and biotechnological exploitation. The essential use of alkaliphilic enzymes is in the detergent industry, for example, an extracellular protease from *Natrialba magadii*, a haloalkaliphilic archaeon, is a solvent tolerant enzyme and suggests a potential application in aqueous-organic solvent biocatalysis (Diego et al., 2007). Recently it was reported that the gene encoding the protease Nep secreted by the *Natrialba magadii* was cloned and sequenced (DeCastro et al., 2008). The study was announced the molecular characterization of a halolysin-like protease from alkaliphilic haloarchaea and the description of a recombinant system that facilitated high-level secretion of a haloarchaeal protease. Alcohol dehydrogenase is a key enzyme in production and utilization of ethanol. The gene encoding for ADH of the haloalkaliphilic archaeon *Natronomonas pharaonis* was cloned and expressed in *Escherichia coli* (Cao et al., 2008). The enzyme was haloalkaliphilic and thermophilic, being most active at 5 M NaCl or 4 M KCl and 70°C, respectively. The optimal activity was observed at pH 9.0 and significantly inhibited by Zn^{2+}. It was concluded that the physiological role of this enzyme is likely related to the oxidation of ethanol to acetaldehyde.

4. Acidophilic Archaea

Both natural and man-made acidic environments on the Earth are commonly found in the sites like pyrite ores, solfatara fields and marine volcanic vents; the microorganisms that inhabited these areas are called acidophiles and have a pH optimum for growth pH <3 (Baker-Austin & Dopson, 2007). Acidophiles are most widely distributed in the bacterial and archaeal domains and contribute to numerous biogeochemical cycles including the iron and sulfur cycles.

Acidophiles might have played some crucial function in the evolution because metabolic processes might have originated on the surface of sulfide minerals (Wachtershauser, 2006) and structuring of the genetic code could have occurred in an intracellular environment with acidic pH (Di Giulio, 2005). Acidophiles optimally grow in low pH and metal-rich environments which might be quite resemble to volcanic aqueous conditions during Archaean and early Proterozoic periods. Therefore, it was suggested that acidophiles could represent primordial form of life from which more complex life have evolved (Baker-Austin & Dopson, 2007). Acidophiles are mostly found in isolated and inaccessible environments like geothermal vents. These environments generally have an impassable physical barrier which reduces the growth of neutrophiles. Recent bioinformatic analysis of several thermoacidophile archaeal genomes have implied that the similarities between these organisms are greater than expected when compared with other more closely related organisms. Therefore, acidic environments could establish an old and genetically distinct niche of life in which ecological closeness disregards phylogenetic nearness (Futterer et al., 2004).

The ongoing exploration of the Earth has led to continued discoveries of life in environments that were previously considered uninhabitable. Thus, interest in the biodiversity and ecology of extreme environments and their inhabitants has grown over the past several years. In this regard, the study of extremely acidic environments is taking too much attention, because environmental acidity is often a consequence of microbial activity (Hallberg & Johnson, 2001). Highly acidic environments are relatively less common on Earth and are generally associated with mining activities. One of the main sources of acidity is the natural oxidation and dissolution of sulfidic minerals exposed to oxygen and water and this process can be greatly enhanced by microbial metabolism (Nordstrom & Alpers, 1999). Jhonson (1998) claimed that microorganisms that is involved in the generation of acidic metalliferous wastes cause widespread environmental pollution.

The extremely thermoacidophilic archaea are a group of interesting microorganisms in that they have to simultaneously cope with biologically extreme pHs and temperatures in their natural environments (Auernik et al., 2008). The current studies of the thermophilic and mesophilic acidophilic archaea have implied that there might be a stronger association between tetra-ether lipids and the tolerance to acid gradients than previously thought (Macalady & Banfield, 2003). Archaeal cell membrane including tetra-ether lipids instead of bacterial ester linkages is an example of highly impermeable cell membrane. These compounds were identified in *Thermoplasma acidophilum* (Shimada et al., 2002). Their expanding biotechnological significance relates to their role in biomining of base and precious metals and their unique mechanisms of survival in hot acid, at both the cellular and biomolecular levels. Extreme thermoacidophiles are microorganisms that are characterized by having an optimal growth temperature $\geq$ 60 °C and an optimal pH of $\leq$4. A majority of the extremely thermoacidophilic species studied to date belongs to the archaeal orders of *Sulfolobales* and *Thermoplasmatales* (Auernik et al., 2008). *Acidianus infernus* which is the most thermophilic of the extreme thermoacidophiles grows at temperatures up to 95 °C and at pHs as low as 1.0. On the other hand, *Picrophilus* species, member of the *Thermoplasmatales* (euryarchaeon), are the most acidophilic organisms growing at pHs as low as 0 and at temperatures up to 65 °C (Huber & Stetter, 2006).

The application of recent molecular genetic systems and genome sequence data have given new clues in understanding of heavy metal tolerance, implementation of a genetic system and discovery of a new carbon fixation pathway. As a consequence, new insights into the molecular mechanisms that define extreme thermoacidophily have been gained. Extreme thermoacidophiles have evolved mechanisms for tolerating heavy metals which are physiologically toxic to most microorganisms (Salzano et al., 2007). These mechanisms cover to their capacity to recover from metal-induced damage (similar to oxidative stress) and to stop the accumulation of effective toxic metal concentration. In some cases, extreme thermoacidophiles could reduce or oxidize metals to less toxic forms by metabolic pathways. In other cases, metal chelation or complexation can perform the same job. Other mechanisms are based on the exporting toxic metal ions via P-type ATPases, instead of direct or indirect metal transformation (Ettema et al., 2006).

Acidophiles have some important biotechnological applications like metal recovery from ores, known as biomining (Rawling, 2002). Biomining is becoming increasingly important in mining because of its lower and containable pollutant outputs comparing to thermal processes. Efficacy in biomining environments demands tolerance to high levels of toxic heavy metals and the ability to incorporate inorganic carbon, as organic carbon can be scarce in this environment (Auernik et al., 2008). The future success of the biomining industry will depend upon the cellular biocatalysts with favorable features. The knowledge coming from genomic research of extreme thermoacidophiles will open new applications in this area.

5. Methanogens

Methane (CH_4) is an important greenhouse gas and its atmospheric abundance has been increasing by about 0.5% per year. Methanogenesis is a common process in many anaerobic environments such as digesters, rumen, rice fields, oil wells, landfills, and a range of extreme habitats (Garcia et al., 2000). Microorganisms are considered to be responsible for about the 50% of total methane production in the world. Methanogens, CH_4-producing microorganisms, are strict anaerobes in the *Euryarchaeota* and can produce CH_4 from a limited number of substrates: CO_2 and H_2, formate, methanol, methylamines or acetate. As terminal oxidizers in complex microbial communities, they are vital to the anaerobic microbial degradation of organic compounds in natural environments and probably also in defined ecological niches of the human body (Cavicchioli et al., 2003). Since methanogens coexist and closely interact with anaerobic bacteria at certain sites (e.g., human colon or dental plaque) they could be implicated in mixed anaeorobic infections. In fact, methanogens have recently been linked to periodontal disease (Lepp et al., 2004), a polymicrobial infection that affects the gums and supporting structures of the teeth and is characterized by periodontal pockets. Methanogens are phylogenetically diverse organisms although they rely on few metabolic substrates (Yavitt et al., 2011).

Methanogens are located in the domain *Archaea* and the phylum *Euryarchaeota*. Unlike *Bacteria*, some methanogens have pseudomurein in their cell wall instead of peptidoglycan (Balch et al., 1979). The glycan strand of pseudomurein is composed of alternating β-(1→3)-linked N-acetyl-D-glucosamine and β-(1→3)-linked N-acetyl-L-talosaminuronic acid (Kandler & Konig, 1998). The most closely related bacterial homologs to pseudomurein have been found in anaerobic Gram-positive or δ-proteobacterial lineages. It is reported that these organisms share the same ecological niche, and could also be the donors of peptide ligase homologs (MurC, MurE and MurF) which are probably responsible for pseudomurein biosynthesis in the *Methanobacteriales* (Graham & Huse, 2008). While pseudomurein is a common cell wall component in *Methanobrevibacter* and *Methanobacterium*, heteropolysaccharide is in *Methanosarcina*, and protein is in *Methanococcus* and *Methanocalculus*. Methanogens are also characterized by including coenzyme F420, which is a cofactor necessary for certain enzyme activities such as hydrogenase and formate dehydrogenase. Another coenzyme typical to methanogens is coenzyme M, which is either produced by the methanogens, such as *Methanobacterium*, or is obtained from an external source, which is the case for *Methanobrevibacter ruminantium* (Ashby et al., 2001).

Methanogens were the first organisms to be identified as archaea and classified as a separate domain. Six orders of methanogens exist: *Methanobacteriales*, *Methanopyrales*, *Methanococcales*, *Methanomicrobiales*, *Methanosarcinales* and *Methanocellales*. All of these orders contain a wide diversity of taxa that display great variance in their morphological and physiological characteristics. However, they all retain in common an anaerobic lifestyle and the ability to produce methane metabolically (Bapteste et al., 2005). Complete genome sequences for representatives of all of these orders are available. Initial analyses of the *Methanocaldococcus* - formerly *Methanococcus jannaschii* genome have revealed that the archaea have many metabolic characteristics in common with bacteria, but that the genes used for information processing are more similar to equivalent systems in eukaryotes (Walters & Chong, 2010).

Following the suggestion that the *Archaea* are a distinct taxonomic group, it was considered that the domain would be divided along phenotypic lines and that the methanogenic archaea would be monophyletic (Bapteste et al., 2005). The 16S rRNA gene comparisons, however, indicated that this was not the case; the *Methanomicrobiales* were more closely related to extremely halophilic archaea (*Halobacteriales*) than to other methanogens (Woese et al 1990). The isolation of *Methanopyrus kandleri* and the sequencing of its 16S rRNA gene sequences showed that it was unrelated to any other known methanogens and originated at the base of the euryarchaeal branch of the tree, implied that the ancestor of euryarchaeal species might have been a methanogen (Burggraf et al., 1991). Recent studies based on transcriptional and translational proteins and 16S rRNA sequences have approved the absence of monophyly of methanogens and strongly suggested a close phylogenetic relationship between *Methanopyrus*, *Methanococcus* and *Methanothermobacter* (Brochier et al., 2004).

As most of archaea have been found living at the extreme sites on this planet, they have often been proposed to resemble the life outside the earth if exists (Jarell et al., 2011). The expectations of Earth-like organisms that could exist on other planets has varied, with methanogens often mentioned because of their adaptation to anaerobic niches with little or no organic carbon (Moissl-Eichinger, 2011), and especially with respect to the possible biogenic formation of the methane on Mars. Methanogenesis is a process occurring in many anaerobic environments. For instance methanogenesis in cold marine sediments has a global significance by leading to methane hydrate deposits, cold seeps, physical instability of sediment, and atmospheric methane emissions. The analysis of cultivation-independent archaeal community revealed that uncultivated microbes of the kingdoms *Euryarchaeota* and *Crenarchaeota* are present and that methanogens comprised a small proportion of the archaeal community. Methanogens were cultivated from depths of 0 to 60 cm in the sediments, and several strains related to the orders *Methanomicrobiales* and *Methanosarcinales* were isolated (Kendall et al., 2007). Microbial diversity was examined in the cold marine sediments (Orphan et al., 2002), however, the function that these microbes carry out during geochemical processes is still not clear. In addition to methanogens, uncultivated lineages of other *Archaea* have been also identified in marine sediments (Vetriani et al., 1999).

Unlike that of bacteria, the diversity of gut methanogenic archaea seems to be well understood and limited to two species belonging to *Methanobacteriales*, one of the five methanogenic orders defined to date: *Methanobrevibacter smithii* and occasionally *Methanosphaera stadtmanae* (Miller & Volin, 1982). In a current study the diversity of methanogenic Archaea from the gut of humans was analyzed by targeting mcrA, a molecular metabolic marker of methanogenesis (Mihajlovski et al., 2008). They reported the presence of *Methanobacteriales, Methanobrevibacter smithii, Methanosphaera stadtmanae* and a distant phylotype that did not cluster with any of the methanogenic orders. Their results were also supported by 16S archaeal sequences retrieved from the same volunteer, strongly suggests there may be a sixth order and hence potential underestimation of the role of methanogens in gut physiology. There is a link between the physiology of cultured methanogenic Archaea and their phylogenetic closeness based on 16S rRNA sequences (Zinder, 1993). For example, while most species of the *Methanobacteriaceae* and *Methanomicrobiaceae* prefer H_2 and CO_2 (or formate) as substrates for methanogenesis, *Methanosaeta*, a genus within the *Methanosarcinaceae*, is known to generate energy only from acetate fermentation. Most of the other *Methanosarcinaceae* preferentially use methanol and related methyl-substrates for the generation of CH_4 (Kleikemper et al., 2005).

It has been considered that methanogenesis may function in the mineralization of petroleum hydrocarbons in contaminated aquifers (Chapelle et al., 2002). However, petroleum hydrocarbons can not be degraded directly by methanogenic microorganisms (Zengler et al., 1999). Methanogens metabolizing H_2 and CO_2 involve indirectly into PHC degradation by maintaining H_2 concentrations low by that way fermentation of PHC becomes exergonic and fermenting microrganisms can grow (Garcia et al., 2000). On the other hand, methanogens metabolizing acetate or methanol can degrade PHC directly by cleaving end products of fermentation. However, the role of different metabolic groups of methanogens with respect to overall methanogenic activity in PHC-contaminated aquifers is reported to be unclear (Kleikemper et al., 2005). Aceticlastic methanogenesis was assumed to be the final step of hydrocarbon degradation in a PHC-contaminated aquifer, but this was not confirmed with activity measurements (Dojka et al., 1998). The research for alternative forms of energy, including recovery of methane via anaerobic digestion of wastes, has becoming very popular since successive petroleum crisis in 1970s.

6. Thermophilic Archaea

In the domain Archaea, hyperthermophilic and extremely thermophilic archaea are determined extensively over the phyla *Crenarchaeota* and *Euryarchaeota*. Hyperthermophiles, growing optimally at ≥ 80 °C, have been acknowledged since 1981 (Zillig et al., 1981). They represent the upper temperature limit of life and are found in environments characterized with high temperature. In their basically anaerobic environments, they generally obtain energy by inorganic redox reactions. Stetter (2006) claimed that in his lab so far about 50 new species of hyperthermophiles have been isolated and characterized, among them representatives of the novel bacterial genera *Thermotoga, Thermosipho, Aquifex, Thermocrinis* and archaeal genera *Acidianus, Metallosphaera, Stygiolobus, Thermoproteus, Pyrobaculum, Thermofilum, Desulfurococcus, Staphylothermus, Thermosphaera, Ignicoccus, Thermodiscus, Pyrodictium, Pyrolobus, Thermococcus, Pyrococcus, Archaeoglobus, Ferroglobus, Methanothermus, Methanopyrus* and *Nanoarchaeum*. The earliest archaeal phylogenetic lineage is represented by the extremely tiny members of the novel kingdom of *Nanoarchaeota* (Stetter, 2006). Hyperthermophiles occupy all the short deep branches closest to the root within the universal phylogenetic tree. In rRNA-based phylogenetic tree constructed by Woese et al. (1990), all extremely short and deeply branching-off lineages within the archaea and bacteria are exclusively represented by hyperthermophiles (Fig. 1), indicating a slow rate of evolution (Stetter, 2006).

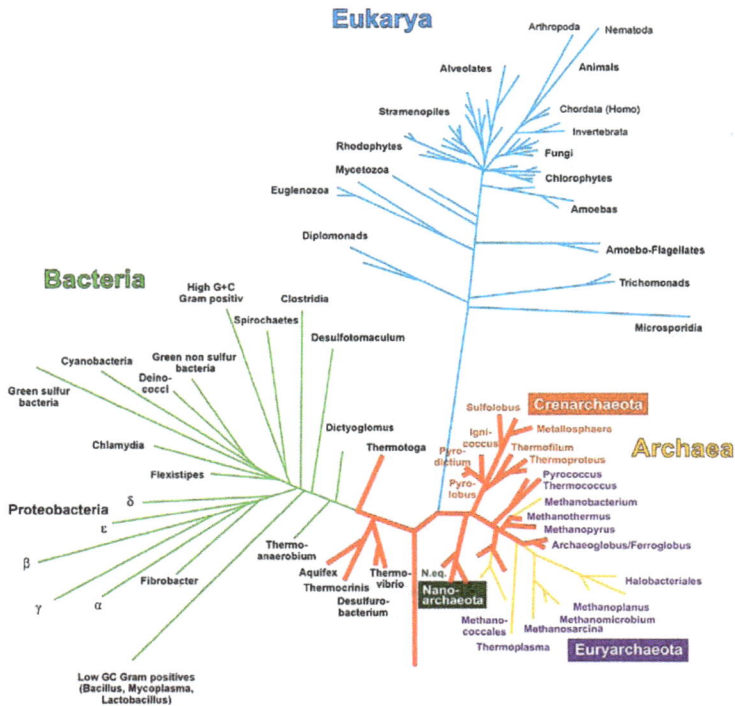

Fig. 1. Universal phylogenetic tree constructed based on rRNA sequence comparison. Hyperthermophiles represented with thick red lineages (Stetter, 2006).

Until now, all *Crenarchaeota* species that have been isolated are either hyperthermophilic or extremely thermophilic, although the existence of mesophilic and psychrophilic *Crenarchaeota* is also suggested by culture-independent molecular phylogenetic analyses. The cultured *Crenarchaeota* are composed of five orders: *Acidilobales*, *Fervidicoccales*, *Desulfurococcales*, *Sulfolobales* and *Thermoproteales*, which are well-supported by 16S rDNA sequence data and by phenotypic properties, such as cell morphology and lipid composition (Burggraf et al., 1997; Reysenbach, 2001; Chaban et al., 2006; Prokofeva et al., 2009). Some of the crenarchaeotas are sequenced from low-temperature environments (Dawson et al., 2001). So far only one member of the groups, *Crenarchaeum symbiosum*, from non-thermophilic environments has been cultivated in axenic culture (Preston et al., 1996). The information about non-thermophilic *Crenarchaeota* is, therefore, solely based on sequence data collected from various low temperature environments like soil, freshwater, deep drillings, and seawater.

The analyses about the energy sources of hyperthermophiles have revealed that most species are chemolithoautotroph (Amend & Shock, 2001; Stetter, 2006). Mode of Respiration is anaerobic -nitrate, sulphate, sulphur and carbon dioxide respiration- and aerobic, CO_2 is the solely carbon source required to synthesize organic cell material. While molecular hydrogen generally serves as main electron donor, sulphide, sulphur, and ferrous iron are other electron donors. Oxygen also may serve as an electron acceptor in some hyperthermophiles which are usually microaerophilic. Anaerobic respiration forms are the nitrate-, sulphate-, sulphur- and carbon dioxide respirations. While chemolithoautotrophic hyperthermophiles synthesize organic matter, some obligate heterotrophic hyperthermophiles depend on organic material as energy- and carbon-sources. In addition, a few chemolithoautotrophic hyperthermophiles are opportunistic heterotrophs. They obtain energy either by aerobic or anaerobic respiration or by fermentation (Stetter, 2006).

Recently a novel phylum of Archaea, called *Nanoarchaeota*, was discovered. The phylum is currently represented by a single species named as *Nanoarchaeum equitans*. The species is a nano-sized hyperthermophilic symbiont that grows attached to the surface of an *Ignicoccus* species (Huber et al., 2002). It has a cell diameter of only 400 nm and grows under strictly anaerobic conditions at temperatures between 75 °C and 98 °C. The *N. equitans* has a genome of only 490.885 bp, it is the smallest genome of any Archaea, and the most compact, with 95% of the DNA predicted to encode proteins or stable RNAs. Its DNA encodes the complete machinery for information processing and repair, but lacks genes for lipid, cofactor, amino acid, and nucleotide biosynthesis. The inadequate biosynthetic and catabolic capacity of *N. equitans* considers its symbiotic relationship to its *Ignicoccus* host as a parasitic relationship, and this makes *N. equitans* the only known archaeal parasite (Waters et al., 2003).

7. Biotechnological importance of Archaea

Industries such as food processing, cleaning, biosynthetic processes and environmental bioremediation need efficient biocatalysts which can operate in harsh environments. The sources for these biocatalysts have been animals, plants, fungi and mostly bacteria. Recently, the extremophilic bacteria and archaea have become more popular since the enzymes of these organisms are able to remain catalytically active under extremes of temperature, salinity, pH and pressure (Synowiecki et al., 2006; Ozcan et al., 2009). In biotechnology for an efficient application, one has to determine the most suitable enzymes and best reaction conditions. Nowadays there are two main strategies for obtaining enzymes with desired properties, namely the genetic engineering of currently known enzymes and the search for new activities in previously uncharacterized microorganisms. Within the second approach, the search for enzymes in extremophiles (called extremozymes) seems to be particularly promising since the enzymes of these organisms have particular adaptations to increase their stability in adverse environments, which can potentially also increase their stability in the harsh environments in which they are to be applied in biotechnology (Oren, 2002).

The domain archaea is particularly under extensive research for their potential biotechnological applications. The domain includes the extreme halophiles, the hyperthermophiles, the thermoacidophilic archaea and the psychrophiles that live in the cold waters of the Antarctic. But the most important value of these organisms (along with some of the bacteria that also tolerate extreme environments) is that their enzyme systems work at harsh conditions. For example, many of the restriction enzymes used in gene splicing and cloning are products of extremophiles. In 1993 a report from US National Academy of Sciences noted that world enzyme sales equaled to US$1 billion and it is a market that has been expected to grow about 10% per year. It is expected that enzymes from extremophiles will constitute an important part of this market.

Recently genes encoding several enzymes from extremophiles have been cloned in mesophilic hosts, with the objective of overproducing the enzyme and altering its properties to suit commercial applications (Alqueres et al., 2007). It has been revealed that enzymes derived from extremophilic archaea are in many cases superior to bacterial homologs. They have higher stability towards heat, pressure, detergents and solvents (Egorova & Antranikian, 2005). The archaeal hyperthermophiles, known as heat-stable prokaryotes, have exceptionally high growth temperature limits and unique ether-linked lipids as well as eukaryotic transcription and translation factors. It has been proposed that heat shock proteins which are inducible by supraoptimal temperatures (Kagawa et al., 1995) and reverse gyrase which installs positive supercoils in DNA (Guipaud et al., 1997) could have been involved in unusual adaptive mechanisms of the hyperthermophiles.

DNA polymerase I from the bacterium *Thermus aquaticus*, Taq polymerase, is the first thermostable DNA polymerases in biotechnology and has an extensive use in PCR. For PCR applications, the archaeal DNA polymerases have been found to have some superior features to bacterial ones. For instance, Pwo from Pyrococcus woesei, Pfu from P. furiosus, Deep Vent polymerase from the *Pyrococcus* strain GB-D and Vent polymerase from *Thermococcus litoralis*, reported to have an error rate that is much lower than that of Taq polymerase.

The starch-processing industry needs thermostable enzymes in order to convert starch into more valuable products such as dextrins, glucose, fructose and trehalose, (Egorova & Antranikian, 2005). It is well established that in all starch-converting processes, high temperatures are required to liquefy starch and to make it accessible to enzymatic hydrolysis. The synergetic action of thermostable amylases, pullulanases and α-glucosidases include the advantage of lowering the cost of making sugar syrup, production of new starch-based materials that have gelatin-like characteristics and defined linear dextrins that can be used as fat substitutes, texturizers, aroma stabilizers and prebiotics. Recently a variety of starch-degrading enzymes from extremophilic archaea has been published. They determined the optimal temperatures for the activity of the archaeal amylases in a range between 80 °C and 100 °C. Especially, the high thermostability of the extracellular α-amylase from *Pyrococcus* sp. (retain its activity even at 130°C) makes these enzymes ideal candidates for industrial application (Egorova & Antranikian, 2005).

Lipases (carboxyl ester hydrolases) are ubiquitous in nature, produced by animals, plants, and fungi, as well as bacteria. Recently halophilic archaeal organisms are taking more attention due to their lipolytic enzymes. For instance, *Haloarcula marismortui*, a halophilic archaeon whose genome was sequenced contains genes encoding for putative esterase and lipase. These genomic predictions have been verified recently (Camacho et al. 2009). They reported that *Haloarcula marismortui* displays esterase and lipase activity intracellularly and extracellularly. While lipase was accumulated mainly extracellular, esterase was generally accommodated intracellular (Camacho et al., 2009). The esterase and lipase genes of *Pyrococcus furiosus*, a hyperthermophilic archaeon, have been cloned in *E. coli* and functional properties have been determined. The archaeal enzyme reported as the most thermostable and thermoactive esterase known to date (Ikeda and Clark 1998). In our laboratory, the halophilic archaeal isolates, grown best in a range of 2-5 M NaCl, produced high lipolytic activity in the range of 3-4.5 M NaCl (Ozcan et al., 2009). It was found that the lipolytic activity dropped at 5 M NaCl. Therefore these enzymes can be classified as not only salt dependent but many can be also thermostable (Oren, 2002; Ozcan et al., 2009).

Heat-stable proteases have some utilities in biotechnology, especially in the detergent industry. It has been revealed that most proteases from extremophilic Archaea belong to the serine type and are stable at high temperatures, even in the presence of high concentrations of detergents and denaturing agents (Antranikian et al., 2005). Proteases are also applied for peptide synthesis using their reverse reaction, mainly because of their compatibility with organic solvents (Egorova & Antranikian, 2005). Recently a protease from *Thermococcus kodakarensis* has been characterized (Foophow et al., 2010). It has been shown that a β-jelly roll domain is not directly involved in proteolytic activity, however, takes role in the extreme thermostability of the enzyme.

One of the important strategies in searching the alternative energy sources could be the conversion of biomass to fuel products (biofuels). Both chemical and biological processes are being explored for the production of bioethanol, biodiesel, biobutanol, biomethane and biohydrogen. For the production of biohydrogen some heterotrophic hyperthermophiles particularly have been found useful because of their abilities to produce molecular hydrogen (Atomi et al., 2011). It is covalently linked to Lys216 in the chromophore by Schiff base action.

Habitats with harsh environmental conditions are often populated by archaea that are specialized to live in water near salt saturation also specialized to utilize solar energy. Light of distinct wavelengths is used to fuel primary transport of both protons and chloride ions as well as for phototaxis. The halophilic archaeon makes use of light for both energy and sensory transduction by exploiting a family of light-sensitive proteins called rhodopsin. Bacteriorhodopsin is a 25-kDa integral membrane protein that carries a retinal group (cromophore) and it is covalently linked to lysine-216 by Schiff-base action. Its function was discovered in the early 1970s during studies of the purple membrane, patches of membrane that contain only bacteriorhodopsin and lipids, found within the cell membrane of *Halobacterium salinarum* (Oren, 2008). The excellent thermodynamic and photochemical stability of bacteriorhodopsin has led to many uses in technical applications like holography, spatial light modulators, artificial retina, neural network optical computing, and volumetric and associative optical memories (Alqueres et al., 2007).

Archaeosomes, a kind of liposomes, are produced from natural lipids found in *Archaea* or from synthetically synthesized compounds that share the unique structural features of archaeal lipids. The isoprenoid glycerolipid membrane of archaeosomes can develop into a bilayer, a monolayer, or a combination of mono- and bilayers made from bipolar and monopolar archaeal lipids (Jacquemet et al., 2009). The archaeal lipids formulations have shown to exhibited relatively higher physico-chemical stabilities to oxidative stress, high temperature, alkaline pH, action of phospholipases, bile salts and serum media (Brard et al., 2007). These properties contribute to their efficacy as self-adjuvant vaccine delivery vesicles. Additionally, archaesomes were found to be safe and not toxic in mice both in-vitro and -vivo studies (Omri et al., 2003). Thus, it has been claimed that the biocompatibility and the superior stability properties of archaeosomes in several conditions give advantages over conventional liposomes in the manufacture and the use in biotechnology including vaccine and drug delivery (Jacquemet et al., 2009).

8. Conclusion

The majority of the living beings thrive in environments having physically and geochemically temperate conditions. The extreme environments found on the planet are generally inhabited by microorganisms, which belong to the Archaeal and Bacterial domains of life. Archaea, members of the third domain of life, are prokaryotes that harbour many unique genotypic and phenotypic properties. Two archaeal phyla are presently recognized; the *Euryarchaeota* and the *Crenarchaeota* (Woese et al., 1990).

Archaea exhibit a wide diversity of phenotypes. The first phenotypes to be recognized were the methanogens which are strict anaerobes and methane producers. Many archaea are extremophiles which are capable of growth at high tempereture, salinity and extremes of pH. In common usage, cultivated *Archaea* were, without exception, considered to be extremophiles. Extreme environments comprise sites of extreme temperature, pH, pressure and salinity. *Archaea* exist in a broad range of habitats, and as a major part of global ecosystems, may contribute up to 20% of earth's biomass (Delong and Pace, 2001).

Thermophilic archaea inhabit environments like hot springs, ocean vents, and geysers which are inhospitable to many other organisms. These habitats not only have extremely high temperatures but have high concentrations of dissolved minerals and low concentrations of oxygen (Egorova and Antranikian, 2005). In particular, enzymes from thermophilic and hyperthermophilic Archaea have industrial relevance. Enzymes of thermophilic archaea are superior to the traditional catalysts because they can perform industrial processes even under harsh conditions (Egorova and Antranikian, 2005).

The extreme halophilic archaea require at least 1.5 M NaCl. Most strains grow best at 3.5–4.5 M NaCl. Members of the halophilic archaea are dominant microorganisms in hypersaline environments worldwide including salt lakes, crystallizer ponds of solar salterns, salt mines, as well as hypersaline soda lakes (Oren, 2002). Halophilic archaea have a number of useful applications in biotechnological processes and potential new applications are being investigated. For instance, they produce bacteriorhodopsin (used in information processing and ATP generation), novel extracellular polysaccharides, exoenzymes (amylase, cellulase, xylanase, lipase and protease) and poly-hydroxyalkanoate (used in biodegradable plastic production), and a protein from *Halobacterium salinarum* has significance in cancer research. Some members are extremely alkaliphilic as well as being halophilic. They grow optimally in alkaline environment with pH values above 9 such as soda lakes and carbonate-rich soils but cannot grow or grow only slowly at the near-neutral pH value (6.5). On the other hand alkalithermophilic members of the Archaea were isolated besides halophilic. These oorganisms have economic value because alkaliphilic enzymes have been used in different industries (Horikoshi, 1999).

Acidophilic archaea have a pH optimum for growth of less than pH 3. They contribute to numerous biogeochemical cycles including the iron and sulfur cycles. Acidophiles have different biotechnological applications. Firstly metal extraction from ores and this sustainable biotechnological process is becoming increasingly important because of its reduced and containable pollutant outputs. Acidophiles could also be a source of gene products; for example, acid-stable enzymes with applications as lubricants and catalysts (Baker-Austin and Dopson, 2007).

Metabolism of methanogens is unique in that energy is obtained by the production of methane (natural gas). Biological methanogenesis is applied to the anaerobic treatment of sewage sludge, and agricultural, municipal and industrial wastes, where the maintenance of a desired methanogenic flora is achieved by inoculation (Schiraldi et al., 2002).

Many archaea colonize extreme environments. Because extremophilic microorganisms have unusual properties, they are a potentially valuable resource in the development of novel biotechnological processes. Especially, based on the unique stability of archaeal enzymes at high temperature, salt and extremes of pH, they are expected to be a very powerful tool in industrial biotransformation processes that run at harsh conditions.

The growing demand for more effective biocatalysts has been satisfied either by improving the properties of existing proteins or by producing new enzymes. The majority of the industrial enzymes known to date have been extracted mostly from bacteria and fungi. Until now, only a few archaeal enzymes have been found to be useful in industrial applications. Based on the unique stability of archaeal enzymes at high temperature, salt and extremes of pH, they are expected to be a very powerful tool in industrial biotransformation processes that run at harsh conditions. Owing to the unique features of *Archaea*, their potential applications in biotechnology are far reaching, ranging from bioremediation potential of nitrite and nitrate from groundwater, oil pollutions, toxic compounds, metal polluted sites and biomining and nitrate removing from brines (Schiraldi et al., 2002).

9. References

Alquéres, S.M.C., Almeida, R.V., Clementino, M.M., Vieira, R.P., Almeida, W.I., Cardoso, A.M., Martins, O.B. (2007) Exploring the biotechnologial applications in the archaeal domain Brazilian Journal of Microbiology 38:398-405

Amend, J. P., Everett, L. (2001) Shock Energetics of overall metabolic reactions of thermophilic and hyperthermophilic Archaea and Bacteria. *FEMS Microbiology Reviews*. Vol. 25, pp. 175-243

Antranikian G, Vorgias C, Bertoldo C. (2005) Extreme environments as a resource for microorganisms and novel biocatalysts. Adv Biochem Eng Biotechnol. 96, 219-62.

Ashby, K. D., Casey, T. A., Rasmussen, M. A., and Petrich, J. W. (2001) Steady-state and time-resolved spectroscopy of F420 extracted from methanogen cells and its utility as a marker for fecal contamination. *Journal of Agricultural and Food Chemistry*. Vol. 49 (3), pp. 1123–1127

Atomi, H., Sato, T., and Kanai, T. (2011) Application of hyperthermophiles and their enzymes Current Opinion in Biotechnology, 22, 618–626.

Auernik KS, Maezato Y, Blum PH, Kelly RM. (2008) The genome sequence of the metal-mobilizing, extremely thermoacidophilic archaeon *Metallosphaera sedula* provides insights into bioleaching-associated metabolism. Appl Environ Microbiol, 74, 682-692.

Balch, W. E., Fox, G. E., Magrum, L. J., Woese, C. R., and Wolfe, R. S. (1979) Methanogens: reevaluation of a unique biological group. *Microbiol. Rev.* Vol. 43, pp. 260-296

Baker-Austin, C. and Dopson, M. (2007) Life in acid: pH homeostasis in Acidophiles. Trends in Microbiology, Vol.15 No.4.

Bapteste, E., Brochier, C., and Boucher, Y. (2005) Higher-level classification of the Archaea: evolution of methanogenesis and methanogens. *Archaea*. Vol. 1, pp. 353–363

Begon, M., Harper, J. L., and Townsend, C. R. (1986) Ecology: individuals, populations, and communites, First edit., Sinauer Associates, Sunderland, MA

Belay, N., Mukhopadhyay, B., Conway, D. M., Galask, R., and Daniels, L. (1990) Methanogenic bacteria in human vaginal samples. *J. Clin. Microbiol.* Vol.28, pp. 1666–1668

Benlloch, S., Acinas, S.G., Anton, J., Lopez-Lopez, S.P., Luz, F., Rodriguez-Valera, F., (2001) Archaeal biodiversity in crystallizer ponds from a solar saltern: culture versus PCR. *Microb. Ecol.* Vol. 41, pp. 12-19.

Bonete, M. J., Martínez-Espinosa, M. R., Pire, C., Zafrilla, B., and Richardson, D. J. (2008) Nitrogen metabolism in haloarchaea. Saline Systems, Vol.4(9), pp.1-12

Boone, D. R., Whitman, W. B., and Koga, Y. (2001) Order III. Methanosarcinales ord. nov. In: *Bergey's Manual of Systematic Bacteriology, (The Archaea and the deeply branching and phototrophic Bacteria)*. Bone D. R., Castenholz, R. W., and Garrity, G. M. (Eds), Second edit., Vol. 1, p.268, Springer-Verlag, New York.

Brard M., Laine C., Rethore G., Laurent I., Neveu C., Lemiegre L., Benvegnu T. (2007) Synthesis of archaeal bipolar lipid analogues: a way to versatile drug/gene delivery systems, J. Org. Chem. 72, 8267–8279.

Brochier, C., Forterre, P., and Gribaldo, S. (2004) Archaeal phylogeny based on proteins of the transcription and translation machineries: tackling the *Methanopyrus kandleri* paradox. *Genome Biol.* 5. R17

Brochier-Armanet, C., Boussau, B., Gribaldo, S., and Forterre, P. (2008) Mesophilic Crenarchaeota: Proposal for a third archaeal phylum, the Thaumarchaeota. *Nat. Rev. Microbiol.* Vol.6, pp. 245–252

Buckley, D. H., Graber, J. R., and Schmidt, T. M. (1998) Phylogenetic analysis of nonthermophilic members of the kingdom Crenarchaeota and their diversity and abundance in soils. *Appl. Environ. Microbiol.* Vol.64, pp. 4333–4339

Burggraf, S., Huber, H. & Stetter, K. O. (1997). Reclassification of the crenarchaeal orders and families in accordance with 16S rRNA sequence data. *Int J Syst Bacteriol.* Vol. 47, pp. 657–660

Burggraf, S., Stetter, K.O., Rouvière, P., and Woese, C.R. (1991) *Methanopyrus kandleri*: an archaeal methanogen unrelated to all other known methanogens. *Syst. Appl. Microbiol.* Vol. 14, pp. 346–351

Camacho, R.M., Mateos, J.C., Gonzalez-Reynoso, O., Prado, L.A., and Cordova, (2009) Production and characterization of esterase and lipase from Haloarcula marismortui, J. Ind. Microbiol. Biotechnol., 36, pp.901-909.

Cao, Y., Liao, L., Xu, X., Oren, A., Wang, C., Zhu X., and Wu, M. (2008) Characterization of alcohol dehydrogenase from the haloalkaliphilic archaeon Natronomonas pharaonis. Extremophiles. Vol. 12(3), pp. 471-476

Castro, H., Ogram, A., and Reddy, K.R. (2004) Phylogenetic characterization of methanogenic assemblages in eutrophic and oligotrophic areas of the Florida Everglades. *Appl. Environ. Microbiol.* Vol. 70, pp. 6559–6568

Castro, R. E. D., Ruiz, D. M., Giménez, M. I., Silveyra, M. X., Paggi, R. A., and Maupin-Furlow, J. A. (2008) Gene cloning and heterologous synthesis of a haloalkaliphilic extracellular protease of *Natrialba magadii* (Nep). Extremophiles. Vol. 12(5), pp. 677-687

Cavicchioli, R., Curmi, P. M. G., Saunders, N., and Thomas, T. (2003) Pathogenic archaea: do they exist? *BioEssays*, Vol.25, pp.1119–1128

Chaban, B., Ng, S. Y. M., and Jarrell, K. F. (2006) Archaeal habitats-from the extreme to the ordinary. *Can. J Microbiol.* Vol.52, pp.73–116

Chapelle, F. H., Bradley, P. M., Lovley, D. R., O'Neill, K., and Landmeyer, J. E. (2002) Rapid evolution of redox processes in a petroleum hydrocarbon- contaminated aquifer. *Ground Water.* Vol. 40, pp. 353–360

Clermont, D., Diard, S., Motreff, L., Vivier, C., Bimet, F., Bouchier, C., Welker, M., Kallow, W., Bizet, C. (2009) Description of *Microbacterium binotii* sp. nov.,isolated from human blood. *Int. J. Syst. Evol. Microbiol.* Vol.59, pp.1016–1022.

Dawson, S., DeLong, E. and Pace, N.R. (2001) Phylogenetic and Ecological Perspectives on Uncultured Crenarchaeota and Korarchaeota, in: *The Prokaryotes*, Dworkin, M. (Ed), Springer- Verlag, Release 3.7.

DeLong E. F.(2005) Microbial community genomics in the ocean. *Nat Rev Microbiol* Vol.3, pp. 459-469.

Delong, E. F. and Pace, N.R. (2001) Environmental diversity of bacteria and archaea. *Systematic Biology* Vol.50, pp. 470–478.

Di Giulio, M. (2005) Structuring of the genetic code took place at acidic pH. J. Theoret. Biol. 237, 219–226.

Egorova K. and Antranikian G. (2005) Industrial relevance of thermophilic Archaea Current Opinion in Microbiology, 8, 649–655.

Elkins, J. G., Podar. M., Graham, D. E., Makarova, K. S., Wolf, Y., Ranau, L., Hedlund, BP., Brochier-Armanet, C, Kunin, V., and Anderson, I. (2008) A korarchaeal genome reveals insights into the evolution of the Archaea. *Proc Natl Acad Sci* USA Vol.105, pp. 8102-8107.

Elshahed, M. S., Najar, F. Z., Roe, B. A., Oren, A., Dewers, T. A., and Krumholz, L. R. (2004) Survey of archaeal diversity reveals an abundance of halophilic *Archaea* in a low salt, sulfide- and sulfur-rich spring. *Appl. Environ. Microbiol.* Vol.70, pp. 2230–2239

Ettema TJ, Brinkman AB, Lamers PP, Kornet NG, de Vos WM, van der Oost J. (2006) Molecular characterization of a conserved archaeal copper resistance (cop) gene cluster and its copperresponsive regulator in Sulfolobus solfataricus P2.Microbiology 152,1969-1979.

Euzeby, J.P (2011). http://www.bacterio.cict.fr/archaea.html

Falz, K. Z., Holliger, C., Grosskopf, R., Liesack, W., Nozhevnikova, A. N., Muller, B., Wehrli, B., and Hahn, D. (1999) Vertical distribution of methanogens in the anoxic sediment of Rotsee (Switzerland). *Appl. Environ. Microbiol.* Vol.65 pp. 2402–2408

Fierer, N., Breitbart, M., Nulton, J., Salamon, P., Lozupone, C., Jones, R., Robeson, M., Edwards, A. R., Felts, B., Rayhawk, S., Knight, R., Rohwer, F., and Jackson, R. B. (2007) Metagenomic and Small-Subunit rRNA Analyses Reveal the Genetic Diversity of Bacteria, Archaea, Fungi, and Viruses in Soil. *Appl. Environ. Microbiol.*, Vol. 73., p. 7059–7066

Foophow T, Tanaka S, Angkawidjaja C, Koga Y, Takano K, Kanaya S. (2010) Crystal structure of a subtilisin homologue, Tk-SP, from Thermococcus kodakaraensis: requirement of a Cterminal b-jelly roll domain for hyperstability. J Mol Biol, 400, 865-877.

Forterre, P. (2002) Evolution of the Archaea *Theoretical Population Biology* Vol.61, pp. 409–422

Futterer, O. et al. (2004) Genome sequence of Picrophilus torridus and its implications for life around pH 0. Proc. Natl. Acad. Sci. U. S. A. 101, 9091–9096.

Guipaud, O., Marguet, E., Noll, K.M., de la Tour, C.B., and Forterre, P. (1997). Both DNA gyrase and reverse gyrase are present in the hyperthermophilic bacterium Thermotoga maritima. Proc. Natl. Acad. Sci. USA. 94: 10606-10611.

Garcia, J. L., Patel, B. K. C., and Ollivier, B., (2000) Taxonomic phylogenetic and ecological diversity of methanogenic Archaea. Anaerobe Vol. 6, pp. 205–226.

Gibson, J. A. E., Miller, M. R., Davies, N. M., Neill, N. P., Nichols, D. S., and Volkmann, J. K. (2005) Unsaturated diether lipids in the psychrotrophic archaeon Halorubrum lacusprofundi. Syst. Evol. Microbiol. Vol.28, pp.19–26

Graham, D., Holly, E., and Huse, K. (2008) Methanogens with pseudomurein use diaminopimelate aminotransferase in lysine biosynthesis. *FEBS letters.* Vol. 582(9), pp. 1369-1374

Grant, W. D., Kamekura, M., McGenity, T. J., and Ventosa, A. (2001) Class III. *Halobacteria* class. nov. In: *Bergey's Manual of Systematic Bacteriology.* Boone, D. R. and Castenholz, R.W. (Eds), Second edit., Vol. 1, pp. 294, Springer, New York

Hallberg, K.B. and Johnson, D.B. (2001) Biodiversity of acidophilic prokaryotes, Adv. Appl. Microbiol. 49, 37–84.

Hedi, A., Sadfi, N., Fardeau, M.L., Rebib, H Cayol, J.C., Ollivier, B., and Boudabous, A. (2009). Studies on the Biodiversity of Halophilic Microorganisms Isolated from El-Djerid Salt Lake (Tunisia) under Aerobic Conditions. *Int. J Microbiol.* V. 2009, pp.1-17

Heidi, H. K.,(2011) Skin microbiome: genomics-based insights into the diversity and role of skin microbes. Trends Mol. Med. Vol. 17(6), pp.320-328

Horikoshi, K. (1999) Alkaliphiles: Some Applications of Their Products for Biotechnology. Microbiology and Molecular Biology Reviews, Vol. 63(4), pp. 735–750

Huber, H., Hohn, M. J., Rachel, R., Fuchs, T., Wimmer, V. C., Stetter, K. O. (2002) A new phylum of Archaea represented by a nanosized hyperthermophilic symbiont. *Nature.* Vol. 417, pp. 63–67

Huber, R., and Stetter, K. O., (2001) Order I. *Methanopyrales* ord. nov. In: *Bergey's Manual of Systematic Bacteriology (The Archaea* and the deeply branching and phototrophic *Bacteria)*, Boone, D. R., Castenholz, R. W., and Garrity, G. M. (Eds), Second edit., vol. 1, Springer-Verlag, New York

Huber H, Stetter KO (2006) Thermoplasmatales. In The Prokaryotes, edn 3. Edited by Dworkin M, Falkow S, Rosenberg E, Schleifer K, Stackebrandt E.Springer, 101-112.

Ikeda, M., and Clark, D.S. (1998) Molecular cloning of extremely thermostable esterase gene from hyperthermophilic archaeon Pyrococcus furiosus in Escherichia coli, Biotechnol. Bioeng., vol.57, pp.624-629.

Itoh, T., Yamaguchi, T., Zhou, P., Takashina, T. (2005) *Natronolimnobius baerhuensis* gen. nov., sp. nov. and *Natronolimnobius innermongolicus* sp. nov., novel haloalkaliphilic archaea isolated from soda lakes in Inner Mongolia, China. Extremophiles. Vol. 9(2), pp.111-6

Jacquemet, A., Barbeau J., Lemie`gre, L., Thierry Benvegnu, T. (2009) Archaeal tetraether bipolar lipids: Structures, functions and applications. Biochimie 91, 711–717.

Jarell, K. F., Walters, A. D., Bochiwal, C., Borgia,2, J. M., Dickinson,T., and Chong, J. P. J. (2011) Major players on the microbial stage: why Archaea are important. *Microbiology.* Vol. 157, pp. 919–936

Johnson, D.B. (1998) Biodiversity and ecology of acidophilic microorganisms, FEMS Microbiol. Ecol. 27 307–317.

Kagawa, H.K., Osipiuk, J., Maltsev, N., Overbeek, R., Quaite-Randall, E., Joachimiak, A., and Trent, J.D. (1995) The 60 kDa heat shock proteins in the hyperthermophilic archaeon Sulfolobus shibatae. J. Molec. Biol. 253, 712-725.

Kamekura, M., Dyall-Smith, M. L., Upasani, V., Ventosa, A., and Kates, M. (1997) Diversity of alkaliphilic Halobacteria: Proposals for transfer of Natronobacterium vacuolatum, Natronobacterium magadii, and Natronobacterium pharaonis to Halorubrum, Natrialba, and Natronomonas gen. nov., respectively, as Halorubrum vacuolatum comb. nov., Natrialba magadii comb. nov., and Natronomonas pharaonis comb. nov., respectively. Int. J. Syst. Bacteriol. Vol. 47, pp. 853-857

Kandler, O., and Konig, H. (1998) Cell wall polymers in Archaea (Archaebacteria), *Cell Mol. Life Sci.* Vol.54, pp. 305–308.

Karr, E. A., Ng, J. M., Belchik, S. M., Sattley, W. M., Madigan, M. T., and Achenbach, L. A. (2006) Biodiversity of Methanogenic and Other Archaea in the Permanently Frozen Lake Fryxell, Antarctica *Appl Environ Microb*, pp. 1663–1666

Kates, M., (1993) Membrane lipids of Archaea, In:"*The Biochemistry of Archaea (Archaebacteria)*", Kates, M., Kushnen, D. J., and Matheson, A. T. (Eds.), pp. 261–295, Elsevier, Amsterdam

Kendall, M., Wardlaw, M. G. D., Tang, C. F., Bonin, A. S., Liu, Y., and Valentine, D. L. (2007) Diversity of Archaea in Marine Sediments from Skan Bay, Alaska, Including Cultivated Methanogens, and Description of Methanogenium boonei sp. nov. *Applied And Environmental Microbiology.* Vol. (Jan.), pp. 407–414

Kevbrin, V. V., Lysenko, A. M., and Zhilina, T. N. (1997) Physiology of the alkaliphilic methanogen Z-7936, a new strain of Methanosalsus zhilinaeae isolated from Lake Magadi. Microbiology. Vol. 66, pp. 261–266

Kleikemper, J., Pombo, S. A., Schroth, M. H., Sigler, W. V., Pesaro, M., and Zeyer, J. (2005) Activity and Diversity of Methanogens in a Petroleum Hydrocarbon-Contaminated Aquifer. *Appl. Environ. Microbiol.* Vol. (jan), p. 149–158

Knittel, K., Losekann T., Boetius, A., Kort, R., and Amann, R.(2005) Diversity and distribution of methanotrophic archaea at cold seeps. *Appl Environ Microbiol* ,Vol.71, pp.467-479

Konneke, M., Bernhard, A.E., de la Torre, J.R., Walker, C.B., Waterbury, J.B., and Stahl, D.A. (2005) Isolation of an autotrophic ammonia-oxidizing marine archaeon. *Nature* Vol.437 pp.543–546

Kushner, D. J. (1964) Lysis and dissolution of cells and envelopes of an extremely halophilic bacterium. *J Bacteriol.* Vol. 87, pp.1147-1156.

Kushner, D. J. (1978) Life in high salt and solute concentrations. In: *Microbial Life in Extreme Environments.* Kushner, D. J. (Ed.), Academic Press, London

Lepp, P. W., Brinig, M. M., Ouverney, C. C., Palm, K., Armitage, G. C., and Relman, D. A. (2004). Methanogenic Archaea and human periodontal disease. *Proc Natl Acad Sci U S A* Vol.101, pp. 6176–6181

Ludwig, W. and Klenk, H. P. (2001) Overview: a phylogenetic backbone and taxonomic framework for prokaryotic systematics. In: *Bergey's Manual of Systematic Bacteriology,* Boone, D.R., Castenholz, R.W., and Garrity, G.M. (Eds), Second edit., pp. 49–65, Springer-Verlag, New York

Ma, Y., Galinski, E.A., Grant, W. D., Oren A., and Ventosa, A. (2010) Halophiles 2010: Life in Saline Environments. *Appl Environ Microb.* Vol. 76(21), pp.6971–6981

Macalady, J. and Banfield, J.F. (2003) Molecular geomicrobiology: genes and geochemical cycling. Earth Planet. Sci. Lett. 209, 1–17

Maturrano, L., Santos, F., Rosello-Mora, R., Anton, J., (2006). Microbial diversity in Maras salterns, a hypersaline environment in the Peruvian Andes. *Appl. Environ. Microb.* Vol.72, pp. 3887-3895

McGenity, T. J., and Grant, W. D. (1993) The haloalkaliphilic archaeon (archaebacterium) Natronococcus occultus represents a distinct lineage within the Halobacteriales, most closely related to the other haloalkaliphilic lineage (Natronobacterium). Syst. Appl. Microbiol. Vol. 16, pp. 239–243

Mihajlovski, A., Alric, M., Bruge`re J. F. (2008) A putative new order of methanogenic Archaea inhabiting the human gut, as revealed by molecular analyses of the mcrA gene. *Research in Microbiology* Vol.159, pp. 51-521

Miller, T.L., and Wolin, M.J. (1982) Enumeration of *Methanobrevibacter smithii* in human feces. *Arch. Microbiol.* Vol. 131, pp. 14-18

Mills, H. J., Martinez, R. J., Story, S., and Sobecky, P. A. (2005) Characterization of microbial community structure in Gulf of Mexico fas hydrates: comparative analysis of DNA- and RNA-derived clone libraries. *Appl Environ Microbiol* Vol.71, pp.3235-3247

Moune, S., Caumette, P., Matheron, R., and Willison J. C. (2003) Molecular sequence analysis of prokaryotic diversity in the anoxic sediments underlying cyanobacterial mats of two hypersaline ponds in Mediterranean salterns. *FEMS Microbiol. Ecol.,* Vol. 44(1), pp. 117–130

Nagaoka, S., Minegishi, H., Echigo, A., Shimane, Y., Kamekura, M., And Usami, R. (2011) Halostagnicola alkaliphila sp. nov., an alkaliphilic haloarchaeon from commercial rock salt. Int. J. Syst. Evol. Microbiol. Vol. 61, pp. 1149-1152

Nordstrom, D.K. and Alpers, C.N. (1999) Negative pH, efflorescent mineralogy, and consequences for environmental restoration at the Iron Mountain Superfund site, California, Proc. Natl. Acad. Sci. USA 96, 3455–3462.

Nunoura, T., Takaki, Y, Kakuta, J, Nishi, S., Sugahara, J., Kazama, H., Chee, G.J., Hattori, M., Kanai, A., and Atomi, H. (2011) Insights into the evolution of Archaea and eukaryotic protein modifier systems revealed by the genome of a novel archaeal group. *Nucleic Acid Res* Vol.39, pp. 3204-3223

Omri, A. Agnew, B.J., Patel, G.B. (2003) Short-term repeated-dose toxicity profile of archaeosomes administered to mice via intravenous and oral routes, Int. J. Toxicol. 22, 9–23.

Oren, A. (2002) Diversity of halophilic microorganisms: Environments, phylogeny, physiology, and applications. *J Indust. Microbiol. Biotechnol.* Vol.1, pp. 56–63

Oren, A. (2008) Microbial life at high salt concentrations: phylogenetic and metabolic diversity Saline Systems, Vol.4(2), pp. 1-13

Orphan, V. J., House, C. H., Hinrichs, K.-U., McKeegan, K. D., and DeLong, E. F. (2002) Multiple archaeal groups mediate methane oxidation in anoxic cold seep sediments. *Proc. Natl. Acad. Sci.* Vol. 99, pp.7663–7668

Oxley, A. P. A., Lanfranconi, M. P., Würdemann D., Ott, S., Schreiber, S., McGenity, T. J., Timmis, K. N., and Nogales, B. (2010) Halophilic archaea in the human intestinal mucosa. *Environ. Microbiol.* Vol.12(9), pp. 2398–2410

Ozcan B., Ozyilmaz G., Cokmus C., Caliskan M. (2009) Characterization of extracellular esterase and lipase activities from Wve halophilic archaeal strains J Ind Microbiol Biotechnol 36:105–110.

Preston, C. M., Wu, K., Molinski, T. F., and Delong, E. F. (1996) A psychrophilic crenarchaeon inhabits a marine sponge: *Cenarchaeum symbiosum* gen. nov., sp. nov. *Proc. Natl. Acad. Sci. USA* Vol.93, pp. 6241-6246

Prokofeva, M. I., Kostrikina, N. A., Kolganova, T. V., Tourova, T. P., Lysenko, A. M., Lebedinsky, A. V., and Bonch-Osmolovskaya, E. A. (2009) Isolation of the anaerobic thermoacidophilic crenarchaeote Acidilobus saccharovorans sp. nov. and proposal of Acidilobales ord. nov., including Acidilobaceae fam. nov. and Caldisphaeraceae fam. nov. *Int. J. Syst. Evol. Microbiol.* Vol. 59, pp. 3116-3122

Purdy, K. J., Cresswell-Maynard, T. D., Nedwell, D. B., McGenity, T. J., Grant, W.D., Timmis, K. N., and Embley, T.M. (2004) Isolation of haloarchaea that grow at low salinities. *Environ. Microbiol.* Vol.6, pp. 591–595

Reysenbach, A.L., (2001) Order I. *Thermoplasmatales* ord. nov. In: *Bergey's Manual of Systematic Bacteriology (The Archaea and the deeply branching and phototrophic Bacteria)*, Boone, D.R., Castenholz, R.W., and Garrity, G.M. (Eds), Second edit., Vol. 1, p. 335, Springer-Verlag, New York

Ruiz, D. M., and Castro, R. E. D. (2007) Effect of organic solvents on the activity and stability of an extracellular protease secreted by the haloalkaliphilic archaeon Natrialba magadii. Journal of Industrial Microbiology & Biotechnology. Vol.34(2), pp. 111-115.

Salzano AM, Febbraio F, Farias T, Cetrangolo GP, Nucci R, Scaloni A, Manco G. (2007) Redox stress proteins are involved in adaptation response of the hyperthermoacidophilic archaeon Sulfolobus solfataricus to nickel challenge. Microb Cell Fact, 6, 25.

Savage, K. N., Krumholz, L. R., Oren, A., and Elshahed, M. S. (2008) *Halosarcina pallida* gen. nov., sp. nov., a halophilic archaeon isolated from a low salt, sulfide-rich spring. *Int. J Syst. Evol. Microbiol.* Vol. 58, pp.856-860

Savage, K. N., Krumholz, L. R., Oren, A., Elshahed, M. S. (2007) *Haladaptatus paucihalophilus* gen. nov., sp. nov., a halophilic archaeon isolated from a low-salt, sulfide-rich spring. Int. J. Syst. Evol. Microbiol. Vol.57, pp.19-24.

Schiraldi C., Giulliano M, DeRosa M. (2002). Perspectives on biotechnological applications of archaea. Archaea 1, 75–86.

Shimada, H. et al. (2002) Complete polar lipid composition of *Thermoplasma acidophilum* HO-62 determined by high-performance liquid chromatography with evaporative light-scattering detection. J. Bacteriol. 184, 556–563.

Shimane, Y., Hatada, Y., Mınegıshı, H., Echıgo,A., Nagaoka, S., Mıyazakı, M., Ohta, Y., Maruyama, T., Usamı, R., Grant, W.D. and Horıkoshı, K. (2011) *Salarchaeum japteronicum* gen. nov., sp. nov., an aerobic, extremely halophilic member of the *Archaea* isolated from commercial salt. *Int. J. Syst. Evol. Microbiol.*, Vol.61, pp.2266-2270

Stetter, K. O. (2006) History of discovery of the first hyperthermophiles. *Extremophiles*. Vol. 10, pp. 357–362

Synowiecki, J., Grzybowska, B., and Zdzieblo, A. (2006) Sources, properties and suitability of new thermostable enzymes in food processing, rit. Rev. Food Sci. Nutrit. vol.46, pp.197-205.

Takai, K. and Nakamura K. (2011) Archaeal diversity and community development in deep-sea hydrothermal vents. *Curr. Opın. Mıcrobıol.*, Vol.14, pp. 282–291

Taxonomic outline of the Bacteria and Archaea, Available online at http://www.taxonomicoutline.org/

Tebbe, A., Klein, C., Bisle, B., Siedler, F., Scheffer, B., Garcia-Rizo, C., Wolfertz, J., Hickmann, V., Pfeiffer, F., and Oesterhelt, D. (2005). Analysis of the cytosolic proteome of *Halobacterium salinarum* and its implication for genome annotation. *Proteomics* Vol.5, pp.168-179

Tehei, M., Franzetti, B., Maurel, M. C., Vergne, J., Hountondji, C., and Zaccai G., (2002.) The search for traces of life: the protective effect of salt on biological macromolecules. *Extremophiles*, Vol. 6(5), pp. 427–430

Ventosa, A., Nieto, J. J., and Oren, A. (1998) Biology of moderately halophilic aerobic bacteria. *Microbiol. Mol. Biol. Rev.* Vol. 62, no. 2, pp. 504–544.

Vetriani, C., Jannasch, H.W. MacGregor, B.J. Stahl, D.A. and Reysenbach, A.-L. (1999) Population structure and phylogenetic characterization of marine benthic archaea in deep-sea sediments. Appl. Environ. Microbiol. 65:4375–4384.

Vianna, M. E., Conrads, G., Gomes, B. P. F. A, and Horz, H. P. (2006) Identification and Quantification of Archaea Involved in Primary Endodontic Infections. *J. Clın. Mıcrobıol.*, Vol. 44(4), pp. 1274–1282

Wachtershauser, G. (2006) From volcanic origins of chemoautotrophic life to Bacteria, Archaea and Eukarya. Phil. Trans. R. Soc. Biol. Sci. 361, 1787-1806.

Walters, A. D. and Chong, J. P. J. (2010) An archaeal order with multiple minichromosome maintenance genes. *Microbiol.* Vol. 156, pp. 1405–1414

Wang, S., Yang, Q., Liu, Z. H., Sun, L., Wei, D., Zhang, J. Z., Song, J. Z., And Yuan, H. F. (2010) *Haloterrigena daqingensis* sp. nov., an extremely haloalkaliphilic archaeon isolated from a saline–alkaline soil. Int. J. Syst. Evol. Microbiol. Vol. 60, pp. 2267-2271

Waters E., Hohn M. J., Ahel I., Graham D. E., Adams M. D., Barnstead, M., Beeson, K. Y., Bibbs, L., Bolanos, R., Keller, M., Kretz, K., Lin, X., Mathur, E., Ni, J., Podar, M., Richardson, T., Sutton, G. G., Simon, M., So, D., Stetter, K. O., Short, J. M., Noordewier, M. (2003) The genome of *Nanoarchaeum equitans*: Insights into early archaeal evolution and derived parasitism. *Proc Natl Acad Sci.* Vol.100, pp. 12984–12988

Welkera, M. and Edward R. B. (2011) Moore, Applications of whole-cell matrix-assisted laser-desorption/ionization time-of-flight mass spectrometry in systematic microbiology. *Syst. Appl. Microbiol.* Vol. 34, pp. 2–11

Whittaker, R. H. (1969). New concepts of kingdoms of organisms. Evolutionary relations are better represented by new classifications than by the traditional two kingdoms. *Science* Vol.163, pp.150–160

Woese C. R., ,Kandler, O., and Wheelis, M. L. (1990) Towards a natural system of organisms: Proposal for the domains Archaea, Bacteria, and Eucarya. *Proc. Nati. Acad. Sci. USA,* Vol. 87, pp. 4576-4579

Xu, Y., Wang,1 Z., Xue, Y., Zhou, P., Ma, Y., Ventosa, A., and Grant,W. D. (2001) Natrialba hulunbeirensis sp. nov. and Natrialba chahannaoensis sp. nov., novel haloalkaliphilic archaea from soda lakes in Inner Mongolia Autonomous Region. China International Journal of Systematic and Evolutionary Microbiology. Vol. 51, pp. 1693-1698

Xu, Y., Zhou, P. J., and Tian, X. Y. (1999) Characterization of two novel haloalkaliphilic archaea Natronorubrum bangense gen. nov., sp. nov., and Natronorubrum tibetense gen. nov., sp. nov. Int. J. Syst. Bacteriol. Vol. 49, pp.261– 266

Xue, Y., Fan, H., Ventosa, A., Grant, W. D., Jones, B. E., Cowan, D. A., Ma, Y. (2005) Halalkalicoccus tibetensis gen. nov., sp. nov., representing a novel genus of haloalkaliphilic archaea. Int J Syst Evol Microbiol. Vol. 55(6), pp. 2501-5

Yarza, P., Ludwig, W., Euzeby, J., Amannd, R., Schleifer, K. H., Glöcknerd, F.O., and Rossello-Moraa, R. (2010) Update of the All-Species Living Tree Project based on 16S and 23S rRNA sequence analyses. *Syst. Appl. Microbiol.,* Vol.33, pp.291–299

Yavitt, J., Yashiro, E., Quiroz, H. C., and Zinder S. (2011) Methanogen diversity and community composition in peatlands of the central to northern Appalachian Mountain region, North America. *Biogeochemistry.* Vol. (24 September), pp. 1-15

Youa, J., Dasa, A., Dolanb, E.M., and Hua, Z. (2009) Ammonia-oxidizing archaea involved in nitrogen removal. *Water Research,* Vol. 43(7), pp.1801-1809

Zhilina, T. N., and Zavarzin, G. A. (1994) Alkaliphilic anaerobic community at pH 10. Curr. Microbiol. Vol. 29, pp. 109–112

Zillig, W., Holz, I., Klenk, H. P., Trent, J., Wunderl, S., Janekovic, D., Imsel, E. and Haas, B. (1987) *Pyrococcus woesei,* sp. nov., an ultra-thermophilic marine *Archaebacterium,* representing a novel order, *Thermococcales. Syst. Appl. Microbiol.,*Vol.9, pp. 62-70

Zillig, W. (1991) Comparative biochemistry of Archaea and Bacteria, *Curr. Opin. Genet. Dev.* Vol.1, pp.544–551

Zillig. W., Stetter. K. O., Schäfer, W., Janekovic, D., Wunderl. S., Holz, f., and Palm, P.: (1981) Thermoproteales: a novel type of extremely thermoacidophilic anaerobic archaebacteria isolated from Icelandic solfataras. *Zbl. Bakt. Hyg. I. Abt. Orig.* Vol.C2, pp. 205–227

Zinder, S. H. (1993) Physiological ecology of methanogens, In: *Methanogenesis: ecology, physiology, biochemistry and genetics,* Ferry, J. G., (Ed), pp. 128–206, Chapman & Hall, Inc., New York.

3

DNA Based Techniques
for Studying Genetic Diversity

Ahmed L. Abdel-Mawgood
Faculty of Agriculture
El-Minia University
Egypt

1. Introduction

Genetic diversity is a study undertaken to classify an individual or population compared to other individuals or populations. This is a relative measure, as the distance between any pair of entries in the study is greater or lesser depending on all pairwise comparisons that can be made in the study. However, genetic fingerprinting is the unambiguous identification of an individual (based on the presence or absence of alleles at different markers) or a population (based on frequencies of alleles of the markers). This is an absolute measure and does not change depending on other individuals or populations under study. Both Genetic diversity and fingerprinting studies are done using molecular markers.

Determining genetic diversity can be based on morphological, biochemical, and molecular types of information (Mohammadi & Prasanna 2003; Sudre et al., 2007; Goncalves et al., 2009). However, molecular markers have advantages over other kinds, where they show genetic differences on a more detailed level without interferences from environmental factors, and where they involve techniques that provide fast results detailing genetic diversity (Binneck et al., 2002; Garcia et al., 2004; Saker et al., 2005; Goncalves et al., 2008; Souza et al., 2008). Moreover, the discovery of high throughput platforms increases number of data per run and reducing the cost of the data and increasing map resolution.

1.1 DNA based markers for genetic diversity studies

Molecular markers are segments of chromosomes which don't necessarily encode any traits and are not affected by the environment but which are inherited in a Mendelian fashion. Some segments of the chromosome change faster than others (i.e. coding vs. non coding DNA). As a result it is recommended to use fast changing markers for closely related individuals and slow changing markers for less related individuals (different species). Different marker types therefore have different usefulness in fingerprinting individuals and populations. Moreover; a good marker for fingerprinting studies will be cheap to run, or gives a lot of information per run; very repeatable between assays; experience very low error rate and easy, unambiguous to score; and contain many alleles (high information content). The following techniques are those most used in genetic diversity studies and listed in chronological order: RFLP (restriction fragment length polymorphism) (Botstein et al., 1980), SSR (simple sequence repeats or just microsatellites) (Tautz, 1989), RAPD (randomly amplified polymorphic DNA) (Williams et al., 1990) or AP-PCR (arbitrarily primed PCR) (Welsh & McClelland 1990), ISSR (inter-simple sequence repeats) (Zietkiewicz et al., 1994), AFLP (amplified fragment length polymorphism) (Vos et al., 1995), SNPs (single nucleotide polymorphisms) (Chen & Sullivan, 2003) and, more recently, DarT (diversity array technology) (Kilian et al., 2005) and other high throughput platforms. These different types of molecular markers are also different as to their potential to detect differences between individuals, their cost, facilities required, and consistency and replication of results (Schlotterer 2004; Schulman, 2007; Bernardo, 2008). A review summarizes various tools of DNA markers technology for application in molecular diversity analysis with special emphasis on wildlife conservation was presented by Arif et al., 2011. However, authors reviewed only mitochondrial DNA based markers including ribosomal DNA (12S and 16S rDNA), mitochondrial protein coding genes, non-coding or control region sequences and nuclear DNA based markers including random amplified polymorphic DNA, Amplified fragment length polymorphism, and microsatellites or simple sequence repeats.

As a laboratory methodology, fingerprinting and diversity studies require the following steps: a) isolation of DNA, b) digestion, hybridization, and/or amplification of DNA into specific fragments, c) sizing /or separation of DNA fragment combinations or patterns into a set of individual DNA fingerprints, d) comparison of DNA fingerprints from different individuals e) calculation of similarity (or dissimilarity) coefficients for all pairs of entries in the genetic study, f) creation of a dendrogram or graph to visualize the differences.

1.2 Restriction Fragment Length Polymorphism (RFLP)

Restriction fragment length polymorphism (RFLP) has much greater power and was originally developed for mapping human genes than anything previously available (Botstein et al., 1980). This technique quickly proved their utility in virtually all species. O'Brien, 1991 groups genetic markers into two types: Type I markers are associated with a gene of known function, and Type II markers are associated with anonymous gene segments of one sort or another. For now, RFLP remain the most common Type I marker presently used in many eukaryotic organisms.

Variations in the characteristic pattern of a RFLP digest can be caused by base pair deletions, mutations, inversions, translocations and transpositions which result in the loss or gain of a recognition site resulting in a fragment of different length and polymorphism. Only a single base pair difference in the recognition site will cause the restriction enzyme not to cut. If the base pair mutation is present in one chromosome but not the other, both fragment bands will be present on the gel, and the sample is said to heterozygous for the marker. Only co-dominant markers exhibit this behavior which is highly desirable, dominant markers exhibit a present/absent behavior which can limit data available for analysis. RFLP has some limitations since it is time consuming. Moreover, in some organisms such as wheat, RFLP is of low frequency which is attributed to the polyploidy nature of wheat, and large genome size. However; in the past RFLP was used for several purposes including genome mapping, varietal identification, identification of wheat rye recombinants, and identification of homologous chromosome arm (Tankesley at al., 1989). RFLP was also used for varietal identification, for example, it has been used for mapping different storage protein loci. Set of polymorphic probes was used to identify 54 common wheat cultivar, mostly Italian type (Vaccino et al., 1993).

2. PCR based methods

With the beginning of studies that led to the development of polymerase chain reaction (PCR) technology (Saiki et al., 1985; Mullis & Faloona, 1987), there were amazing advances in the refinement of techniques to obtain specific or non-specific DNA fragments, relevant mainly to research in genetic diversity.

2.1 Randomly Amplified Polymorphic DNA (RAPD)

RAPD was the first PCR based molecular marker technique developed and it is by far the simplest (Williams et al., 1990). Short PCR primers (approximately 10 bases) are randomly and arbitrarily selected to amplify random DNA segments throughout the genome. The resulting amplification product is generated at the region flanking a part of the 10 bp priming sites in the appropriate orientation. RAPD products are usually visualized on agarose gels stained with ethedium bromide.

RAPD markers are easily developed and because they are based on PCR amplification followed by agarose gel electrophoresis, they are quickly and readily detected. RAPD technique was used extensively in studying genetic diversity between plant species. For example, it was used to study genetic structure and diversity among and between six populations of Capparis deciduas in Saudi Arabia (Abdel-Mawgood et al. 2010). As a result, RAPD's may permit the wider application of molecular maps in plant science. Most RAPD markers are dominant and therefore, heterozygous individuals cannot be distinguished from homozygotes. This contrasts with RFLP markers which are co-dominant and therefore, distinguish among the heterozygote and homozygotes. Thus, relative to standard RFLP markers, and especially VNTR loci, RAPD markers generate less information per locus examined. One disadvantage of using RAPD technique is the reproducibility between different runs which is due to the short primer length and low annealing temperature.

2.2 Inter-simple sequence Repeat (ISSR)

The Inter-simple sequence repeat (ISSR) are semiarbitrary markers amplified by polymerase chain reaction (PCR) in the presence of one primer complementary to a target microsatellite. Each band corresponds to a DNA sequence delimited by two inverted microsatellites (Zietkiewicz et al., 1994; Tsumara et al., 1995; Nagaoka & Ogihara, 1997). It does not require genome sequence information; it leads to multilocus, highly polymorphous patterns and produces dominant markers (Mishra et al., 2003). ISSR PCR is a fast, inexpensive genotyping technique based on variation in the regions between microsatellites. This method has a wide range of uses, including the characterization of genetic relatedness among populations, genetic fingerprinting, gene tagging, detection of clonal variation, cultivar identification, phylogenetic analysis, detection of genomic instability (for example, it was used in human quantification of genomic instability to estimate of prognosis in colorectal cancer (Brenner, 2011), and assessment of hybridization.

ISSRs have been used in genetic diversity studies in different crop plants (Nagaraju et al.,39 2002; Reddy et al., 2002; Obeed et al. 2008). ISSR markers are also suitable for the identification and DNA fingerprinting (Gupta et al., 2002; Gupta & Varshney, 2000). This method has several benefits over other techniques: first, it is known to be able to discriminate between closely related genotypes (Fang & Roose, 1997; Hodkinson et al., 2002) and second, it can detect polymorphisms without any previous knowledge of the crop's DNA sequence.

ISSRs are like RAPDs markers in that they are quick and easy to handle, but they seem to have the reproducibility of SSR markers because of the longer length of their primers. However, ISSR is more informative than RAPD in wheat, fruit plants (strawberry, apple and Ribes species) and the common bean for the evaluation of genetic diversity (Korbin et al., 2002; Rakoczy-Trojanowska et al., 2004). It was proven to be reproducible, and quick for characterization many cultivars like poplar (Gao et al., 2006).

2.3 Simple Sequence Repeats (SSR)

Simple sequence repeat (SSR) markers are repeats of short nucleotide sequences, usually equal to or less than six bases in length, that vary in number (Rafalski et al., 1996; Reddy et al., 2002). SSR are becoming the most important molecular markers in both animals and plants. They are also called microsatellites. SSR are stretches of 1 to 6 nucleotide units repeated in tandem and randomly spread in eukaryotic genomes. SSR are very polymorphic due to the high mutation rate affecting the number of repeat units. Such length-polymorphisms can be easily detected on high resolution gels (e. g. sequencing gels). It is suggested that the variation or polymorphism of SSRs are a result of polymerase slippage during DNA replication or unequal crossing over (Levinson & Gutman, 1987). SSRs are not only very common, also are hypervariable for numbers of repetitive DNA motifs in the genomes of eukaryotes (Vosman & Arens, 1997; Rallo et al., 2000; van der Schoot et al., 2000).

SSR have several advantages over other molecular markers. For example, (i) microsatellites allow the identification of many alleles at a single locus, (ii) they are evenly distributed all over the genome, (iii) microsatellites can offer more detailed population genetic insight than maternally inherited mitochondrial DNA (mtDNA) because of the high mutation rate and bi-parental inheritance (iv) they are co-dominant, (v) highly polymorphic and specific (Jones et al., 1997). (vi) very repeatable (vii) little DNA is required and (viii) so cheap and easy to run (ix) need a small amount of medium quality DNA and (x) the analysis can be semi-automated and performed without the need of radioactivity (Gianfranceschi et al., 1998; Guilford et al., 1997), (xi) with the advance of DNA isolation technology, it was possible to identify loci in highly degraded ancient DNA (aDNA), where traditional enrichment procedures have been unsuccessful (Allentoft et al., 2009), (xii) with the development of high-throughput sequencing platforms, such as the GS-FLX (Roche, Branford, CT, USA) SSR has recently become fast and efficient (Abdelkrim et al., 2009; Allentoft et al., 2009; Santana et al., 2009). SSRs are typically codominant and multiallelic, with expected heterozygosity frequently greater than 0.7, allowing precise discrimination even of closely related individuals. However, since genomic sequencing is needed to design specific primers, it is not very cost effective and also requires much discovery and optimization for each species before use.

For the past two decades they have been the markers of choice in a wide range of forensic profiling, population genetics and wildlife-related research. The importance and applicability of these markers are confirmed by observing an excess of 2,450,000 hits on the word "microsatellite" on the Web of Science database (accessed October-2011).

Searches of EST databases for microsatellite containing sequences have been useful for a number of species including humans (Haddad et al., 1997), catfish (Serapion et al., 2004), rice (Cho et al., 2000) and barley (Thiel et al., 2003). In Rainbow trout, cDNAs and expressed sequence tags (ESTs) available in public databases offer an in silico approach to marker development at virtually no cost. Marker development in salmonids is complicated by the evolutionarily recent genome duplication event which often results in multiple copies of loci in the haploid genome (Venkatesh, 2003). The AFLP was also used to test for purity of three inbred lines by examining the pattern of 5 individuals from each of inbred tested (Ismail et al. 1999). Characterization of loci including copy number is important when conducting analyses of genetic variability in genomic regions under control of different evolutionary constraints. Two linkage maps have been published for rainbow trout using AFLP and microsatellite markers with an average marker spacing of 10 cM (Sakamoto et al., 2000; Nichols et al., 2003).

In natural plant populations, microsatellites have great potential for helping to understand what determines patterns of genetic variation, particularly when used in concert with chloroplast DNA (cpDNA) markers. Their utility has been demonstrated in studies of genetic diversity (Morand *et al.*, 2002; Zeid et al., 2003), mating systems (Durand *et al.*, 2000), pollination biology (White et al., 2002) and seedling establishment (Dow & Ashley 1996). ISSR was also used for hybrid identification in maize (Abdel-Mawgood et al 2006), DNA fingerprinting of wheat genotypes in conjunction with RAPD (Abdel-Mawgood et al 2007). However, few studies have been carried out using microsatellites in analysis of population structure of polyploid species because the polyploidy complicates the results of the SSR. This is likely due to the problems in analysing polyploid data as well as difficulties in amplifying loci, possibly because of differences in the parental genomes of polyploids (Roder et al., 1995). A number of studies have demonstrated that microsatellite alleles of the same size can arise from mutation events which either interrupt repeat units or occur in the regions flanking the repeat region. This has been shown to occur both within (Angers & Bernatchez, 1997; Viard *et al.*, 1998) and among populations (Estoup et al., 1995; Viard et al., 1998) and closely related species (Peakall et al., 1998; van Oppen et al., 2000). One approach to minimizing the risk of misinterpretation of genetic information is to characterize different electromorphs by sequencing, particularly in cases in which other genetic data (e.g. chloroplast or mitochondrial sequences) suggest strong levels of genetic structuring that is not being detected by microsatellite analysis.

2.4 Amplified Fragment Length Polymorphisms (AFLP)

Amplified Fragment Length Polymorphisms (AFLP) based genomic DNA fingerprinting is a technique used to detect DNA polymorphism. AFLP is a polymerase chain reaction (PCR) based technique, (Vos et al., 1995) has been reliably used for determining genetic diversity and phylogenetic relationship between closely related genotypes. AFLP analysis combines both the reliability of restriction fragment length polymorphism (RFLP) and the convenience of PCR-based fingerprinting methods. AFLP markers are generally dominant and do not require prior knowledge of the genomic composition. AFLPs are produced in great numbers and are reproducible

The AFLP is applicable to all species giving very reproducible results. It was also used in microbial population: in studying genetic diversity of human pathogenic bacteria (Purcell & Hopkins, 1996), microbial taxonomy (Vaneechonette, 1996) and in characterizing pathovars of plant pathogenic bacteria (Bragard et al., 1997). In that regards it has the advantage of the extensive coverage of the genome under study. In addition the complexity of the bands can be reduced by adding selective bases to the primers during PCR amplification. It was also used in studying genetic diversity of human pathogenic bacteria (Bragard et al., 1997). After the completion of the genome sequencing of E. coli, it was possible to predict the band pattern of the AFLP analysis of E. coli. This indicates the power of this technique. In higher plants AFLP was used in variety of applications which includes examining genetic relationship between species (Hill et al., 1996), investigating genetic structure of gene pool (Tohme et al., 1996), and assessment of genetic differentiation among populations (Travis et al., 1996; Paul et al., 1997).

Disadvantages of this technique are that alleles are not easily recognized, has medium reproducibility, labor intensive and has high operational and development costs (Karp et al., 1997). Moreover, AFLP require knowledge of the genomic sequence to design primers with specific selective bases. Dominant markers such as RFLP as well as RAPD are very limited in their ability to precisely determine parentage. They can readily be used to establish that two individuals are not the same, but the statement, that two individuals are identical is usually only approximate and no formal statistics can be attached to this assertion. There are several advantage of this technique a) no need for prior knowledge of any sequence information, b) multiple bands are produced per each experiment, c) these bands are produced from all over the genome, d) the technique is reproducible (Blears et al., 1998; Vos & Kuiper, 1997), e) have highly discriminatory power, and f) the data can be stored in database like AmpliBASE MT (Majeed et al., 2004) for comparison purposes.

3. Single nucleotide polymorphism SNP's

Single nucleotide polymorphism SNP's, represent sites in the genome where DNA sequence differs by a single base when two or more individuals are compared. They may be individually responsible for specific traits or phenotypes, or may represent neutral variation that is useful for evaluating diversity in the context of evolution. SNPs are the most widespread type of sequence variation in genomes discovered so far. About 90% of sequence variants in humans are differences in single bases of DNA (Collins, 1998).

Several disciplines such as population ecology and conservation and evolutionary genetics are benefitting from SNPs as genetic markers. There is widespread interest in finding SNP's because they are numerous, more stable, potentially easier to score than the microsatellite repeats currently been used in gene mapping in human. Within coding regions there are on average four SNPs per gene with a frequency above 1%. About half of these cause amino acid substitutions: termed non-synonymous SNPs (nsSNPs) (Cargill et al., 1999).

Because of the importance of the SNP's in the discovery of DNA sequence variants, the National Human Genome Research Institute (NHGRI) of NIH along with the Center for

Disease Control and Prevention and several individual investigators have assembled a DNA Polymorphism Discovery Resource of samples from 450 U.S. residents (Collins et al., 2011). This DNA variant discovery will help in finding SNP's that are deleterious to gene function or likely to be disease associated.

In plants, SNPs are rapidly replacing simple sequence repeats (SSRs) as the DNA marker of choice for applications in plant breeding and genetics because they are more abundant, stable, amenable to automation, efficient, and increasingly cost-effective (Duran et al., 2009; Edwards & Batley, 2010; Rafalski, 2002a). Generally, SNPs are the most abundant form of genetic variation in eukaryotic genomes. Moreover; they occur in both coding and non-coding regions of nuclear and plastid DNA (Kwok et al., 1996). As in the case of human genome, SNP-based resources are being developed and made publicly available for broad application in rice research. These resources include large SNP datasets, tools for identifying informative SNPs for targeted applications, and a suite of custom-designed SNP assays for use in marker-assisted and genomic selection. SNPs are widely used in breeding programs for several applications such as a) marker assisted and genomic selection, b) association and QTL mapping, positional cloning, c) haplotype and pedigree analysis, d) seed purity testing and d) variety identification e) monitoring the combinations of alleles that perform well in target environments (Bernardo, 2008; Jannink et al., 2010; Kim et al., 2010; Moose & Mumm, 2008; Xu & Crouch, 2008; McCouch et al., 2010).

Although SNP's have several advantages over other technology, there are limitations to the discovery of SNP's in the non-model organism. This is due to the expenses and technical difficulties involved in the currently available SNP isolation strategies (Brumfield, 2003; Seddon et al., 2005). Typical direct SNP discovery strategies involve sequencing of locus-specific amplification (LSA) products from multiple individuals or sequence determination of expressed sequence tags (EST-sequencing) (Suhn & Vijg, 2005; Twyman, 2004). Other direct strategies include whole genome (WGSS) and reduced representation (RRSS) shotgun sequencing approaches. If comparative sequence data are available in public or other databases, various sequence comparison algorithms that identify nucleotide differences provide an alternative means to empirically discover SNPs (Guryev et al., 2005). In rice, for example, SNPs discovery on a genome wide basis was based on using either the genomic-scale re-sequencing approaches or Sanger sequencing-based strategies. However, the later approach require the design of specific primer pairs, which are generally located in exons and may span intronic as well as exonic regions (Caicedo et al., 2007; Ebana et al., 2010; Tung et al., 2010; Yamamoto et al., 2010; McCouch et al., 2010).

A direct analysis of sequence difference between many individuals at a large number of loci can be achieved by the next generation sequencing. Re-sequencing is used to identify genetic variation between individuals, which can provide molecular genetic markers and insights into gene function. The process of whole-genome re-sequencing using short-read technologies involves the alignment of a set of literally millions of reads to a reference genome sequence. Once this has been achieved, it is possible to determine the variation in nucleotide sequence between the sample and the reference. There does not appear to be a major disadvantage to using short-read technology for SNP discovery where a reference genome is available. The approach is also low cost, at approximately $0.25 per SNP, compared with about $2.95 using Sanger sequencing. Where a draft reference genome is not available, it may be possible to combine the long- and short-read next-generation sequencing technologies, and use 454 sequencing to generate an assembly against which to align the short reads. Re-sequencing has proved to be a valuable tool for studying genetic variation and, with the advent of James Watson's genome being sequenced using this method (Wheeler et al., 2008), the challenge of whole-genome re-sequencing has largely been conquered. Whole-genome re-sequencing for SNP discovery has been demonstrated in Caenorhabditis elegans. Solexa technology was used to sequence two C.elegans strains, which were then compared with the reference genome sequence for SNP and indel identification. The software applications PyroBayes and Mosaik were used to differentiate between true polymorphisms and sequence errors (Hillier et al., 2008). Although whole-genome re-sequencing can be very useful, there are some drawbacks with this approach. First, a reference genome sequence is required and the quality of the re-sequenced genome is highly dependent on the quality of the reference sequence. In addition, for very large and complex plant genomes, a vast amount of sequence data is required to confidently call SNPs, with SNPs in repetitive sequences being particularly difficult to call. However, as sequencing technology continues to improve, it is expected that whole-genome re-sequencing of crop genomes will become common.

The new development in technologies that collect high-throughput data contribute substantially in the progress in evolutionary genomics (Gilad et al., 2009). The next-generation sequencing technologies has the potential to revolutionize genomic research and enable us to focus on a large number of outstanding questions that previously could not be addressed effectively (Rafalski, 2002b). The next generation sequencing (NGS) provides the capacity for high-throughput sequencing of whole genomes at low cost. They have advantage of improving the capacity to finding novel variations that are not covered by genotyping arrays. The efficiency of NGS-mediated genotyping has recently been improved through employing amplicon libraries of long-range PCR, which encompass discrete genomic intervals. However, at present, pitfalls of next-generation sequencing data is challenging, in particular because most sequencing platforms provide short reads, which are difficult to align and assemble. The next-generation sequencing technology' can produce very large amounts; typically millions of short sequence reads (25–400 bp). However, these large numbers of relatively short reads are usually achieved at the expense of read accuracy (Imelfort et al., 2009). Moreover, only little is known about sources of variation that are associated with next-generation sequencing study designs (Morozova & Marra, 2008).

The first commercially available next-generation sequencing system was developed by 454 and commercialized by Roche (Basel, Switzerland) as the GS20, capable of sequencing over 20 million base pairs, in the form of 100-bp reads, in just over 4h. The GS20 was replaced during 2007 by the GS FLX model, capable of producing over 100 million base pairs of sequence in a similar amount of time. Roche and 454 continue to improve data production with the expectation of 4–500 Mbp of sequence per run, and an increase to 500-bp reads with the release of their Titanium system towards the end of 2008. Two ,alternative ultrahigh-throughput sequencing systems now compete with the GS FLX: Solexa technology, commercialized by Illumina (San Diego, California, USA), and the SOLiD system from Applied Biosystems (AB) (Carlsbad, California, USA). A rapid and effective method for high-throughput SNP discovery for identification of polymorphic SNP alleles in the oat genome was developed based on high resolution melting and high-throughput 454 sequencing technology. The developed platform for SNP genotyping is a simple and highly-informative and can be used as a model for SNP discovery and genotyping in other species with complex and poorly-characterized genomes (Oliver et al., 2011).

Indirect SNP discovery strategies include prescreening methods which detect heteroduplexes on the basis of mismatch-induced altered DNA characteristics such as: single-strand conformational polymorphism (SSCP) which exploits the physical differences (Orita et al., 1989) and altered melting behavior of mismatch-containing DNA fragment (Fisher & Lerman, 1979; Vijg & van Orsouw, 1999; Xiao & Oefner, 2001), any reagent that specifically recognizes and cleaves mismatched DNA can be used for the SNP's detection (Goldrick, 2001), heteroduplex-cleaving chemicals (Ellis et al., 1998), heteroduplex-cleaving proteins (Till et al., 2004) and bacteriophage Mu DNA transposition (Yanagihara & Mizuuchi, 2002).

4. Array based platforms

Several different types of molecular markers have been developed over the past three decades (Kumar, 1999; Gupta & Rustgi 2004), motivated by requirements for increased throughput, decreased cost per data point, and greater map resolution. Recently, oligonucleotide-based gene expression microarrays have been used to identify DNA sequence polymorphisms using genomic DNA as the target (Hazen & Kay, 2003).

4.1 Diversity arrays technology (DArT)

Diversity arrays technology (DArT) is a microarray hybridization based technique that permits simultaneous screening thousands of polymorphic loci without any prior sequence information. The DArT methodology offers a high multiplexing level, being able to simultaneously type several thousand loci per assay, while being independent of sequence information. DArT assays generate whole genome fingerprints by scoring the presence versus absence of DNA fragments in genomic representations generated from genomic DNA samples through the process of complexity reduction. DArT has been developed as a hybridisation-based alternative to the majority of gel-based marker technologies currently in use It can provide from hundreds to tens of thousands of highly reliable markers for any species as it does not require any precise information about the genome sequence (Jaccoud et al., 2001). Moreover, DArT was recently shown to provide good genome coverage in wheat and barley (Wenzl et al., 2004; Akbari et al., 2006). An important step of this technology is a step called "genome complexity reduction" which increasing genomic representation by reducing repetitive sequence that is abundant in eukaryotes. With DArT platform, comprehensive genome profiles are becoming affordable for virtually any crop, genome profiles which can be used in management of bio-diversity, for example in germplasm collections. DArT genome profiles enable breeders to map QTL in one week. DArT profiles accelerate the introgression of a selected genomic region into an elite genetic background (for example by marker-assisted backcrossing). In addition, DArT profiles can be used to guide the assembly of many different regions into improved varieties (marker assisted breeding). The number of markers DArT detects is determined primarily by the level of DNA sequence variation in the material subjected to analysis and by the complexity reduction method deployed (Kilian et al., 2003). Another advantage of DArT markers is that their sequence is easily accessible compared to amplified fragment length polymorphisms (AFLPs) making DArT a method of choice for non-model species (James et al., 2008). DArT has been also applied to a number of animal species and microorganisms (The Official Site of Diversity Arrays Technology (DArT P/L).

4.2 Restriction Site-Associated DNA (RAD)

Another high throughput method is restriction site-associated DNA (RAD) procedures which involved digesting DNA with a particular restriction enzyme, ligating biotinylated adapters to the overhangs, randomly shearing the DNA into fragments much smaller than the average distance between restriction sites, and isolating the biotinylated fragments using streptavidin beads (Miller et al., 2007a). RAD specifically isolates DNA tags directly flanking the restriction sites of a particular restriction enzyme throughout the genome. More recently, the RAD tag isolation procedure has been modified for use with high-throughput sequencing on the Illumina platform (Baird et al., 2008; Lewis et al., 2007). In addition, Miller et al., 2007b demonstrate that RAD markers, using microarray platform, allowed high-throughput, high-resolution genotyping in both model and nonmodel systems.

4.3.Single Feature Polymorphism (SFP)

A third high throughput method is single feature polymorphism (SFP) which is done by labeling genomic DNA (target) and hybridizing to arrayed oligonucleotide probes that are complementary to indel loci. The SFPs can be discovered through sequence alignments or by hybridization of genomic DNA with whole genome microarrays. Each SFP is scored by the presence or absence of a hybridization signal with its corresponding oligonucleotide probe on the array. Both spotted oligonucleotides and Affymetrix-type arrays have been used in SFP. Borevitz et al. 2003 coined the term "single feature polymorphism" and demonstrated that this approach can be applied to organisms with somewhat larger genomes, specifically Arabidopsis thaliana with a genome size of 140 Mb. Similarly, whole-genome DNA-based SFP detection has been accomplished in rice (Kumar et al., 2007), with a genome size of 440 Mb, barley, which has a 5300 Mb genome composed of more than 90% repetitive DNA, (Cui et al., 2005). Thus SFPs have become an attractive marker system for various applications including parental polymorphism discovery, which.

The development of DNA based technologies such as SFP , DArT and RAD which are based on microarray have the merits of SNP;s without going through sequencing. These technologies have provided us platforms for medium- to ultra-high-throughput genotyping to discover regions of the genome at a low cost, and have been shown to be particularly useful for genomes, where the level of polymorphism is low (Gupta et al., 2008). These array based technologies are expected to play an important role in crop improvement and will be used for a variety of studies including the development of high-density molecular maps, which may then be used for QTL interval mapping and for functional and evolutionary studies. Some of these arrays are available from Illumina and Affymetrix.

5. Internal Transcribed Spacer (ITS)

Eukaryotic ribosomal RNA genes (known as ribosomal DNA or rDNA) are parts of repeat units that are arranged in tandem arrays. They are located at the chromosomal sites known as nucleolar organizing regions (NORs). Nuclear ribosomal DNA has 2 internal transcribed spacers: ITS-1 that is located between the small subunit (16s-18s) and 5.8S rRNA cistronic regions, and the ITS-2 which is located between the 5.8S and large subunit (23S-28S) rRNA cistronic regions. The 2 spacers and the 5.8S subunit are collectively known as the internal transcribed spacer (ITS) region.

The ITS regions of rDNA (600-700 bp) repeats are believed to be fast evolving and therefore may vary in length and sequences. The regions flanking the ITS are highly conserved and was used to design universal PCR primers to enable easy amplification of ITS region. Although the biological role of the ITS spacers is not well understood, the utilization of yeast models has definitely shown their importance for production of the mature rRNA.

The number of copies of rDNA repeats is up to-30000 per cell (Dubouzet & Shinoda, 1999). This makes the ITS region an interesting subject for evolutionary and phylogenetic investigations (Baldwin et al., 1995) as well as biogeographic investigations (Baldwin, 1993). The sequence data of the ITS region has also been studied earlier to assess genetic diversity in cultivated barley (Petersen & Seberg 1996). Generally, the ITS has become an important nuclear locus for systematic molecular investigations of closely related taxa. This is because the ITS region is highly conserved intraspecifically, but variable between different species (Bruns et al., 1991; Hillis & Dixon, 1991). Furthermore, the ITS region evolves much more rapidly than other conserved regions of rDNA (Baldwin et al., 1995). Thus, phylogenetic studies based on nrDNA, ITS sequences have provided novel insights into plant evolution and hybridization in various plant species (Sang et al., 1995; Wendel et al., 1995; Buckler & Holtsford 1996a; Quijada et al., 1998; Semerikov & Lascoux, 2003). The ITS region and trnL intron are the most widely used markers in phylogenetic analyses of the Brassicaceae (Koch et al., 2003a; Koch et al., 2003b). The ITS sequences is one of the most successfully used of nuclear genome in studying phylogenetic and genomic relationships of plants at lower taxonomic levels (Baldwin et al., 1995; Wendel et al., 1995).

The genetic diversity studies using ITS can be used by either direct sequencing of the region from different individuals followed by tree consctruction based on sequence comparison. The other method is by measuring sequence variation by restriction digestion of the ITS region and separate the digested fragments on high concentration agarose or acrylamide gel electrophoresis. Sequence variation produced by the latter method called restriction fragment length polymorphisms (PCR-RFLPs), which can be used for taxonomic goals. The ITS region was successfully used for the diagnostics and quick identification of cyst nematodes. Comparisons of PCR-RFLP profiles and sequences of the ITS-rDNA of unknown nematodes with those published or deposited in GenBank facilitate quick identification of most species of cyst nematodes (Subbotin et al., 2001 & Subbotin et al., 2000).

Because of the assumed conservation in secondary structure, it is also referred to as a double-edged tool for eukaryote evolutionary comparison (Tippery & Les, D.H., 2008; Young & Coleman, 2004). This is because the ITS region is highly conserved intraspecifically, but variable between different species (Bruns et al., 1991; Hillis & Dixon, 1991). For many organisms, the ITS2 in premature rRNA is organized around a preserved central core of secondary structure from which four helices emerge (Coleman, 2003). Structural ITS2 database contains more than 288,000 pre-calculated structures for the currently known ITS2 sequences and provides new possibilities for incorporating structural information in phylogenetic studies (http://its2.bioapps.biozentrum.uni-wuerzburg.de). Global pairwise alignments from about ribosomal RNA (rRNA) internal transcribed spacer 2 (ITS2) sequences - all against all - have been generated in order to model ITS2 secondary structures based on sequences with known structures. Via 60,000 known ITS2 sequences that fit a common core of the ITS2 secondary structure described for the eukaryotes homology based modeling (Wolf et al. 2005) and reannotation procedures revealed in addition more than 150,000 homologous structures that could not be predicted by standard RNA folding programs. This database was used for studying gene phylogeny of the bioassay alga "Selenastrum capricornutum" (Buchheim, et al., 2011; Krienitz et al., 2011).

Although ITS is widely used in phylogenetic studies, there is a reported case -for the genus Corylus- in which the ITS region failed to explain the genetic relationship between species (Erdogan & Mehlenbacher, 2000). Moreover, Concerted evolution of ribosomal DNA repeats (including the ITS region) may be a problem if there are instances of allopolyploid speciation within the group (Wendel et al., 1995). Despite this potential problem, ITS is generally considered to be of great utility for phylogenetic analysis among closely related species (Baldwin, 1992; Baldwin et al., 1995),

6. Chloroplast DNA as source of genetic diversity

The cytoplasm DNA consisting of the chloroplast and mitochondrial genomes which became very useful for studying genetic diversity and other taxonomic studies specially after the development of molecular markers. The analysis of the chloroplast organelle provides information on genetic diversity of plants that is complementary to that obtained from the nuclear genome. The chloroplast genome is highly conserved and has a much lower mutation rate than plant nuclear genomes. Consequently, chloroplast DNA has been extensively exploited in studies of plant genetic diversity. Restriction site analysis of chloroplast (cp) DNA has been widely used for interspecific studies and in some cases the magnitude of intraspecific variation has been ssufficient to allow population based studies (Soltis et al., 1992). In addition, Restriction fragment length polymorphism (RFLP) analysis of cpDNA was used to study genetic diversity of Douglas-fir in British Columbia from coastal, interior and transition zones (Ponoy et al., 1994). The RFLP is the result of length mutation or rearrangements. Such polymorphisms are often associated with localized hotspots that contain repetitive DNA. Because of their mutational complexity and lack of representativeness of the genome, they provide biased estimate of nucleotide diversity and thus may also give rise to incorrect estimates of genetic subdivision.

6.1 Chloroplast Transfer RNAs (tRNAs)

Transfer RNAs (tRNAs) are ancient macromolecules that have evolved under various environmental pressures as adaptors in translation in all forms of life but also towards alternative structures and functions (Muller-Putz et al., 2010). In other words, the tRNA world presents a large diversity in terms of function (which includes cell wall synthesis, porphyrin biosynthesis for heme and chlorophyll, N-terminal modification of proteins, initiation of reverse transcription in retroviruses, and lipid remodelling in addition to ribosome-dependent protein synthesis) as well as in terms of structure. Several recent reviews illustrate this diversity (Hopper & Phizicky, 2003; Ryckelynck et al., 2005; Dreher 2009; Roy & Ibba, 2009). Another feature of this region is that it has highly conserved nature which allows structural changes, in the form of indels and repeat sequences, as well as base substitutions to be considered phylogenetically informative (Palmer, 1991; Raubson & Jansen 2005). Low base pair substitution rates within the chloroplast genome has lead to the use of indels in population level studies, with small structural changes (< 10 bp) being useful for increasing phylogenetic resolution and increasing the ability to discriminate within species variation (Mitchell-Olds et al., 2005).

The trnT–trnF region is located in the large single copy region of the chloroplast (cp) genome. The plant trnL-trnF intergenic spacer is less than 500-bp long. From the conserved region, two primers designed (Taberlet et al., 1991) can be used to amplify this spacer in various plant species . In addition, there is generally a high degree of polymorphism exists in the spacer between species. For example, spacer sequences of Acer pseudoplatanus and A. platamoides, two closely related species, are different (Taberlet et al., 1991). Sequence differences between different species within this spacer region can be detected using polyacrylamide gel electrophoresis, polymerase chain reaction –single stranded conformation polymorphism (PCR-SSCP), or even agarose gel electrophoresis. Therefore, the existence of the two universal primers and the high degree of polymorphism in the trnL-trnF inergenic spacer makes it a good marker for paternal analysis in many plant species, studies (Baker et al. 1999; Chen et al., 2002). The noncoding regions of the intron of *trnL* (UAA) and the intergenic spacer of *trnL* (UAA)– *trnF* (GAA) was also extensively utilized for evolutionary analysis in plants and for developing markers for identifying the maternal donors of polyploids with additional capacity to reveal phylogenetic relationships of related species (Sang et al., 1997; Xu & Ban, 2004).

Chloroplast genome is more stable since larger structural changes, and complex structural rearrangements, such as inversions, translocations, loss of repeats and gene duplications are not common. However, there are occasional reports suggesting structural changes in the chloroplast genome (such as pseudogene which are non functioning duplications of functional genes formation) (Ingvarsson et al., 2003). Two main mechanisms have been suggested to account for this structural changes: intramolecular recombination between two similar regions on a single double-stranded DNA (dsDNA) molecule that results in the excision of the intermediate region, yielding a shorter dsDNA molecule and a separate circularised dsDNA molecule. It has been suggested that this will be mediated by short repeats formed by slip-strand mispairing (Ogihara et al., 1988), and intermolecular recombination between two genome copies of the plastids because the plastid contains a large number of copies of the chloroplast genome. Recombination by this method could act to increase pseudogene variation through uneven crossing over (Dobes, 2007). The difference in copy number of pseudogenes can be used in genetic diversify studies as well as classification. In the Brassicaceae, for example, the *trnL*-F region from at least 20 genera contain vaiable copy number of a duplicated *trnF* pseudogene (Ansell et al., 2007; Koch & Matschinger, 2007; Schmickl et al., 2008). These copies are thought to be non-functional, and are made up of partial *trnF* gene fragments ranging between 50-100 bp in length. Tedder et al. 2010 suggested that the pseudogene duplications can serve as a useful lineage marker within the Brassicaceae, as they are absent from many genera, including *Brassica, Draba,* and *Sinapis* (Koch et al., 2006), and in some cases the copy number can be increased to 12 in certain genera (Schmickl et al., 2008). This study requires amplification by PCR using the 'E' and 'F' primers of Taberlet et al., 1991 and sequencing of the *trnL*(UAA)–*trnF*(GAA) gene region. This system was used to study the phylogeography of European populations of *Arabidopsis lyrata* (Ansell, 2007, Ansell et al., 2010), which were considered as *A. lyrata* ssp. *petraea.* Taxonomy of this genus has subsequently been altered, and now both subspecies are considered as the *Arabidopsis lyrata* complex (Schmickl et al., 2008).

6.2 Chloroplast SSR

Nuclear simple sequence repeats SSR is considered a popular marker for population genetics because they are considered selectively neutral, highly polymorphic, co-dominant and inherited in Mendelian mode (Sunnuck, 2000). SSR or microsatellites (Tautz, 1989) are sequence of repetitive DNA where a single motif consisting of one to six base pairs is repeated tandomely or number of times. The cpSSR differs from nuclear SSR in that they are consisted of mononucleotide motif that is repeated 8-15 times in comparison to one to six nucleotide repeats for nuclear SSR. Chloroplast SSR evolves faster than the other gene regions in the chloroplast genome (Jakob et al., 2007). It has been identified in number of genomes.

SSR from organelle genomes, mitochondrial SSR have little impact comparing to chloroplasts. Moreover, chloroplst genomes are transimitted in multiple copies during mitosis and miosis and for that reason they are subject to random drift between and within individuals. The uniparentally inherited, haploid and non-recombinanat nature of chloroplast genome make them very useful tool in evolutionary studies (Petit et al., 2005). Chloroplast SSR (cpSSR) have been used in several plant species such as conifer (Vendramin et al., 1996), graminae (Provan et al., 2004). The potential value of cpDNA markers in complementing nuclear genetic markers in population genetics is widely recognized (Provan et al., 1999; Provan et al., 2001; Petit et al., 2005). The cpSSR evolves faster than any other part of the chloroplast genome. When they are located in noncoding region of cpDNA they show intraspecific variation in repeat numbers. Moreover the effective population size for haploid cpDNA is smaller than diploid nuclear genes so it is considered to be stronger in differentiation between species. Quintela-Sabaris et al., 2010 used cpSSR markers to analyze the colonization pattern of Cistus ladanifer. The cpSSR is sensitive to drift and was used in studying differentiation due to genetic drift (Comes & Kadereit, 1998), in genetic structure and gene flow (Hansen et al., 2005), in bottleneck phenomena (Echt et al., 1998), detection of reduction in genetic diversity within M population of Silene paradoxab (Mengoni et al., 2001). They can also be used to monitor the transmission of chloroplast genomes during hybridization and introgression in wild or breeding populations or to characterize plastid genome type for breeding purposes (Flannery et al., 2006). A comprehensive review for the technical resources, applications, and recommendations for expanding DNA discovery in wide array of plant species is presented by Ebert & Peakall, 2009.

7. Conclusion

Different marker types have different usefulness in studying genetic diversity. Fast changing markers can be used in studying closely related species. RFLP was the first molecular marker to be used in genetic diversity. Although reproducible, it is time consuming and in ploidy species it is of low frequency. RAPD marker is easy to perform, however; it has inherited problem of reproducibility. ISSR is more reproducible and polymorphic than RAPD. The SSR have been the marker of choice for the last two decades especially before the discovery of SNP's. SNP are the most widespread sequence variation in the genome. They are numerous, more stable and easier to score than SSR. AFLP, although having high discriminatory power, it has medium reproducibility and alleles are not easily recognized. On the other hand the development of microarray based technologies such as

SFP, DArT and RAD which have the merits of SNP's without going through sequencing. They are medium- to ultra-high-throughput genotyping at a low cost. They have been shown to be particularly useful for genomes, where the level of polymorphism is low. They are expected to play an important role in crop improvement and will be used for a variety of studies including the development of high-density molecular maps, which may then be used for QTL interval mapping and for functional and evolutionary studies.

One important characteristics of ITS region is that it is highly conserved intraspecifically, but variable between different species. It provides novel insights into plant evolution and hybridization and considered one of the most successfully used of nuclear genome in studying phylogenetic and genomic relationships of plants. The noncoding regions of the intron of *trnL* (UAA) and the intergenic spacer of *trnL* (UAA)– *trnF* (GAA) was extensively utilized for evolutionary analysis in plants and for developing markers for identifying the maternal donors of polyploids with additional capacity to reveal phylogenetic relationships of related species. In addition to the high degree of polymorphism, the intergenic spacer of *trnL* (UAA)– *trnF* (GAA) can be amplified from several plant species using two universal primers. On the other hand, cpSSR evolves faster than any other part of the chloroplast genome It has been used in several plant species.

8. References

Abdelkrim J., Robertson B.C., Stanton, J.A.L., Gemmell, N.J. (2009). Fast, cost effective development of species-specific microsatellite markers by genomic sequencing. Biotechniques 46: 185–192.

Abdel-Mawgood, A.L. (2007). DNA fingerprinting studies of some bread wheat (Triticum aestivum L.) genotypes using RAPD and ISSR techniques. Alex. J. Agric. Res. 52:57-61.

Abdel-Mawgood, A.L. , Jakse, J., Al-Doss, A.A & Assaeed, A.M. (2010). Genetic Structure and Diversity Within and Among Six Populations of Capparis decidua (Forssk.) Edgew., in Saudi Arabia. African Journal of Biotechnology 9: 6256-

Abdel-Mawgood, A.L., Ahmed, M.M.M. & Ali, B.A. (2006). Application of molecular markers for hybrid maize identification. Journal of Food Agriculture and Envirnoment 4:176-178.

Akbari, M., Wenzl, P., Caig, V., Carling, J., Xia, L., Yang, S., Uszynski, G., Mohler, V., Lehmensiek. A., Kuchel, H., Hayden, M., Howes, N., Sharp, P., Vaughan, P., Rathmell, B., Huttner, E., Kilian, A. (2006). Diversity arrays technology (DArT) for high-throughput profiling of the hexaploid wheat genome. Theor Appl Genet, 113:1409-1420.

Allentoft M.E., Schuster, S., Holdaway, R.N., Hale, M.L., McLay, E. (2009). Identification of microsatellites from an extinct moa species using highthroughput (454) sequence data. Biotechniques 46: 195–200.

Angers, B., Bernatchez, L. (1997). Complex evolution of a salmonid microsatellite locus and its consequences in inferring allelic divergence from size information. Molecular Biology and Evolution, 14, 230–238.

Ansell, S.W., Schneider, H., Pedersen, N., Grundmann, M., Russell, S.J., Vogel, J.C. (2007). Recombination diversifies chloroplast trnF pseudogenes in Arabidopsis lyrata. J Evol Biol.

Ansell, S.W., Stenoien, H.K., Grundmann, M., Schneider, H., Hemp, A., Bauer, N., Russell, S.J., Vogel, J. (2010). Population structure and historical biogeography of European Arabidopsis lyrata. Heredity 105:543-53

Arif, I., Khan, H.A., Bahkali, A.H., Al Homaidan, A.A., Al Farhan, A.H., Al Sadoon, M. Shobrak, M. (2011). DNA marker technology for wildlife conservation. Saudi Journal of Biological Sciences. 18:219-225.

Baird N.A., Etter, P.D., Atwood, T.S., Currey, M.C., Shiver, A.L., Lewis, Z.A., Selker, E.U., Cresko, W.A., Johnson, E.A. (2008). Rapid SNP discovery and genetic mapping using sequenced RAD markers. PLoS One. 3(10):e3376.

Baker, W.J., Asmussen, C.B., Barrow, S.C., Dransfield, J., Hedderson, T.A. (1999). A phylogenetic study of the palm family (palmae) based on chloroplast DNA sequenxes from the trnL-trnF region. Plant syst. Evol. 219. 111-126.

Baldwin, B.G. (1992). Phylogenetic utility of the internal transcribed spacers of nuclear ribosomal DNA in plants: an example from the Compositae. Mol Phylogenet Evol 1:3–16.

Baldwin, B.G. (1993). Molecular phylogenetics of Calcydenia (Compositae) based on ITS sequences of nuclear ribosomal DNA: Chromosomal and morphological evolution reexamined. Am. J. Bot. 80:222-238.

Baldwin, B.G., Sanderson M.J., Porter, Wojciechowski, M.F., Campbell, J.S., Donoghue, M.J. (1995). The ITS region of nuclear ribosomal DNA: a valuable source of evidence on angiosperm phylogeny. Annals of the Missouri Botanical Garden. 82: 247-277.

Bernardo, R. (2008). Molecular markers and selection for complex traits in plants: learning from the last 20 years. Crop Sci. 48: 1649-1664.

Binneck, E., Nedel, J.L., Dellagostin, O.A. (2002). RAPD analysis on cultivar identification: a useful methodology? Rev. Bras. Sem. 24: 183-196.

Blears, M.J., De Grandis, S.A., Lee, H., Trevors, J.T. (1998). Amplified fragment length polymorphism (AFLP): review of the procedure and its applications. J. Ind. Microbiol. Biotechnol. 21:99–114.

Borevitz, J.O., Liang, D., Plouffe, D., Chang, H.S., Zhu, T., Weigel, D., Berry, C.C., Winzeler, E., Chory, J. (2003). Large-scale identification of single-feature polymorphisms in complex genomes. Genome Research, 13:513-523.

Botstein, D., White, R.L., Skolnick, M., Davis, R.W. (1980). Construction of a genetic linkage map in man using restriction fragment length polymorphisms. American Journal of Human Genetics 32, 314–31.

Bragard, C., Singer, E., Alizadeh, A., Vauterin, L, Maraite, H., Swings, J. (1997). Xanthomonas translucens from small grains: diversity and phylopathological relevance. Phytopathology 87:1111-1117.

Brenner, B.M., Swede, H., Jones, B.A., Anderson, G.R., Stoler, D.L.(2011). Genomic Instability Measured by Inter-(Simple Sequence Repeat) PCR and High-Resolution Microsatellite Instability are Prognostic of Colorectal Carcinoma Survival After Surgical Resection. Ann Surg Oncol.
(http://www.annsurgoncol.org/journal/10434/0/0/1708/0/).

Brumfield, R.T., Beerli, P., Nickerson, D.A., Edwards, S.V. (2003). The utility of single nucleotide polymorphisms in inferences of population history. Trends Ecol. Evol., 18, 249–256.

Bruns,T.D., White, T.J., Taylor, J.W. (1991). Fungal molecular systematics. Annu. Rev. Ecol. Syst. 22: 525-564.

Buchheim, M., Keller, A., Koetschan, C., Forster, F., Merget, B., Wolf, M. (2011). Internal Transcribed Spacer 2 (nu ITS2 rRNA) Sequence-Structure Phylogenetics: Towards an Automated Reconstruction of the Green Algal Tree of Life. PLoS One 6: 2. e16931.

Buckler, E.S., Holtsford T.P. 1996 Zea systematics: Ribosomal ITS evidence. Molecular Biology and Evolution. 13: 612-622

Caicedo, A.L., Williamson, S.H., Hernandez, R.D. Boyko, A., Fledel-Alon, A. York, T.L. Polato N.R., Olsen, K.M., Nielsen, R., McCouch, S.R. (2007). Genome-wide patterns of nucleotide polymorphism in domesticated rice. PLoS Genet. 3: e163.

Cargill, M., et al. (1999) Characterization of single-nucleotide polymorphisms in coding regions of human genes. Nat. Genet., 22, 231–238.

Chen, J., Tauer, C.G., Huang, Y. (2002). Paternal chloroplast inheritance patterns in pine hybrids detected with trnL-trnF intergenic region polymorphism. Theor. Appl. Genet. 104, 1307-1311.

Chen, X., Sullivan, P.F. (2003). Single nucleotide polymorphism genotyping: biochemistry, protocol, cost and throughput. Pharmacogenomics J. 3: 77-96.

Cho, Y.G., Shii, T., Temnyk, S., Chen, X., Lipovich, L., McCouch, S.R., Park, W.D., Ayres, N., Cartinhour, S. (2000). Diversity of microsatellites derived from genomic libraries and GenBank sequences in rice (Oryza sativa L.). Theoretical and Applied Genetics 100, 713–22.

Coleman, A.W. 2003. ITS2 is a double-edged tool for eukaryote evolutionary comparisons. Trends Genet. 19: 370–375

Collins, F., Brooks, L., Chakaravarti, A. (2011). A DNA polymorphism discovery resources for human genetic variation. Genome Research. 8:1229-1231.

Collins, F.S., Brooks, L.D., Chakravarti, A. (1998). A DNA polymorphism discovery resource for research on human genetic variation. Genome Res., 8, 1229–1231.

Comes, H., Kadereit, J. (1998). The effect of quaternary climatic changes on plant distribution and evolution. Tredns in Plant Science. 3:432-438.

Cui, X.P., Xu, J., Asghar, R., Condamine, P., Svensson, J.T, Wanamaker ,S., Stein, N., Roose, M., Close, T.J. (2005). Detecting single-feature polymorphisms using oligonucleotide arrays and robustified projection pursuit. Bioinformatics 21:3852-3858.

Dobes, C., Kiefer, C., Kiefer, M., Koch, M.A. (2007). Plastidic trnF(UUC) pseudogenes in North American genus Boechera (Brassicaceae): Mechanistic aspects of evolution. Plant Biol. 9: 502-515.

Dow, B.D., Ashley, M.V. (1996). Microsatellite analysis of seed dispersal and parentage of saplings in bur oak, Quercus macrocarpa. Molecular Ecology 5:615-627.

Dreher, T.W. (2009). Role of tRNA-like structures in controlling plant virus Giegé, R. (2008) Toward a more complete view of tRNA biology. Nat. Struct. Hao G, Yuan YM,

Dubouzet, J.G., Shinoda, K. (1999). Relationships among old and New world Alliums according to ITS DNA sequence analysis. Theor. Appl. Genet. 98: 422-433

Duran, C.N., Appleby, T., Clark, D., Wood, M., Batley, I.J., Edwards, D. (2009). AutoSNPdb: an annotated single nucleotide polymorphism database for crop plants. Nucleic Acids Res. 37: D951-953.

Durand, J., Garnier, L., Dajoz, I., Mousset, S., Veuille, M. (2000). Gene flow in a facultative apomictic Poacea, the savanna grass Hyparrhenia diplandra. Genetics, 156, 823-831.

Ebana, K., Yonemaru, J.-I., Fukuoka, S.H., Iwata, H., Kanamori, N., Namiki, H., Nagasaki, M. Yano, M. (2010). Genetic structure revealed by a whole-genome single-nucleotide polymorphism survey of diverse accessions of cultivated Asian rice (Oryza sativa L.). Breed. Sci. 60: 390-397.

Ebert, D. and Peakall, R. (2009). Ltd Chloroplast simple sequence repeats (cpSSRs): technical resources and recommendations for expanding cpSSR discovery and applications to a wide array of plant species. Molecular Ecology Resources 9: 673-690.

Echt, C., DeVerno, L., Anzide, M., Vendramin, G. (1998). Chloroplast microsatellites reveal population genetic diversity in red pine, Pinus resinosa Ait. Molecular Ecology 7:307-316.

Edwards, D., Batley, J. (2010). Plant genome sequencing: applications for plant improvement. Plant Biotech. J. 8: 2-9.

Ellis, T.P., Humphrey, K.E., Smith, M.J., Cotton, R.G.H. (1998). Chemical cleavage of mismatch: a new look at an established method. Hum. Mutat. 11:345-353.

Erdogan,V., Mehlenbacher S.A. (2000). Phylogenetic relationships of Corylus species (Betulaceae) based on Nuclear Ribosomal DNA ITS region and Chloroplast matK gene sequences. Systematic Botany. 25: 727-737.

Estoup, A., Tailliez, C., Cornuet, J.M., Solignac, M. (1995). Size homoplasy and mutational processes of interrupted microsatellites in two bee species, Apis mellifera and Bombus terrestris (Apidae). Molecular Biology and Evolution, 12, 1074-1084.

Fang, D., Roose, M.L. (1997). Identification of closely related citrus cultivars with inter-simple sequence repeat markers. Theor. Appl. Genet. 95: 408-417.

Fisher, S., Lerman, L.S. (1979). Length-independent separation of DNA restriction fragments in two-dimensional gel electrophoresis. Cell 16:191-200.

Flannery, M.L., Mitchell, F.J.G., Coyne, S., Kavanagh, T.A., Burke, J.I., Salamin, N. (2006). Plastid genome characterisation in Brassica and Brassicaceae using a new set of nine SSRs. Theor Appl Genet 113: 1221-1231.

Gao, J., Zhang, S., Qi, L., Zhang, Y., Wang, C., Song, W., Han, S. (2006). Application of ISSR Markers to Fingerprinting of Elite Cultivars (Varieties/Clones) From Different Sections of the Genus Populus L. Silvae Genetica 55(1): 1-6.

Garcia, A.A.F., Benchimol, L.L., Barbosa, A.M.M., Geraldi, I.O. (2004). Comparison of RAPD, RFLP, AFLP and SSR markers for diversity studies in tropical maize inbred lines. Genet. Mol. Biol. 27: 579-588.

Gilad, Y., Pritchard, J., Thornton, K. (2009). Characterizing natural variation using next-generation sequencing technologies. Trends in Genetics 25:463-471

Goldrick, M.M. (2001). RNase cleavage-based methods for mutation/SNP detection, past and present. Hum. Mutat., 18: 190-204.

Goncalves, L.S., Rodrigues, R., do Amaral. Junior, A.T., Karasawa, M. (2009). Heirloom tomato gene bank: assessing genetic divergence based on morphological, agronomic and molecular data using a Ward-modified location model. Genet. Mol. Res. 8: 364-374.

Goncalves, L.S.A., Sudre, C.P., Bento, C.S., Moulin, M.M. (2008). Divergencia genetica em tomate estimada por marcadores RAPD em comparacao com descritores multicategóricos. Hortic. Bras. 26: 364-370.

Guilford, K., Prakash, S., Zhu, J., Gardiner, S., Bassettte, H., Forster, R. (1997). Microsatellites in Malus x domestica (apple), abundance, polymorphism and cultivar identification. Theor Appl Genet 94 : 249-254.

Gupta PK, Rustgi S, Mir RR: (2008). Array-based high-throughput DNA markers for crop improvement. Heredity, 101:5-18.

Gupta, P.K., Rustgi, S. (2004). Molecular markers from the transcribed/expressed region of the genome in higher plants. Funct. Integr. Genomics 4: 139–162.

Gupta, P.K., Varshney, R.K. (2000). The development and use of microsatellite markers for genetic analysis and plant breeding with emphasis on bread wheat, Euphytica 113:163–185.

Gupta, P.K., Varshney, R.K., Prasad, M. (2002). Molecular markers: principles and methodology, in: S.M. Jain, B.S. Ahloowalia, D.S. Brar (Eds.), Molecular Techniques in Crop Improvement, Kluwer Academic Publishers, The Netherlands, , pp. 9–54.

Guryev, V., Berezikov, E., Cuppen, E. (2005). CASCAD: a database of annotated candidate single nucleotide polymorphisms associated with expressed sequences. BMC Genomics, 6, 10.

Haddad, L.A., Fuzikawa, A.K., Pena, S.D. (1997). Simultaneous detection of size and sequence polymorphisms in the transcribed trinucleotide repeat D2S196E (EST00493). Human Genetics 99, 796–80.

Hansen, O.K., Kjær, E.D., Vendramin, G.G. (2005). Chloroplast microsatellite variation in Abies nordmanniana and simulation of causes for low differentiation among populations. Tree Genetics & Genomes 1: 116–123.

Hazen, S.P., Kay, S.A. (2003). Gene arrays are not just for measuring gene expression. Trends Plant Sci. 8: 413–416.

Hill, M., Witsenboer, H., Zabeau, M., Vos, P., Kesseli, R., Michelmore, R. (1996). PCR-based fingerprinting using AFLPs as a tool for

Hillier, L.W., Marth, G.T., Quinlan, A.R., Dooling, D., Fewell, G., Barnett, D., Fox, P., Glasscock, J.I., Hickenbotham, M., Huang, W.C.,Magrini, V.J., Richt, R.J., Sander, S.N., Stewart, D.A., Stromberg, M., Tsung, E.F., Wylie, T., Schedl, T., ilson, R.K. and Mardis, E.R. (2008). Whole-genome sequencing and variant discovery in C. elegans . Nat. Methods 5: 183–188.

Hillis, D.M., Dixon, M.T. (1991). Ribosomal DNA: molecular evolution and phylogenetic inference. Quat. Rev. Biol. 66: 411-453.

Hodkinson, T.R., Chase, M.W., Renvoize, S.A. (2002). Characterization of a genetic resource collection for Miscanthus (Saccharinae, Andropogoneae, Poaceae). Annals of Botany. 89:627-636.

Hopper, A.K. Phizicky, E.M. (2003). TRNA transfers to the limelight. Genes Dev. 17, 162–180.

Hu, C.M., Ge, X.J., Zhao, N.X. (2004). Molecular phylogeny of. Lysimachia (Myrsinaceae) based on chloroplast trnL-F and nuclear ribosomal ITS sequences. Molecular Phylogenetics and Evolution 31:323–339.

Imelfort, M., Duran, C., Batley, J., Edwards, D. (2009). Discovering genetic polymorphisms in next-generation sequencing data. Plant Biotechnology Journal. 7: 312–317.

Ingvarsson, P.K., Ribstein, S., Taylor, D.R. (2003). Molecular Evolution of Insertions and Deletion in the Chloroplast Genome of Silene. Mol. Biol. Evol. 20: 1737-1740

Ismail A. A., Ellingboe, A. H., & Abdel-Magood, A.L. (1999). Genetic diversity between three inbred lines of maize as revealed by AFLP technique. Egypt J. Plant Breed. 3:329-335.

Jaccoud, D., Peng, K., Feinstein, D., Kilian, A (2001). Diversity Arrays: a solid state technology for sequence information independent genotyping. Nucleic Acids Res 29:e25.

Jakob, S.S., Ihlow, A., Blattner, F.R. (2007). Combined ecological niche modeling and molecular phylogeography istory of Hordeum marinum (Poaceae) – niche differentiation, loss of genetic diversity, and speciation in Mediterranean Quaternary refugia. Molec Ecol 16: 1713–1727.

James, K,E,, Schneider, H., Ansell, S.W., Evers, M., Robba, L., Uszynski, G., Pedersen, N., Newton, A.E., Russell, S.J., Vogel, J.C., Kilian, A. (2008). Diversity Arrays Technology (DArT) for pan-genomic evolutionary studies of non-model organisms. PLoS ONE 2008, 3:e1682.

Jannink, J., Lorenz, A.J., Iwata, H. (2010). Genomic selection in plant breeding: from theory to practice. Brief. Funct, Genomics 9: 166–177.

Jones, M.P., Dingkuhn, M., Aluko, G.K., Semon, M. (1997). Interspecific Oryza sativa L. x O. glaberrima Steud. progenies in upland rice improvement. Euphytica, 92: 237-246.

Karp, A., Kresovich S., Bhat, K.V., Ayada, W.G., Hodgkin, T. (1997). Molecular tools in plant genetic resources conservation: a guide to the technologies. IPGRI technical bulletin no 2. International Plant Genetic Resources Institute, Rome, Italy.

Kilian, A., Huttner, E., Wenzl, P., Jaccoud, D. (2005). The Fast and the Cheap: SNP and DArT-based Whole Genome Profiling for Crop Improvement. In: Proceedings of the International Congress in the Wake of the Double Helix: From the Green Revolution to the Gene Revolution (Tuberosa R, Phillips RL and Gale M, eds.). May 27-31, Bologna, 443-461.

Kilian, A., Huttner, E., Wenzl, P., Jaccoud, D., Carling, J., Caig, V., Evers, M., Heller-Uszynska, K., Cayla, C., Patarapuwadol S., Xia, L., Yang, S., Thomson, B. (2003). The fast and the cheap: SNP and DArT-based whole genome profiling for crop improvement. In the Wake of the Double Helix: From the Green Revolution to the Gene Revolution; 27–31May Bologna 2005:443-461.

Kim, J.S., Ahn, S.G., Kim, C.K., Shim, C.K. (2010). Screening of rice blast resistance genes from aromatic rice germplasms with SNP markers. Plant Pathol. J. 26: 70–79.

Koch, M., Dobea, C., Bernhardt, K.G., Kochjarova, J. (2003a). Cochlearia macrorrhiza: A bridging species between Cochlearia taxa from the Eastern Alps and the Carpathians. Plant Syst. Evol. 242: 137-147

Koch, M., Dobea, C., Mitchell-Olds, T. (2003b). Multiple hybrid formation in natural populations: Concerted evolution of the internal transcribed spacer of nuclear ribosomal DNA (ITS) in North American Arabis divaricarpa (Brassicaceae). Mol. Biol. Evol. 20: 338-350.

Koch, M.A., Matschinger, M. (2007). Evolution and genetic differentiation among relatives of Arabidopsis thaliana. Proc Natl Acad Sci USA 104:6272–6277.

Korbin, M., Kuras, A., Zurawicz, E. (2002). Fruit Plant Germplasm characterization using molecular markers generated in RAPD and ISSR-PCR. Cellular and Molecular Biology Letters. 7(2B): 785–794.

Krienitz, L., Bock, C., Nozaki, H., Wolf, M. (2011). SSU rRNA gene phylogeny of morphospecies affiliated to the bioassay alga "Selenastrum capricornutum" recovered the polyphyletic origin of crescent-shaped. Chlorophyta Journal of Phycology 47: 880-893.

Kumar, L.S. (1999). DNA markers in plant improvement: An overview. Biotechnol. Adv. 17: 143–182.

Kumar, R., Qiu, J., Joshi, T., Valliyodan, B., Xu, D., Nguyen, H.T. (2007). Single feature polymorphism discovery in rice. PLoS One, 2(3):e284

Kwok, P.Y., Q. Deng, H. Zakeri, S.L. Taylor, D., Nickerson (1996). Increasing the information content of STS-based genome maps: identifying polymorphisms in mapped STSs. Genomics 31: 123–126.

Levinson, G., Gutman, G.A. (1987). Slipped-strand mispairing: Amajor mechanism for DNA sequence evolution. Mol. Biol. Evol. 4:203−221.

Lewis, Z.A., Shiver, A.L., Stiffler, N., Miller, M.R., Johnson, E.A,, Selker, E.U. (2007). High density detection of restriction site associated DNA (RAD) markers for rapid mapping of mutated loci in Neurospora. Genetics. 177(2):1163-1171.

Majeed, A.A., Ahmed, N., Rao, K.R., Ghousunnissa, S., Kauser, F., Bose, B., Nagarajaram, H.A., Katoch, V.M., Cousins, D.V., Sechi, L.A., Golman, R.H. & Hasnain, S.E. (2004). AmpliBASE MT: a Mycobacterium tuberculosis diversity knowledgebase. Bioinformatics 20:989-992.

McCouch, S., Zhao, K., Wright, M., Tung, C., Ebana, K., Thomson, M., Reynolds, A., Wang, D., DeClerck, G., Ali, M., McClung, A., Eizenga, G., Bustamante, C. (2010). Development of genome-wide SNP assays for rice. Breeding Science 60: 524–535.

Miller, M.R., Atwood, T.S., Eames, B.F., Eberhart, J.K., Yan, Y.L., Postlethwait, J.H., Johnson, E.A. (2007b). RAD marker microarrays enable rapid mapping of zebrafish mutations. Genome Biol. 2007;8(6):R105.

Miller, M.R., Dunham, J.P., Amores, A., Cresko, W.A., Johnson, E.A. (2007a). Rapid and cost-ffective polymorphism identification and genotyping using restriction site associated DNA (RAD) markers. Genome Research. 17(2):240-248.

Mishra, P.K., Fox, R.T.V., Culhamm, A. (2003). Inter-simple sequence repeat and aggressiveness analyses revealed high genetic diversity, recombination and long-range dispersal in Fusarium culmorum. School of Plant Sciences, The University of

Reading, Whiteknights, Reading RG6 6AS, UK, 2003 Association of Applied Biologists.

Mitchell-Olds, T.; Al-Shehbaz, I., Koch, M., Sharbel, T. (2005). Crucifer evolution in the post-genomic era. In Plant Diversity and Evolution: Genotypic and Phenotypic Variation in Higher Plants; Henry, R.J., Ed.; CABI Publishing: Oxon, UK; pp. 119-136.

Mohammadi, S.A. and B.M. Prasanna, 2003. Analysis of genetic diversity in crop plants-salient statistical tools and considerations. Crop Sci., 43: 1235-1248.

Mohammadi, S.A., Prasanna, B.M. (2003). Analysis of genetic diversity in crop plants - salient statistical tools and considerations. Crop Sci. 43: 1235-1248.

Moose, S.P., Mumm, R.H. (2008). Molecular plant breeding as the foundation for 21st century crop improvement. Plant Physiol. 147:969–977.

Morand, M.E., Brachet S, Rossignol P, Dufour J, Frascaria-Lacoste N (2002) A generalized heterozygote deficiency assessed with microsatellites in French common ash populations. Molecular Ecology, 11, 377–385.

Morozova, O., Marra, M. (2008). Applications of next-generation sequencing technologies in functional genomics. Genomics 92: 255-264.

Muller-Putz, G. R., Kaiser V., Solis-Escalante, T., Pfurtscheller G. (2010). Fast set-up asynchronous brain-switch based on detection of foot motor imagery in 1-channel EEG. Med. Biol. Eng. Comput. 48, 229–233.

Mullis, K.B., Faloona, F.A. (1987). Specific synthesis of DNA in vitro via a polymerase-catalyzed chain reaction. Methods Enzymol. 155: 335-350.

Nagaoka, T., Ogihara, Y. (1997). Applicability of inter-simple sequence repeat polymorphisms in wheat for use as DNA markers in comparison to RFLP and RAPD markers. Theor Appl Genet 94: 597-602.

Nagaraju, J., Kathirvel, M., Kumar, R., Siddiq, E.A., Hasnain, S.E. (2002). Genetic analysis of traditional and evolved Basmati and non-Basmati rice varieties by using fluorescence-based ISSR-PCR and SSR markers. Proc.Natl.Acad.Sci., 99-9. 5836-5841.

Nichols, K.M., Young, W.P., Danzmann, R.G. (2003). A consolidated linkage map for rainbow trout (Oncorhynchus mykiss). Animal Genetics 34, 102–15.

O'Brien, S.J. (1991). Molecular genome mapping: lessons and prospects. Curr. Opin. Genet. Dev. 1:105–111.

Obeed, R.S., Harhash, M.M. and Abdel-Mawgood, A.L. (2008). Fruit properties and genetic diversity of five ber (Ziziphus mauritiana Lamk) cultivars. Pakistani J. Biological Sciences: 11:888-893.

Ogihara, Y., Terachi, T., Sasakuma, T. (1988). Intramolecular Recombination of Chloroplast Genome Mediated by Short Direct-Repeat Sequences in Wheat Species. Proc. Nat. Acad. Sci.USA 85:8573-8577.

Oliver, R., Lazo, G., Lutz, J., Rubenfield, M,, Tinker, N., Anderson J, Morehead, N, Adhikary, D, Jellen, E., Maughan, P, Guedira G, Chao, S, Beattie, A, Carson, M, Rines, H, Obert D, Bonman J, JacksonE. (2011). Model SNP development for complex genomes based on hexaploid oat using high-throughput 454 sequencing technology. BMC Genomics 2011, 12:77.

Orita, M., Iwahana, H., Kanazawa, H., Hayashi, K., Sekiya, T. (1989). Detection of polymorphisms of human DNA by gel electrophoresis as single-strand conformation polymorphisms. Proc. Natl. Acad. Sci. U.S.A., 86: 2766–2770.

Palmer, J.D. (1991). Plastid chromosomes: structure and evolution. In Cell Culture and Somatic CellGenetics in Plants: The Molecular Biology of Plastids; Bogorad, L., Vasil, I.K., Eds.; Acedemic Press: San Deigo, CA, USA, Volumn 7, pp. 5-53.

Paul, S., Wachira, F.N., Powel, W. & Waugh, R. (1997). Diversity and genetic differentiation among populations of Indian and Kenyan tea (Camellia sinensis (L.) O. Kuntze) revealed by AFLP markers. Theoretical and Applied Genetics 94:255-263.

Peakall, R., Gilmore, S., Keys, W., Morgante, M., Rafalski, A. (1998). Cross-species amplification of soybean (Glycine max) simple sequence repeats (SSRs) within the genus and other legume genera: implications for the transferability of SSRs in plants. Molecular Biology and Evolution, 15, 1275–1287.

Petersen, G., Seberg, O. (1996). ITS regions highly conserved in cultivated barleys. Euphytica 90: 233-234.

Petit RJ, Duminil J, Fineschi S, Hampe A, et al. (2005). Comparative organization of chloroplast, mitochondrial and nuclear diversity in plant populations. Mol. Ecol. 14:689-701.

Ponoy, B., Hongy, P., Woods, J., Jaqoub, B., Carlson, J. (1994). Chloroplast DNA diversity of Douglas-fir in British Columbia. Can. J. For. Res. 24: 1824-1 834.

Provan, J.W., Powell, P.M., Hollingsworth, A. (2001). Chloroplast microsatellites: new tools for studies in plant ecology and evolution. Tends in Ecol. Evol. 16(3): 142-147

Provan, J., Soranzo, N., Wilson, N.J., Goldstein, D.B. (1999). A low mutation rate for chloroplast microsatellites. Genetics 153: 943-947.

Provan,J., Biss, P.M., McMeel, D., Mathews, M. (2004). Universal primers for the amplification of chloroplast microsatellites in grasses (Poaceae). Molecular Ecology Notes 4: 262-264

Purcell, A.H., Hopkins, D.L. (1996). Fastidious xylem-limited bacterial plant pathogens. Annual Review of Phytopathology 1996;34:131-151.

Pütz, J., Giege, R., Florentz, C. (2010). Diversity and similarity in the tRNA world: Overall view and case study on malaria-related tRNAs. FEBS Letters 584:350–358

Pyke, K.A. (1991). Plastid Division and Development. Plant Cell 11:549-556.

Quijada, A., Liston, A., Delgado, P., Lobo, A.V., Alvarez-Buylla, E.R. (1998). Variation in the nuclear ribosomal DNA internaltranscribed spacer (ITS) region of Pinus rzedowskii revealed by PCR-RFLP. Theor. Appl. Genet. 96: 539-544.

Quintela-Sabarís, C., Vendramin, G., Castro-Fernández, D., Fraga, M. (2010). Chloroplast microsatellites reveal that metallicolous populations of the Mediterranean shrub Cistus ladanifer L have multiple origins. Plant Soil 344:161-174

Rafalski, A. (2002) Applications of single nucleotide polymorphisms in crop genetics. Curr. Opin. Plant Biol. 5: 94–100.

Rafalski, A. (2002b). Novel genetic mapping tools in plants: SNPs and LD-based approaches. Plant Science. 162: 329-333.

Rafalski, J., Morgante, M., Powell, J., Tinggey, S. (1996). Generating and using DNA markers in plants. In: Birren, and E. Lai (rds), Analysis of non-mammalian Genomes: a practical Guide, 75-134. Academic press, Boca Raton.

Rakoczy-Trojanowska, M., Bolibok, H. (2004). Characteristics and comparison of three classes of microsatellite-based markers and their application in plants. Cell. Mol. Biol. Lett. 9: 221–238.

Rallo, P., Dorado, G., Martin, A. (2000). Development of simple sequence repeats (SSRs) in olive tree (Olea europaea L.). Theor. Appl.Genet. 101: 984–989.

Raubson, L.A., Jansen, R.K. (2005). Chloroplast Genomes of Plants. In Plant Diversity and Evolution:Genotypic and Phenotypic Variation in Higher Plants; Henry, R.J., Ed.; CABI Publishing:Cambridge, MA, USA, pp. 45-68.

Reddy, P.M., Sarla, N., Siddiq, E.A. (2002). Inter simple sequence repeat (ISSR) polymorphism and its application in plant breeding. Euphytica, 128-2: 9-17

Roder, M.S., Plashke, J., Konig, S.U. (1995). Abundance, variability, and chromosomal location of microsatellites in wheat. Molecular and General Genetics, 246: 327–333.

Roy H, Ibba M (2009). Broad range amino acid specificity of RNA dependent lipid remodeling by multiple peptide resistance factors. J. Biol. Chem. 284: 29677–29683.

Ryckelynck, M., Giege, R., Frugier, M. (2005). TRNAs and tRNA mimics as Schmickl, R.; Jorgensen, M.; Brysting, A.; Koch, M., Phylogeographic implications for the North

Saiki, R.K., Scharf, S., Faloona, F., Mullis, K.B. (1985). Enzymatic amplification of beta-globin genomic sequences and restriction site analysis for diagnosis of sickle cell anemia. Science 230: 1350-1354.

Sakamoto, T., Danzmann, R.G., Gharbi, K. (2000). A microsatellite linkage map of rainbow trout (Oncorhynchus mykiss) characterized by large sex-specific differences in recombination rates. Genetics 155, 1331–45.

Saker, M.M., Youssef, S.S., Abdallah, N.A., Bashandy, H.S. (2005). Genetic analysis of some Egyptian rice genotypes using RAPD, SSR and AFLP. Afr. J. Biotechnol. 4: 882-890.

Sang, T., Crawford, D., Stuessy, T. (1997). Chloroplast phylogeny, reticulate, evolution, and biography of Paeonia (Paeoniaceae). American Journal of Botany 84: 1120–1136.

Sang, T., Crawford, D.J. Stuessy, T.F. (1995). Documentation of reticulate evolution in peonies (Paeonia) using internal transcribed spacer sequences of nuclear ribosomal DNA: implications for biogeography and concerted evolution. Proc.Natl. Acad. Sci. USA. 92: 6813-6817.

Santana, Q.C., Coetzee, M.P.A., Steenkamp, E.T., Mlonyeni, O.X., Hammond, G.N.A. (2009). Microsatellite discovery by deep sequencing of enriched genomic libraries. Biotechniques 46: 217–223

Schlotterer, C. (2004). The evolution of molecular markers - just a matter of fashion? Nat. Rev. Genet. 5: 63-69.

Schmickl, R., Kiefer, C., Dobes, C., Koch, M. (2008). Evolution of trn F(GAA) pseudogenes in cruciferous plants. Plant Syst. Evol. 282:229-240.

Schulman, A.H. (2007). Molecular markers to assess genetic diversity. Euphytica 158: 313-321.

Scribner, K.T., Prince, H. H. (2001). Definition of spatial genetic structure in mid-continent Canada geese. In Management of Urban Canada Geese Symposium, W. Rapley (ed.). Toronto Univ.

Seddon, J.M., Parker, H.G., Ostrander, E.A., Ellegren, H. (2005). SNPs in ecological and conservation studies: a test in the Scandinavian wolf population. Mol. Ecol., 14, 503–511.

Semerikov, L.V., Lascoux, M. (2003). Nuclear and Cytoplasmic variation within and between Eurasian Larix species. American Journal of Botany. 90: 1113-1123.

Serapion, J., Kucuktas, H., Feng, J., Liu, Z. (2004). Bioinformatic mining of type I microsatellites from expressed sequence tags of channel catfish (Ictalurus punctatus). Marine Biotechnology 6, 364–77.

Soltis, D.E., Soltis, P.S., Milligan, B.G. (1992). in Molecular Systematics of Plants, eds. Soltis, P. S., Soltis, D. E. & Doyle, J. J. (Chapman and Hall, New York), pp. 117-150.

Souza, S.G.H., Carpentieri-Pípolo, V., Ruas, C.F., Carvalho, V.P. (2008). Comparative analysis of genetic diversity among the maize inbred lines (Zea mays L.) obtained by studying genetic relationships in Lactuca spp. Theoretical and Applied Genetics 93: 1202–1210.

Subbotin, S.A., Halford, P.D., Warry, A., Perry, R. (2000). Variations in ribosomal DNA sequences and phylogeny of Globodera parasitising Solanaceae. Nematology 2: 591-604.

Subbotin, S.A., Vierstraete, A., De Ley, P., Rowe, J., Waeyenberge, L., Moens, M., Vanfleteren, J. R. (2001). Phylogenetic relationships within the cyst-forming nematodes (Nematoda, Heteroderidae) based on analysis of sequences from the ITS regions of ribosomal DNA. Molecular Phylogenetics and Evolution 21: 1-16.

Sudre, C.P., Leonardecz, E., Rodrigues, R., Amaral Júnior, A.T. (2007). Genetic resources of vegetable crops: a survey in the Brazilian germplasm collections pictured through papers published in the journals of the Brazilian Society for Horticultural Science. Hortic. Bras. 25: 496-503.

Suh, Y., Vijg, J. (2005). SNP discovery in associating genetic variation with human disease phenotypes. Mutat Res., 573, 41–53.

Sunnucks, P. (2000). Efficient genetic markers for population biology. Trends in Ecology and Evolution, 15, 199–203.

Taberlet, P., Gielly, L., Pautou, G., and bouvet, J. (1991). Universal Primers for amplification of three non-coding regions of chloroplast DNA. Plant Mol. Biol. 17,1105-1109.

Tanksley, S. D., Young, N. D., Paterson, A.H., Bonierbale, M.W. (1989). RFLP mapping in plant breeding-new tools for an old science. Biotechnology 7: 257-264.

Tanksley, S., Young, A., Patterson, H., Bonierbale, M. (1989). RFLP mapping in plant breeding- new tools for an old science. Bio/Technology 7:257-264.

Tautz, D. (1989). Hypervariability of simple sequences as a general source for polymorphic DNA markers. Nucleic Acids Res. 17: 6463-6471.

Tedder , A., Hoebe, P.N., Ansell, S.W., Mable, B.K. (2010). Using chloroplast trnF pseudogenes for phylogeography in Arabidopsis Lyrata. Diversity 2 : 653-678.

The Official Site of Diversity Arrays Technology (DArT P/L). [http://www.diversityarrays.com]

Thiel, T., Michalek,W., Varshney, R.K., Graner, A. (2003). Exploiting EST databases for the development and characterization of genederived SSR-markers in barley (Hordeum vulgare L.). Theoretical and Applied Genetics 106, 414–22.

Till, B.J., Burtner, C., Comai, L., Henikoff, S. (2004). Mismatch cleavage by single-strand specific nucleases. Nucleic Acids Res., 32, 2632–2641.

Tippery, N.P., Les, D.H. (2008). Phylogeneic analysis of the internal transcriped spacer in Menyanthaceae using predicted secondary structure. Molecular Phylogenetic and Evolution 49:528-537.

Tohme, J., Gonzalez, D.O., Beebe, S., Duque, M.C. (1996). AFLP analysis of gene pool of a wild bean core collection. Crop Sci, 36: 1375–1384.

Travis, S.E., Maschinski, J., Keim, P. (1996). An analysis of genetic variation in Astrgalus cremnophylax var. Cremnophylax, a critically endangered plant, using AFLP markers. Mol Ecol.;6:735–745.

Tsumura, Y., K. Yoshimura, N. Tomaru, and K. Ohba. K. (1995). Molecular phylogeny of conifers using PCR-RFLP analysis of chloroplast genes. Theor. Appl. Genet. 91:1222–1236

Tsumura, Y., Ohba, K. Strauss, S.H. (1996). Diversity and inheritance of inter-simple sequence repeat olymorphisms in Douglas-fir (Pseudotsuga menziesii) and sugi (Cryptomeria japonica). Theor Appl Genet 92: 40-45.

Twyman, R.M. (2004). SNP discovery and typing technologies for pharmacogenomics. Curr. Top. Med. Chem., 4, 1423–1431.

Vaccino P, Accerbi M, Corbelini, A 1993. Cultivar identification in T. aestivum using highly polymorphic RFLP probes. Theor. Appl. Genet. 86:833-836.

van der Schoot, J., Pospiskova, M., Vosman, B. (2000). Development and characterization of microsatellite markers in black poplar (Populus nigra L.). Theor. Appl. Genet. 101: 317–322.

van Oppen, M.J.H., Rico, C., Turner, G.F., Hewitt, G.M. (2000). Extensive homoplasy, nonstepwise mutations, and shared ancestral polymorphism at a complex microsatellite locus in lake Malawi cichlids. Molecular Biology and Evolution, 17, 489–498.

Vaneechonette, M. (1996). DNA fingerprinting techniques for microorganisms: a proposal for classification and nomenclature. Mol. Biotechnol 6:115-142.

Vendramin, G., Lelli, L., Rossi, P., Morgante, M. (1996). A set of primers for the amplificaions of 20 chloroplast microsatellites in Pinaceae. Molecular Ecology. 5:595-598.

Venkatesh, B. (2003). Evolution and diversity of fish genomes. Current Opinion in Genetics and Development 13, 588–92.

Viard, F., Franck, P., Dubois, M-P., Estoup, A., Jarne, P. (1998). Variation of microsatellite size homoplasy across electromorphs, loci, and populations in three invertebrate species. Journal of Molecular Evolution, 47, 42–51.

Vijg, J., van Orsouw, N.J. (1999). Two-dimensional gene scanning: exploring human genetic variability. Electrophoresis, 20:1239–1249.

Vos, P., Hogers, R., Bleeker, M., Reijans, M., Van der Lee, T., Hornes, M., Frijters, A., Pot, J., Peleman J, Kuiper M and Zabeau M (1995). AFLP: a new technique for DNA fingerprinting. Nucleic Acids Res 23: 4407-4414.

Vos, P., Kuiper, M. (1998). AFLP analysis. In: DNA markers, protocols, applications and overviews (G. Caetano-Arnolles & P.H. Gresshoff, eds.) pp 115-131. John Wiley & Sons, New York.

Vosman, B., Arens, P. (1997). Molecular characterization of GATA/GACA microsatellite repeats in tomato. Genome 40: 25–33.

Welsh, J., McClelland, M. (1990). Fingerprinting genomes using PCR with arbitrary primers. Nucleic Acids Res. 18: 7213-7218.

Wendel, J.F., Schnabel, A., Seelanan, T. (1995). Bi-directional interlocus concerted evolution following allopolyploid speciation in cotton (Gossypium). Proc. Natl. Acad. Sci. USA.92: 280-284.

Wenzl, P., Carling, J., Kudrna, D., Jaccoud, D., Huttner, E., Kleinhofs, A., Kilian, A. (2004). Diversity Arrays Technology (DArT) for wholegenome profiling of barley. Proc Natl Acad Sci USA 2004, 101:9915-9920.

Wheeler, D.A., Srinivasan, M., Egholm, M., Shen, Y., Chen, L., McGuire, A., He, W., Chen, Y.-J., Makhijani, V., Roth, G.T., Gomes, X., Tartaro, K., Niazi, F., Turcotte, C.L., Irzyk, G.P., Lupski, J.R., Chinault, C., Song, X.-Z., Liu, Y., Yuan, Y., Nazareth, L., Qin, X., Muzny, D.M., Margulies, M., Weinstock, G.M., Gibbs, R.A. and Rothberg, J.M. (2008). The complete genome of an individual by massively parallel DNA sequencing. Nature 452: 872–876.

White, G.M., Boshier, D.H., Powell, W. (2002). Increased pollen flow counteracts fragmentation in a tropical dry forest: an example from Swietenia humilis Zuccarini. Proceedings of the National Academy of Sciences of the USA, 99, 2038–2042.

Williams, J.G., Kubelik, A.R., Livak, K.J., Rafalski, J.A. (1990). DNA polymorphisms amplified by arbitrary primers are useful as genetic markers. Nucleic Acids Res. 18: 6531-6535.

Williams, J.G.K., Kubelik, A.R., Livak, K.J., Rafalski, J.A., Tingey, S.V. (1990). DNA polymorphisms amplified by arbitrary primers are useful as genetic markers. Nucleic Acids Res 18: 6231-6235.

Wolf, M., Achtziger, M., Schultz, J., Dandekar, T., Muller, T. (2005). Homology modeling revealed more than 20,000 rRNA internal transcriped spacer 2 (ITS2) secondary structures. RNA 11:327-336.

Xiao, W., Oefner, P.J. (2001). Denaturing high-performance liquid chromatography: a review. Hum. Mutat., 17, 439–474.

Xu, D.H., Ban, T. (2004). Phylogenetic and evolutionary relationships between Elymus humidus and other Elymus species based on sequencing of noncoding regions of cpDNA and AFLP of nuclear DNA. Theoretical and Applied Genetics 108: 1443–1448.

Xu, Y., Crouch, J.H. (2008). Marker-assisted selection in plant breeding: from publications to practice. Crop Sci. 48: 391–407.

Yamamoto, T., Nagasaki, H., Yonemaru, J., Ebana, K., Nakajima, M., Shibaya, T., Yano, M. (2010), Fine definition of the pedigree haplotypes of closely related rice cultivars by means of genome-wide discovery of single-nucleotide polymorphisms. BMC Genomics 11: 267.

Yanagihara, K., Mizuuchi, K. (2002). Mismatch-targeted transposition of Mu: a new strategy to map genetic polymorphism. Proc. Natl. Acad. Sci. U.S.A., 99, 11317–11321.

Young, I., Coleman, A.W. (2004). The advantages of the ITS2 region of the nuclear rDNA cistron for analysis of phylogenetic relationships of insects: A Drosophila example. Mol. Phylogenet. Evol. 30: 236–242.

Zeid, M., Schon, C., Link, W. (2003). Genetic diversity in recent elite faba bean lines using AFLP markers. Theor. Appl. Genet. 107:1304-1314.

Zietkiewicz, E., Rafalski, A., Labuda, D. (1994). Genome fingerprinting by simple sequence repeat (SSR)-anchored polymerase chain reaction amplification. Genomics 20: 176-183.

4

Genotyping Techniques for Determining the Diversity of Microorganisms

Katarzyna Wolska[1] and Piotr Szweda[2]
[1]University of Natural Science and Humanities n Siedlce,
[2]Gdansk University of Technology,
Poland

1. Introduction

Typing of microbial pathogens, or identifying bacteria at the strain level, is particularly important for diagnosis, treatment, and epidemiological surveillance of bacterial infections. This is especially the case for bacteria exhibiting high levels of antibiotic resistance or virulence, and those involved in nosocomial or pandemic infections. Strain typing also has applications in studying bacterial population dynamics. The part that molecular methods have to play in elucidating bacterial diversity is increasingly important. The shortcomings of phenotypically based typing methods (generally these methods are viewed as being too time consuming and lacking in sufficient resolution amongst related strains) have led to the development of many DNA – based techniques. A suitable typing method must have high discrimination power combined with good to moderate inter- and intra-laboratory reproducibility. In addition, it should be easy to set up, to use and to interpret, and inexpensive (Olive & Bean, 1999). In this chapter, we review the current bacterial genotyping methods and classify them into six main categories: (1 and 2) DNA banding pattern-based methods, which classify bacteria according to the size of fragments generated respectively by enzymatic digestion of genomic/plasmid DNA, and PCR amplification, (3) DNA hybridization–based methods using nucleotidic probes, (4) DNA sequencing-based methods, which study the polymorphism of DNA sequences, (5) differentiation of isolates on the basis of presence or absence of particular genes and (6) high resolution melting analysis–real–time monitoring of melting process of PCR amplified polymorphic DNA fragment. We described and compared the applications of genotyping methods to the study of bacterial strain diversity. We also discussed the selection of appropriate genotyping methods and the challenges of bacterial strain typing and described the current trends of genotyping methods.

2. Description of current microbial genotyping methods

2.1 DNA banding pattern-based methods, which classify bacteria according to the size of fragments generated by enzymatic digestion of genomic or plasmid DNA

2.1.1 Pulsed-Field Gel Electrophoresis – PFGE

Pulsed–field gel electrophoresis (PFGE) was developed in 1984 and has since become the "gold standard" of molecular typing methods. PFGE is extensively used in research laboratories that specialize in analyzing specimens sent in by hospitals or state laboratories. Clinically, it is an invaluable tool in detecting the occurrence of an outbreak and trying to determine its source (Tenover et al., 1995). The procedural steps in PFGE are the following: embed organisms in agarose plugs→enzyme digestion→restriction endonuclease digestion→electrophoresis→gel staining→interpretation. The bacterial suspension is combined with molten agarose and mixed with a protease (an enzyme that disrupts the cell membrane by attacking the membrane proteins) and with SDS (a detergent that unfolds proteins). The enzyme-detergent mixture denatures the cell membrane proteins thus forming holes in the cell through which the chromosomal DNA is released. The agarose keeps the DNA embedded in its gel matrix. Next, the plug is washed several times to remove cell debris and proteases. These diffused out of the agarose gel matrix more easily than the large DNA molecules. It is important to remove the proteases so as not to harm the restriction enzymes that cleave the DNA in the next step of the process. A small piece of the plug is cut off and added to a restriction endonuclease(s) mixture which cleaves DNA at a specific sequence resulting in 10-30 DNA fragments ranging from 0.5 to 1000 kb. The large DNA fragments are then separated by size by pulsed-field gel electrophoresis. PFGE facilitates the migration of the large DNA fragments (>600 kb) through the agarose gel by regularly changing the direction of the electrical field during electrophoresis, allowing the fragments to maneuver through the agarose. The smaller DNA fragments move faster through the agarose than the larger fragments and the result is a pattern of DNA fragments. Migration distances are compared to reference standards of known molecular weight and a profile for each strain/isolate is obtained. The PFGE pattern from one isolate can be compared to other patterns to determine whether the samples may have originated from a common source. The electrophoretic patterns are visualized following staining of the gels with a fluorescent dye such as ethidium bromide. Gel results can be photographed, and the data can be stored by using one of the commercially available digital systems. Data analysis can be accomplished by using any of a number of commercially software packages (Olive & Bean, 1999; Tenover et al., 1994, 1995, 1997).

Use of PFGE has been greatly facilitated by the incorporation of standard methods of analysis suggested by Tenover et al. (1995), which has led to its widespread adoption. In their scheme, bacterial isolates yielding the same PFGE pattern are considered the same strain. Bacterial isolates differing by a single genetic event, reflected as a difference of one to three bands, are closely related. Isolates differing by four to six bands, representing two independent genetic changes, are possibly related. Bacterial isolates containing six or more band differences, representative of three or more genetic changes, are considered unrelated. These criteria are applicable to small, local studies in which genetic variability is presumed to be limited.

PFGE is one of the most reproducible and highly discriminatory typing techniques available and is the most reliable technique for analysis of variety of foodborne pathogens: *Escherichia coli* O157:H7 (Arbeit, 1995), *Salmonella* Typhimurium (Tsen et al., 2002), *Salmonella* Enteritidis (Thong et al., 1995), *Campylobacter jejuni* (Eyles et al., 2006; Nebola & Steinhauserova, 2006), *Listeria monocytogenes* (Okwumabua et al., 2005) as well as nosocomial pathogens: methicilin-resistant *Staphylococcus aureus* (MRSA) (Saulnier et al., 1993; Tenover et al., 1994), vancomycin-resistant enterococci (Barbier et al., 1996), *Klebsiella pneumoniae* (Vimont et al., 2008), *Serratia marcescens* (Shi et al., 1997), *Acinetobacter calcoaceticus-Acinetobacter baumannii* complex (Liu et al., 1997), *Pseudomonas aeruginosa* (Grundmann et al., 1995), *Mycobacterium avium* (Arbeit, 1994), *Neisseria gonorrhoeae* (Poh et al., 1996), *Neisseria meningitidis* Serogroup C (Shao et al., 2007), *Stenotrophomonas maltophilia* (Valdezate et al., 2004) and *Legionella pneumophilla* (Tenover et al., 1995). PFGE has also proved to be discriminatory and reproducible for typing *Clostridium difficile*, the main etiologic agent of nosocomial *diarrhoea*. Nevertheless, a high proportion of strains are non-typable by this technique due to the degradation of the DNA during the process. The introduction of several modifications in the PFGE standard procedure proposed by Alonso and coworkers (2005) increased typability from 40% (90 isolates) to 100% (220 isolates) while maintaining the high degree of discrimination and reproducibility of the technique.

The PFGE characterize with some advantages. The method can be easily applied to different species, all the strains can be typed with good reproducibility, restriction profiles are easily read and interpreted, patterns consistent within and between laboratories (strict adherence to standard conditions is necessary), PFGE generally yields a high amount of pattern diversity (Olive & Bean, 1999; Tenover et al., 1997).

However it is necessary to remember about some drawbacks of PFGE. The technique is labor-intensive, relatively slow (approximately 2 to 3 days to completion), complex patterns challenging for inter-laboratory pattern comparisons, one mutation can yield differences in several fragments (PFGE can not determine phylogenic relationships), the agarose gel used for PFGE has a lower resolving power compared to those of polyacrylamide-urea gels (this may be a disadvantage for reliable comparison of patterns and computer analysis) and high initial cost of the equipment can also be an important limitation for many laboratories and investigators (Olive & Bean,1999; Tenover et al., 1997).

2.1.2 Restriction Enzyme Analysis of Plasmid

The development of novel genotypic methods, characterized by high discriminatory power, universality and reproducibility, caused that the REAP method (Restriction Enzyme Analysis of Plasmid) is currently not so popular. However it is one of the first method of molecular biology which was applied for differentiation of pathogenic bacteria such as *Yersinia enterocolitica* (Kapperud et al., 1990) or methicilin resistant *S. aureus* (Harstein et al., 1989). From the technical point of view, the REAP method is easy and includes five relatively simple steps: isolation and analysis of presence of plasmid/s DNA in the cells →digestion of the plasmid/s with a selected endonuclease→ electrophoresis→gel staining→interpretation.

In the REAP method, strains are distinguished due to presence, number and size of plasmids, and size of fragments generated in the result of plasmid digestion with a restriction enzyme. The most important limitation of this method is that it can be used only for strains carrying plasmids. The differentiation power of the method can be modified by selection of restriction enzymes, or by using a combination of several of them. Although the method is not very popular, it still should be considered an important tool of microorganisms' differentiation. The REAP method can be very useful for investigation of spreading of plasmid - dependent antibiotic resistance or virulence of pathogenic microorganisms, or analysis of transmission of strains harboring plasmids. Plasmids belong to the mobile genetic elements and very often contain genes encoding virulence factors and genes, expression of which results in antibiotic resistance. Some bacteria eg. staphylococci and *Enterobacteriaceae* are very prone to the plasmid DNA transfer. It is possible that two strains of bacteria with identical chromosomal DNA, differ in the virulence potential or sensitivity to some antibiotics, only because of the presence of a plasmid DNA. The plasmid DNA transfer often leads to the creation of endemic strains in hospital environment. For example the REAP analysis performed by Trilla and coworkers revealed that 95% of MRSA strains isolated after an outbreak of MRSA nosocomial infections in Hospital Clínic in Provincial of Barcelona presented an unique homologous pattern of DNA fragments (Trilla et al., 1993). High virulence potential and decreased susceptibility towards antibiotics and chemoterapeutics are very often characteristic for such endemic strains.

2.2 DNA banding pattern-based methods, which classify bacteria according to the size of fragments generated by a PCR amplification

2.2.1 PCR ribotyping and PCR – RFLP

Rybotyping is based on the analysis of sequences of a rRNA operone, a very important element of genomes of all bacteria (Fig. 1). The method was developed by Kostman et al. (1992) and Gurtler (1993) in the 1990s as a response in part to the need in the clinical microbiology laboratory setting for expeditious epidemiological discrimination among pathogenic microorganisms without the use of probes, thus making the analysis more widely applicable. The genes coding for 16S rRNA, 23S rRNA and 5S rRNA are absolutely essential for existence of each bacterial cell due to its involvement in protein synthesis, and their sequences are generally highly conserved among bacteria, however some variable regions are present within these sequences. The sequence of 16S rRNA gene is conserved even at the strain level. As a result sequencing of 16S rRNA gene is at present one of the most important method used for bacterial species identification. However also less advanced techniques such as analysis of restriction fragments length polymorphism of PCR amplified fragments of 16S rRNA gene (PCR – RFLP) was successfully applied for identification of bacteria species. Okhavir et al. (2000) performed PCR – RFLP analysis on 53 strains of 14 bacterial species (eight Gram-positive and five Gram-negative) collected from both *keratitis* and *endophthalmitis* patients. Two pairs of oligonucleotide primers based on the 16S rDNA gene were used to PCR-amplify 1.2- and 1.0-kb fragments of bacterial genomic DNA. All bacteria tested could be identified and speciated using RFLP analysis except for *E. coli* and *S. marcescens*, which could not be interdifferentiated using PCR – RFLP.

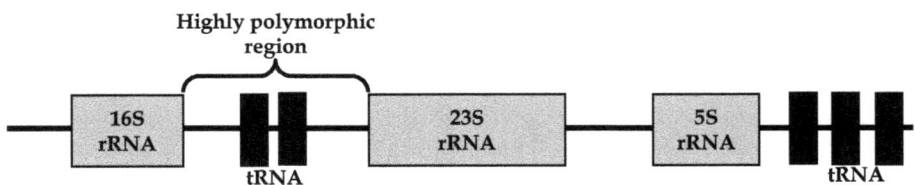

Fig. 1. Organization of bacterial rRNA operon

From the point of view of analysis of genetic differentiation of microorganism especially interesting is the fragment of the operon located between 16S and 23S, called internal transcribed spacer (ITS) or intergenic space region (ISR). In the case of some kinds of bacteria ITS characterize with surprisingly high polymorphism of both length and sequence even at the species level. Additionally some of the species of bacteria contain in the genome more than one copy of the operon. *In silico* analysis of ISR length variability in 27 genomically sequenced bacterial species revealed that while in some species ISR length is variable within and between isolates, in others ISR lengths are limited to one or two sizes, usually dependent on the number of tRNAs present (Christensen et al., 1999). For example, while the multiple copies of the *S. enterica* serovar Typhimurium, serovar Infantis, and serovar Derby ITS fragments are polymorphic, they are conserved in other *Salmonella* species, as well in *Listeria, Streptococcus*, and certain species of *Staphylococcus* (Giannino et al., 2003; Gurtler & Stanisich, 1996; Lagatolla et al., 1996; Marsou et al., 1999). As consequence depending of the construction of rRNA operon amplification of its noncoding fragments can be a very useful technique for identifying bacteria to the species level (if the sequence is conserved) or for their differentiation if the target fragment of gene characterize with high polymorphism. As it was mentioned above the sequences of 16S and 23S rRNA are generally conserved among bacteria thus one universal pair of PCR primers can be used for amplification of a target polymorphic region of rRNA operon of different species of bacteria. If the examined strains contain several polymorphic copies of rRNA operon several products of ITS – PCR amplification are generated, and strain specific patterns of DNA bands is obtained after electrophoresis. The discriminatory power of the ITS – PCR method can be improved by digestion of PCR products, and analysis of restriction fragment length polymorphism (PCR – RFLP) after agarose or polyacrylamide electrophoresis. The differences in sequences of the amplified fragments of intergenic spacer regions of the same length can be alternatively analyzed by using the single sequence conformation polymorphism (SSCP) (Daffonchio et al., 1998).

The PCR ribotyping method has proven to be suitable for the routine investigation of foodborne outbreaks of *Shigella* isolates (De Paula et al., 2010). Cho & Tiedje (2000) successfully used a PCR ribotyping approach to define the biogeography of fluorescent pseudomonad isolates, and found it as discriminating as the BOX rep-PCR. The other species with which this methodology has found some success is *C. difficile*, having a total of 11 ribosomal operons, with differing tRNAs and ISR lengths found among ribosomal operons of the same organism (Sebaihia et al., 2006)   Almost 43% of all published PCR ribotyping studies have been performed with *C. difficile* (Bidet et al., 2000; Kikkawa et al., 2007; Rupnik et al., 1998; Stubs et al., 1999). An analysis of 45 isolates of *C. difficile* from which the ITS had been sequenced did not reveal striking sequence length diversity; however, it appears that the selection of PCR ribotyping as a typing technique relies more on the comparative ease of the technique and repeatability rather than on the discriminatory ability (Bidet et al., 2000), due to the occurrence of DNA degradation in *C. difficile* interfering with PFGE analysis.

Beside ITS fragment of rRNA operon, there are also some other targets which can be used for PCR – RFLP genotyping of different species of bacteria. The popular criterion often used for examining genetic variability of staphylococci is RFLP analysis of polymorphism of size an sequence of coagulase gene (Goh et al., 1992). Nebola and Steinhauserowa (1996) used PFGE technique and PCR – RFLP analysis of the flagellin gene (*fla*) for differentiation of 92 poultry and 110 human strains of *C. jejuni*, and obtained results were comparable.

And lately the restriction analysis of PCR amplified gene coding for major outer membrane protein was successfully applied for genotyping *Chlamydia trachomatis* isolated from cervical specimens in New Delhi (India) (Gita et al., 2011).

PCR ribotyping and RFLP technique is easy in comparison to other molecular methods and consists of: PCR amplification→restriction endonuclease digestion→agarose gel electrophoresis→gel staining→interpretation. The technique is not time-consuming and inexpensive, but the most important drawback is fact that only limited fragment of genome is analyzed as a result its discriminatory power is usually poorer than for example PFGE, RAPD, ERIC or BOX rep-PCR (Bouchet et al., 2008; Kikkawa et al., 2007, Van den Berg et al., 2007).

2.2.2 Random Amplified Polymorphic DNA – RAPD

The random amplified polymorphic DNA (RAPD) assay, also referred to as arbitrary primed (AP) PCR and DNA amplification fingerprinting (DAF), is a powerful tool for genetic studies. It was first described by Williams et al. (1990) and Welsh & McClelland (1990). RAPD analysis is useful as a screening genotyping method (Speijer et al., 1999). RAPD is fast, simple and less labor than the usual fingerprinting method with non radioactive isotopes used (Leal et al., 2004). It detects differences along the entire bacterial genome, not only in particular sequences. Thus, this system is helpful in characterizing bacteria isolates over long periodes (Ortiz-Herrera et al., 2004). RAPD can generate various fingerprint profiles with unlimited number of primers (Leal et al., 2004). A simple short primers are used without the need of prior knowledge of the template DNA. The selection of an appropriate primer and optimization of PCR conditions are the important factor in RAPD analysis (Blixt et al., 2003). The primers that work for some bacteria may fail for others and because of that, the screening process needs to determine the appropriate primers (Shangkuan & Lin, 1998). RAPD uses oligonucleotide (9 to 10 bases in length) primers with arbitrary sequence, which hybridize with sufficient affinity to chromosomal DNA sequences at low annealing temperatures such that they can be used to initiate amplification of regions of the bacterial genome. If two RAPD primers anneal within a few kilobases of each other in the proper orientation, the result is a PCR product with a molecular length corresponding to the distance between the two primers results. The number and location of these random primer sites vary for different strains of a bacterial species. Thus, following separation of the amplification products by agarose gel electrophoresis, a pattern of bands which, in theory, is characteristic of the particular bacterial strain results (Caetano et al, 1991; Meunier & Grimont, 1993; Welsh & McClelland 1990; Williams et al., 1990). The relationship between strains may be determined by comparing their unique fingerprint information (Leal et al., 2004).

Several investigators found poor reproducibility with RAPD. However, the technique is reliable if the PCR conditions are optimized (Benter et al., 1995; Leelayuwet et al., 2000; Ortiz-Herrera et al., 2004;). Leelayuwet et al. (2000) have optimized the RAPD conditions using eighteen deca-oligo nucleotide primers with 70% GC content, eight 60%GC RAPD primers, and four random deca oligomers to produce reproducible but complex patterns showing a high degree of variation between strains of *Burkholderia pseudomallei* isolated from five patients with localized and four with septicemic *melioidosis*. They found that reproducible RAPD patterns are dependent upon the optimal concentrations of DNA in accordance with Taq polymerase and magnesium as well as PCR cycling conditions. Thus, DNA samples should be quantified, and the same lot of Taq enzyme should be used for the entire study. If a new batch of Taq is introduced, re-optimization is required. Furthermore, a ramping time of 7 min is essential in obtaining reproducible RAPD patterns. The RAPD patterns were analyzed by high resolution polyacrylamide gel electrophoresis using a laser based automated fragment analyzer, GS2000.

RAPD is considered as a simple (PCR amplification with single or two primers→agarose gel electrophoresis→gel staining→interpretation), rapid, highly discriminating, less costly and simple technique for molecular typing of various microorganisms (Van Belkum et al., 1995). It has been successfully applied in the genetic differentiation of *Salmonella* (Tikoo, 2001), *E. coli* (Renqua-Mangia et al., 2004), *A. baumannii* (Reboli et al., 1994), *S. marcescens* (Debast et al., 1995), *Proteus mirabilis* (Bingen et al., 1993), *Enterobacter cloacae* (Hou et al., 1997), *Haemophilus somnus* (Myers et al., 1993), *Leptospira* species (Ralph et al., 1993), *L. pneumophilia* (Tram et al., 1990, Van Belkum et al., 1993), *B. pseudomallei* (Haase et al., 1995, Leelayuwet et al., 2000), *Aeromonas salmoniciola* (Miyata et al., 1995), *Aeromonas hydrophila* (Miyata et al., 1995), *Vibrio cholerae* (Leal et al., 2004), *S. aureus* (Saulnier et al., 1993), *Lactobacillus plantarum* (Elegado et al., 2004; Lawrence et al., 1993), *Bacillus cereus* (Svensson et al., 2004), *Listeria monocytogenes* (Hiroshi et al., 2007), *Candida albicans* (Lehmann et al., 1992, Robert et al., 1995), *Histoplasma capsulatum* (Kersulyte et al., 1992) and *Cryptococcus neoformans* (Yamamoto et al., 1995). RAPD is highly discriminatory and therefore especially useful in the investigation of short term or local outbreaks of disease. Several authors have used RAPD fingerprinting of material obtained by gastric biopsy (Akopyanz et al., 1992) or gastric aspirates (Konno et al., 2005) to investigate transmission of *Helicobacter pylori*, supporting the hypothesis of intra-familial transmission. RAPD assay can be also useful in identification of *Proteus penneri* strains (Kwil et al., 2002). This typing presented a high discriminatory power between strains of *Malassezia furfur* and can be applied in epidemiological investigation of skin disease caused by these bacteria (Gandra et al., 2006). *Lactobacillus*-probiotic strains from 5 *Lactobacillus* species (*L. brevis*, *L. reuteri*, *L. gallinarium*, *L. salivarius* and *L. panis*) were specifically, rapidly, immediately and conveniently differentiated after optimization of the RAPD parameters such as $MgCl_2$, Taq polymerase, primer concentration and type of primer (Manan et al., 2009). It is a potentially useful assay for the rapid characterization of neonatal infections associated with group B streptococci (Zhang et al., 2002). Random amplified polymorphic DNA (RAPD) and enterobacterial repetitive intergenic consensus (ERIC) can be applied successfully in study of genetic distribution and epidemiology of *Vibrio parahaemolyticus* (Zulkifli et al., 2009). This technique utilizing a universal typing primer was successfully used for genotyping the isolates *Arcobacter butzleri*, *A. cryaerophilus* and *A. skirrowii* (Atabay et al., 2008).

RAPD-PCR can be practically applied in most laboratories since it requires no special and/or complex equipment and takes less time and is less labourous as compared with some other genotyping methods such as PFGE and AFLP (Atabay et al., 2008). Vila et al. (1996) and Speijer et al. (1999) found that the RAPD assay was more discriminating than RFLP analysis of either the 16S rRNA genes or the 16S-23S rRNA spacer region but less discriminating than PFGE, AFLP and Rep-PCR analysis in the case of studing of *P. aeruginosa* and *A. calcoaceticus-A. baumannii* complex.

2.2.3 Rep–PCR

Versalovic et al. (1991) described a method for fingerprinting bacterial genomes by examining strain or subtype-specific patterns obtained from PCR amplification of repetitive DNA elements present within bacterial genomes. There are three main sets of repetitive DNA elements used for typing purposes. The 35-40 bp repetitive extragenic palindromic (REP) elements are palindromic units, which contain a variable loop in the proposed stem-loop structure (Stern et al., 1984). The 124-127 bp enterobacterial repetitive intergenic consensus (ERIC) sequences also known as intergenic repeat units (IRUs) are characterized by central, conserved palindromic structures. They are present in many copies in the genomes of many enterobacteria (Hulton et al., 1991). The position of ERIC elements in enterobacterial genomes varies between different species and has been used as a genetic marker to characterize isolates within a bacterial species (Son et al., 2002; Versalovic et al., 1991). The 154 bp BOX elements consist of differentially conserved subunits, namely boxA, boxB, and boxC (Martin et al., 1992). Only the boxA-like subunit sequences appear highly conserved among diverse bacteria (Versalovic et al., 1994). BOX elements were the first repetitive sequences identified in a Gram-positive organism (*Streptococcus pneumoniae*) (Martin et al., 1992). REP- and ERIC-sequences were originally identified in Gram-negative bacteria (*E. coli* and *Salmonella* Typhimurium) and then found to be conserved in all related Gram-negative enteric bacteria and in many diverse, unrelated bacteria from multiple phyla (Olive & Bean, 1999; Versalovic et al., 1994). The repetitive elements may be present in both orientations, and oligonucleotide primers have been designed to prime DNA synthesis outward from the inverted repeats in REP and ERIC, and from the boxA subunit of BOX, in the polymerase chain reaction (PCR) (Versalovic et al., 1994). The use of these primer(s) and PCR leads to the selective amplification of distinct genomic regions located between REP, ERIC or BOX elements (Scheme 5). The corresponding protocols are referred to as REP-PCR, ERIC-PCR and BOX-PCR genomic fingerprinting respectively, and rep-PCR genomic fingerprinting collectively (Versalovic et al., 1991, 1994). The amplified fragments can be resolved in a gel matrix, yielding a profile referred to as a rep-PCR genomic fingerprint (Versalovic et al., 1994). These fingerprints resemble "bar code" patterns analogous to UPC codes used in grocery stores (Lupski et al., 1993). The rep-PCR genomic fingerprints generated from bacterial isolates permit differentiation to the species, subspecies and strain level.

The procedural steps in rep-PCR are: PCR amplification with REP or ERIC or BOX primers→agarose gel electrophoresis→gel staining→interpretation. Rep-PCR can be performed with DNA extracted from bacterial colonies or by a modified method using unprocessed whole cells (Woods et al., 1993). Studies carried out by many laboratories on a variety of different bacterial genera and species have revealed that, at fine taxonomic resolution, phylogenetic trees derived from BOX-, ERIC- and REP-PCR genomic fingerprinting are not always identical. This is to be expected, since different numbers of bands may be generated with each primer set, the annealing conditions vary between primers or sets, and the prevalence/distribution of the target repetitive elements in question may vary. To compare multiple different gels with each other, the best results are obtained when all experimental parameters are standardized as much as possible. This is especially important, when large databases are to be generated or, data generated by different laboratories need to be compared. The standardized conditions used should include sample preparation and processing, use of similar growth conditions, the same DNA isolation methods and use of the same rep-PCR conditions. Moreover the use of standardized electrophoresis conditions and size markers is essential (Versalovic et al., 1994). REP or ERIC amplification can be performed with a single primer, a single set of primers, or multiple sets of primers. ERIC patterns are generally less complex than REP patterns, but both give good discrimination at the strain level. Application of both REP and ERIC-PCR to samples to be typed increases the discriminatory power over that of either technique used alone. BOX-PCR is the most robust of the three rep-PCR methods. BOX-PCR patterns are not affected by the culture age of the strain to be analyzed (Kang & Dunne, 2003) and fingerprinting output can be easily analyzed by computer assisted methods (Ni Tuang et al., 1999). These features make BOX-PCR a frequently used tool in biogeography studies in environmental microbiology (Cherif et al., 2003; Dombek 2000; Oda et al., 2003; Singh et al., 2001). The highest data point scatter was observed with ERIC-PCR fingerprint similarity values, which are more sensitive to disturbances.

Rep-PCR genomic fingerprinting protocols have been developed in collaboration with the group led by Dr. J.R. Lupski at Baylor College of Medicine (Houston, Texas) and have been applied successfully in many medical, agricultural, industrial and environmental studies of microbial diversity (Versalovic et al., 1994). In addition to studying diversity, rep-PCR genomic fingerprinting has become a valuable tool for the identification and classification of bacteria, and for molecular epidemiological studies of human and plant pathogens (Louws et al., 1996; Versalovic et al., 1997). It has been applied in the classification and differentiation of strains of many Gram-positive and -negative bacteria including *Bartonella* (Rodriguez-Barrados et al., 1995), *Bacillus subtilis* (Pinna et al., 2001), *B. sporothermodurans* (Herman & Heyndrickx, 2000), *E. coli* (Leung et al., 2004; Panutdaporn et al., 2004; Silveira et al., 2003), *Citrobacter diversus* (Woods et al., 1992), *Enterobacter aerogenes* (Georghiou et al., 1995), *Salmonella* (Chmielewski et al., 2002; Kerouanton et al., 1996; Millemann et al., 1996; Rasschaert et al., 2005), *Vibrio cholerae* (Colombo et al., 1997; Rivera et al., 1995), *Pseudomonas corrugata* (Achouak et al., 2000), *Vibrio parahaemolyticus* (Khan et al., 2002; Son et al., 1998), *Pseudomonas syringae-Pseudomonas viridiflava* group (Marques et al., 2008), *Aeromonas* spp. (Taco et al., 2005), *Xanthomonas* (Rademaker et al., 2000), *Rhizobium meliloti* (De Bruijn, 1992; Niemann et al., 1997), *Pandoraea apista* (Atkinson et al., 2006), methicillin-resistant *S. aureus* (Van Belkum et al., 1992), *S. pneumoniae* (Versalovic et al., 1993), *A. baumanii* (Dijkshoorn et al., 1996), *Burkholderia cepacia* (Hamill et al., 1995), *B. pseudomallei* (Currie et al., 2007), *L. pneumophilia* (Georghiou et al., 1994), *Helicobacter pylori* (Kwon et al., 1998), *N. gonorrhoeae* (Poh et al., 1996), *N. meningitidis* (Woods et al., 1996), *Enterococcus* spp. (Svec et al., 2005), *Paenibacillus larvae* subsp. *larvae* (Genersch & Otten, 2003) and *Lactobacillus* spp. (Gevers et al., 2001). Rep-PCR is emerging as a potential tool for identification of the source of environmental *E. coli* populations owing to its success in classifying the correct host source, reproducibility, cost-effectiveness and easy operational procedures (Baldy-Chudzik et al., 2003; Carson et al., 2003; Dombek et al., 2000; Johnson et al. 2004; Mohapatra et al., 2007). Dombek et al. (2000) compared the ability of REP-PCR and BOX-PCR to discriminate 154 *E. coli* isolates of 7 source groups (human, duck, geese, chicken, pig, sheep and cow) and concluded from the ARCC results that the discriminatory efficacy of BOX-PCR (ARCC Ľ 93.3%) was superior to REP-PCR (ARCC Ľ 65.8%). Leung et al. (2004) documented that ERIC-PCR was not an effective tool in distinguishing *E. coli* isolates between animal and human sources. In contrast to above study, the ERIC sequence was used successfully for differentiation of *E. coli* isolates obtained from patients showing clinical signs of urinary tract infection (UTI) (Dalal et al., 2009). Baldy-Chudzik et al. (2003) have evaluated REP-PCR and ERIC-PCR and found greater discriminatory power for REP-PCR than ERIC-PCR for genotyping of 93 aquatic *E. coli* isolates. Results of Wieser & Busse (2000) demonstrated that ERIC- and BOX-PCR are excellent tools for rapid identification of *Staphylococcus epidermidis* strains at the species level. For species differentiation ERIC-PCR appeard to be more suitable than BOX-PCR, however, the combination of these two PCR methods provide more reliable results in classifying and identifying staphylococcal isolates. ERIC-PCR and BOX-PCR were rapid, highly discriminatory and reproducible assays that proved to be powerful surveillance screening tools for the typing of clinical *P. aeruginosa* isolates (Dawson et al.

2002; Syrmis et al., 2004; Wolska & Szweda, 2008; Yang et al., 2005). Rep-PCR was also used to study the epidemiology of *Vibrio parahaemolyticus* isolated from cockles in Padang, Indonesia. For the ERIC primer, it produced bands ranged from 3-15 with sizes from 0.1 - 5.0 kb and twenty seven different ERIC patterns (Zulkifli et al., 2009).

Rep-PCR has been shown to have similar or even better strain differentiation power, as well as to be easier, quicker and cheaper to perform, than ribosomal intergenic spacer analysis (RISA), restriction fragment length polymorphism (RFLP), amplified fragment length polymorphism (AFLP), random amplified polymorphic DNA (RAPD) and other techniques (Chmielewski et al., 2002; Niemann et al., 1997; Olive & Bean, 1999). Several studies have shown Rep-PCR to have good correlation with PFGE results but, in general, with slightly less discriminatory power and less reproducible (Kidd et al., 2011).

The rep-PCR technique, in which the amplified products are separated by agarose gel electrophoresis, suffers from several limitations like poor band resolution and run standardization for comparison of the different profiles in different gels. To overcome these limitations separation of fluorescent labelled products in automated DNA sequencer can be used (Del Vecchio et al., 1995; Versalovic et al., 1995). This method allows consistent pattern formation and storage of the data in a database as a digitized image. Unknown strains can be compared against the stored database for identification purposes.

The recently introduced DiversiLab (DL) system (bio-Mérieux) is based on the repetitive-sequence-based PCR. The DL system is a semiautomated rep-PCR with a high level of standardization, in particular for the electrophoresis step by using a Bioanalyzer (Agilent Technologies, Inc., Santa Clara, CA). This reduces reproducibility problems due to variation in assay conditions. The analysis software allows the comparison of individual amplification product patterns (peak patterns), which enables easier interpretation of the patterns, but a virtual gel image is also generated. The patterns can be stored in a database and used for comparison. An important advantage of DL is that a result can be obtained in 1 day starting from a pure culture. DL is a useful tool to help identify hospital outbreaks of *Acinetobacter* spp., *S. maltophilia*, the *Enterobacter cloacae* complex, *Klebsiella* spp., and, to a somewhat lesser extent, *E. coli*. DL was inadequate for *P. aeruginosa*, *Enterococcus faecium*, and MRSA (Fluit et al., 2010).

2.2.4 Ligation mediated–PCR

This is a group of highly sensitive methods widely used for detection of DNA polymorphism of both prokaryotic and eukaryotic organisms. Several techniques classified to this group have been described to date. The procedure of all of them is based on five steps: digestion of the total cellular DNA with restriction enzymes (one or more, depending of the type of LM-PCR) → ligation of restriction half-site specific adaptors to all restriction fragments generated during digestion of the cellular DNA → selective amplification of some of the DNA fragments generated in the ligation step → visualization of the amplified fragments of DNA →interpretation.

2.2.4.1 Amplified Fragment Length Polymorphism – AFLP

The AFLP-PCR (Amplified Fragment Length Polymorphism) is the first described (Vos et al., 1995) and the most popular method belonging to the LM-PCR (Ligation mediated) group. In the first step the total cellular DNA is digested with a combination of two restriction endonucleases, a frequent cutting enzyme with a 4 – base recognition and a rare-cutting enzyme of 6 to 8 - base recognition. Three types of restriction fragments are generated following digestion:

a. with a cohesive ends left by a rare cutter at both ends,
b. with a cohesive ends left by a frequent cutter at both ends,
c. with two distinct ends left by each of used enzymes.

More than 90% of fragments are expected to have frequent – cutter sites at both ends (A) (Blears et al., 1998). The short double - stranded fragments of synthesized DNA (10-30 base pairs long), called adapters, complementary to the corresponding restriction site, are ligated to the generated fragments of DNA of analyzed organisms. The adapters serve as a primer binding sites for PCR amplification carried out in the next step. The ligation of adapters and restriction products is performed in the presence of both restriction enzymes used previously for the digestion of the template DNA. It prevents ligation of restriction fragments obtained in the first step. The adaptor is designed to ensure that ligation of a fragment to an adaptor does not reconstitute the restriction site and the obtained products are not digested with the endonucleases. The products of ligation are amplified with primers consisting of an adapter sequences at the 5' ends and extended with a variable number of 3' nucleotides, usually 1 to 3. The addition of nucleotides at the 3' end of the primers reduces a number of amplified fragments of DNA. The number of amplified fragments is reduced approximately four fold with each additional selective base, assuming a random base distribution (Vos et al., 1995). The limitation of a number of amplified fragments is necessary to make the analysis of DNA fragments pattern easier or rather possible. Even after using the extended primers, the number of amplicons is still very high, in a range of 50 to 100. As a result, special methods have to be used for analysis of the results of amplification, such as high resolution electrophoresis with silver staining or autoradiography. The fluorescent AFLP method (FAFLP), which uses fluorescently labeled primers for PCR amplification and an automated DNA sequencer for fragment detection can be used as a solution to such problems. In some cases, less complex patterns can be obtained by AFLP protocols employing just one restriction enzyme, which can be analyzed by simple agarose gels, as it has been demonstrated for *Chlamydia psittaci*, *H. pylori* and *L. pneumophila* (Boumedine & Rodolakis, 1998; Gibson et al., 1998; Jonas et al., 2000; Valsangiacomo et al., 1995). Each time, the AFLP – PCR must be carried out under stringent conditions, permitting only a selective amplification of those genomic fragments that are perfectly complementary to the 3' ends of the primer sequences. The stringent PCR conditions used in AFLP – PCR lead to highly reproducible results that are readily comparable among different samples.

This method has been successfully applied for genotypic analysis of different species of bacteria, yeast, fungi and plants. A model system for AFLP is genus of *Xanthomonas* (Rademaker et al., 2000). Lomonaco and coworkers have shown that AFLP-PCR can be a very useful tool for the subtype of *L. monocytogenes* isolated from environmental and food samples (Lomonaco et al., 2011). Also coagulase-positive and coagulase-negative staphylococci from different origins: veterinary (Cuteri et al., 2004; Piessens, 2010) and clinical samples (Sloos et al., 1998) were differentiated using the AFLP technique. Analyzing the group of over 50 clinical isolates of *S. aureus*, Melles and coworkers (2007) obtained similar results for three genotypic methods: AFLP, PFGE and MLTS, and AFLP was shown to be more reproducible than PFGE. Lan and Reeves (2007) successfully applied radioactively and fluorescent dye-labeled AFLP method for analysis of a genetic polymorphism of one of most dangerous species of foodborn pathogens – *Salmonella*. An interesting modification of the AFLP method has been recently proposed and applied for typing of 70 clinical isolates of *E. coli* by Brillowska-Dabrowska and coworkers (2008). The authors digested the template DNA with TspRI, the enzyme which recognizes and digests 9 base degenerated sequences, generating the DNA fragments with degenerated cohesive ends. The long adaptor containing the primer binding site was cloned to DNA fragments containing only one selected sequence of a cohesive end. The remaining cohesive ends were covered with short - lacking primer site adaptors with degenerated cohesive ends. In the result, only the selected TspRI digested DNA fragments were amplified. The analysis of results was performed using a classical polyacrylamide gel electrophoresis and the grouping obtained was identical with that resulting from REA–PFGE. Identification and epidemiological examination of *Candida* species isolated from clinical samples using the AFLP method were carried out by several authors eg. Ball and coworkers (2004). Satisfactory results were also obtained in the case of analysis of genetic similarity between 55 isolates of *Aspergillus fumigatus* obtained from 15 different patients suffering from the proven invasive aspergillosis (De Valk et al., 2007). The AFLP can be also applied for analysis of more complicated higher organisms: plants, animals and human. For example the method has been used to study human DNA samples in forensic investigations and in paternity tests (Brinkman et al, 1991), however the analysis of diversity of higher organisms is not a subject of this chapter and will not be discussed in details.

In addition to their widespread use in DNA fingerprinting, the AFLP-based approaches have also been used to produce gene expression fingerprints. AFLP gene expression fingerprints are generated using cDNA (rather than genomic DNA) as the PCR template. With this approach, researchers can study gene expression from multiple *loci* as a means of comparison between two different individuals or populations (Rice, 2010).

The most important advantage of AFLP – PCR is the high discriminatory power of this method, that is comparable with that of PFGE – the method currently considered to be "the gold standard" for epidemiological studies and can be easily determined by combination of a pair of restriction enzymes and by design of primers. The selection of restriction enzymes is usually based on the in *silico* analysis (San Millan et al., 2005), however the combination of MseI and EcoRI enzymes is the most popular in the literature. If fact, the AFLP analysis can be performed without any prior knowledge about DNA sequence of a tested organism. However, on the other hand pre - isolation of the pure culture is necessary. The analyzed DNA cannot be contaminated with any plasmid or chromosomal DNA of another organism. Therefore the method is not suitable for analysis of DNA samples isolated directly from an infected tissue, food or environmental sample, which in some applications is an obvious limitation.

2.2.4.2 Amplification of DNA Surrounding Rare Restriction Sites ADSRRS

At the beginning of the first decade of the 21st - century Masny and Plucienniczak described two new methods of LM–PCR, that are called respectively ADSRRS (Amplification of DNA Surrounding Rare Restriction Sites) and PCR MP (Melting Profiles) (Masny and Plucienniczak, 2001, 2003). In comparison to the AFLP technique, the main advantage of both proposed methods is reduction of number of amplified fragments of DNA, that leads to much easier analysis of obtained results. In this procedure, similary like in the AFLP technique, the products of digestion of the template DNA with a pair of restriction enzymes (one frequent and one rare cutting), are ligated with adapters. Adapter design is the crucial element of the ADSRRS method. The adapter complementary to the cohesive ends of rare cutter is short and not really rich in GC pairs, in contrast to the adapter complementary to the ends generated by the other enzyme, which is long and contains a lot of G and C nucleotides. It is important to mention that adapters are not phosphorylated and in fact only one (shorter) strand is ligated to the fragments of digested template DNA. The other strand of adapter, called helper, is only necessary for binding of the adapter with complementary cohesive end before ligation. The helper is than thermally dissociated and the ends of DNA fragments ligated with one strand of adapter are filled with thermostable DNA polymerase. In the effect, three sets of DNA fragments are generated:

a. containing sequences of a long adapter at both ends. These are the products of digestion of the template DNA with only frequent cutter. As it was mentioned they constitute about 90% of the obtained population of DNA fragments,

b. containing sequences of short adapter at both ends,

c. with ends containing sequences of two different adaptors at their ends.

In consequence of ligation of the same adaptors to the both ends, the fragments belonging to the groups A and B contain in fact the terminal complementary sequences within each of a single DNA strand. It leads to the creation of secondary DNA structures, called sometimes the tennis racket structures, during the PCR procedure and inhibition of amplification of these fragments of DNA. The PCR amplification is carried out with primers complementary to the sequences of ligated adaptors. Especially effective is suppression of amplification of fragments belonging to the group A, which contain the long and GC rich complementary sequences at both ends of each DNA strand. Amplification of fragments of group B is not inhibited so effectively. The terminal complementary sequences are shorter and do not contain so much GC so that they are unstable in the conditions of PCR reaction. The fragments of group C are amplified exponentially. Finally, in the electrophoretic analysis one can find only the products of amplification of fragments belonging to the groups B and C, which constitute less than 10% of fragments generated after digestion of template DNA. Usually the number of amplified fragments range between 20 and 30, so the interpretation of results is much easier than in the case of AFLP, which usually generates 50 to 200 products of amplification. In spite of reduction of the number of amplicons, the ADSRRS method has high discriminatory power. The method still is not very popular, however it has been successfully applied for genotyping of several important species of bacteria: *Enterococcus faecium* (Krawczyk et al, 2003 a), *S. marcescens* (Krawczyk et al., 2003 b), *K. pneumonia* (Krawczyk et al., 2005), *S. aureus* (Krawczyk et al., 2007a), *Corynobacterium pseudotuberculosis* (Stefanska et al., 2008) and *C. jejuni* and *C. coli* (Krutkiewicz & Klimuszko, 2010). The above authors have revealed that the discriminatory power of the ADRRS method is comparable with that of a gold standard RAE-PFGE.

2.2.4.3 PCR Melting Profiles – PCR MP

In the case of PCR MP, the template DNA is digested with only one enzyme and generated DNA fragments are ligated with only one adaptor. The mechanism of ligation is identical as in the ADSRRS method. The adaptor is not phosphorylated and only one strand is ligated with fragments of the digested template DNA. After thermal dissociation of helper sequences, the ends of DNA fragments are filled with a thermostable polymerase. Finally, the terminal complementary sequences within a single DNA strand are present in the whole population of the obtained DNA fragments. The reduction of number of amplified fragments in this method is achieved by using low denaturation temperature of template DNA during PCR reaction. Typically during PCR, denaturation temperatures around 94 – 95°C are applied, whilst in PCR MP the temperature of this step is reduced to the level of 80 – 86°C, depending of analyzed species of microorganism. As a result most of the fragments of DNA, especially long and containing a high number of GC pairs, are not melted and are not amplified because only single – stranded DNA may serve as a template in PCR. In other words, the primers cannot effectively bind to the complementary sequences of template DNA fragments. Additionally, as in the case of AFLP, the limitation of the number of amplified fragments is possible by using extended primers with a variable number of 3' nucleotides (Masny & Plucienniczak, 2003). The PCR MP has been widely investigated in the group of Krawczyk and coworkers, who revealed a comparable discriminatory power of this method with RAE PFGE in differentiation of staphylococci (Krawczyk et al., 2007 a), vancomycin – resistant *E. faecium* (Krawczyk et al., 2007 b), and *C. albicans* (Krawczyk et al., 2009), and higher discriminatory power than ITS – PCR in genotypic analysis of *Klebsiella oxytoca* (Stojowska et al., 2009).

2.2.5 Cleavase Fragment Length Polymorphism – CFLP

Cleavase fragment length polymorphism (CFLP) is a subtyping system based on the single-stranded DNA patterns resulting from digestion with the enzyme cleavase, a structure-specific, thermostable nuclease. This enzyme recognizes and cleaves secondary structures that consist of double-stranded hairpin regions interspersed with single-stranded regions of DNA and that are formed after denaturation and cooling to an intermediate temperature, in a pattern unique to the nucleotide sequence. CFLP analysis is a method with the capacity for direct assignment of alleles based on the nucleotide sequences of genes. CFLP can be applied to the rapid screening of a large number of strains during investigations of outbreaks and/or surveillance systems (Tondella et al., 1999).

The use of a thermostable endonuclease, Cleavase I (engineered endonuclease which consists of the 5'-nuclease domain of *Thermus aquaticus* DNA polymerase), allows the cleavage reactions to be performed at elevated temperature, which is fundamental to realizing the full benefit of this assay. For example, if a particularly stable secondary structure is assumed by the DNA, a single nucleotide change is unlikely to significantly alter that structure or the cleavage pattern it produces. Elevated temperatures can be used to bring structures to the brink of instability, so that the effects of small changes in sequence are maximized and revealed as alterations in the cleavage pattern. Furthermore, the use of high temperature reduces long-range interactions along the DNA strands, thereby increasing the likelihood that observed cleavage differences will reflect the locations of the base changes. The potential for multiple localized pattern changes in response to a single sequence change presents an additional advantage of this method of mutation detection over other methods in use. Direct sequencing of this region would provide a single variant peak on which to base a conclusion. In contrast, the multiplicity of effects seen here provides redundant confirmation of the base change, allowing multiple checkpoints within the pattern to be compared. This is especially useful in the cases of heterozygosity, or the presence of drug-sensitive and drug-resistant bacteria in a single sample. While the analyses shown above have been done with nonisotopic labels, the method is fully compatible with radiolabeled DNAs and standard autoradiography. When staining or uniform labeling are used, the patterns produced are more complex, as all fragments are visible, and the localization feature is lost because bands cannot be measured from a discrete labeled end, yet variations in the cleavage patterns are readily detected (Brow et al., 1996).

CFLP can be performed on differently labeled wild-type and mutant alleles in the same reaction, with no discernible effect of the patterns of each. When fragment size analyses are performed on a fluorescence based DNA sequencer, the wild-type pattern can serve as an internal control for enzyme digestion and as an internal reference that allows accurate discrimination of pattern differences in the mutant samples. When calibrated against a sequence ladder or other known markers, the wild-type pattern also allows precise sizing of the mutant bands. With the CFLP patterns normalized in this way, experimental samples can be compared day-to-day and lab-to-lab, with a high degree of confidence. The ability to rapidly perform an independent analysis of both DNA strands with the use of differently labeled PCR primers facilitates complementary confirmatory information analogous to sequencing both DNA strands. Analysis of DNAs through the probing of their structures with the Cleavase I enzyme promises to be a powerful tool in the comparison of sequences (Tondella et al., 1999). The procedural steps in CFLP are the following: PCR amplification→exonuclease digestion→spin column purification→cleavase digestion→eectrophoresis→overnight transfer to nylon membrane→probe hybridization→interpretation.

CFLP can rapidly screen for the major epidemic clones of serogroups B and C *N. meningitides*, e.g., clonal complexes ET-5 and ET-37 and also, more importantly, can provide guidance regarding the choice of genes for the DNA sequence-based approach to the molecular subtyping of *N. meningitides (Tondella et al., 1999).*

The low cost, rapidity, and reliability of the analysis make this method very suitable for the rapid screening of DNAs, not only for the examination of disease-associated mutations but also in applications such as tissue typing, genetic identity, bacterial and viral typing, and mutant screening in genetic crosses. The method is able to detect single base changes in DNA molecules over 1 kb long. It has been able to apply CFLP analysis to detection of mutations responsible for rifampin resistance in *M. tuberculosis* (Brow et al., 1996) and for determining the genotype of HCVs and have successfully analyzed DNAs as long as 2.7 kb (Marschall et al., 1997; Sreevtsatsan et al., 1998). The potential of the CFLP method was examined for differentiating many bacteria. In order to provide the greatest utility, a bacterium-typing method should be able to differentiate both species and strains. The genotype of the 16S rRNA genes, while useful for general classification, would not be expected to necessarily allow identification to the species and strain levels. For these purposes, a 550-bp amplicon derived from the intergenic region of several isolates of *Shigella* and *Salmonella* spp. was examined. The structural fingerprints generated by Cleavase I treatment clearly differentiated all species of *Salmonella* and *Shigella*. Again, a familial resemblance is clear in all of these evolutionarily related bacteria, yet differences in the structural fingerprints of each genus and species are clearly visible. The power of CFLP analysis is even more striking when the analysis of *Shigella* spp. was examined. *Shigella sonnei* was clearly differentiated from the three *Shigella dysenteriae* isolates examined. Furthermore, although serotypes 1, 2, and 8 of *S. dysenteriae* yielded related structural fingerprints, the antisense patterns for the individual serotypes could, however, be distinguished (Tondella et al., 1999).

A similar variation to CFLP analysis may occur with restriction-enzyme-dependent fingerprinting methods like RFLP, ARDRA and AFLP, depending on the restriction enzyme that is used (Tondella et al., 1999).

2.3 Hybridization – Based method

Southern blotting is a technique named after its inventor and developer, the British biologist Edwin M. Southern in 1975 (Southern et al., 1975). Southern blotting is a technique which allows the detection of a specific DNA sequence (gene or other) from a variety of prokaryotic and eukaryotic organisms. This classic method has been adapted to the differentiation of bacterial strains on the basis of the observation that the locations of various restriction enzyme recognition sites within a particular genetic locus of interest can be polymorphic from strain to strain, resulting in gel bands that differ in size between unlike strains. Thus, the name restriction fragment length polymorphism (RFLP) refers to the polymorphic nature of the locations of restriction enzyme sites within defined genetic regions. Only the genomic DNA fragments that hybridize to the probes are visible in RFLP analysis, which simplifies the analysis greatly.

The procedural steps in RFLP-Southern blotting are: **restriction endonuclease digestion**→agarose gel electrophoresis→DNA fragments transfer to nitrocellulose or nylon membrane→probe hybridization→interpretation. DNA (genomic or other source) is digested with one or more restriction enzymes and the resulting fragments are separated according to size by electrophoresis through an agarose gel. The DNA is then denatured into single strands by incubation with NaOH and transferred to a solid support (usually nitrocellulose filter or nylon membrane). The relative positions of the DNA fragments are preserved during their transfer to the filter. The membrane-bound nucleic acid is then hybridized to one or more labeled probes homologous to the gene to be examined. The probe is labeled before hybridization either radioactively or enzymatically (e.g. alkaline phosphatase or horseradish peroxidase). Finally, the location of the probe is detected by directly exposing the membrane to X-ray film or chemiluminescent methods. The optimum size range of DNA fragments detected this way is 2 – 25 kb. Although RFLP analysis does not require knowledge of the genome sequence, RFLP studies can be extremely time consuming and challenging in the absence of such data. When sequence data is not available, the experimenter must physically clone a region of the genome under study that is large enough to be cut up by enzymes, and this process requires a great deal of time and resources (Campbell, 2001; Olive & Bean, 1999).

Genetic specific probes have been used to subtype *Brucella* species (Grimont et al., 1992), *L. pneumophilla* (Tram et al., 1990) and *P. aeruginosa* (Loutit & Tompkins, 1991). RFLP was commonly found in the *tcdB* gene of *C. difficile* (Rupnik et al., 1998). It was earlier the gold standard for *M. tuberculosis* complex species genotyping - *IS6110*. One of the factors that have limited the use of RFLP was it being laborious and time consuming and the resulting complex banding patterns made interlaboratory comparisons difficult (Van Embden et al., 1993). Ribotyping, a variation of RFLP-Southern blotting in which the probes are derived from the 16S and 23S rRNA genes, has been applied successfully in many studies for differentiating strains of *Salmonella* (Landers et al., 1998), *P. aeruginosa* (Dawson et al., 2002) and *Pseudomonas fluorescens* (Dawson et al., 2002). Ribotyping results in only a small number of bands (one to four varing between about 6 and 9 kb), which simplifies the interpretation. However, this also limits the ability of the technique to distinquish between closely related strains (Olive & Bean, 1999). Analysis of the 16S rRNA nucleotide sequence, the 16S to 23S rRNA intergenic spacer region (ISR) sequence, and the 5'-terminal region of 23S rRNA allowed the strains of *Bacillus thuringensis* to be subdivided into three groups based on the pattern of nucleotide substitutions (Boulygina et al., 2009). Furthemore, a phylogenetic approach based on 16S rRNA genes (rDNA) has been applied to investigate the diversity of cultivable and non-cultivable species in the human oral cavity without cultivation (Sakamoto et al., 2003).

Advantages of RFLP-Southern blotting: all strains carrying loci homologous to probe are typeable; they are reproducible, have good ease of interpretation. Disadvantages of RFLP-Southern blotting: the discriminatory power depends on the choice of probes; the process requires costly reagents and equipment besides being labour intensive (Campbell, 2001).

Both RFLP-Southern blotting probe sets yielded information about the evolutionary relationships of the strains examined, yet neither technique was as efficient at discriminating strain differences as PFGE (Olive & Bean, 1999).

2.4 Sequencing methods

The results of "classical" genotypic methods are presented as specific patterns of DNA fragments after electrophoresis. However, the profiles of DNA fragments generated in PFGE or AFLP techniques are very complicated and consist of at least 50 bands. It is very difficult to use these results for long – term or interlaboratory investigations. This problem can be solved by application of the newest group of genotypic methods applying nucleotide sequencing approach. The results of sequencing are presented as a sequence of combination of four letters: A, T, G, C, corresponding to the sequence of nucleotides, respectively: adenine, thymine, guanine and cytosine. The phylogenetic classification of microorganisms is based on differences of sequences of a particular fragment of the genome. Various types of differences among DNA sequences may be identified: insertion, deletion, duplication and a situation when a single nucleotide A, T, G, or C differs between members of a given species. This form of results is much easier to analyze and can be used for studying long-term or global pathogen epidemiology, such as the worldwide distribution and frequency of bacterial lineages, virulence properties associated with certain lineages, etc. The data, i.e. sequences can be easily stored in the online accessible databases and compared among laboratories. Some of the most important databases containing sequences widely used for global investigation of genetic variability of microorganisms are presented below. Although the cost of sequencing decreased substantially during the least 20 years it is still very high, what is the biggest drawback of sequencing – based methods. At present there are two main strategies for DNA sequencing: the traditional Sanger method and the newly developed pyrosequencing method. The key principle of the Sanger method is using of dideoxy nucleoside triphosphates (ddNTPs) as DNA chain terminators. Dideoxynucleotides are essentially the same as nucleotides except they contain a hydrogen (H) atom on the 3' carbon instead of a hydroxyl group (OH). These modified nucleotides, when integrated into a sequence, prevent the addition of further nucleotides. Each of the four ddNTPs is also labeled with specific fluorescent dye for detection in automated sequencing machines. After the PCR amplification, a set of DNA fragments is generated. Each of them is labeled with a specific dye depending on the last nucleotide. The newly synthesized and labeled polynucleotides are separated by capillary electrophoresis according to their size and particular ddNTPs are recognized by a fluorescence detector. The results of sequencing are presented as a fluorescent peak tracing.

Another technique of DNA sequencing called pyrosequencing was developed by Pal Nyren and Mostafa Ronaghi at the Royal Institute of Technology in Stockholm in 1996. In this method, a DNA fragment of interest (sequencing primer hybridized to a single-stranded DNA template immobilized on solid matrice) is incubated with DNA polymerase, ATP sulfurylase, firefly luciferase, and a nucleotide-degrading enzyme. Repeated cycles of deoxynucleotide addition are performed. A deoxynucleotide will only be incorporated into the growing DNA strand if it is complementary to the base in the template strand. The synthesis of DNA is accompanied by release of PP_i in molar concentration equal to that of the incorporated deoxynucleotide. Thereby, real-time signals are obtained by the enzymatic inorganic pyrophosphate detection assay. In this assay, the released PP_i is converted to ATP by ATP sulfurylase and the concentration of ATP is then sensed by the luciferase. The amount of light produced in the luciferase-catalyzed reaction can readily be estimated by a suitable light-sensitive device such as a luminometer or a CCD (charge-coupled device) camera. The sequence of solutions of deoxynucleotide which produce chemiluminescent signals allows the determination of the sequence of the template (Ronaghi et al., 1998).

Recently developed sequencing – based methods of differentiation of microorganisms represent a real progress because of their accuracy, reproducibility, versality and exchangeability of the results. Their discriminatory power depends of the size of the fragment of the genome that is sequenced. Theoretically, the analysis of the bigger fragment should guarantee the higher discriminatory power. However there are also some examples of single-locus-sequence-based (SLST) typing methods which give unexpectedly good results. Probably the most widely used method of the SLST group, called the *spa* typing, is sequencing of the polymorphic X region of the protein A gene (*spa*) of *S. aureus*, which contains a variable number of 24-bp-repeat regions flanked by well-conserved fragments (Harmsen et al, 2003). Several authors have demonstrated fairly good correlation between the clonal groupings classification of MRSA obtained by *spa* typing and those obtained by other typing techniques including PFGE (Hallin et al, 2007, Strommenger et al, 2008). At the moment it is possible to compare the sequence of fragment X of the *spa* gene of any analyzed strain with 9073 *spa* sequence types, from 17,9659 strains isolated in 81 countries (August 2011), which are stored on the Ridom Spa Server (http://spa.ridom.de). The SpaServer was developed to collate and harmonize data from various geographic regions. This WWW site is freely accessible for internet users and the *spa*-repeat sequences and the -types can be downloaded. Chromatograms of new *spa*-repeats and/or -types can be submitted online for inclusion into the reference database. Another example of a SLST method that can be used for typing staphylococci is analysis of the polymorphism of sequence of the *coa* gene encoding *S. aureus* coagulase. Short review of other genes which can be used for bacterial strain typing was recently presented by Li and coathors (Li et al, 2009). Sequence analysis of the *ompt* gene, which encodes a major antigenic outer membrane protein of *Rickettsia* species, was demonstrated to be an efficient tool for identification and subtyping of these bacteria (Fournier et al, 1998). Karjalainen and coworkers revealed that serotyping of *C. difficile* can be replaced by sequencing of a variable fragment of *slpA* gene, that encodes a surface layer protein of this bacteria (Karjalainen et al, 2002). Great potential for genotyping of bacteria represents also the *rpoB* gene encoding β – subunit of RNA polymerase. The gene exist in only one copy in most bacteria and its sequence polymorphism offers higher discriminatory power than the 16S rRNA gene. The analysis of its polymorphism enabled genotyping of *Bacillus licheniformis* strains from different sources (De & De, 2004; Li et al, 2009).

To date, for most species of bacteria, yeast and fungi it is still impossible to select one appropriate gene which could be used for differentiation of isolates with enough high discriminatory power. More detailed analysis of genetic diversity of these microorganisms can be performed by the MLST method (Multilocus sequence typing), which was developed for *N. meningitidis* as the model species (Maiden et al, 1998). MLST is an unambiguous procedure for characterizing isolates of bacterial species using the sequences of internal fragments of several, usually seven house-keeping genes coding for proteins that are required for maintenance of basic cellular function. Approximately 450 - 500 bp internal fragments of each gene are used, as these can be accurately sequenced on both strands using an automated DNA sequencer. Depending on the species of microorganism, different sets of housekeeping genes are selected as targets for MLTS. For example the *S. aureus* MLST scheme uses internal fragments of the following seven house-keeping genes: *arc* (carbamate kinase), *aro* (shikimate dehydrogenase), *glp* (glycerol kinase) , *gmk* (guanylate kinase), *pta* (phosphate acetyltransferase), *tpi* (triosephosphate isomerase) and *yqi* (acetyl coenzyme A acetyltransferase) (Enright et al, 2000a), whilst *S. pneumoniae* scheme uses: *aroE* (shikimate dehydrogenase), *gdh* (glucose-6-phosphate dehydrogenase), *gki* (glucose kinase), *recP* (transketolase), *spi* (signal peptidase I), *xpt* (xanthine phosphoribosyltransferase) and *ddl* (D-alanine-D-alanine ligase) (Enright et al., 2000b). For each house-keeping gene, the different sequences present within a bacterial species are assigned as distinct alleles and each strain is assigned a seven – number allelic profile designated as a sequence type (ST) (Maiden et al, 1998). MLST is an especially useful tool for long - term investigation of bacterial population structures and is rather rarely used for analyzing local, short - term epidemics. The results of MLST of 27 microorganisms species (August 2011) are available on the MLST homepage (http://www.mlst.net), and the DNA sequences of all housekeeping genes of all analyzed strains and all necessary software are available online, that facilitates the global epidemiology survey of infections caused by these groups of pathogens (Li et al, 2009).

Besides sequences coding for particular proteins, different forms of RNA or sequences important for regulation of expression of some genes, the genomes of all living organisms contain surprisingly large amounts of non-coding DNA sequences, i.e. intergenic spacers. With high probability it can be assumed that intergenic spacers are a subject of much weaker selection pressure than genes and their variability is higher than that of genes. Therefore the intergenic spacers have a great potential as targets for genotyping. A particularal example of noncoding sequence of prokaryotic organisms useful for their genotyping is a genomic region separating 16S and 23S rRNA genes called the internal transcribed spacer (ITS). Possibilities of application of analysis of polymorphism of size and sequences of this target by classical genotypic techniques was presented in the earlier parts of this chapter. The presence of more than one copy of the rRNA operone in most bacterial species cause that it is not really good target for sequencing analysis. In order to achieve a good discriminatory power usually several intergenic spacers are analyzed (MTS –

Multispacer typing). The crucial point of the MTS method is selection of non-coding sequences characterized by the high level of variability. Amplification and sequencing of selected spacers provide unique sequence types (ST) for each of them. Finally combination of STs from each studied spacer provides a multispace genotype (MTS genotype) (Li et al., 2009). First successful analysis of polymorphism of sequences of several selected highly variable intergenic spacers was presented by Drancourt and coworkers (2004). The authors performed subtyping of *Y. pestis* from dental pulp collected from remains of eight persons who likely died during two historical plagues caused probably by this species of bacteria. The MST system was also applied with success to strains of other human pathogens, including *Bartonella henselae* (Li et al., 2007), *M. tuberculosis* (Djelouadji et al., 2008), *Rickettsia conorii* (Wenjun et al., 2009) and *M. avium* (Cayrou et al., 2010).

As a result of the presented advantages with great certainty it can be assumed that sequencing – based methods will gain popularity in the future. It should also be stressed that all sequencing – based methods do not require isolation of a pure culture of analyzed microorganism. The biggest drawback of this group of methods is the fact that they are time demanding and still too expensive to be used in routine genotyping in most laboratories. However a huge progress in the sequencing techniques has been observed during the last years and it is quite realistic that in not too remote future the analysis of polymorphism of the whole genomes will become a routine genotypic method, at least in some specialized institutions.

2.5 Detection of presence or absence of particular genes

Another PCR – based genotypic method relies on detection of presence or absence of sets of genes, usually coding for virulence factors or genes involved in antibiotic resistance. However the main goal of this kind of search is identification of genes being important factors in particular types of infection. They can be also used for characterisation of genetic differentiation of isolated pathogens. More or less detailed analysis of genetic features has been carried out and described for probably all known groups of microorganism, and it is impossible to present all of them in this short review. *S. aureus,* common and important human and animal pathogen may serve as a model for description of this genotyping method. One of the most spectacular, classical analyse of genetic features of staphylococci was presented by the group of Ote , who determined the genetic profiles of 229 strains of *S. aureus* collected from bovine cases of mastitis. The presence of about forty virulence-associated genes was investigated by specific polymerase chain reaction (PCR) amplification, and a high number of genetic subtypes were observed (Ote at al., 2011). This method does not require using of expensive and sophisticated equipment, but on the other hand, this is an extremely time consuming and laborious approach for analysis of genetic diversity of microorganisms. The mentioned authors had to carry out about ten thousand of PCR reactions, assuming that they were not repeated. The time required for analysis of presence of selected genes can be dramatically reduced by using macro or microarrays. These technologies can provide detailed, clinically relevant information on the isolate by detecting the presence or absence of a large number of virulence-associated genes simultaneously in a single assay. Trad and coworkers designed and prepared the DNA macroarray containig 465 intergenic amplicons of genes characteristic for *S. aureus*. The genes selected included those encoding *S. aureus*-specific proteins, staphylococcal and enterococcal proteins mediating antibiotic resistance and factors involved in their expression, putative virulence proteins and factors controlling their expression, and finally proteins produced by mobile elements. The macroarray was hybridized with the labeled cellular DNAs of 80 *S. aureus* clinical isolates. In a gene content dendrogram, the isolates were distributed into 52 clusters, and outbreak-related isolates were linked in the same or a closely related cluster(s) (Trad et al., 2004). The most spectacular example of application of macroarrays, which in fact does not concern staphylococci but in our opinion should be presented because of its importance and popularity is spoligotyping of mycobacteria (Groenen et al., 1993). Spoligotyping (spacer oligonucleotide typing) is a method that can be simultaneously used for detection as well as typing of the *M. tuberculosis* complex. This is a PCR-based method, which depends on the amplification of a highly polymorphic Direct Repeat (DR) locus in *M. tuberculosis* genome. The DR region contains direct repeat sequences of 36 bp, that are interspersed by the non-repetitive DNA spacers of 35-41 bp in length. One DR and its neighboring non repetitive spacer is termed as Direct Variant Repeat (DVR). Spoligotyping can be used for detecting the presence or absence 43 spacers of known sequences in DR region by hybridizing the PCR amplified spacer DNA to the set of immobilized oligonucleotides, representing each of the unique spacer sequences. In the effect, a distinct spoligotyping pattern is generated.

Use of microarrays – miniaturized version of macroarrays enables performance even more detailed analysis. Fitzgerald and coworkers demonstrated that 2,198 (78%) of the 2,817 of chromosomal open reading frames (ORFs) of reference COL *S. aureus* strain represented on a DNA microarray were shared by the 36 analyzed *S. aureus* isolates from various sources and geographical regions. The investigation also revealed that 10 out of 18 large regions of difference carry genes that encode putative virulence factors and proteins that mediate antibiotic resistance (Fitzgerald et al., 2001). The microarrays which are used for detection of presence or absence of particular genes are called cDNA microarrays. The microarray technology seems to be also a very promising tool for analysis of SNP, however a modified construction of array, called oligonucleotide microarray, is required. The most important difference of both kinds of arrays is the size of the DNA probe that is immobilized on surface of the chip. The probes of the oligonucleotide microarray are definitely shorter (20 – 70 nuclotides), in comparison to cDNA arrays. In consequence of using shorter DNA probes, the hybridization on the oligonucleotide microarray is more stringent than that on the cDNA microarray. The detection of SNP within the sequences of analyzed genes can be achieved by preparing and immobilization on the chip surface of short nucleotide sequences complementary to the fragments of genes of reference strain (without mutations) and also a number of modifications of each of these nucleotide sequences containing all possibilities at the potential polymorphic sites including short deletions, insertions and replacement of particular nucleotides. The hybridization of labeled DNA of analyzed strain of microorganism and the immobilized DNA probes is the most efficient in the case when their sequences are identical. The special equipment and software enable quantitative analysis of hybridization of analyzed DNA to the oligonucleotides spotted on the surface of the chip. At present it is possible to prepare a microarrays containing up to 1 000 000 oligonuclotides per chip, so that the actual sequences of a large number of genes can be deduced simultaneously in one experiment. The advantages of macro or microarrays are undeniable, however with some exceptions like the presented above spoligotyping of mycobacteria, they are still rarely used for routine genotyping of viruses, bacteria, yeast or fungal pathogens. It is mainly caused by the high cost of these methods – especially microarrays and some technical problems related with their preparation and analysis of obtained results. In addition it can be difficult to achieve satisfactory quality control in DNA microarray analysis because many factors affect nucleic acid hybridization reaction (Li et al., 2009). The macroarrays can be self-prepared by the methodology presented for example by Trad and colleagues (2004), however it requires specific equipment and advanced bioinformatic and laboratory skills. The microarrays are specially treated microscope slides (chips) that carry an ordered mosaic of sequences of selected genes of analyzed species of microorganism. As it was mentioned, the analysis of presence of two - three thousands of genes is not a problem, so that in the case of the most pathogenic bacteria it is possible to analyze the presence of all possible open reading frames. The preparing of microarrays can be definitely classified as a hi - tech technology and can be performed only in specialized institutions. The particular microarray can be prepared and supplied by these institutions, however performance of experiment and then analysis of results is definitely more complicated than in the case of macroarrays, it requires using specific equipment and software and therefore can be done in specialized laboratories only. In spite of all presented actual limitations, there are no doubts that macro and especially micro array technologies will gain in popularity and in near future they will become routine techniques of genotyping of microorganisms.

2.6 High Resolution Melting analysis – HRM

High Resolution Melting (HRM) analysis is a next powerful technique for the detection of epigenetic differences in double-stranded DNA samples. It was discovered and developed in 2003 by the group of Carl T Wittwer from University of Utah in collaboration with Idaho Technology (Gundry et al., 2003; Wittwer et al., 2003). The method is based on the real – time monitoring of the melting process of PCR amplified fragment of a DNA region in which the potential mutation of interest lies. The measurement is carried out in the range of temperatures from about 50 to 95°C. This can be achieved by using fluorescent dyes, that specifically bind to double - stranded DNA. The complexes of the double - stranded DNA and the dye characterize with high value of fluorescence. At the beginning of the HRM analysis there is a high level of fluorescence in the sample because of the billions of copies of the amplicon present. Heating cause progressive melting of the double - stranded DNA and in consequence a reduction of fluorescence of the sample is observed. Using a Real-time PCR machinery enables execution of the whole analysis consisting of PCR amplification and melting monitoring in one tube in the time period shorter than 1 hour. The kinetics of DNA melting and reduction of fluorescence level is presented as a fluorescence *vs.* temperature curve. The melting profile of a given DNA fragment and a shape of the obtained curve depend on length and sequence, especially the GC content of the analyzed PCR product. The method is very sensitive and even a single nucleotide difference can be detected. Thus the method is very popular for detection of SNP. Several authors successfully applied the HRM method for genotyping important human pathogens. The HRM analysis of 35 - fold repetitive *B1* gene was successfully used for genetic differentiation of the protozoan parasite *Toxoplasma gondi* (Costa et al., 2011). HRM analysis of 44 diverse MRSA isolates carried out by Stephens and coworkers generated 20 profiles from 22 *spa* sequence types. The two unresolved HRM *spa* types differed by only 1 bp (Stephens et al., 2008). Surprisingly good results of genotyping of *C. jejuni* were obtained by HRM scanning of polymorphism of several targets: a regularly interspaced short-palindromic-repeat (CRISPR) locus (Price et al., 2007), highly variable fragment of flagellin-encoding *flaA* gene (Merchant – Patel et al.,

2010). Levesquote and coworkers revealed that HRM can complement full MLST characterization of *C. jejuni* by identifying the most common alleles more rapidly and at the lower cost (Levesquote et al., 2011). The group of Levesquote performed the HRM analysis of polymorphism of seven house - keeping genes which are used in MLTS analysis of *C. jejuni*: *aspA* (aspatase), *glnA* (glutamine synthetase), *gltA* (citrate synthase), *glyA* (serine hydroxymethyl transferase), *pgm* (phosphoglucomutase), *tkt* (transketolase), *uncA* (ATP synthase alpha subunit).The HRM method also provides a rapid, robust, and inexpensive way to detect the dominant mutations known to confer MDR in *M. tuberculosis* strains and offers evident advantages over current molecular and culture-based techniques. First of all, the time of assay is very short. This method was used to screen 252 *M. tuberculosis* clinical isolates, including 154 rifampicin - resistant strains and 174 isonizaid - resistant strains based on the agar proportion method of drug susceptibility testing (DST). The rifampicin resistance determinant region (RRDR) of *rpoB* and specific regions of *katG* and the *inhA* promoter were HRM scanned for the detection of mutations conferring rifampicin (RIF) and isoniazid (INH) resistance, respectively. Of the 154 RIF-resistant strains, 148 were also resistant to INH and therefore classified as multidrug resistant (MDR). The assay demonstrated sensitivity and specificity of 91% and 98%, respectively, for the detection of RIF resistance and 87% and 100% for the detection of INH resistance (Ramirez et al., 2010). The HRM scanning of PCR amplified fragments of *rpoB, katG and inhA genes* was also successfully applied for rifampicin and isoniazid resistance among 217 clinical isolates of *M. tuberculosis* from South Korea (Choi et al., 2010) and lately Wang and coworkers revealed usefulness of the HRM method for detection of mutations of *rplS* gene resulting in streptomycin resistance of *M. tuberculosis* (Wang et al., 2011). Because of some important adavantages: rapidity, relatively low costs, possibility of differentiation pathogens without isolation of a pure culture and simplicity of interpretation of results it can be expected that HRM method will be more and more popular. However it is necessary to remember that expensive PCR – real time machinery is required and first of all it can be difficult to use the results in interlaboratory investigations.

3. Conclusion

DNA-based strain typing techniques are a remarkably useful set of tools for complementing the epidemiological analysis of nosocomial outbreaks. This chapter highlights the strengths of molecular typing methods. The ideal technique to assess genetic diversity should meet the following criteria. First of all, should be typed, that is, give the possibility to assign tested isolates for genotypic group of microorganisms. It should also have appropriate high discriminatory power, which will show the differences between unrelated microorganisms and the similarity between organisms isolated from the same source. The other important criterion is the reproducibility, the possibility of obtaining the same results for the same tested microorganisms, with each test and using the same procedure, reagents and apparatus. For epidemiological studies it is necessary that the method was quick and easy to interpret, not require specialized equipment or expensive reagents, and which is extremely important, gave the opportunity to be tested at the same time many trials and did not require specialized and expensive training of medical staff (could be used by routine microbiology laboratory personnel). So far there has not been found a universal and ideal typing method, each has its advantages and disadvantages. Moreover, each method has its pluses and minuses with regard to ease of application, reproducibility, requirement for equipment and level of resolution (Akkermans et al., 1995).

Table 1 shows the characteristics of the methods described above, in the context of their suitability for genetic differentiation of microorganisms. Currently, the gold standard for microbial typing methods is the PFGE technique and methods based on DNA-DNA hybridization. However, they are labour intensive and limited by high cost and extended turnaround times. The methods based on PCR appear to be more common and available. A simple and fast technique with high discriminatory power, and low cost is rep-PCR. But its disadvantage is a relatively small reproducibility in scale inter-laboratory. AFLP analysis has also a highly reproducible discriminatory power however the large number of band differences obtained by use primers without a selective base may be confusing and will usually not give additional information about whether strains are epidemiologically related. The other method, CFLP has the average discriminatory power and its need for DNA purification, careful optimization, and complex interpretation and its questionable reproducibility can be difficult to implement in a clinical laboratory. RAPD analysis is characterized by the lowest discriminatory power and poor reproducibility, however it gives the fastest typing results with the least hands-on time. It everything stimulates the search for new solutions, without the mentioned drawbacks. Most modifications are aimed towards the improvement of verified and reliable methods of analysis or applications of the combined techniques, which aims to increase the power of differentiation, and thus the universality of the method.

No.	Methodo-logy	Discrimina-tion power	Intralabora-tory reproducibi-lity	Interlabora-tory reproducibi-lity	Ease of use	Ease of interpreta-tion	Time to result (days)	Setup cost	Cost per test
1	PFGE	High	Good	Good	Moderate	Moderate	3	High	High
2	REAP	Moderate	Good	Good	Easy	Easy	1	Low	Low
3	PCR rybotyping	Moderate	Good	Good	Easy	Easy	1	Moderate	Low
4	RAPD	High	Moderate	Poor	Easy	Moderate	1	Moderate	Low
5	Rep-PCR	High	Good	Moderate	Easy	Easy	1	Moderate	Low
6	AFLP	High	Good	Good	Moderate	Moderate	2	High	Moderate
7	ADSRRS	High	Good	Good	Moderate	Easy	2	Moderate	-*
8	PCR-MP	High	Good	-*	Moderate	Easy	2	Moderate	-*
9	CFLP	Moderate	Good	Poor	Moderate	Moderate	2	Moderate	High
10	IRS-PCR	High	Good	Good	Moderate	Easy	2	Moderate	-*
11	Sequencing	High	Good	Good	Difficult	Moderate	2	High	High
12	HRM	High	Good	Moderate	Moderate	Moderate	1	High	Moderate

*(-) – not shown

Table 1. Summary of the characteristics of the various molecular typing methods

4. Acknowledgment

We sincerely thank Mrs. Irena Latacz and Professor Slawomir Milewski for checking the text in terms of language.

5. References

Achouak, W., Thiery, J. M., Rouband, P., & Heulin, T. (2000). Impact of crop management on intraspecific diversity of *Pseudomonas corrugata* in bulk soil. *FEMS Microbiology Ecology*, Vol.31, No.1 (January 2000), pp. 11-19, ISSN 1574-6941

Akopyanz, N., Bukanov, N.O., Westblom, T.U., Kresovich S., & Berg, D.E. (1992). DNA diversity among clinical isolates of *Helicobacter pylori* detected by PCR-based RAPD fingerprinting. *Nucleic Acids Research*, Vol.20, No.19 (September 1992), pp. 5137-5142, ISSN 0305-1048

Akkermans, A.D.L., van Elsas, J.D., & de Bruijn, F.J. (1995). *Molecular Microbial Ecology Manual*. Kluwer Academic Publishers, ISBN 90-6960-097-8, Dordrecht, The Netherlands

Alonso, R., Martin, A., Pelaez, T., Martin, M., Rodriquez-Creixe, M., & Bouza, E. (2005). An improved protocol for pulsed-field gel electrophoresis typing of *Clostridium difficile*. *Journal of Medical Microbiology*, Vol.54, No.2 (February 2005), pp. 155-157, ISSN 0022-2615

Arbeit, R.D., Maslow, J.N., & Mulligan, M.E. (1994). Polymerase chain reaction-mediated genotyping in microbial epidemiology. *Clinical Infectious Diseases*, Vol.18, No.6 (June 1994), pp. 1018-1019, ISSN 1058-4838

Arbeit, R.D. (1995). Laboratory procedures for the epidemiologic analysis of microorganisms, In: *Manual of clinical microbiology*, P.R. Murray, Baron, E. J., Pfaller, M.A., Tenover, F.C., & Yolken, R.H. (Eds.), pp. 90-208, 6th ed. ASM Press, ISBN 15-5581-086-1, Washington, D.C., USA

Atabay, H.I., Unver, A., Otlu, S., & Taner Kalaycioglu, A. (2008). Genotyping of various *Arcobacter* species isolated from domestic geese by randomly amplified polymorphic DNA (RAPD) analysis. *Archiv für Lebensmittelhygiene*, Vol.9, No.4 (2008), pp. 84-88, ISSN 0003-925X

Atkinson, R.M., LiPuma, J.J., Rosenbluth, D.B., & Dunne, W.M. Jr. (2006). Chronic colonization with *Pandoraea apista* in cystic fibrosis patients determined by repetitive-element-sequence PCR. *Journal of Clinical Microbiology*, Vol.4, No.3 (March 2006), pp. 833-836, ISSN 0095-1137

Baldy-Chudzik, K., Niedbach, J., & Stosik, M. (2003). Rep-PCR fingerprinting as a tool for the analysis of genomic diversity in *Escherichia coli* strains isolated from an aqueous/fresh water environment. *Cellular & Molecular Biology Letters*, Vol.8, No.3 (June 2003), pp. 793-798, ISSN 1689-1392

Ball, L.M., Bes, M.A., Theelen, B., Boekhout, T., Egeler, R.M., & Kuijper, E.J. (2004). Significance of amplified fragment length polymorphism in identification and epidemiological examination of *Candida* species colonization in children undergoing allogeneic stem cell transplantation. *Journal of Clinical Microbiology*, Vol.42, No.4 (April 2004), pp. 1673-1679, ISSN 0095-1137

Barbier, N., Saulnier, P., Chachaty, E., Dumontier, S., & Andremont, A. (1996). Random amplified polymorphic DNA typing versus pulsed-field gel electrophoresis for epidemiological typing of vancomycin-resistant enterococci. *Journal of Clinical Microbiology*, Vol.34, No.5 (May 1996), pp. 106-109, ISSN 0095-1137

Benter, T., Papadopoulos, S., Pape, M., Manns, M., & Poliwada, H. (1995). Optimization and reproducibility of random amplified polymorphic DNA in human. *Analytical Biochemistry*, Vol.230, No.1 (September 1995), pp. 92–100, ISSN 0003-2697

Bidet, P., Lalande, V., Salauze, B., Burghoffer, B., Avesani, V., Delmee, M., Rossier, A., Barbut, F., & Petit, J.C. (2000). Comparison of PCR-ribotyping, arbitrarily primed PCR, and pulsed-field gel electrophoresis for typing *Clostridium difficile*. *Journal of Clinical Microbiology*. Vol. 38, No.7 (July 2000), pp. 2484-2487, ISSN 0095-1137

Bingen, E., Boissinot, C., Desjardins, P., Cave, H., Lambert-Zechovsk, N., Denamur, E., Blot, P., & Elion, J. (1993). Arbitrary primed polymerase chain reaction provides rapid differentiation of *Proteus mirabilis* isolates from a pediatric hospital. *Journal of Clinical Microbiology*, Vol.31, No.10 (October 1993), pp. 1055-1059, ISSN 0095-1137

Blears, M.J., De Grandis, S.A., Lee, H., & Trevors, J.T. (1989). Amplified fragment length polymorphism (AFLP): a review of the procedure and its applications. *Journal of Industrial Microbiology & Biotechnology*, Vol.21, No.3 (June 1989), pp. 99-114, ISSN 1367-5435

Blixt, Y., Knutsson, R., Borch, E., & Radstrom, P. (2003). Interlaboratory random amplified polymorphic DNA typing of *Yersinia enterocolitica* and *Y. enterocolitica* like bacteria. *International Journal of Food Microbiology* Vol.83, No.1 (June 2003), pp. 15-26, ISSN 0168-1605

Bouchet, V., Huot H., & Goldstein R. (2008). Molecular genetic basis of ribotyping. *Clinical Microbiology Reviews*, Vol. 21, No.2 (April 2008), pp. 262-273, ISSN 0893-8512

Boulygina, E.S., Ignatov, A.N., Tsygankova, S.V., Korotkov, E.V., & Kuznetsov, B.B. (2009). Interspecies relations between *Bacillus thuringiensis* strains studied by AP-PCR and sequence analysis of ribosomal operon regions. *Microbiology*, Vol.78, No.6 (July 2009), pp. 703-710, ISSN 0026-2617

Boumedine, K.S., & Rodolakis, A. (1998). AFLP allows the identification of genomic markers of ruminant *Chlamydia psittaci* strains useful for typing and epidemiological studies. *Research in Microbiology*, Vol.149, No.10 (November 1998), pp. 735-744, ISSN 0923-2508

Brillowska-Dabrowska, A., Wianecka, M., Dabrowski, S., Mladenovska, Z., Kur, J., & Ahring, BK. (2008). ALIS-FLP: amplified ligation selected fragment-length polymorphism method for microbial genotyping. *Scandinavian Journal of Clinical and Laboratory Investigation*, Vol.68, No.8 (August 2008), pp. 720-730, ISSN 0036-5513

Brinkman, B. (1991). Population studies on selected AMP-FLPs and their use in the investigation of mixtures of body fluids. *Crime Laboratory Digest*, Vol.18, No.2 (June 1991), pp. 153-155, ISSN 0743-1872

Brow, M.A.D., Oldenburg, M.C., Lyamichev, V., Heisler, L.M., Lyamicheva, N., Hall, J.G., Eagan, N.J., Olive, D.M., Smith, L.M., Fors, L., & Dahlberg, J.E. (1996). Differentiation of bacterial 16S rRNA genes and intergenic regions and *Mycobacterium tuberculosis* katG genes by structure-specific endonuclease cleavage. *Journal of Clinical Microbiology*, Vol.34, No.12 (December 1996), pp. 3129-3137, ISSN 0095-1137

Caetano-Anollés, G., Bassam, B.J., & Gresshoff, P.M. (1991). DNA amplification fingerprinting using very short arbitrary oligonucleotide primers. *Biotechnology Nature Publishing Company*, Vol.9, No.6 (July 1991), pp. 553-557, ISSN 0968-0004

Campbell M. (2001). Department of Biology, Davidson College, Davidson, NC 28036, 05.09.2011, Available from http://www.bio.davidson.edu/.../Southernblot.html

Carson, C.A., Shear, B.L., Ellersieck, M.R., & Schnell, J.D. (2003). Comparison of ribotyping and repetitive extragenic palindromic-PCR for identification of fecal *Escherichia coli* from humans and animals. *Applied and Environmental Microbiology*, Vol.69, No.3 (March 2003), pp. 1836–1839, ISSN 0099-2240

Cherif, A., Brusetti, L., Borin, S., Rizzi, A., Boudabous, A., Khyami-Horani, H., & Daffonchio, D. (2003). Genetic relationship in the "*Bacillus cereus* group" by rep-PCR fingerprinting and sequencing of a *Bacillus anthracis*-specific rep-PCR fragment. *Journal of Applied Microbiology*, *Vol.*94, No.6 (June 2003), pp. 1108-1119, ISSN 1365-2672

Chmielewski, R., Wieliczko, A., Kuczkowski, M., Mazurkiewicz, M., & Ugorski M. (2002). Comparison of ITS profiling, REP- and ERIC-PCR of *Salmonella enteritidis* isolates from Poland. *Journal of Veterinary medicine B*, *Vol.49, No.4 (May 2002), pp. 163-168, ISSN 0931-1793*

Cho, J.C. & Tiedje, J.M. (2000). Biogeography and degree of endemicity of fluorescent *Pseudomonas* strains in soil. *Applied and Environmental Microbiology*, Vol.66, No.2, No. (December 2000), pp. 5448-5456, ISSN 0099-2240

Choi, G.E., Lee, S.M., Yi, J., Hwang, S.H., Kim, H.H., Lee, E.Y., Cho, E.H., Kim, J.H., Kim, H.J., & Chang, C.L. (2010). High-resolution melting curve analysis for rapid detection of rifampin and isoniazid resistance in *Mycobacterium tuberculosis* clinical isolates. *Journal of Clinical Microbiology*, Vol.48, No.11 (November 2010), pp. 3893-3898, ISSN 0095-1137

Christensen, H., Jorgensen, K., & Olsen, J.E. (1999). Differentiation of *Campylobacter coli* and *C. jejuni* by length and DNA sequence of the 16S-23S rRNA internal spacer region. *Microbiology*, Vol. 145, No.1 (January 1999), pp. 99-105, ISSN 1553-619X

Colombo, M.M., Mastrandrea, S., Leite, F., Santona, A., Uzzau, S., Rappelli, P., Pisano, M., Rubino, S., & Cappuccinelli, P. (1997). Tracking of clinical and environmental *Vibrio cholerae* O1 strains by combined analysis of the presence of the toxin cassette, plasmid content and ERIC PCR. *FEMS Immunology and Medical Microbiology*, Vol.19, No.1 (September 1997), pp. 33-45, ISSN 1574-695X

Costa, J.M., Cabaret, O., Moukoury, S., & Bretagne, S. (2011). Genotyping of the protozoan pathogen *Toxoplasma gondii* using high-resolution melting analysis of the repeated B1 gene. *Journal of Microbiological Methods*, *Vol.86, No.3 (March 2011), pp 357-63, ISSN 0167-7012

Currie, B. J., Gal, D., Mayo, M., Ward, L., Godoy, D., Sptatt, B.G., & LiPuma, J.J. (2007). Using BOX-PCR to exclude a clonal outbreak of *melioidosis. BMC Infectious Diseases*, Vol.7, No.1 (June 2007), pp. 68-72, ISSN 1471-2334

Cuteri, V., Marenzoni, M.L., Mazzolla, R., Tosti, N., Merletti, L., Arcioni, S., & Valente, C. (2004). *Staphylococcus aureus*: study of genomic similarity of strains isolated in veterinary pathology using amplified fragment length polymorphism (AFLP). *Comparative Immunology, Microbiology & Infectious Diseases*, Vol. 27, No.4 (July 2004), pp. 247-253, ISSN 0147-9571

Cayrou, C., Turenne, C., Behr, M.A., & Drancourt, M. (2010). Genotyping of *Mycobacterium avium* complex organisms using multispacer sequence typing. *Microbiology*, Vol.15, No.3 (March 2010), pp. 687-694, ISSN 1350-0872

Daffonchio, D., Borin, S., Consolandi, A., Mora, D., Manachini, P.L., & Sorlini, C. (1998). 16S-23 rRNA internal transcribed spacers as molecular markers for the species of the 16S rRNA group I of the genus *Bacillus*. *FEMS Microbiology Letters*, Vol.163, No.2 (June 1998), pp. 229-236, ISSN 0378-1097

Dalal, S., Lindsay N., Marrs, C. F., Zhang L., Harding, G., & Foxman, B. (2009). Long-term *Escherichia coli* asymptomatic bacteriuria among women with diabetes *mellitus*. *Clinical Infectious Diseases*, Vol.49, No.4 (June 2009), pp. 491-497, ISSN 1058-4838

Dawson, S.L., Fry, J.C., & Dancer, B.N. (2002). A comparative evaluation of five typing techniques for determining the diversity of fluorescent pseudomonads. *Journal of Microbiological Methods*, Vol.50, No.1 (June 2002), pp. 9-22, ISSN 0167-7012

De, C.E., & De, V.P. (2004). Gentypic diversity of *Bacillus licheniformis* strains from various sources. *FEMS Microbiology Letters*, Vol.231, No.1 (Januray 2006), pp. 91-98, ISSN 1574-6968

Debast, S.B., Melchers, W.J., Voss, A., Hoogkamp-Korstanje, J.A., & Meis, J.F. (1995). Epidermiological survey of an outbreak of multi-resistant *Serratia marcescens* by PCR fingerprinting. *Infection*, Vol.23, No.2 (September-October 1995), pp. 267–271, ISSN 0300-8126

De Bruijn, F.J. (1992). Use of repetitive (repetitive extragenic palindromic and enterobacterial repetitive intergenic consensus) sequences and the polymerase chain reaction to fingerprint the genomes of *Rhizobium meliloti* isolates and other soil bacteria. *Applied and Environmental Microbiology*, Vol.58, No.7 (July 1992), pp. 2180-2187, ISSN 0099-2240

Del Vecchio, V.G., Petroziello, J.M. , Gress, M.J. , McClesky, F.K., Melcher, G.P., Crouch, H.K., & Lupski, J.R. (1995). Molecular genotyping of methicillin-resistant *Staphylococcus aureus* via fluorophore-enhanced repetitive-sequence PCR. *Journal of Clinical Microbiology*, Vol.33, No.8 (August 1995), pp. 2141-2144, ISSN 0095-1137

De Paula, C.M.D., Geimba, M.P., do Amaral, P.H., & Tondo, E.C. (2010). Antimicrobial resistance and PCR-ribotyping of *Shigella* responsible for foobborne outbreaks occurred in southern Brazil. *Brazilian Journal of Microbiology*, Vol.41, No.4 (June 2010), pp. 966-977, ISSN 1517-8382

De Valk, H.A., Meis, J.F., de Pauw, B.E., Donnelly, P.J., & Klaassen, C.H. (2007). Comparison of two highly discriminatory molecular fingerprinting assays for analysis of multiple *Aspergillus fumigatus* isolates from patients with invasive *aspergillosis*. *Journal of Clinical Microbiology*, Vol.45, No.5 (May 2007), pp. 1415-1419, ISSN 0095-1137

Dijkshoorn, L., Aucken, H. M., Gerner-Smidt, P. Janssen, P., Kaufman, M.E., Garaizar, J., Ursing, J., & Pitt, T.L. (1996). Comparison of outbreak and nonoutbreak *Acinetobacter baumannii* strains by genotypic and phenotypic methods. *Journal of Clinical Microbiology*, Vol.34, No.6 (June 1996), pp. 1519-1525, ISSN 0095-1137

Djelouadji, Z., Arnold, C., Gharbia, S., Raoult, D., & Drancourt, M. (2008). Multispacer sequence typing for *Mycobacterium tuberculosis* genotyping. *PLoS One*, Vol.18, No.3 (June 2008), e2433, ISSN 1932-6203

Dombek, P.E., Johnson, L.K., Zimmerley, S.T., & Sadowsky, M.J. (2000). Use of repetitive DNA sequences and the PCR to differentiate *Escherichia coli* isolates from human and animal sources. *Applied and Environmental Microbiology*, Vol.66, No.6 (June 2000), pp. 2572-2577, ISSN 0099-2240

Drancourt, M., Roux, V., Dang, L.V., Tran-Hung, L., Castex, D., Chenal-Francisque, V., Ogata, H., Fournier, P.E., Crubezy, E., & Raoult, D. (2004). Genotyping orientalis – like *Yersinia pestis*, and plague pandemics. *Emerging Infectious Diseases*, Vol.10, No.9 (September 2004), pp. 1585-1592, ISSN 1080-6059

Elegado, F.B., Guerra, M.A.R.V., Macayan, R.A., Mendoza, H.A., & Lirazan, M.B. (2004). Spectrum of bacteriocin activity of *Lactobacillus plantarum BS* and fingerprinting by

RAPD-PCR. *International Journal of Food Microbiology*, Vol.95, No.1 (August 2004), pp. 11-18, ISSN 0168-1605

Enright, M.C., Day, N.P., Davies, C.E., Peacock, S.J., & Spratt, B.G. (2000a). Multilocus sequence typing for characterization of methicillin-resistant and methicillin-susceptible clones of *Staphylococcus aureus*. *Journal of Clinical Microbiology*. Vol.38, No.3 (Mar 2000), pp. 1008-1015, ISSN 0095-1137

Enright, M.C., Knox, K., Griffiths, D., Crook, D.W., & Spratt, B.G. (2000b). Molecular typing of bacteria directly from cerebrospinal fluid. *European Journal of Clinical Microbiology & Infectious Diseases*, Vol.19, No.8 (August 2000), pp. 627-630, ISSN 0934-9723

Eyles, R.F., Brooks, H.J., Townsend, C.R., Burtenshaw, G.A., Heng, N.C., Jack, R.W., & Weinstein, P. (2006). Comparison of Campylobacter jejuni PFGE and Penner subtypes in human infections and in water samples from the Taieri River catchment of New Zealand. *Journal of Applied Microbiology*, Vol.101, No.1 (July 2006), pp. 18-25, ISSN 1365-2672

Fitzgerald, J.R., Sturdevant, D.E., Mackie, S.M., Gill, S.R., & Musset, J.M. (2001). Evolutionary genomics of *Staphylococcus aureus*: insights int the origin of methicillin-resistant strains and the toxic shock syndrome epidemic. *Proceedings of the National Acadademy of Sciences of the USA*, Vol.98, No.15 (July 2001), pp. 8821–8826, ISSN 0027-8424

Fluit, A.C., Terlingen, A.M., Andriessen, L., Ikawaty, R., van Mansfeld, R., Top, J., Cohen Stuart, J.W., Leverstein-van Hall, M.A., & Boel, C.H. E. (2010). Evaluation of the DiversiLab System for detection of hospital outbreaks of infections by different bacterial species. *Journal of Clinical Microbiology*, Vol.48, No.11 (November 2010), pp. 3979-3989, ISSN 0095-1137

Fournier, P.E., Roux, V., & Raoult, D. (1998). Phylogenetic analysis of spotted fever group rickettsiae by study of the outer surface protein rOmpA. *International Journal of Systematic Bacteriology*, Vol.48, No.3 (July 1998), pp. 839-849, ISSN 0020-7713

Gandra, R.F., Simão, R.C., Matsumoto, F. E., da Silva, B.C., Ruiz L.S., da Silva, E.G., Gambale, W., & Paula, C.R. (2006). Genotyping by RAPD-PCR analyses of *Malassezia furfur* strains from pityriasis versicolor and seborrhoeic dermatitis patients. *Mycopathologia*, Vol.162, No.4 (October 2006), pp. 273-280, ISSN 0301-486X

Genersch, E., & Otten, C. (2003). The use of repetitive element PCR fingerprinting (rep-PCR) for genetic subtyping of German field isolates of *Paenibacillus larvae* subsp. *larvae*. *Apidologie*, Vol.34, No.4 (July-August 2003), pp. 195-206, ISSN 0044-8435

Georghiou, P.R., Doggett, A.M., Kielhofner, M.A., Stout, J.E., Watson, D.A., Lupski, J.R., & Hamill, R.J. (1994). Molecular fingerprinting of *Legionella* species by repetitive element PCR. *Journal of Clinical Microbiology*, Vol.32, No.12 (December 1994), pp. 2989-2994, ISSN 0095-1137

Georghiou, P.R., Hamill, R.J., Wright, C.E., Versalovic, J., Koeuth, T., Watson, D.A., & Lupski, J.R. (1995). Molecular epidemiology of infections due to *Enterobacter aerogenes*: identification of hospital-associated strains by molecular techniques. *Clinical Infectious Diseases*, Vol.20, No.1 (January 1995), pp. 84-94, ISSN 1058-4838

Gevers, D., Huys, G., & Swings, J. (2001). Applicability of rep-PCR fingerprinting for identification of *Lactobacillus* species. *FEMS Microbiology Letters*, Vol.205, No.1 (November 2001), pp. 31–36, ISSN 0378-1097

Giannino, V., Santagati, M., Guardo, G., Cascone, C., Rappazzo, G., & Stefani, S. (2003). Conservation of the mosaic structure of the four internal transcribed spacers and localisation of the *rrn* operons on the *Streptococcus pneumoniae* genome. *FEMS Microbiology Letters*, Vol.223, No.2 (June 2003), pp. 245-252, ISSN 0378-1097

Gibson, J.R., Slater, E., Xerry, J., Tompkins, D.S., & Owen, R.J. (1998). Use of an amplified-fragment length polymorphism technique to fingerprint and differentiate isolates of *Helicobacter pylori*. *Journal of Clinical Microbiology*, Vol.36, No.9 (September 1998), pp. 2580-2585, ISSN 0095-1137

Gita, S., Suneeta, M., Anjana, S., Niranjan, N., Sujata, M., & Pandey, R.M. (2011). C. *trachomatis* in female reproductive tract infections and RFLP-based genotyping: a 16-year study from a tertiary care hospital. *Infectious Diseases in Obstetrics and Gynecology*, 2011;2011:548219. Epub 2011 Jun 26, ISSN 1064-7449

Goh, S., Byrne, S.K., Zhang, J.L., & Chow, A.W. (1992). Molecular typing of *Staphylococcus aureus* on the basis of coagulase gene polymorphisms. *Journal of Clinical Microbiology*, Vol.30, No.7 (July 1992), pp. 1642-1645, ISSN 0095-1137

Grimont, F., Verger, J.M., Cornelis, P., Limet, J., Lefbvre, M., Grayon, M., Rdgnault, B., Van Broeck, J., & Grimont, P.A.D. (1992). Molecular typing of *Brucella* with cloned DNA probes. *Research in Microbiology*, Vol.143, No.1 (January 1992), pp. 55-65, ISSN 0923-2508

Groenen, P.M.A., Bunschoten, A.E., van Soolingen, D., & van Embden, J.D.A. (1993). Nature of DNA polymorphism in the direct repeat cluster of *Mycobacterium tuberculosis*; application for strain differentiation by a novel typing method. *Molecular Microbiology*, Vol.10, No(5) (December 1993), pp. 1057–1065, ISSN 1365-2958

Grundmann, H., Schneider, C., Hartung, D., Daschner, F. D., & Pitt, T.L. (1995). Discriminatory power of three DNA-based typing techniques for *Pseudomonas aeruginosa*. *Journal of Clinical Microbiology*, Vol. 33, No.3 (March 1995), pp. 528-534, ISSN 0095-1137

Gundry, C.N., Vandersteen, J.G., Reed, G.H., Pryor, R.J., Chen, J., & Wittwer, C.T. (2003). Amplicon melting analysis with labeled primers: a closed-tube method for differentiating homozygotes and heterozygotes. *Clinical Chemistry*, Vol.49, No.3 (March 2003), pp. 396–406, ISSN 0009-9147

Gürtler, V. (1993). Typing of *Clostridium difficile* strains by PCR-amplification of variable length 16S-23S rDNA spacer regions. *Journal of Genetic Microbiology*, Vol.139, No.12 (December 1993), pp. 3089-3097, ISSN 0001-8482

Gürtler, V., & Stanisich A. (1996). New approaches to typing and identification of bacteria using the 16S-23S rDNA spacer region. *Microbiology*, Vol.142, No.1 (January 1996), pp. 3-16, ISSN 1553-619X

Haase, A., Smith, V.H., Meldr, A., Wood, Y., Janmaat, A., Gilfeder, J., Kemp, D., & Curie, B. (1995). Subdivision of *Burkholderia pseudomallei* ribotype into multiple types by random amplified polymorphic DNA analysis provides new insights into epidemiology. *Journal of Clinical Microbiology*, Vol.33, No.7 (July 1995), pp. 1687-1690, ISSN 0095-1137

Hallin, M., Deplano, A., Denis, O., De Mendonça, R., De Ryck, R., & Struelens, M.J. (2007). Validation of pulsed-field gel electrophoresis and spa typing for long-term, nationwide epidemiological surveillance studies of *Staphylococcus aureus* infections. *Journal of Clinical Microbiology*. Vol.45, No.1 (January 2007), pp. 127-133, ISSN 0095-1137

Hamill, R.J., Houston, E.D., Georghiou, P.R., Wright, C.E., Koza, M.A., Cadle, R.M., Goepfert, P.A., Lewis, D.A., Zenon, G.J., & Clarridge, J.E. (1995). An outbreak of *Burkholderia* (formerly *Pseudomonas*) *cepacia* respiratory tract colonization and infection associated with nebulized albuterol therapy. *Annals of Internal Medicine*, Vol.122, No.10 (May 1995), pp. 762-766, ISSN 0003-4819

Harmsen, D., Claus, H., Witte, W., Rothganger, J., Claus, H., Turnwald, D., & Vogel, U. (2003). Typing of methicillin-resistant *Staphylococcus aureus* in a university hospital setting using a novel software for spa-repeat determination and database management. *Journal of Clinical Microbiology.* Vol.41, No.12 (December 2003), pp. 5442-5448, ISSN 0095-1137

Hartstein, A.I., Morthland, V.H., Eng, S., Archer, G.L., Schoenknecht, F.D., & Rashad, A.L. (1989). Restriction enzyme analysis of plasmid DNA and bacteriophage typing of paired *Staphylococcus aureus* blood culture isolates. *Journal of Clinical Microbiology.* Vol.27, No.8 (August 1989), pp. 1874-1879, ISSN 0095-1137

Herman, L., & Heyndrickx, M. (2000). The presence of intragenically REP-like elements in *Bacillus sporothermodurans* in sufficient for REP-PCR typing. *Research in Microbiology,* Vol.151, No.4 (May 2000), pp. 255-261, ISSN 0923-2508

Hiroshi, Hiwaki., Sumi, E., Al., B., & Takahisa, M. (2007). Genotyping of *Listeria monocytogenes* strains by PFGE, RAPD-PCR, ERIC-PCR and PCR-SSCP based on the jap gene. *Journal of Antibacterial and Antifungal Agents,* Vol.35, No.1 (June 2007), pp. 13-22, ISSN 0385-5201

Hou S.-T., Wang, C.-C., & Chu, M.-L. (1997). Ribotyping and random amplification of polymorphic DNA for nosocomial *Enterobacter cloacae* isolates in a pediatric Intensive Care Unit. *Infection Control and Hospital Epidemiology,* Vol.18, No.11 (November 1997), pp. 769-771, ISSN 0899-823X

Hulton, C.J.S., Higgins, C.F., & Sharp, P.M. (1991). ERIC sequences: a novel family of repetitive elements in the genomes of *Escherichia coli, Salmonella typhimurium* and other enterobacteria. *Molecular Microbiology,* Vol.5, No.4 (April 1991), pp. 825-834, ISSN 0950-382X

Johnson, L. K., Brown, M. B., Carruthers, E. A., Ferguson, J. A., Dombek, P.E., & Sadowsky, M.J. (2004). Sample size, library composition, and genotypic diversity among natural populations of *Escherichia coli* from different animals influence accuracy of determining sources of fecal pollution. *Applied and Environmental Microbiology,* Vol.70, No.8 (August 2004), pp. 4478–4485, ISSN 0099-2240

Jonas, D., Meyer, H.-G. W., Matthes, P., Hartung, D., Jahn, B., Daschner, F.D., & Jansen B. (2000). Comparative evaluation of three different genotyping methods for investigation of nosocomial outbreaks of legionnaires´ disease in hospitals. *Journal of Clinical Microbiology,* Vol.38, No.6 (June 2000), pp. 2284-2291, ISSN 0095-1137

Kang, H.P., & Dunne, W.M. (2003). Stability of repetitive-sequence PCR patterns with respect to culture age and subculture frequency. *Journal of Clinical Microbiology,* Vol.41, No.6 (June 2004), pp.2694-2696, ISSN 0095-1137

Kapperud, G., Nesbakken, T., Aleksic, S., & Mollaret, H.H. (1990). Comparison of restriction endonuclease analysis and phenotypic typing methods for differentiation of *Yersinia enterocolitica* isolates. *Journal of Clinical Microbiology.* Vol.28, No.6 (June 1990), pp. 1125-1131, ISSN 0095-1137

Karjalainen, T., Saumier, N., Barc, M.C., Delmee, M., & Collignon, A. (2002). *Clostridium difficile* genotyping based on *slpA* variable region in S-layer gene sequence: an alternative to serotyping. *Journal of Clinical Micrbiology.* Vol.40, No.7 (July 2002), pp. 2452-2458, ISSN 0095-1137

Kerouanton, A., Brisabois, A., Grout, J., & Picard, B. (1996). Molecular epidemiological tools for *Salmonella* Dublin typing. *FEMS Immunology & Medical Microbiology,* Vol.14, No.1 (May 1996), pp. 25-29, ISSN 1574-695X

Kersulyte, D., Woods, J.P., Kreath, E.J., Goldman, W.E., & Berg, D.E. (1992). Diversity among clinical isolates of *Histoplasma capsulatum* detected by polymerase chain reaction with arbitrary primers. *Journal of Bacteriology*, Vol.174, No.22 (November 1992), pp. 7075-7079, ISSN 0021-9193

Khan, A.A, McCarthy, S., Wang, R.F., & Cerniglia, C.E. (2002). Characterization of United States outbreak isolates of *Vibrio parahaemolyticus* using enterobacterial repetitive intergenic consensus (ERIC) PCR and development of a rapid PCR method for detection of O3:K6 isolates. *FEMS Microbiology Letters*, Vol.206, No.2 (January 2002), pp. 209-214, ISSN 0378-1097

Kidd, T.J., Grimwood, K., Ramsay, K.A., Rainey, P.B., & Bell S.C. (2011). Comparison of three molecular techniques for typing *Pseudomonas aeruginosa* isolates in sputum samples from patients with *cystic fibrosis*. *Journal of Clinical Microbiology*, Vol.49, No.1 (January 2011), pp. 263-268, ISSN 0095-1137

Kikkawa, H., Hitomi, S., & Watanabe, M. (2007). Prevalence of toxin A-nonproducing/toxin-B-producing *Clostridium difficile* in the Tsukuba-Tsuchiura district, Japan. *Journal of Infection and Chemotherapy*, Vol. 13, No.1 (February 2007), pp. 35-38, ISSN 1341-321X

Konno, M., Fujii, N., Yokota, S., Sato, K., Takahashi, M., Sato, K., Mino, E., & Sugiyama, M. (2005). Five-year follow-up study of mother-to-child transmission of *Helicobacter pylori* infection detected by a random amplified polymorphic DNA fingerprinting method. *Journal of Clinical Microbiology*, Vol.43, No.5 (May 2005), pp. 2246-2250, ISSN 0095-1137

Kostman, J.R., Edlind, T.D., Lipuma, J.J., & Stull, T.L. (1992). Molecular epidemiology of *Pseudomonas cepacia* determined by polymerase chain reaction ribotyping. *Journal of Clinical Microbiology*, Vol.30, No.8 (August 1992), pp. 2084-2087, ISSN 0095-1137

Krawczyk, B., Lewandowski, K., Bronk, M., Samet, A., Myjak, P., & Kur, J. (2003a). Evaluation of a novel method based on amplification of DNA fragments surrounding rare restriction sites (ADSRRS fingerprinting) for typing strains of vancomycin-resistant *Enterococcus faecium*. *Journal of Microbiological Methods*, Vol.52, No.3 (March 2003), pp. 341-351, ISSN 0167-7012

Krawczyk, B., Naumiuk, L., Lewandowski, K., Baraniak, A., Gniadkowski, M., Samet, A., & Kur, J. (2003b). Evaluation and comparison of random amplification of polymorphic DNA, pulsed-field gel electrophoresis and ADSRRS-fingerprinting for typing *Serratia marcescens* outbreaks. *FEMS Immunology and Medical Microbiology*, *Vol.38, No.3* (October 2003), pp. 241-248, ISSN 0928-8244

Krawczyk, B., Samet, A., Czarniak, E., Szczapa, J., & Kur, J. (2005). Extended-spectrum beta-lactamase-producing *Klebsiella pneumoniae* in a neonatal unit: control of an outbreak using a new ADSRRS technique. *Polish Journal of Microbiology*, Vol.54, No.2 (June 2005), pp. 105-110, ISSN 1733-1331

Krawczyk, B., Leibner, J., Barańska-Rybak, W., Samet, A., Nowicki, R., & Kur, J. (2007a). ADSRRS-fingerprinting and PCR MP techniques for studies of intraspecies genetic relatedness in *Staphylococcus aureus*. Journal of Microbiological Methods, Vol.71, No.2 (November 2007), pp. 114-122, ISSN 0167-7012

Krawczyk, B., Leibner, J., Stojowska, K., Bronk, M., Samet, A., & Kur J. (2007b). PCR melting profile method for genotyping analysis of vancomycin-resistant *Enterococcus faecium* isolates from Hematological Unit patients. *Polish Journal of Microbiology*, Vol.56, No.2 (June 2007), pp. 65-70, ISSN 1733-1331

Krawczyk, B., Leibner-Ciszak, J., Mielech, A., Nowak, M., & Kur J. (2009) PCR melting profile (PCR MP)--a new tool for differentiation of *Candida albicans* strains. *BMC Infectious Diseases, Vol.9. No.177* (November 2009), ISSN 1471-2334

Krutkiewicz, A., & Klimuszko, D. (2010) Genotyping and PCR detection of potential virulence genes in *Campylobacter jejuni* and *Campylobacter coli* isolates from different sources in Poland. *Folia Microbiologica, Vol.55, No.2 (March 2010), pp.* 167-175, ISSN 0015-5632

Kwil, I., Dziadek, J., Babicka, D., Cierniewska-Cieślak, A., & Różalski, A. (2002). Use of Randomly Amplified Polymorphic DNA (RAPD) analysis for identification of *Proteus penneri*. *Advances in Experimental Medicine and Biology*, Vol.485, pp. 321-324, ISSN 0065-2598

Kwon, D.H., El-Zaatari, F.A.K., Woo, J.S., Perng, C.L., Graham, D.Y., & Goo, M.F. (1998). Rep-PCR fragments as biomarkers for differentiating gastroduodenal disease-specific *Helicobacter pylori*. *Digestive Diseases and Sciences*, Vol.43, No.5 (May 1998), pp. 980-987, ISSN 0163-2116

Lagatolla, C., Dolzani, L., Tonin, E., Lavenia, A., Di Michele, M., Tommasini, T., & Monti-Bragadin, C. (1996). PCR ribotyping for characterizing *Salmonella* isolates of different serotypes. *Journal of Clinical Microbiology*, Vol.34, No.10 (October 1996), pp. 2440-2443, ISSN 0095-1137

Lan, R., & Reeves, P.R. (2007). Amplified Fragment Length Polymorphism analysis of *Salmonella enterica*, In: *Salmonella: Methods and Protocols*, H. Schatten, & A. Eisenstark, (Eds), pp. 119-132, ISBN 9781588296191, Humana Press: Totowa Netherlands

Landers, E., Gonzalez-Heria, M.A., Mendoza, M.C. (1998). Molecular epidemiology of *Salmonella* serotype *enteritidis*. Relationships between food, water and pathogenic strains. *International Journal of Food Microbiology*, Vol.43, No.1-2 (August 1998), pp. 81-90, ISSN 0168-1605

Lawrence, L.M., Harvey, J., & Gilmour, A. (1993). Development of a random amplification of polymorphic DNA typing method for *Listeria* monocytogenes. *Applied and Environmental Microbiology*, Vol.59, No.9 (September 1993), pp.3117–3119, ISSN 0099-2240

Leal, N. C., Sobreira, M., Leal-Balbino, T.C., Almeida, A. M. P. Silva, M.J.B., Mello, D.M., Seki, L.M., & Hofer, E. (2004). Evaluation of a RAPD-based typing scheme in a molecular epidemiology study of *Vibrio cholerae* O1, Brazil. *Journal of Applied Microbiology*, Vol.96, No.3 (July 2004), pp. 447-454, ISSN 1364-5072

Leelayuwat, C., Romphruk, A., Lulitanond, A., Trakulsomboon, S., & Thamlikitkul, V. (2000). Genotype analysis of *Burkholderia pseudomallei* using randomly amplified polymorphic DNA (RAPD): indicative of genetic differences amongst environmental and clinical isolates. *Acta Tropica*, Vol.77, No.2 (November 2000), pp. 229-237, ISSN 1013-3119

Lehmann, P.F., Lin, D., & Lasker, B.A. (1992). Genotypic identification and characterization of species and strains within the genus *Candida* using random amplified polymorphic DNA. *Journal of Clinical Microbiology*, Vol.30, No.12 (December 1992), pp. 3249-3254, ISSN 0095-1137

Leung, K.T., Mackereth, R., Tien, Y. & Topp, E. (2004). A comparison of AFLP and ERIC-PCR analyses for discriminating *Escherichia coli* from cattle, pig and human sources. *FEMS Microbiology Ecology*, Vol.47, No.1 (January 2004), pp. 111-119, ISSN 0168-6496

Levesque, S., Michaud, S., Arbeit, R.D., & Frost, E.H. (2011). High-Resolution Melting System to Perform Multilocus Sequence Typing of *Campylobacter jejuni*. *PLoS One*, Vol.6, No.1 (January 2011), e16167, ISSN 1932-6203

Li, W., Raoult, D., & Fournier, P.E. (2009). Bacterial strain typing in the genomic era. *FEMS Microbiology Review*, Vol.33, No.5 (September 2009), pp. 892-916, ISSN 1574-6976

Liu, P.Y.-F., & Wu, W.-L. (1997). Use of different PCR-based DNA fingerprinting techniques and pulsed-field gel electrophoresis to investigate the epidemiology of *Acinetobacter calcoaceticus-Acinetobacter baumannii* complex. *Diagnostic Microbiology and Infectious Disease*, Vol.28, No.1 (September 1997), pp. 19-28, ISSN 0732-8893

Lomonaco, S., Nucera, D., Parisi, A., Normanno, G., & Bottero, M.T. (2011). Comparison of two AFLP methods and PFGE using strains of *Listeria monocytogenes* isolated from environmental and food samples obtained from Piedmont, Italy. *International Journal of Food Microbiology*, In press, ISSN 0168-1605

Loutit, J.S., & Tompkins, L.C. (1991). Restriction enzyme and Southern hybridization analysis of *Pseudomonas aeruginosa* strains from patients with *cystic fibrosis*. *Journal of Clinical Microbiology*. Vol.29, No.12 (December 1991), pp.2897-2900, ISSN 0095-1137

Louws, F.J., Schneider, M., & de Bruijn, F.J. (1996) In: *Nucleic Acid Amplification Methods for the Analysis of Environmental Samples*, G. Toranzos (Ed), ISBN 978-82-8208-004-0, pp. 63-94, Technomic Publishing Co

Lupski, J.R. (1993). Molecular epidemiology and its clinical application. *JAMA*, Vol.270, No.11 (September 1993), pp. 1363-1364, ISSN 0098-7484

Maiden, M.C., Bygraves, J.A., Feil, E., Morelli, G., Russell, J.E., Urwin, R., Zhang, Q., Zhou, J., Zurth, K., Caugant, D.A., Feavers, I.M., Achtman, M., & Spratt, B.G. (1998). Multilocus sequence typing: a portable approach to the identification of clones with populations of pathogenic microorganisms. *Proceedings of the National Acadademy of Sciences of the* USA,Vol.95, No.6 (March 1998), pp. 3140-3145, ISSN 0027-8424

Manan, N., Chin, S., Abdullach, N., & Wan, H. (2009). Differentiation of *Lactobacillus*-probiotic strains by visual comparison of random amplified polymorphic DNA (RAPD) profiles. *The African Journal of Biotechnology*, Vol.8, No.16 (August 2009), pp., ISSN 1684-5315

Marques, A.S.A., Marchaison, A., Gardan, L., & Samson, R. (2008). BOX-PCR-based identification of bacterial species belonging to *Pseudomonas syringae* – *P. viridiflava* group. *Genetics and Molecular Biology*, Vol.31, No.1 (January 2008), pp. 1415-4757, ISSN 1415-4757

Marschall, D.J., Heisler, L.M., Lyamichev, V., Murvine, C., Olive, D.M., Ehrlich, G.D., Neri, B.P., & de Arruda, M. (1997). Determination of *hepatitis* C virus genotypes in the United States by Cleavase Fragment Length Polymorphism Analysis. *Journal of Clinical Microbiology*, Vol.35, No.12 (December 1997), pp. 3156-3162, ISSN 0095-1137

Marsou, R., Bes, M., Boudouma, M., Brun, Y., Meugnier, H., Freney, J., Vandenesch, F., & Etienne, J. (1999). Distribution of *Staphylococcus sciuri* subspecies among human clinical specimens, and profile of antibiotic resistance. *Research in Microbiology*, Vol.150, No.8 (October 1999), pp. 531-541, ISSN 0923-2508

Martin, B., Humbert, O., Camara, M., Guenzi, E., Walker, J., Mitchell, T., Andrew, P., Prudhomme, M., Alloing, G., & Hakenbeck, R. (1992). A highly conserved repeated DNA element located in the chromosome of *Streptococcus pneumoniae*. *Nucleic Acids Research*, Vol.20, No.13 (July 1992), pp. 3479-3483, ISSN 0305-1048

Masny, A., & Plucienniczak, A. (2001). Fingerprinting of bacterial genomes by amplification of DNA fragments surrounding rare restriction sites. *Biotechniques*, Vol.31, No.4 (October 2001), pp.930-934, ISSN 0736-6205

Masny, A., & Płucienniczak, A. (2003). Ligation mediated PCR performed at low denaturation temperatures-PCR melting profiles. *Nucleic Acids Research*, Vol.31, No.18 (September 2003), e114, ISSN 0305-1048

Melles, D.C., van Leeuwen, W.B., Snijders, S.V., Horst-Kreft, D., Peeters, J.K., Verbrugh, H.A., & van Belkum, A. (2007). Comparison of multilocus sequence typing (MLST), pulsed-field gel electrophoresis (PFGE), and amplified fragment length polymorphism (AFLP) for genetic typing of *Staphylococcus aureus*. *Journal of Microbiological Methods*, Vol.69, No.2 (May 2007), pp 371-375, ISSN 0167-7012

Merchant-Patel, S., Blackall, P.J., Templeton, J., Price, E.P., Tong, S.Y., Huygens, F., & Giffard, P.M. (2010) *Campylobacter jejuni* and *Campylobacter coli* genotyping by high-resolution melting analysis of a *flaA* fragment. *Applied and Environmental Microbiology*, Vol.76, No.2 (January 2010), pp. 493-499, ISSN 0099-2240

Meunier, J.R., & Grimont, P.A.D. (1993). Factors affecting reproducibility of random amplified polymorphic DNA fingerprinting. *Research in Microbiology*, Vol.144, No.5 (June 1993), pp. 373-379, ISSN 0923-2508

Millemann, Y., Lesage-Descauses, M.C, Lafont, J.P., & Chaslus-Dancla, E. (1996). Comparison ofrandom amplified polymorphic DNA analysis enterobaterial repetitive Intergenic consensus-PCR for epidemiological studies of *Salmonella*. *FEMS Immunology and Medical Microbiology*, Vol.14, No.2-3 (Juny 1996), pp. 129-134, ISSN 0928-8244

Miyata, M.A., Aoki, T., Inglis, V., Yoshida, T., & Endo, M. (1995). RAPD analysis of *Aeromonas salmonicida* and *Aeromonas hydrophila*. *Journal of Applied Bacteriology*, Vol.79, No.2 (August 1995), pp. 181-185, ISSN 1364-5072

Mohapatra, B. R., Broersma, K., Nordin, R., & Mazumder, A. (2007). Evaluation of repetitive extragenic palindromic-PCR for discrimination of fecal *Escherichia coli* from humans, and different domestic-and wild-animals. *Microbiology Immunology*, Vol.51, No.8 (July 2007), pp. 733-740, ISSN 0902-0055

Mueller, U.G., & Wolfenbarger, L.L. (1999). AFLP genotyping and fingerprinting. *Trends in Ecology & Evolution*, Vol.14, No.10 (October 1999), pp. 389-394, ISSN 0169-5347

Myers, L.E., Silva, S.V.P.S., Procunir, J.D., & Little, P.B. (1993). Genomic fingerprinting of *Haemophilus somnus* isolates using a random-amplified polymorphic DNA assay. *Journal of Clinical Microbiology*, Vol.31, No.3 (March 1993), pp. 512-517, ISSN 0095-1137

Nebola, M., & Steinhauserova, I. (2006). PFGE and PCR/RFLP typing of *Campylobacter jejuni* strains from poultry. *British Poultry Science*, Vol.47, No.4 (August 2006), pp. 456-461, ISSN 0007-1668

Niemann, S, Pühler, A, Ticky, H.V., Simon, R., & Selbitschka, W. (1997). Evaluation of the resolving power of three different DNA fingerprinting methods to discriminate among isolates of a natural *Rhizobium meliloti* population. *Journal of Applied Microbiology*, Vol.82, No.4 (April 1997), pp. 477-484, ISSN 1364-5072

Ni Tuang, F, Rademaker, J.L.W., Alocilja, E.C., Louws, F.J., & de Bruijn, F.J. (1999). Identification of bacterial rep-PCR genomic fingerprints using a backpropagation neural network. *FEMS Microbiology Letters*, Vol.177, No.2 (August 1999), pp. 249-256, ISSN 0378-1097

Oda, Y, Star, B, Huisman, L.A., Gottschal, J.C., & Forney, L.J. (2003). Biogeography of the purple nonsulfur bacterium *Rhodopseudomonas palustris*. *Applied and Environmental Microbiology*, Vol.69, No.9 (September 2003), pp. 5186-5191, ISSN 0099-2240

Okhravi N., Adamson, P., Matheson, M.M., Towler, H.M. A., & Lightman, S. (2000). PCR-RFLP–mediated detection and speciation of bacterial species causing endophthalmitis. *Investigative Ophthalmology & Visual Science*, Vol.41, No.6 (May 2000), pp. 1438–1447, ISSN 1552-5783

Okwumabua, O., O'Connor, M., Shull, E., Strelow, K., Hamacher, M., Kurzynski, T., & Warshauer, D. (2005). Characterization of *Listeria monocytogenes* isolates from food animal clinical cases: PFGE pattern similarity to strains from human *listeriosis* cases. *FEMS Microbiology Letters*, Vol.15, No.2 (August 2005), pp. 275-281, ISSN 0378-1097

Olive M.D., & Bean P. (1999). Principles and applications of methods for DNA-based typing of microbial organisms. *Journal of Clinical Microbiology*, Vol.37, No.6 (June 1999), pp. 1661-1669, ISSN 0095-1137

Ortiz-Herrera, M., Geronimo-Gallegos, A., Cuevas-Schacht, F., Perez-Fernandez, L., & Coria-Jimenez R. (2004). RAPD-PCR characterization of *Pseudomonas aeruginosa* strains obtained from *cystic fibrosis* patients. *Salud Publicade México*, Vol.46, No.2 (April 2004), pp. 149-157, ISSN 0036-3634

Ote, I., Taminiau, B., Duprez, J.N., Dizier, I., & Mainil, J.G. (2011). Genotypic characterization by polymerase chain reaction of *Staphylococcus aureus* isolates associated with bovine mastitis. *Veterinary Microbiology*, in press, ISSN 0378-1135

Panutdaporn, N., Chongsa-nguan, M., Nair, G.B., & Ramamurthy, T. (2004). Genotypes and phenotypes of *Shiga* toxin producing *E. coli* isolated from healthy cattle in Thailand. *Journal of Infection*, Vol.48, No.2 (February 2004), pp. 149-160, ISSN 1532-2742

Piessens, V., Supre, K., Heyndrickx, M., Haesebrouck, F., De Vliegher, S., & Van Coillie, E. (2010). Validation of amplified fragment length polymorphism genotyping for species identification of bovine associated coagulase-negative *staphylococci*. *Journal of Microbiological Methods*, Vol.80, No.3 (March 2010), pp. 287-294, ISSN 0167-7012

Pinna, A., Sechi, L.A., Zanetti, S., Usai, D., Delogu, G., Cappuccirelli, P., & Carta, F. (2001). *Bacillus cereus keratitis* associated with contact lens wear. *Ophthalmology*, Vol.108, No.10 (October 2001), pp. 1830-1834, ISSN

Poh, C.L., Ramachandran, V., & Tasall, J.W. (1996). Genetic diversity of *Neisseria gonorrhoeae* IB-2 and IB-6 isolates revealed through whole-cell repetitive element sequence-based PCR. *Journal of Clinical Microbiology*, Vol.34, No.2 (February 1996), pp. 292-295, ISSN 0095-1137

Price, E.P., Smith, H., Huygens, F., & Giffard, P.M. (2007). High-resolution DNA melt curve analysis of the clustered, regularly interspaced short-palindromic-repeat locus of *Campylobacter jejuni*. *Applied and Environmental Microbiology*, Vol.73, No.10 (May 2007), pp. 3431-3436, ISSN 0099-2240

Rademaker J.L.W., Hoste, B., Louws, F.J., Kersters, K., Swings, J., Vauterin, L., Vauterin, P., & de Bruijn, F.J. (2000). Comparison of AFLP and rep-PCR genomic fingerprinting with DNA–DNA homology studies: *Xanthomonas* as a model system. *International Journal of Systematic and Evolutionary Microbiology*, Vol.50, No.2 (March 2000), pp. 665-677, ISSN 1466-5026

Ralph, D., McClelland, M., Welsh, J., Baranton, G., & Perolat, P. (1993). *Leptospira* species categorized by arbitrarily primed polymerase chain reaction (PCR) and by mapped restriction polymorphisms in PCR-amplified rRNA genes. *Journal of Bacteriology*, Vol.175, No.4 (February 1993), pp. 973-981, ISSN 1098-5530

Ramirez, M.V., Cowart, K.C., Campbell, P.J., Morlock, G.P., Sikes, D., Winchell, J.M., & Posey, J.E. (2010). Rapid detection of multidrug-resistant *Mycobacterium tuberculosis* by use of real-time PCR and high-resolution melt analysis. *Journal of Clinical Microbiology*. Vol.48, No.11 (November 2010), pp. 4003-4009. ISSN 0095-1137

Rasschaert, G., Houf, K., Imberechts, H., Grijspeerdt, K., De Zutter, L., & Heyndrickx, M. (2005). Comparison of five repetitive sequence- based PCR typing methods for molecular discrimination of *Salmonella enterica* isolates. *Journal of Clinical Microbiology*, Vol.43, No.8 (August 2005), pp. 3615–3623, ISSN 0095-1137

Reboli, A.C., Houston, E.D., Monteforte, J.S., Wood, C.A., & Hamill, R.J. (1994). Discrimination of epidemic and sporadic isolates of *Acinetobacter baumannii* by repetitive element PCR-mediated DNA fingerprinting. *Journal of Clinical Microbiology*, Vol.32, No.11 (November 1994), pp. 2635-2640, ISSN 0095-1137

Rengua-Mangia, A.H., Guth, B.C., da Costa Adrada, J.R., Irino, K., Pacheco, A.B.F., Ferreira, L.C.S., Zahner, V., & Teixeira, L.M. (2004). Genotypic and phenotypic characterization of enterotoxigenic *Escherichia coli* (ETEC) strains isolated in Rio de Janeiro city, Brazil. *FEMS Immunology and Medical Microbiology*, Vol.40, No.2 (March 2004), pp. 55-162, ISSN 0928-8244

Rice, W.C. (2010). Application of Amplicon Length Polymorphism to differentiate amongst closely related strains of bacteria. In: *Current Research,Technology and Education Topics Applied Microbiology and Microbial Biotechnology*, A. Mendez-Vilas (Ed), pp. 1509-1517, ISBN 0-12-198370-6, New Caledonia

Rivera, I.G., Chowdhury, M.A.R., Huq, A., Jacobs, D., Martins, M.T., & Colwell, R.R. (1995). Enterobacterial repetitive intergenic consensus sequences and the PCR to generate fingerprints of genomic DNA from *Vibrio cholerae* O1, O139, and Non-O1 Strains. *Applied and Environmental Microbiology*, Vol.61, No.8 (August 1995), pp. 2898-2904, ISSN 0099-2240

Robert, F., Lebreton, F., Bougnoux, M.E., Paugam, A., Wassermann, D., Schlotterer, M., Tourte-Schaefer, C., & Dupouy-Camet, J. (1995). Use of random amplified polymorphic DNA as a typing method for *Candida albicans* in epidemiological surveillance of a burn unit. *Journal of Clinical Microbiology*, Vol.33, No.9 (September 1995), pp. 2366-2371, ISSN 0095-1137

Rodriguez-Barradas, M.C., Hamill, R.J., Houston, E.D., Georghiou, P.R.,. Clarridge, J.E., Regnery, R.L., & Koehler, J.E. (1995). Genomic fingerprinting of *Bartonella* species by repetitive element PCR for distinguishing species and isolates. *Journal of Clinical Microbiology*, Vol.33, No.5 (May 1995), pp. 1089-1093, ISSN 0095-1137

Ronaghi, M., Uhlen, M., & Nyren, P. (1998). A sequencing method based on real-time pyrophosphate. *Science*, Vol.281, No.5379 (July 1998), pp. 363-365, ISSN 0036-8075

Rupnik, M. Avesani, V., Janc, M., von Eichel-Streiber, C., & Delmee, M. (1998). A novel toxinotyping scheme and correlation of toxinotypes with serogroups of *Clostridium difficile* isolates. *Journal of Clinical Microbiology*, Vol.36, No.8 (August 1998), pp. 2 240-2247, ISSN 0095-1137

Sakamoto, M., Takeuchi, Y., Umeda, M., Ishikawa, I., & Benno, Y. (2003). Application of terminal RFLP analysis to characterize oral bacterial flora in saliva of healthy subjects and patients with *periodontitis*. *FEMS Immunology & Medical Microbiology,*. Vol.52, No.1 (January 2003), pp. 79-89, ISSN 1574-695X

Saulnier, P., Bourneix, C., Prevost, G., & Andremont, A. (1993). Random amplified polymorphic DNA assay is less discriminant than pulsed-field gel electrophoresis for typing strains of methicillin-resistant *Staphylococcus aureus*. *Journal of Clinical Microbiology*, Vol.31, No.4 (April 1993), pp. 982-985, ISSN 0095-1137

Sebaihia, M., Preston, A., Maskell, D.J., Kuzmiak, H., Connell, T.D., King, N.D., Orndorff, P.E., Miyamoto, D.M., Thomson, N.R., Harris, D., Goble, A., Lord, A., Murphy, L., Quail, M.A., Rutter, S., Squares, R., Squares, S., Woodward, J., Parkhill, J., Temple, L.M. (2006). Comparison of the genome sequence of the poultry pathogen *Bordetella avium* with those of *B. bronchiseptica*, *B. pertussis*, and *B. parapertussis* reveals extensive diversity in surface structures associated with host interaction. *Journal of Bacteriology*, Vol.188, No.16 (August 2006), pp. 6002-6015, ISSN 1098-5530

San Millán, R., Garaizar, J., & Bikandi, J. (2005). *In silico* simulation of fingerprinting techniques based on double endonuclease digestion of genomic DNA. *In Silico Biology*, Vol.5, No.3 (April 2005), pp. 341-346, ISSN 1434-3207

Shangkuan, Y.H., & Liu, H.C. (1998). Application of random amplified polymorphic DNA analysis to differentiate strains of *Salmonella typhi* and other *Salmonella* species. *Journal of Applied Microbiology*, Vol.85, No.4 (October 1998), pp. 693-702, ISSN 1364-5072

Shao, Z.-J., Ren, H.-Y., Xu, L., Diao, B.-W., Li, W., Li, M.-C., Cui, Z.-G., Liang, X.-F., Li, Y.-X., Liu, D.Q., Yang, M., Zhang, T.-G., Li, M.-S., & Xu, J.-G. Zhonghua Liu Xing Bing Xue Za Zhi. (2007). Molecular typing of *Neisseria meningitidis* serogroup C strains with pulsed field gel electrophoresis in China. *Zhonghua Liu Xing Bing Xue Za Zhi*, Vol. 28, No.8 (August 2007), pp. 756-760, ISSN 1808-0560

Shi, Z.-Y, Liu P. Y.-F., Lau Y.-J., Lin, Y.-H., & Hu, B.-S. (1997). Use of Pulsed-Field Gel Electrophoresis to investigate an outbreak of *Serratia marcescens*. *Journal of Clinical Microbiology*, Vol.35, No.1 (January 1997), pp. 325-327, ISSN 0095-1137

Silveria W.D., Ferreira, A., Lancellotti, M., Barbosa, I.A.G.C.D., Leite, D.S., de Castro, A.F.P., & Brocchi, M. (2003). Clonal relationships among avian *Escherichia coli* isolates determined by enterobacterial repetitive intergenic consensus (ERIC)-PCR. *Veterinary Microbiology*, Vol.89, No.4 (November 2003), pp. 323-328, ISSN 0378-1135

Singh, D.V., Matte, M.H., Matte, G.R., Jiang, S., Sabeena, F., Shukla, B.N., Sanyal, S.C., Huq, A., & Colwell, R.R. (2001). Molecular analysis of *Vibrio cholerae* O1, O139, non-O1, and non-O139 strains: clonal relationship between clinical and environmental isolates. *Applied and Environmental Microbiology*, Vol.67, No.2 (February 2001), pp. 910-921, ISSN 0099-2240

Sloos, J.H., Janssen, P., van Boven, C.P., & Dijkshoorn, L. (1998). AFLP typing of *Staphylococcus epidermidis* in multiple sequential blood cultures. *Research Microbiology*, Vol.143, No.3 (March 1998), pp. 221-228, ISSN 0923-2508

Son, R., Nasreldin, E.H., Zaiton, H., Samuel, L., Rusul, G., & Nimita, F. (1998). Use of randomly amplified polymorphic DNA analysis to differentiate isolates of *Vibrio parahaemolyticus* from cockles (*Anadara granosa*). *World Journal of Microbiology and Biotechnology*, Vol.14, No.6 (November 1998), pp. 895-901, ISSN 1573-0972

Son, R., Micky, V., Kasing, A., Raha, A.R., Patrick, G.B., & Gulam, R. (2002). Molecular characterization of *Vibrio cholerae* O1 outbreak strains in Miri, Sarawak (Malaysia). *Acta Tropica*, Vol.83, No.2 (August 2002), pp. 169-176, ISSN 0001-706X

Southern, E.M. (1975). Detection of specific sequences among DNA fragments separated by gel electrophoresis. *Journal of Molecular Biology*, Vol.98, No.3 (November 1975), pp. 503-517, ISSN 0022-2836

Speijer, H., Savelkoul, P.H.M., Bonten, M.J., Stobberingh, E. E., & Tjhie, J.H. (1999). Application of different genotyping methods for *Pseudomonas aeruginosa* in setting of endemicity in an intensive care unit. *Journal of Clinical Microbiology* Vol.37, No.11 (November 1999), pp. 3654-3661, ISSN 0095-1137

Sreevatsan, S., Bookout, J.B., Ringpis, F.M., Pottathil, M.R., Marshall, D.J., De Arruda, M., Murvine, C., Fors, L., Pottathil, R., & Barathur, R.R. (1998). Algorithmic approach to high-throughput molecular screening for alpha interferon-resistant genotypes in *hepatitis* C patients. *Journal of Clinical Microbiology*, Vol.36, No.7 (July 1998), pp. 1895-1901, ISSN 0095-1137

Stefanska, I., Rzewuska, M., & Binek, M. (2008). Evaluation of three methods for DNA fingerprinting of *Corynebacterium pseudotuberculosis* strains isolated from goats in Poland. *Polish Journal of Microbiology*, Vol.57, No.2 (June 2008), pp. 105-112, ISSN 1733-1331

Stephens, A.J., Inman-Bamber, J., Giffard, P.M., & Huygens, F. (2008). High-resolution melting analysis of the *spa* repeat region of *Staphylococcus aureus*. *Clinical Chemistry*, Vol.54, No.2 (February 2008), pp. 432-436, ISSN 0009-9147

Stern, M.J., Ames, G.F.L., Smith, N.H., Robinson, E.C., & Higgins, C.F. (1984). Repetitive extragenic palindromic sequences: a major component of the bacterial genome. *Cell*, Vol.37, No.3 (July 1984), pp.1015-1026, ISSN 1534-5807

Stojowska, K., Kałuzewski, S., & Krawczyk, B. (2009). Usefulness of PCR melting profile method for genotyping analysis of *Klebsiella oxytoca* isolates from patients of a single hospital unit. *Polish Journal of Microbiology*, Vol.58, No.3 (September 2009), pp. 247-253, ISSN 1733-1331

Strommenger, B., Braulke, C.; Heuck, D., Schmidt, C., Pasemann, B., Nubel, U., & Witte, W. (2008). *spa* typing of *Staphylococcus aureus* as a frontline tool in epidemiological typing. *Journal of Clinical Microbiology*. Vol.46, No.2 (February 2009), pp.574-581, ISSN 0095-1137

Stubbs, S.L., Brazier, J. S., O'Neill, G. L., & Duerden, B. I. (1999). PCR targeted to the 16S-23S rRNA gene intergenic spacer region of *Clostridium difficile* and construction of a library consisting of 116 different PCR ribotypes. *Journal of Clinical Microbiology*, Vol.37, No.2 (February 1999), pp. 461-463, ISSN 0095-1137

Svec, P., Vancanneyt, M., Seman, M., Snauwaert, C., Lefebvre, K., Sedlacek, I., & Swings, J. (2005). Evaluation of (GTG)5-PCR for identification of *Enterococcus* spp. *FEMS Microbiology Letters*, Vol.247, No.1 (June 2005), pp. 59-63, ISSN 0378-1097

Svensson, B., Ekelund, K., Ogura, H., & Christiansson, A. (2004). Characterization of *Bacillus cereus* isolated from milk silo tanks at eight different dairy plants. *International Dairy Journal*, Vol.14, No.1 (January 2004), pp. 17-27, ISSN 0958-6946

Syrmis, M.W., O'Carroll, M.R., Sloots, T.P., Coulter, C., Wainwright, C.E., Bell, S.C., & Nissen, M.D. (2004). Rapid genotyping of *Pseudomonas aeruginosa* isolates harboured by adult and paediatric patients with cystic fibrosis using repetitive-element-based PCR assays. *Journal of Medical Microbiology*, Vol.53, No.11 (November 2004), pp. 1089-1096, ISSN 0022-2615

Taco, M., Alves, A., Saavedra, M.J., & Correia, A. (2005). BOX-PCR is an adequate tool for typing *Aeromonas* spp. *Antoine-van-Leeuwenhoek*, Vol.88, No.2 (August 2005), pp. 173-179, ISSN 1609-6694

Tenover, F.C., Arbeit, R.D., Archer, G., Biddle, J., Byrne, S., Goering, R., Hancock, G., Hébert, G.A., Hill, B., Hollis, R., Jarvis, W.R., Kreiswirth, B., Eisner, W., Maslow, J., McDougal, L.K., Miller, J.M., Mulligan, M., & Pfaller, M.A. (1994). Comparison of traditional and molecular methods of typing isolates of *Staphylococcus aureus*. *Journal of Clinical Microbiology*, Vol. 32, No.2 (February 1994), pp. 407-415, ISSN 0095-1137

Tenover, F.C., Arbeit, R.D., Goering, R.V., Mickelsen, P.A., Murray, B.E., Persing, D.H., & Swaminathan, B. (1995). Interpreting chromosomal DNA restriction patterns produced by pulsed-field gel electrophoresis: criteria for bacterial strain typing. *Journal of Clinical Microbiology*, Vol.33, No.9 (September 1995), pp. 2233-2239, ISSN 0095-1137

Tenover, F.C., Arbeit, R.D., & Goering, R.V. (1997). How to select and interpret molecular strain typing methods for epidemiological studies of bacterial infections: a review for healthcare epidemiologists. Molecular Typing Working Group of the Society for

Healthcare Epidemiology of America. *Infection Control & Hospital Epidemiology*, Vol.18, No.6 (June 1997), pp. 426-39, ISSN 0899-823X

Thong, K.-L., Ngeow, Y.-F., Altweg, M., Navaratnam, P., & Pang, T. (1995). Molecular Analysis of *Salmonella enteritidis* by Pulsed-Field Gel Electrophoresis and Ribotyping. *Journal of Clinical Microbiology*, Vol.33, No.5 (May 1995), pp. 1070-1074, ISSN 0095-1137

Tikoo, A., Tripathi, A.K., Verma, S.C., Agrawal, N., & Nath, G. (2001). Application of PCR fingerprinting techniques for identification and discrimination of *Salmonella* isolates. *Current Science*, Vol.80, No.8 (April 2001), pp. 1049-1052, ISSN 0011-3891

Tondella M.L.C., Reeves, M.W., Popovic, T., Rosenstein, N., Holloway, B.P., & Mayer, L.W. (1999). Cleavase fragment length polymorphism analysis of *Neisseria meningitidis* basic metabolic genes. *Journal of Clinical Microbiology*, Vol.37, No.8 (August 1999), pp. 2402-2407, ISSN 0095-1137

Trad, S., Allignet, J., Frangeul, L., Davi, M., Vergassola, M., Couve, E., Morvan, A., Kechrid, A., Buchrieser, C., Glaser, P., & El-Solh, N. (2004). DNA macroarray for identification and typing of *Staphylococcus aureus isolates*. Microbiology , Vol.42, No.5 (May 2004), pp. 2054-2064, ISSN 0095-1137

Tram, C., Simonet, M., Nicolas, M.-H., Offredo, C., Grimont, F., Lefevre, M., Ageron, E., Debure, A., & Grimont, P.A.D. (1990). Molecular typing of nosocomial isolates of *Legionella pneumophilia* serogroup 3. *Journal of Clinical Microbiology*, Vol.28, No.2 (February 1990), pp. 242-245, ISSN 0095-1137

Trilla, A., Marco, F., Moreno, A., Prat, A., Soriano, E., & Jiménez de Anta, M.T. (1993). Clinical epidemiology of an outbreak of nosocomial infection caused by *Staphylococcus aureus* resistant to methicillin and aminoglycosides: efficacy of control measures. Comité de Control de Infecciones. *Medicina Clínica* (Barcelona), Vol.100, No.6 (February 1993), pp. 205-209. ISSN 0025-7753, In Spanish, abstract in English

Tsen, H.Y., Lin, J.-S., & Hsih H.-Y. (2002). Pulsed field gel electrophoresis for animal *Salmonella enterica* serovar Typhimurium isolates in Taiwan. *Veterinary Microbiology*, Vol.87, No.1 (June 2002), pp. 73-80, ISSN 0378-1135

Valdezate, S., Vindel, A., Martín-Dávila, P., Del Saz, B. S., Baquero, F., & Canton, R. (2004). High genetic diversity among *Stenotrophomonas maltophilia* strains despite their originating at a single hospital. *Journal of Clinical Microbiology*, Vol.42, No.2 (2004 February), pp. 693-699, ISSN 0095-1137

Valsangiacomo, C., Baggi, F., Gaia, V., Balmelli, T., Peduzzi, R., & Piffaretti, J.C. (1995). Use of amplified fragment length polymorphism in molecular typing of *Legionella pneumophila* and application to epidemiological studies. *Journal of Clinical Microbiology*, Vol.33, No.7 (July 1995), pp. 1716-1719, ISSN 0095-1137

Van Belkum, A., Bax, R., Peerbooms, P., Goessens, W.H.F., Van Leeuwen, N., & Quint, W.G.V. (1992). Comparison of phage typing and DNA fingerprinting by polymerase chain reaction for discrimination of methicillin-resistant *Staphylococcus aureus* strains. *Journal of Clinical Microbiology*, Vol.31, No.4 (April 1992), pp. 798-803, ISSN 0095-1137

Van Belkum, A., De Jonckheere, J., & Quint, W.G.V. (1993). Typing *Legionella pneumophilia* strains by polymerase chain reaction-mediated fingerprinting. *Journal of Clinical Microbiology*, Vol.31, No.8 (August 1993), pp. 2198-2200, ISSN 0095-1137

Van Belkum, A., van Leeuwen, W., Kluytmans, J., & Verbrugh, H.A. (1995). Molecular nosocomial epidemiology: high speed typing of microbial pathogens by arbitrary primed polymerase chain reaction assays. *Infection Control & Hospital Epidemiology*, Vol.16, No.11 (November 1995), pp.658-666, ISSN 0899-823X

Van den Berg, R.J., Schaap, I., Templeton, K. E., Klaassen, C.H., & Kuijper, E.J. (2007). Typing and subtyping of *Clostridium difficile* isolates by using multiple-locus variable-number tandem-repeat analysis. *Journal of Clinical Microbiology*, Vol.45, No.3 (March 2007), pp. 1024-1028, ISSN 0095-1137

Van Embden, J.D., Cave, M.D., Crawford, J.T., Dale, J. W., Eisenach, K.D., Gicquel, B., Hermans, P., Martin, C., Mc Adam, R. & Shinnick, T.M. (1993). Strain identification of *Mycobacterium tuberculosis* by DNA fingerprinting: recommendations for a standardized methodology. *Journal of Clinical Microbiology*, Vol.31, No.2 (February 1993), pp. 406–409, ISSN 0095-1137

Versalovic, J., Koeuth, T., & Lupski, J.R. (1991). Distribution of repetitive DNA sequences in eubacteria and application to fingerprinting of bacterial genomes. *Nucleic Acids Research*, Vol.19, No.24 (December 1991), pp. 6823-6831, ISSN 0305-1048

Versalovic, J., Kapur, V., Mason, E.O., Jr., Shah, U., Koeuth, T., Lupski, J.R., &. Musser, J. (1993). Penicillin-resistant *Streptococcus pneumoniae* strains recovered in Houston: identification and molecular characterization of multiple clones. *Journal of Infectious Diseases*, Vol.167, No.4 (April 1993), pp. 850-856, ISSN 0022-1899

Versalovic, J., Schneider, M., de Bruijn, F. J. & Lupski, J.R. (1994). Genomic fingerprinting of bacteria using repetitive sequencebased polymerase chain reaction. *Methods in Molecular and Cellular Biology*, Vol.5, No.2, pp. 25–40, ISSN 0898-7750

Versalovic, J., Kapur, V., Koeuth, T., Mazurek, G.H., Whittam, T., Musser, J., & Lupski, J.R. (1995). DNA fingerprinting of pathogenic bacteria by fluorophore-enhanced repetitive sequence-based polymerase chain reaction. *Archives of Pathology & Laboratory Medicine*, Vol.119, No.1 (January 1995), pp. 23-29, ISSN 0003-9985

Versalovic, J., de Bruijn, F.J. & Lupski, J.R. (1997). In: *Bacterial Genomes: Physical Structure and Analysis*, de Bruijn, F.J., Lupski, J.R., & Weinstock, G.M. (Eds), pp. 437-454, ISBN 978-1-4020-6900-0, Chapman & Hall, New York

Vila, J., Marcos, M.A., & Jimenez de Anta, M.T. (1996). A comparative study of different PCR-based DNA fingerprinting techniques for typing of the *Acinetobacter calcoaceticus-A. baumannii* complex. *Journal of Medical Microbiology*, Vol.44, No.6 (June 1996), pp. 482-489, ISSN 0022-2615

Vimont, S., Mnif, B., Fevre, C., & Brisse, S. (2008). Comparison of PFGE and multilocus sequence typing for analysis of *Klebsiella pneumoniae* isolates. *Journal of Medical Microbiology*, Vol.57, No.10 (October 2008), pp. 1308-1310, ISSN 0022-2615

Vos, P., Hogers, R., Bleeker, M., Reijans, M., van de Lee, T., Hornes, M., Frijters, A., Pot, J.; Peleman, J., Kuiper, M., & Zabeau, M. (1995). AFLP: a new technique for DNA fingerprinting. *Nucleic Acids Research*, Vol.23, No.21 (November 1995), pp. 4407–4414, ISSN 0305-1048

Wang, F., Shen, H., Guan, M., Wang, Y., Feng, Y., Weng, X., Wang, H., & Zhang, W. (2011). High-resolution melting facilitates mutation screening of *rpsL* gene associated with streptomycin resistance in *Mycobacterium tuberculosis*. *Microbiological Research*. Vol.166, No.2 (February 2011), pp. 121-128, ISSN 0944-5013

Welsh, J., & McClelland, M. (1990). Fingerprinting genomes using PCR with arbitrary primers. *Nucleic Acids Research*, Vol.18, No.24 (December 1990), pp. 7213-7218, ISSN 0305-1048

Wenjun, L.I., Mouffok, N., Rovery, C., Parola, P., & Raoult, D. (2009). Genotyping *Rickettsia conorii* detected in patients with Mediterranean spotted fever in Algeria using multispacer typing (MST). *Clinical Microbiology and Infection*, Suppl 2, pp. 281-283 (June 2009), ISSN 1198-743X

Williams, J. G., Kubelik, A.R., Livak, K.J., Rafalsky, J.A., & Tingey, S.V. (1990). DNA polymorphisms amplified by arbitrary primers are useful genetic markers. *Nucleic Acids Research*, Vol.18, No.22 (November 1990), pp. 6531-6535, ISSN 0305-1048

Wittwer, C.T., Gudrun, H.R., Gundry, C.N., Vandersteen, J.G., & Pryor, R.J. (2003). High-Resolution Genotyping by Amplicon Melting Analysis Using LCGreen. *Clinical Chemistry*, Vol.49, No.6 (June 2003), pp. 853–860, ISSN 0009-9147

Wieser, M., & Busse, H.-J. (2000). Rapid identification of *Staphylococcus epidermidis*. *International Journal of Systematic and Evolutionary Microbiology*, Vol.50, No.3 (May 2000), pp. 1087–1093, ISSN 1466-5026

Wolska, K., & Szweda, P. (2008). A comparative evaluation of PCR ribotyping and ERIC PCR for determining the diversity of clinical *Pseudomonas aeruginosa* isolates. *Polish Journal of Microbiology*, Vol.57, No.2, pp. 157-163, ISSN 0137-1320

Woods, C.R., Versalovic, J., Koeuth, T., & Lupski, J.R. (1992). Analysis of relationships among isolates of *Citrobacter diversus* by using DNA fingerprints generated by repetitive sequence-based primers in polymerase chain reaction. *Journal of Clinical Microbiology*, Vol.30, No.11 (November 1992), pp. 2921-2929, ISSN 0095-1137

Woods, C.R., Versalovic, J., Koeuth, T., & Lupski, J.R. (1993). Whole-cell repetitive element sequence-based polymerase chain reaction allows rapid assessment of clonal relationships of bacterial isolates. *Journal of Clinical Microbiology*, Vol.31, No.7 (July 1993), pp. 1927-1931, ISSN 0095-1137

Woods, C.R., Koeuth, T., Estabrook, M.M., & Lupski, J.R. (1996). Rapid determination of outbreak-related strains of *Neisseria meningitidis* by repetitive element-based polymerase chain reaction genotyping. *Journal of Infectious Diseases*, Vol.174, No.4 (October 1996), pp. 760-767, ISSN 0022-1899

Yamamoto, Y., Kohno, S., Koga, H., Kakeya, H., Tomono, K., Kaku, M., Yamazaki, T., Arisawa M., & Hara, K. (1995). Random amplified polymorphic DNA analysis of clinically and environmentally isolated *Cryptococcus neoformans* in Nagasaki. *Journal of Clinical Microbiology*, Vol.33, No.12 (December 1995), pp. 3328–3332, ISSN 0095-1137

Yang, W., Shi, L., Jia, W., Yin, X., Su, J., Kou, Y., Yi, X., Shinoda, S., & Miyoshi, S. (2005). Evaluation of the biofilm-forming ability and genetic Tybing for clinical isolates *Pseudomonas aeruginosa* by enterobacterial intergenic consensus-based PCR. *Microbiology Immunology*, Vol.49, No.12 (September 2005), pp. 1057-1061, ISSN 0902-0055

Zabeau, M., & Vos, P. (1993). Selective restriction fragment amplification: a general method for DNA fingerprinting. *European Patent Application Number: 9240269.7*, Publication Number: 0 534 858 A1.

Zhang, G. W., Kotiw, M., & Daggard, G. (2002). *A RAPD-PCR genotyping assay which correlates with serotypes of Group B streptococci*. Letters in Applied Microbiology, Vol.35, No.3, pp. 247-250, ISSN 0266-8254

Zulkifli, Y., Alitheen, N.B., Son, R., Raha, A.R., Samuel, L. Yeap, S.K., & Nishibuchi, M. (2009). Random amplified polymorphic DNA-PCR and ERIC PCR analysis on *Vibrio parahaemolyticus* isolated from cockles in Padang, Indonesia. *International Food Research Journal*, Vol.16, pp.141-150, ISSN 1985-4668

5

Patterns of Microbial Genetic Diversity and the Correlation Between Bacterial Demographic History and Geohistory

Pei-Chun Liao[1] and Shong Huang[2]
*[1]Department of Biological Science & Technology,
National Pingtung University of Science & Technology
[2]Department of Life Science, National Taiwan Normal University,
Taiwan, R.O.C.*

1. Introduction

Microbial diversity is commonly represented by genotypic frequency of the whole gene pool (metagenomes) of a microbial community. The biological community structure is determined by the environment, species competition, and the evolutionary histories of the species living in the community (Aravalli et al., 1998). Because microorganisms are highly sensitive to environmental changes, they can be used as indicators of the properties of their environment (Aravalli et al., 1998). Therefore, the demographic history of a microbial population may indicate changes that have occurred in the local habitat.

Traditionally, the 16S ribosomal RNA genes (16S rRNA) are widely used as genetic barcodes for identifying and recording the microbial organisms of a specific "microbial community" (waters, soils, digestive tracts, etc.) (Liao et al., 2007; Kulakov et al., 2011). Characteristic of the 16S rRNA gene in species differentiation provides as good genetic tool for ecological survey. The comparison of the genome data of two *Prochlorococcus* ecotypes revealed a genetic differentiation in niches which is also reflected in the 16S rRNA differentiation (Rocap et al., 2003). Recently, the advanced (meta)-genomic survey provides more novel insights into the microbial ecology and niche differentiation (Rocap et al., 2003; Shanks et al., 2006; Staley, 2006; Avarre et al., 2007; Biddle et al., 2008; Kalia et al., 2008; Banfield et al., 2010; Benson et al., 2010; Wang et al., 2010; Morales & Holben, 2011). However, the price for the metagenomic survey even by the next-generation sequencing technologies (e.g. the 4 (Chistoserdova, 2010). Therefore, studies for the goals of microbial diversity of a community are still favoring the 16S rRNA genes as genetic barcodes.

In the past decades, the rapid development of the population genetic and phylogeographic analyses based on the coalescent theory leads advancement of the field of molecular evolution in eukaryotes (Avise, 2009; Hickerson et al., 2010). Until recently, the coalescent theory is found to be used for testing the microbial spatiotemporal hypothesis (Gray et al., 2011). The coalescence theory that was firstly proposed by Kingman (1982) provides a practical framework to model genetic variation in a population. This involves tracing backward through time in order to identify events that occurred since the most recent common ancestor (MRCA) of the samples (Fu & Li, 1999; Kingman, 2000). This theory is sample-based and the speculation of evolutionary processes is more relevant than the classical population genetics theory that describes the properties of the entire population (Fu & Li, 1999). Three essential concepts comprise the coalescent process (Kingman, 2000): (1) the idea of identity-by-descent (Nagylaki, 1989), (2) selective neutrality and a constrained population size (Ewens, 1972, 1972), and (3) independent mutations of genealogy (Kingman, 1980). Along with advanced molecular techniques, approaches developed from the coalescence model can provide a sketch of the demographic history of microorganisms (Perez-Losada et al., 2007). Studies in this field have increasingly supported the reliability of molecular dating by microbial genetic analyses, such as estimating the TMRCA and the radiating time of bacteria and archaea by comparing 16S rRNA gene sequences (Sheridan et al., 2003) and exploring the early evolutionary history of phototrophy and methanogenesis in prokaryotes by the use of a relaxed molecular clock (Battistuzzi et al., 2004). There have also been reports of the successful use of DNA viruses to track recent and ancient local human histories (Pavesi, 2003, 2004, 2005; Kitchen et al., 2008).

2. The problematic definition of species in microbial diversity

Although the genetic diversity inferred by 16S rRNA gene or by genome data well display the entire microbial diversity of a community, the degree of biodiversity that just considers the appearance (birth) and the extinction (death) of "lineages" in phylogeny, i.e. the diversification rate, and ignores the concept of "species diversity" is still difficult to be accepted by traditional biologists. Therefore, species definition (species concept) of microorganisms, especially in prokaryotes, is still perplexed many microbial ecologists and environmental microbiologists, although some of them usually skirt this knotty problem.

However, the use of coalescent theory cannot prevent to discuss this knotty issue because the "monophyletic" species is necessary to be defined firstly (due to the assumption of identity-by-descent) to confirm the accurate coalescent inferences. Several papers discussed the species concept of prokaryotes in different points of view but most of them adopted a concept of "genetic similarity" as the principle (Ward, 1998; Vellai et al., 1999; Rossello-Mora & Amann, 2001; Cohan, 2006; Konstantinidis et al., 2006; Staley, 2006; Wilkins, 2006; Ward et al., 2008; Zimmer, 2008; Doolittle & Zhaxybayeva, 2009; Ereshefsky, 2010; Klenk et al., 2010) rather than the concept of monophyly. Achtman and Wagner (Achtman & Wagner, 2008) summarized five categories of species concepts in microbiology in which none of each has been generally accepted:

1. Monophyletic and genomically coherent cluster of organisms showing a high degree of overall similarity (Rossello-Mora & Amann, 2001)
2. An irreducible cluster of organisms of a common ancestor (Staley, 2006)
3. Having much greater degree of lateral gene transfer between each other than between other groups (Wilkins, 2006; Doolittle & Zhaxybayeva, 2009; Ereshefsky, 2010)
4. Forming a natural cluster (Nesbø et al., 2006)
5. Metapopulation lineages (Ereshefsky, 2010)

The first concept for delimiting species by monophyly should be used for the application of coalescent theory, such as inferring the demographic history (Fu & Li, 1999; Rosenberg & Nordborg, 2002; Rosenberg, 2003). In addition, the degree of genetic similarity is also a key factor to determine the length of coalescent time. In other words, lineages of a clade with shorter genetic distance reflect relatively recent coalescent history (of this clade). In contrast, the clade composed of lineages with long genetic distance can reflect relatively long-historical demography but with wider variance and larger inaccuracy.

In general, the genetic similarity of microbes ≤98.7% estimated by 16S rRNA genes are considered as different species but the opposite is not necessary true, i.e. the genetic similarity ≥98.7% might not be the same species (Stackebrandt & Ebers, 2006; Achtman & Wagner, 2008). This value matches to the threshold of 70% reassociation value in DNA-DNA hybridization (Stackebrandt & Ebers, 2006). The value <98.7% (or to round off <99%) identity in 16S rRNA gene overturns the old threshold of <97% identity for delineating species of microorganisms. Therefore the definition of "a species" in microbiology, for the purpose for phylogeographic or demographic inferences by the coalescent theory, is concluded as an integration of monophyly and genetic distance lower than 1%.

3. A case study: Simple microbial composition of a volcanic pond and a demographic association of demographic history of microbes with geographic history

In this case study, the microbial composition of an acidic sulfidic lake located in the northern Taiwan is exampled by genetic barcodes to represent the microbial diversity of a community and the microevolution of the dominant bacteria is further explored by the application of the coalescent theory. The sulfate lake is a special water type, since only certain chemical autotrophic microbes are able to utilize sulfides or sulfates as energy source (Moreira & Amils, 1997). Our study site, the "Niunai (Milky) Lake", is a small crater pond composed of sulfate substrates, located in a volcanic mountain, Mt. Datun, in the Yangmingshan National Park (YMSNP) in the northern part of Taipei County in Taiwan (Fig. 1). The source of the water is a mineral spring and the abundant rainfall from the northeast monsoon in North Taiwan (4526.4 mm per year on average at Chutzuhu Station in YMSNP, Table 1). This lake has never dried up since records have been kept. Due to the neutralization by rainfall, the water is mildly acidic (approximate pH 6–7) and the water temperature is approximately 38–40°C. The crater took shape during a volcanic eruption in the Quaternary Period and two major eruptions have made the recent topography of the Datun Volcano Groups. The first eruption was 2.8 to 2.5 mega annual before present (Ma BP) and the last time a Datun volcano erupted was approximately 0.8–0.2 Ma BP. The volcanoes have been reposed since. Similar to Yellowstone National Park, sulfur is rich in the nearby soil, rocks, and waters in the Datun Volcanoes. It has been reported that the endolithic microbes preserved the geological history of Yellowstone National Park (Walker et al., 2005). Very few natural microbial communities, especially those in extreme environments like the Datun Volcanoes, have been reported in Taiwan, which is a young island that took shape less than 5 Ma BP (Shen, 1996). Therefore, the sulfate-rich pool, Niunai Lake, serves as an excellent template to explore the microbial community structure and the evolutionary history of the dominant species in the volcanic mountains of Taiwan. Through the analyses of the phylogenetic community structure, which can assess the community assemblages (Kraft et al., 2007), and the population genetic structure, we present

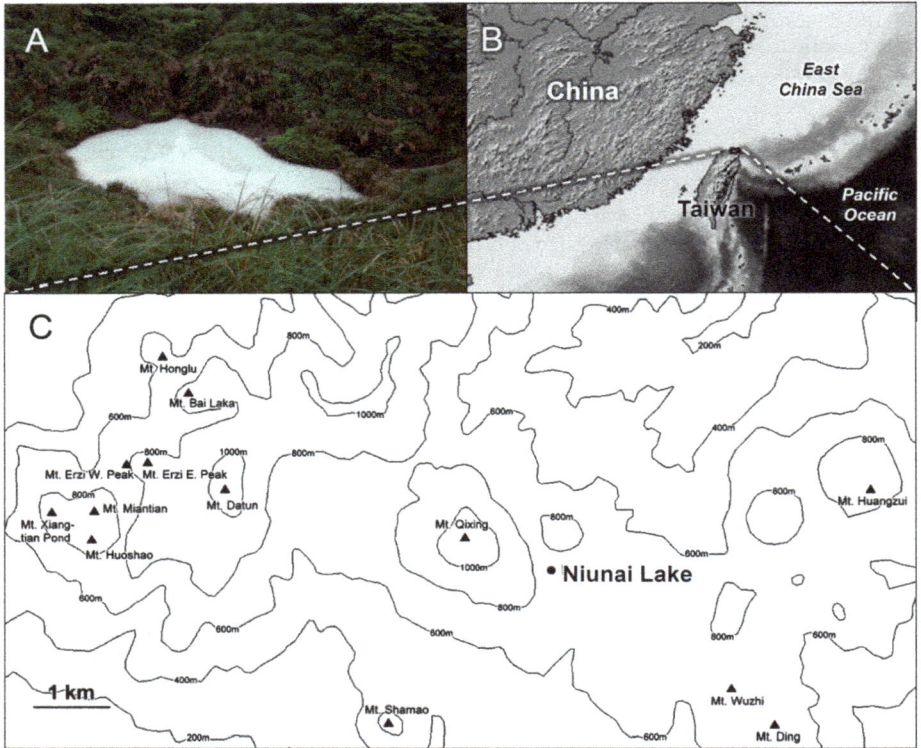

Fig. 1. Picture and map of the Niunai Lake. (A) The panorama of the Niunai Lake; (B) the location of the Niunai Lake in the northern Taiwan; (C) the contour map of the Yangmingshan National Park where the Niunai Lake located.

here the microbial community and population structures of a sulfur-rich environment. Several well-known studies that were based on culture-independent approaches (because less than 1% of microbial species are cultivable) have indicated that several unknown and unexpected taxa were discovered in the microbial communities (Giovannoni & Stingl, 2005). For example, 37% of the clones isolated from an extremely acidic (~pH 1) endolithic microhabitat in the Yellowstone geothermal environment were identified as *Mycobacterium* spp., which are pathogens of humans. (Walker et al., 2005). Additionally, while the long-considered dominant cyanobacteria comprised less than 5% of the clones isolated from the microbial community of the stromatolites of Hamelin Pool in Shark Bay, Western Australia, unknown proteobacteria comprised 28% of the clones. (Papineau et al., 2005). These studies suggest an unknown field of environmental microbiology that requires further investigation. According to the model proposed by Stackebrandt and Ebers (2006) and Acinas et al. (2004), 16S rRNA sequences of microorganisms that were more than 98.7% ~ 99% similar could be treated as an operational taxonomic unit (OTU). These OTUs, as defined by Stackebrandt and Ebers (2006) and Acinas et al. (2004), are 'microdiverse ribotype clusters' and are considered an important differentiation unit in natural bacterial communities. In other words, 99% similarity delineates different microbial species in nature. Therefore we assumed that the observed ribotypes may represent the species or the categories of the microorganisms in the collected samples and that their colony frequencies are indicative of the composition of the microbial species. As such, instead of using microbial culture methods, we used culture-independent techniques (Perez-Losada et al., 2007) to examine the composition of the microbial community in Niunai Lake.

	Jan	Feb	Mar	Apr	May	Jun	Jul	Aug	Sep	Oct	Nov	Dec	Mean
Accumulated precipitation (mm)	269.3	277.3	240.3	207.8	275.3	294.7	248.3	446	588.1	837.3	521.9	320.1	377.2
Precipitation days	20	18	18	15	16	14	10	13	15	19	21	20	16.6
Mean temp. (°C)	11.7	12.2	14.6	18.1	20.9	23.5	24.8	24.5	22.7	19.8	16.4	13.3	18.5
Maximum temp. (°C)	15.3	15.8	18.8	22.4	24.9	27.5	29.4	29	26.9	23.4	19.7	16.8	22.5
Minimum temp. (°C)	9.2	9.6	11.7	15.1	18.2	20.9	21.9	21.8	20.3	17.7	14.2	10.9	16
Relative humidity (%)	88	89	88	87	87	87	84	84	85	87	88	88	86

Table 1. Statistical records (1971 ~ 2000) of weather at the Chutzuhu Station in YMSNP, sources from Central Weather Bureau, Taiwan (http://www.cwb.gov.tw/).

In this case study, we report the results of our study examining the composition of the microbial community in Niunai Lake using a 16S rRNA gene library. In addition, the long-term demographic history of the dominant taxon (*Thiomonas* sp.) of this community and the factors that influenced it are presented. We try to make a connection between the geological and demographic history of microbes in terms of molecular ecology and this should be helpful in understanding the relationship between environmental factors and the microbial composition.

3.1 Methods

3.1.1 Sampling and molecular techniques

The water samples were collected one meter below the water surface from Niunai Lake (25°10'00"N, 121°33'52"E) in the Datun Volcano Group in Yangmingshan National Park (YMSNP) in Taipei, Taiwan. The weather records in YMSNP are listed in Table 1. The Niunai Lake is a sulfate-rich (20–40% sulfide) pond at an altitude of 700 m. Water samples were collected in sterile bottles and immediately incubated on ice until filtration and metagenomic DNA extraction.

DNA extraction was immediately carried out in order to prevent a bias from the foraging of microfauna. The DNA extraction protocols have been previously described (Liao et al., 2001). The water samples were pre-filtered through Nuclepore PE filters with pore size of 11 µm. The filtered water was then passed through 0.22 µm filters and these filters were cut into pieces, soaked in the extraction buffer (200 mM Tris-Cl pH 7.5, 250 mM NaCl, 25 mM EDTA, and 0.5% SDS), and shaken for homogenization. The metagenomic DNA was extracted with phenol-chloroform-isoamyl alcohol buffer and the total extracted DNA was dissolved in ddH$_2$O for subsequent analysis.

16S rRNA gene fragments were selectively amplified from the genomic DNA using the following two PCR primers (Field et al., 1997): forward primer 27F (5'-AGAGTTTGATCMTGGCTCAG-3', nucleotides 8-27 relative to the *Escherichia coli* 16S rRNA gene) and reverse primer 1522R (5'-AAGGAGGTGATCCANCCRCA-3', nucleotides 1522-1541 relative to *E. coli*). This primer set is universal in amplifying most bacterial 16S rRNA genes. The PCR reactions (50 µL) contained 0.4 µL of extracted DNA sample, 1 µM each primer, 0.1 mM each deoxynucleoside triphosphate, 20 µg BSA, and 2.5 U of Super *Taq* polymerase (Violet) in 10X PCR buffer (Violet). The Super *Taq* polymerase was used for enlarging the yields of PCR product and decreasing the rate of PCR error. PCR parameters were as follows: an initial denaturation at 94°C for 5 min, followed by 35 cycles of denaturation (94°C for 60 sec), annealing (55°C for 60 sec), and extension (72°C for 90 sec), with a final extension step (72°C for 7 min). The PCR products were resolved in an ethidium bromide-stained 1% agarose gel in TBE. DNA fragments of the expected size were purified from the gel using the Gel Extraction System Kit (Viogene).

The 16S rRNA gene library was constructed by cloning the amplified PCR products into the *y*T&A vector (Yeastern Biotech). Competent DH5α cells (*E. coli*) were transformed with the vector by heat shock transformation at 42° for 45 sec. The transformed cells were spread onto LB agar plates containing ampicillin, 5-bromo-4-chloro-3-indolyl-β-D-galactopyranoside, and isopropyl-β-D-thiogalactopyranoside (LB Ampicillin/X-gal/IPTG). Positive clones (which contained 16S rRNA gene PCR inserts) were confirmed using the M13F and M13R primers and were picked for further sequence analysis. Both directional sequencing was done at Genomics BioScience & Tech Co., Ltd. The sequences obtained in this study were deposited in GenBank under the accession numbers DQ145964-DQ146147.

3.1.2 Data analyses

The 16S rRNA gene sequences obtained in the study and those of known microorganisms in the NCBI database were aligned using the program Clustal X (Thompson et al., 1997), and then manually edited with the program BioEdit (Hall, 1999). All sequences were tested for possible chimeric artifacts using the Bellerophon software (Kelly et al., 2007). Putative chimeras were excluded from further analyses. Phylogenetic analysis (neighbor-joining (NJ) method) of the aligned data sets was then performed using TOPALi version 2.17 (Milne et al., 2004). The substitution model and rate model used for constructing the NJ tree were F84 (transition/transversion = 1.10) and gamma distribution (alpha = 0.69), respectively. The bootstrap analysis was conducted with 1,000 replications. From the analysis of the phylogenetic tree, the 16S rRNA gene clones of Niunai Lake were classified and the relative frequencies of taxa were counted. The species affinity of the 16S rRNA gene clones were identified through comparison with the Ribosomal Database Project (http://rdp.cme.msu.edu/).

After barcoding with the molecular characteristics by phylogenetic method (Liao et al., 2007), the number of microbial species in the Niunai Lake was estimated using the definition of microbial species proposed by Acinas et al. (2004) and Stackebrandt & Ebers (2006). Additionally, the "expected" richness was estimated using the S_{Chao1} index (Chao, 1984) and the rarefaction was estimated using the Rarefaction Calculator (http://www2.biology.ualberta.ca/jbrzusto/rarefact.php). The S_{Chao1} index is a nonparametric estimator of species richness (Chao, 1984). In addition, the genetic polymorphisms of the dominant microbial species, *Thiomonas* sp. in the Niunai Lake were calculated. The haplotype diversity (Hd) and the π (i.e., the average number of pairwise nucleotide differences) and θ ($\theta = 4N\mu$, where N is the effective population size and μ is the mutation rate estimated by the total number of mutations) indices of nucleotide diversity (Nei, 1987) were calculated in order to understand the style of genetic variation in this microorganism.

In order to infer the demographic history, the mismatch distribution of the 16S rDNA of *Thiomonas* sp. was calculated using DnaSP 4.0 (Rozas et al., 2003). The Tajima's D (Tajima, 1989) and Fu's Fs (Fu, 1997) tests were used to assess the effect of demographic changes and were calculated using Arlequin 3.11 (Excoffier et al., 2005). Generally speaking, these tests are based on the frequency distributions of variations. With the exception of these tests, the demographic inferences were carried out using the Bayesian skyline plot (BSP) analysis (Drummond et al., 2005) and the software BEAST 1.4.8 (Drummond & Rambaut, 2007) in order to estimate fluctuations in the effective population size. This method estimates a posterior distribution of effective population sizes backward through time until the most recent common ancestor using MCMC procedures. The constant population size coalescent model was the basic assumption used for this approach. The model developed by Jukes and Cantor (Jukes & Cantor, 1969) was used for distance matrix correction. A uniform rate across all branches (strict molecular clock) and the general time reversible substitution model were used for this calculation. In order to obtain the correct parameters and a higher effective sample size for BSP analysis, six pre-runs were performed and the parameters were modified according to the suggestions of the runs. Markov chains were run for 1×10^7 generations for pre-runs and 3×10^7 generations for the final run and were sampled every 1,000 generations, with the first 10% of the samples discarded as burn-in. The other parameters were set as default. The TRACER v1.4 program (Rambaut & Drummond, 2007) was used to visualize the posterior probabilities of the Markov chain statistics and to calculate a statistical summary of the genetic parameters.

3.2 Results

3.2.1 Phylogenetic assemblage of the microbial community

A 1625 base pair, after sequence alignment, partial 16S rRNA gene sequence was used in the analysis. Among these 1625 sites (characters), 790 were constant, 171 were variable characters that were parsimony-uninformative, and the remaining 664 sites were parsimony-informative characters. A total of 181 haplotypes (considering gaps) or 148 haplotypes (not considering gaps) were obtained from the 184 clones in our 16S rRNA gene library. The 16S rRNA gene library derived from Niunai Lake samples was analyzed by NJ comparisons (Fig. 2) in which 13 species were identified using the species definition proposed by Acinas et al. (2004) and Stackebrandt & Ebers (2006). One microbial species belonging to the genus *Thiomonas* (Burkholderiales; 91.85% in abundance), four species belonging to the genus *Thiobacillus* (Burkholderiales; 2.72% in abundance), one species belonging to the genus *Acidiphilium* (Acetobacteraceae; one clone), and one species belonging to the genus *Escherichia* (Enterobacteraceae; one clone) were identified. Additionally, eight clones from unknown species were identified (Fig. 3). One of the eight unknown taxa belonged to the epsilonproteobacteria and the others were betaproteobacteria, based on the grouping of the neighbor-joining tree (Fig. 2).

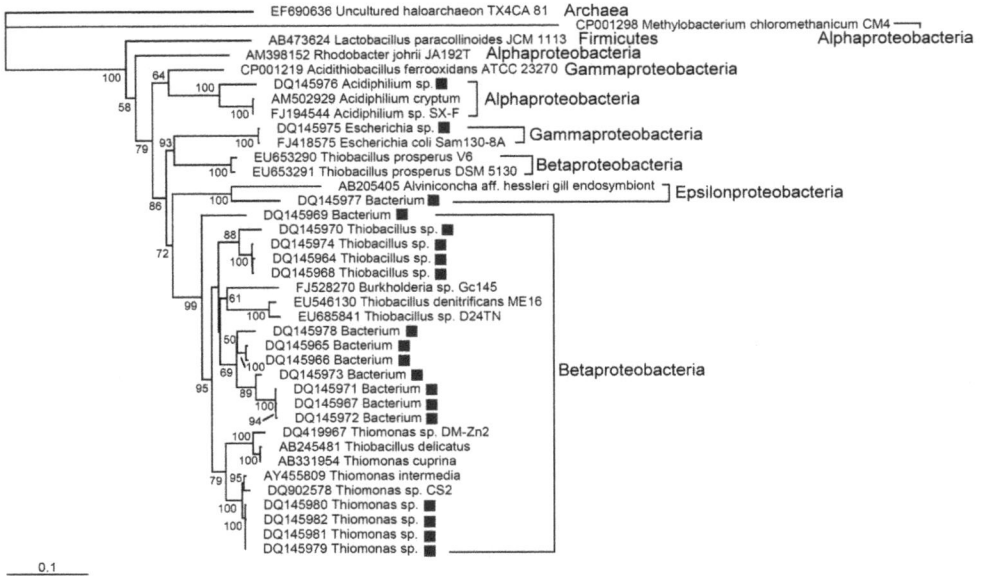

Fig. 2. Phylogenetic analysis of the 16S rRNA genes obtained from the Niunai Lake-derived clones and from the NCBI database. The reference sequences from NCBI were obtained by BLAST search. The tree was constructed by the neighbor joining method with the F84+G model and 1,000 bootstrapping replicates. The paraphyletic grouping is due to the genetic similarity of the sequences and cannot be explained as phylogenetic affinity. The lineages indicated in squares are sequences obtained in this study. The scale bar is the expected substitutions per site.

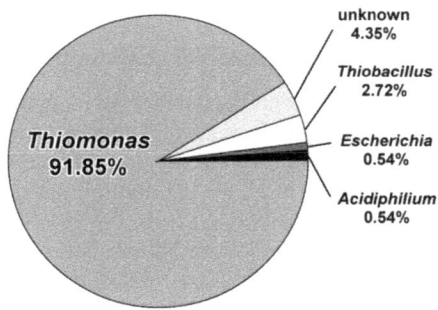

Fig. 3. Microbial species composition of the Niunai Lake community. The frequencies were estimated by the relative frequencies of the colony sizes of the 16S rRNA gene library.

Even though seven taxa of known genera were identified in the microbial community of Niunai Lake, they were all undescribed species. Additionally, although the relative sample size of the unknown taxa was small (8/184) compared to the known genera (176/184), almost half of the taxa identified in this research were previously unknown (6/13). Acinas et al. (2004) have reported that for microorganisms, there was a 70% decrease in the number of OTUs (from 1633 to 520 OTUs), when only 99% sequence identity (as opposed to 100% sequence identity) was used as the cut-off. The dramatic decrease in the OTU number using the cut-off of 99% 16S rRNA gene sequence similarity may be due to variations within species and PCR errors. Based on the definition proposed by Acinas et al. (2004), the number of OTUs versus cluster similarity of our findings is shown in Fig. 4. The similarity in genetic composition was also demonstrated by a pairwise comparison where most variations fell into a 0.02% genetic distance (Fig. 4, inset). This similarity in microbial composition indicates that microbial species were selected by the acidic environment of Niunai Lake and that the composition of microbial community was simple.

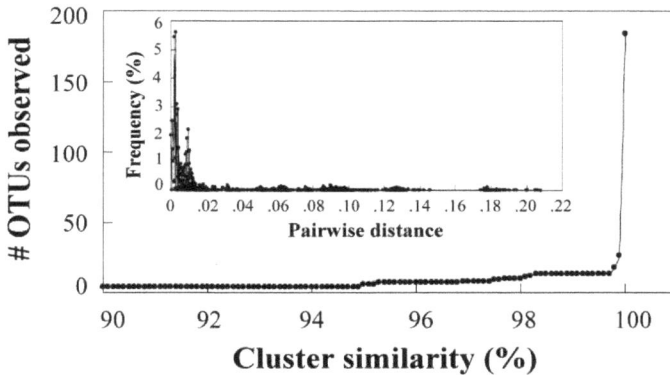

Fig. 4. Cluster-similarity curve of OTUs based on Nei's (1987) genetic distance. The uplift in 99% similarity indicates most microbial organisms in Niunai Lake belong to a single species. The inner plot indicates the frequency distribution of pairwise distance that demonstrates that most haplotypes are similar to each other within 0.02 genetic distances.

In our 16S rRNA gene library, there could be an inevitable amplification bias due to primer specificity. As a result, the unamplified 16S ribotypes would not be seen in this study (Moore et al., 1998; Morris et al., 2002). In order to estimate the probable species richness, both the S_{Chao1} and species-accumulation-curve methods were used. Both analyses indicated a higher number of species than what was detected in the 16S rRNA gene library. The S_{Chao1} index was 26.5 ± 9.418 species and the species-accumulation curve suggested a maximum number of 19.5 species (Fig. 5). Therefore, we could expect a greater microbial richness in Niunai Lake, even in such a harsh environment full of acidic, sulfurous, and *Thiomonas*-rich competitive stresses.

3.2.2 Genetic diversity and demographic history of *Thiomonas* sp.

The dominant bacterium *Thiomonas* sp. had a haplotype diversity (Hd) of 0.993, which indicates that most of the clones are different in genetic composition. The pairwise diversity (π) and genetic diversity index (θ) estimated from the segregating site (S) are 0.0146 ± 0.0077 and 0.0340 ± 0.0065, respectively (Table 2). Both Tajima's D and Fu's Fs indices, which reflect the demography, have significant negatives (D = -2.648, P < 0.00001; Fs = -23.746, P = 0.004,

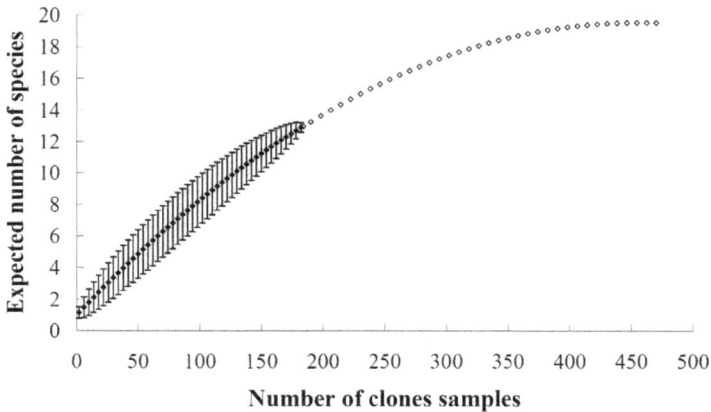

Fig. 5. Cumulative numbers of OTUs (the rarefaction analysis) as a function of the number of clones sequenced. The species-accumulation curve (black diamond dots) is not saturated, which suggests that the estimated number of microbial species of Niunai Lake is underestimated. The curve following the black diamond curve is a simulated curve that achieves a maximum number of species of 19.5 at 454 clones.

respectively) and indicate population expansion events (Table 2). We then used these nucleotide differences to calculate the distribution of pairwise differences (mismatch distribution). There was a left skew curve in the mismatch distribution plot, which had higher observed values than the expected values (Fig. 6). The differences in allelic frequency indicate that the observed values of the first four differences are lower than the expected values (Fig. 6, inset). In addition, clones of *Thiomonas* sp. are most pairwise different from each other in 5–10 nucleotide differences and illustrate an event of rapid population growth in the recent past, but not at the present.

Hd	π ($\times 10^3$)	S.D. ($\times 10^3$)	θ ($\times 10^3$)	S.D. ($\times 10^3$)	Tajima's D	P	Fu's Fs	P
0.993	14.617	7.678	33.974	6.463	-2.648	<0.00001	-23.746	0.004

Table 2. Genetic diversity of *Thiomonas* sp. in Niunai Lake. *Hd*: haplotype diversity (Nei, 1987); π: nucleotide diversity (Nei, 1987); θ: nucleotide diversity estimated by total number of mutations (Nei, 1987).

Except for the mismatch distribution, the BSP analysis was performed to depict the demographic history of *Thiomonas* sp. in Niunai Lake. Initially, the time to MRCA (TMRCA) of the *Thiomonas* sp. of the Niunai Lake was calculated. The strict-molecular-clock mode was selected because the known nodes for suitable molecular-dating were not acquired. Thus, the commonly used substitution rate of 4.5×10^{-9} per nucleotide site per year for prokaryotic SSU rDNA (estimated from *Escherichia coli*) suggested by Ochman and colleagues (Ochman & Wilson, 1987; Ochman et al., 1999) was used for calculating the TMRCA of Thiomonas sp. The coalescent time was estimated to be 7 Ma BP (6–9 Ma BP at the 95% confidence interval; Table 3), which is shorter than the coalescent time of 12 Ma BP (7.6–16.4 Ma BP at the 95% confidence interval) for whole *Thiomonas* species. In addition to the TMRCA, the demographic

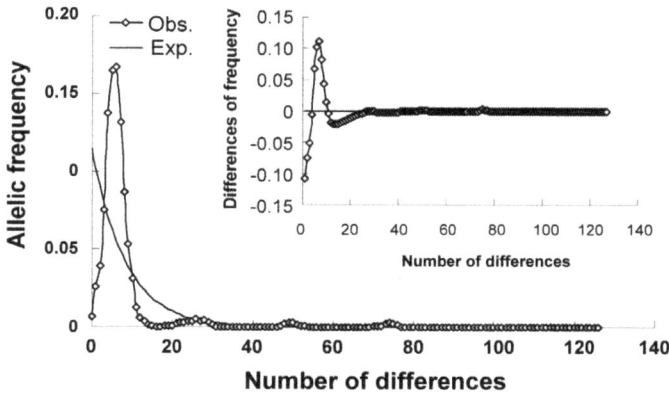

Fig. 6. Mismatch-distribution plot of pairwise differences of nucleotides. The right-skewed peak from the expected curve indicates a past-population-expansion of *Thiomonas* sp. in Niunai Lake. The diamond-line is the observed allelic frequency from the obtained sequences; the straight-line is the expectation by letting $\theta_{initial}$ = 8.971 (per seq) and infinite θ_{final}. The inner plot indicates the differences of the observed- and expected-allelic-frequency curves.

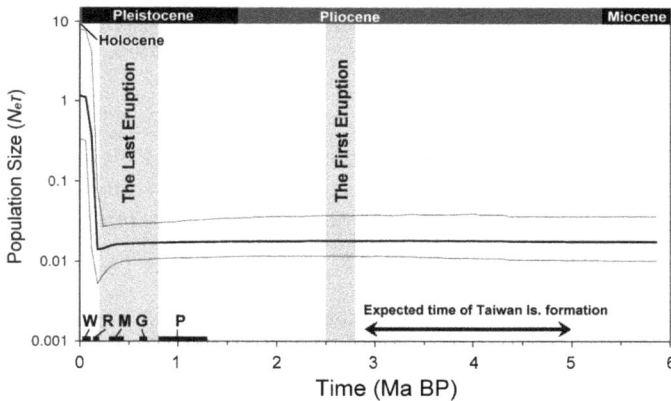

Fig. 7. Bayesian skyline plot for the population of *Thiomonas* in Niunai Lake. The plot was estimated by using the 16S rRNA sequences and was generated with a mutation rate of 4.5 × 10^{-9} per site per year (Ochman et al., 1999). The background effective population size of the *Thiomonas* population before the rapid expansion that occurred approximately 0.18 Ma BP was a result of the ancestral polymorphisms of *Thiomonas*. The x-axis is the time (Ma BP) and the y-axis is the scaled effective population size. The black line represents the medium value and the 95% confidence interval is shown by the gray lines. The geological time scale is presented in the top of graph. The shaded area indicates the eruptions of the Datun Volcanoes (Kim et al., 2005). The bars indicated as W, R, M, G, and P on the x-axis represent the Pleistocene glacial epochs Würm, Riss, Mindel, Günz, and Pre-Pastonian Stages, respectively, named according to the Alps glaciation events.

history was estimated by the BSP analysis. A constant population size through time was estimated until 0.35 Ma BP. The population size of *Thiomonas* sp. declined slightly from 0.35 Ma to 0.18 Ma BP and a rapid population growth followed until approximate 60 millennia BP when the growth rate gradually decreased (Fig. 7). The rapid population growth is approximate at two orders of magnitude.

3.3 Discussion

In this case study, two aspects were discussed: the microbial composition of the Niunai Lake community and the demographic history of the dominant bacteria *Thiomonas* sp. of Niunai Lake. We will try to illustrate the interactions between microorganisms and geologic history in terms of both of these aspects.

	Likelihood		TMRCA			
	Thiomonas sp. (Niunai Lake)	*Thiomonas* (NCBI)	*Thiomonas* sp. (Niunai Lake)		*Thiomonas* (NCBI)	
Mean	-5329.602	-1902.352	0.0332	(7.38)	0.0537	(11.93)
Stdev of mean	0.564	0.202	4.96×10^{-5}	(11.03×10^{-3})	3.80×10^{-4}	(8.45×10^{-3})
Median	-5329.21	-1901.978	0.0330	(7.33)	0.0522	(11.60)
95% HPD lower	-5347.411	-1911.269	0.0266	(5.91)	0.0343	(7.62)
95% HPD upper	-5312.535	-1894.13	0.0400	(8.89)	0.0737	(16.38)
Effective sample size (ESS)	253.488	504.472	4821.903		765.687	

Table 3. Summary statistics of maximum likelihood values for Bayesian skyline plot (BSP) analysis and the time to most recent common ancestor (TMRCA) estimated by BSP analysis. The unit of TMRCA is substitutions per nucleotide site and the parentheses are the dating by dividing the substitution rate (unit: Ma BP).

3.3.1 Microbial composition in the Niunai Lake community

The species composition in the microbial community was estimated using the phylogenetic approach. Even though the genera of most of the microbes were identified, the definite species are unknown, similar to other studies using this method (Blank et al., 2002; Walker et al., 2005; Omoregie et al., 2008). In spite of this drawback, the phylogenetic approach is still reliable for microbial identification and for inferring the microbial composition of the sample. In this study, a severe paraphyletic grouping was obtained (Fig. 2). The nucleotide compositions of these microbes, especially the rRNA nucleotide composition (Rudi, 2009), were not only shaped by heredity but could be affected by the properties of environment that they inhabited (Foerstner et al., 2005; Rudi, 2009). This may result in a similar genetic composition (e.g., GC content), a close genetic distance of distantly related microbes, and the disordered grouping of the phylogenetic tree.

The species composition of the microbial community of Niunai Lake is simpler than other environments (Table 4). Only 13 microbial species were detected in the 16S rDNA library,

Site	Area	Environmental properties	pH	Temp.	Dominant microbes	Species richness[a]	Reference
Chefren mud volcano, Nile Deep Sea Fan	Eastern Mediterranean	Iron- and Sulfide-rich	-	-	*Candidatus* Arcobacter sulfidicus (24%) in white mats; neutrophilic Fe(II)-oxidizing betaproteobacterium *Leptothrix ochracea* in orange mats	Very diverse	(Omoregie et al., 2008)
Soap Lake, lower Grand Coulee in E Washington State	USA	Saline and alkaline	9.8	7.3° ~ 16.3°C	Proteobacteria	508 (653)	(Dirnitriu et al., 2008)
Stromatolites of Hamelin Pool in Shark Bay	Western Australia	Hypersaline	-	17° ~ 27°C	Novel proteobacteria (28%), planctomycetes (17%), and actinobacteria (14%)	71 (178) of surface, 124 (505) of interior, and 90 (566) of irregular sampling	(Papineau et al., 2005)
Danshui River Estuary, mangrove ecosystem	Taiwan	Salinity 7 ~ 25 PSU, dissolved nitrogen 0.15–6.59 (2.88 on average) mg L^{-1}; dissolved phosphorus 0.02–1.53 (0.28 on average) mg L^{-1}; suspended solid 42.75 mg L^{-1}	~7.7	12.9° ~ 32°C (18°C on average)	Rhodobacteraceae (28.65%)	84 (447)	(Liao et al., 2007)
Rowley River in Plum Island Sound salt marsh, NE Massachusetts	USA	Marsh grass *Spartina alterniflora*, 20‰ ~ 34‰ salinity	-	-	Desulfobacteriaceae, Desulfobulbaceae and Desulfovibrionaceae (46.91%)	200 (332)	(Klepac-Ceraj et al., 2004)
Charca Verde pond, the forested university campus of Orsay	France	Freshwater suboxic pond, particles of organic material rich	7.07 ~ 7.8	7.8° ~ 9.9°C	Candidate division OD1 and beta-Proteobacteria	100 ~ 120 (170 ~ 198)	(Briée et al., 2007)
Volcanic lake in Deception Island in	Antarctica	140.1 S cm^{-1} of conductivity, low soluble reactive	6.7 ~ 7.7	1.1° ~ 6.5°C	Bacillariophyceae (58.70%)	46	(Llames & Vinocur, 2007)

Site	Area	Environmental properties	pH	Temp.	Dominant microbes	Species richness[a]	Reference
South Shetland Islands		phosphorus (69.2 g L[-1]) but high total phosphorus (418 g L[-1]), low dissolved inorganic nitrogen (20 µg L[-1]) and total nitrogen (75 µg L[-1])					
Norris Geyser Basin, Yellowstone National Park	USA	Geothermal, high concentrations of sulfuric acid, metals, and silicates	~1	~35°C	*Mycobacterium* spp. (37%) and *Cyanidium* spp. (26%)	~40	(Walker et al., 2005)
Seafloor of Sagami Bay, Hatsushima Island	Japan	Cold-seep sediments in the deep sea (1168 ~1174 m in depth); hypersaline	-	-	*Calyptogena* spp. (64%)	>24	(Fang et al., 2006)
Niunai Lake, Datun Volcanoes	Taiwan	Sulfide-rich (20%–40% sulfide)	~6.5	38° ~ 40°C	*Thiomonas* sp. (91.85%)	13 (26.5)	This study
Saline mud volcano at San Biagio-Belpasso, Mt. Etna	Italy	High salinity brines (up to 100 g/L) with high concentrations of Na[+] and Cl[-] ions (93–95%); CO_2-rich gases (85–87% of total gas composition); lower amounts of methane (7 ~ 10%), nitrogen (1.8 ~ 2.3%), and oxygen (0.3 ~ 0.5%)	6.58	15.8°C	*Marinobacter* sp. (20%), *Propionibacterium acnes* (18.57%), and *Methylomicrobium alcaliphilum* (15%)	19	(Yakimov et al., 2002)
Norris Basin, Yellowstone National Park	USA	Arsenite-oxidizing acidic thermal spring with arsenic concentration ~33 µM	3.1	58° ~ 62°C	*Hydrogenobacter acidophilus* (79% ~84%)	>8	(Jackson et al., 2001)

[a] Numbers in parentheses are the expected species richness estimated by the Chao1 (Chao, 1984) equation.

Table 4. Comparisons of the dominant species and species richness of the Niunai Lake microbial community with other environments. This table is ordered by the species richness.

while 19.5 or 26.5 species were calculated to be expected. We compared the environmental properties and microbial species richness between the Niunai Lake and other locations (Table 4). In contrast to the highly complex microbial communities in nutrient-rich bodies of water (e.g., the Soap Lake in eastern Washington State (Dirnitriu et al., 2008), Charca Verde pond in the University of Orsay campus in France (Briée et al., 2007), and the mangrove ecosystem of the Danshui River estuary in northern Taiwan (Liao et al., 2007)), most of the harsh environments have a lower species richness. When compared to the microbial communities in these volcanic or geothermal environments (e.g., the saline mud volcano at Mt. Etna in Italy (Yakimov et al., 2002) and the acidic thermal spring in Yellowstone National Park (Jackson et al., 2001)), the small species richness of Niunai Lake is not surprising. In addition, the dominant microbes metabolically correspond to the chemical properties of the environment. The simple microbial communities could be a consequence of long-term environmental selection and these dominant microbial species could be bio-indicators of the strict environments.

In the 16S rRNA gene library from Niunai Lake, 169 of 184 sequences (91.85%) were contributed to the genus *Thiomonas*, each as a different genotype. A similar dominance by a single microbial taxon in a strict environment has been reported. For example, 79–84% of the identified strains from the acidic thermal spring in the Norris Basin in Yellowstone National Park were *Hydrogenobacter acidophilus* (Jackson et al., 2001) and 64% of the strains from the deep-sea cold-seep sediments in the seafloor of Sagami Bay in Japan were *Calyptogena* spp. (Fang et al., 2006). However, the abundance of *Thiomonas* sp. (91.85%) is higher than these cases. This suggests that a series of periodic selection events purged the diversity of the microbial community (Cohan, 2006) or just the conditions that favor *Thiomonas* sp. from the onset, and the remnant *Thiomonas* sp. may play an important ecofunctional role in Niunai Lake. The slightly diversified ecotypes could be the descendants of the surviving variant of the last selection event (Cohan, 2006). The existence of multiple ecotypes, or forms, is a general phenomenon in the microbial world for adaptation and survival in a broad range of extreme environments (Moore et al., 1998). For example, in a study by Walker et al. (Walker et al., 2005) that examined the composition of the microbial endolithic community in a geothermal environment (~35°C, pH 1) in Yellowstone National Park, the abundant and diversified microbes (37% *Mycobacterium* spp. and 26% *Cyanidium* spp. in abundance), and most of other microbes, were those that could adapt to the acidic and thermal environment. Although the geologic condition of YMSNP is not as well defined as that of Yellowstone National Park, the microbial communities in Niunai Lake may still be a representative of those dwelling in a volcanic environment similar to YMSNP. The most abundant microbes in Niunai Lake, *Thiomonas* sp., occur widely with the presence of thiosulfate, tetrathionate, H_2S, and elemental sulfur. This genus also contains facultative chemolithoautotrophs (Moreira & Amils, 1997). The genus *Thiomonas* was discovered and classified by Moreira and Amils in 1997 (Moreira & Amils, 1997) and was split from *Thiobacillus* because of characteristics identified in the phylogenetic analysis. However, *Thiomonas* also share some physiological features with Thiobacillus, such as the capability of oxidizing sulfides to sulfuric acid as metabolic products (Temple & Colmer, 1951). The detailed metabolic properties of *Thiomonas* spp. were described by Kelly et al. (2007). Niunai Lake is a sulfate pond in a stratum of volcano. Due to the high sulfate level in the substrate, the dissolved sulfate or sulfide in Niunai Lake results in a special and extreme environment that would select for the microbes that have the ability to gain energy from the oxidation of a reduced sulfur compound. Another taxon present in Niunai Lake is *Acidiphilium* sp. (or related taxon), which plays an important role in sulfur-oxidization similar to *Thiobacillus* and *Thiomonas*. *Escherichia* is another genus found in Niunai Lake. While we cannot exclude the probability of contamination, the existence of *Escherichia* may also be due to the activities of wildlife around the Niunai Lake.

3.3.2 Climatic change, volcanism, and demographic history of *Thiomonas* sp.

In this part of the study, the evolutionary history of *Thiomonas* sp. was explored instead of the short-term population dynamics. The short-term population dynamics of microbial organisms, which are influenced by nutrients, toxins, temperature, and other biotic and abiotic factors, are commonly described through empirical studies (Tang et al., 1997; Lee et al., 2007; Lee et al., 2008; Ying et al., 2008). However, the long-term population demographies of bacteria and viruses are relatively few and most are relative to the evolutionary history of the host (Pavesi, 2003, 2004, 2005; Kitchen et al., 2008). However, some studies of molecular dating on microorganisms have been reported (Franzmann, 1996; Sheridan et al., 2003; Battistuzzi et al., 2004; Acquisti et al., 2007) and these studies provide excellent details on the molecular clock of microbial organisms.

Three independent analyses all indicate a population-growth event of *Thiomonas* sp. in Niunai Lake. The significant negatives of both Tajima's D and Fu's Fs indices, especially the high sensitivity in population expansion of negative Fu's Fs (Fu, 1997), suggest an increase in the population size. In addition, the mismatch distribution also demonstrates a left-skewed curve with a higher difference of frequency than expected, which is commonly explained by a population growth after a bottleneck event (Hwang et al., 2003; Cheng et al., 2005; Johnson et al., 2007). However, the number of differences is lower in the observed than the expected. This indicates that the event of population growth occurred in the recent past but not very recently. This speculation was supported by the Bayesian calculation according to the coalescent theory. The BSP analysis illustrates a similar demographic history of a bottleneck event by mismatch distribution resulting in a slight population decline (bottleneck effect) approximately 0.35 Ma to 0.18 Ma BP that was followed by a rapid expansion of population size until 60 millennia BP. From this point, the growth rate of the population decreased.

Similar to other studies of demographic histories (Flagstad & Roed, 2003; Carnaval & Bates, 2007; Lin et al., 2008), when compared with the paleoclimatic change, we noticed that the timing of the population decline seems to match the Mindel glacial epoch in the middle Pleistocene and the population expansion started at the beginning of Riss glacial epoch. Moreover, the timing of the gradual decrease in the population growth rate started at the beginning of the Würm glacial epoch. Therefore, the demographic history of *Thiomonas* sp. in Niunai Lake seems to perfectly match the glacial cycles. However, the effect of climatic changes on the demographic history of bacteria is full of paradoxes. First, it is difficult to explain the relationship between population growth and the Riss glacial period. Logically, the postglacial population expansion was reasonable, like other studies (Mikheyev et al., 2008), instead of expansion during the cold glacials. Second, the lowlands of SE Asia did not ice during the glacial periods, but became drier and colder. There is no evidence of the ice-covering of Niunai Lake, which is located at a low altitude (~700 m above sea level) in northern Taiwan, during the glacials. Thus the influence of the glacial cycles on the demographies of microorganisms was less clear. The optimal temperature for the growth of *Thiomonas* spp. ranges from 20°C to 53°C (estimated from six *Thiomonas* species) but they can adapt to a wide range of temperatures (some strains can slowly grow at less than 4°C and up to 65°C) (Kelly et al., 2007). Although the temperature records of Mt. Datun during the glacials are not available, the temperature during the last glacial maximum was estimated to be approximately 8°C on average (-4.1±0.68°C in January and 20.3±0.31°C in July) at Jih-Yueh Tan (approximate 750 m in altitude, 23°52′N lat., 120°55′E long., in central Taiwan) (Tsukada, 1966). Therefore, even in the relatively cold temperatures during the glacial maximum, *Thiomonas* spp. still have the ability to survive. For this reason, we do not think that the population size of *Thiomonas* sp. would be greatly influenced by climatic change. Despite some studies suggesting that the demographic histories of other microbes are consistent with climatic changes, they are host-dependent and influenced by the demographies of the hosts, which were affected by climatic changes (e.g. Mikheyev et al., 2008).

Alternatively, the activity of the Datun Volcanoes could have influenced the demographic history of *Thiomonas* sp. in Niunai Lake. According to the geological history, the Datun Volcanoes began to erupt approximately 2.8–2.5 Ma ago in a compressional tectonic environment (Song et al., 2000). These volcanic events are thought to have ceased during the late Pliocene and early Pleistocene and were followed by a second major eruption approximately 0.8–0.2 Ma BP. The Datun Volcanoes developed gradually during the episodic volcanic events that occurred between 2.8 and 0.2 Ma ago (Kim et al., 2005). While the volcanoes are believed to be currently inactive (Song et al., 2000), a recent study has suggested that these volcanoes may still be active because of continuing hydrothermal activities and gas fumaroles (Kim et al., 2005). The last eruption (approximately 0.8 Ma–0.2 Ma BP) formed large volcanoes in this area (Kim et al., 2005). The rapid expansion of the *Thiomonas* sp. population occurred approximately 0.18 Ma BP, just after the last eruption of the Datun Volcanoes. The geological events of the Datun Volcanoes match the demographic history of *Thiomonas* sp. Therefore, a more likely explanation for the rapid increase in the population size of *Thiomonas* sp. is that the volcanism created a sulfide-rich environment around Mt. Datun. Although the demographic changes affected by the earlier eruption events were difficult to examined because most ancestral variations were eliminated during the catastrophes, the population size change after the last eruptions was recorded in the genetic variations. The sulfide-rich lake water would have provided a hotbed for the growth of thiobacteria, like *Thiomonas* spp., which have ability to catabolize sulfur-containing compounds.

Of these two potential explanations for the demographic history of *Thiomonas* sp. (i.e., long-term climatic changes and volcanism), we prefer the model of volcanism affecting the bacterial evolution. This is because substrates in Niunai Lake could be directly used as materials for the catabolism of thiobacteria and the chemical properties of the environment are directly decide the microbial composition (Munster et al., 1998; Mills et al., 2003). The sulfide-rich environment of the Datun Volcanoes has been maintained for very long time since the last volcanic eruption. This extremely harsh substrate could be a stress for other microbes but could be relatively suitable for the growth of thiobacteria (e.g., *Thiomonas*, *Thiobacillus*, and *Acidiphilium*). Although many cases of eukaryotic demographic histories have been reported to be influenced by glacial periods, especially studies detailing postglacial expansion (Bartish et al., 2006; Aoki et al., 2008; Lin et al., 2008), and although the short-term population dynamics of microbes are easily influenced by temperature (Grisi et al., 1998; Skirnisdottir et al., 2000), the climatic change model seems inappropriate for explaining the long-term demographic history of *Thiomonas* sp. in this case.

In summary, based on a survey of the 16S rRNA gene, a very simple microbial composition was detected in the sulfide-rich Niunai Lake in a volcanic mountain of the northern Taiwan. Only 13 microbial taxa were detected and these were predominantly *Thiomonas* species (over 90%) with a small amount of *Thiobacillus*, *Escherichia*, and *Acidiphilium* species and approximately 4% unknown proteobacteria. While the expected microbial species richness was greater than what was observed (19.5 or 26.5 taxa), the species richness was still less than other bodies of water and revealed a simple microbial community structure. The dominant bacteria (belonging to genera *Thiomonas*, *Thiobacillus*, and *Acidiphilium*) function as sulfur oxidizers and may contribute to some of the lake's physical and chemical properties. In addition, we used population genetic approaches to explore the long-term demographic changes of the dominant species, *Thiomonas* sp. The observed significant negatives of both Tajima's D index and Fu's Fs index suggest a rapid population expansion. This suggestion was further supported by the left-skewed curve of the differences in allelic frequency between the observed and expected values in pairwise comparisons of sequences (mismatch distribution). The result of the mismatch distribution indicates a past event of rapid population growth. The Baysian skyline plot that was analyzed according to the coalescent theory also suggests a bottleneck event followed a rapid increase of population size of *Thiomonas* sp. in Niunai Lake. The time of population expansion was estimated to be approximately 0.18 Ma BP until 60 millennia BP and this timing approximately matches the end of the last eruption of the Datun Volcanoes (~ 0.2 Ma BP), where Niunai Lake is located. We eliminated the hypothesis of glacial cycles influencing the bacterial demography and prefer volcanism as the underlying mechanism for the observed bacterial demography. While the sulfide-rich substrates created by the volcanism could directly accelerate the growth of thiobacteria, the climate change model could not account for the population growth during the cold Riss glacial epoch. In conclusion, the periodic selection by the sulfide-rich environment simplified the microbial community and resulted in the dominance of *Thiomonas* sp. Additionally, the geological history corresponds to the demographic history of *Thiomonas* sp. in Niunai Lake.

4. Conclusion

In conclusion, there is a correlation between microbial composition and environmental change (Aravalli et al., 1998). Thus, the environmental properties can directly affect the microbial composition (Ptacnik et al., 2008) and even shape their genetic composition (Foerstner et al., 2005; Rudi, 2009). A harsh environment simplifies the composition of the microbial communities by strong selection forces (i.e., ecotypes of microbes are recurrently purged of their diversity by periodic selection for a long time) (Cohan, 2006). This kind of periodic selection under thermal, acidic, and sulfide-rich conditions causes a high abundance of *Thiomonas* sp. with greater than 99% genetic identity that limits the microbial species richness and simplifies the microbial community in Niunai Lake. The long-term periodic selection of the microbes could last as long as 0.18 Ma. Since the end of the last volcanic eruption of the Datun Volcanoes (0.2 Ma BP), the level of the sulfide-rich substrates steadied and were a hotbed for the growth of thiobacteria. The periodic selection since 0.2

Ma BP resulted in the current microbial communities of the Datun Volcanoes. This study represents the tight connection between environmental-microbial demographic history and the *in situ* geologic characteristic and geohistory. In this case study, the 16S rRNA gene was used as the genetic marker for tracing the demographic history of microorganisms. Recently, the rapid development of the genomic technologies and theories and models for eliminating the effects of recombination or horizontal gene transfer helps for decreasing the variance of coalescence estimation. The use of multilocus (or genomic) makers for exploring the microevolution of microorganisms is expectedly going to be a future issue soon.

5. Acknowledgment

The authors thank Chun-Hsiung Chen (Yangmingshan National Park) for his assistance in water sampling. This study were supported by grants from the Office of Research and Development, NTNU to S. Huang and from the National Science Council, R.O.C (NSC 99-2621-B-020-002-MY3) to P.-C. Liao.

6. References

Achtman, M. & Wagner, M. (2008). Microbial diversity and the genetic nature of microbial species. *Nature Reviews Microbiology*, Vol.6, No.6, pp. 431-440

Acinas, S.G., Klepac-Ceraj, V., Hunt, D.E., Pharino, C., Ceraj, I., Distel, D.L. & Polz, M.F. (2004). Fine-scale phylogenetic architecture of a complex bacterial community. *Nature*, Vol.430, No.6999, pp. 551-554

Acquisti, C., Kleffe, J. & Collins, S. (2007). Oxygen content of transmembrane proteins over macroevolutionary time scales. *Nature*, Vol.445, No.7123, pp. 47-52

Aoki, K., Kato, M. & Murakami, N. (2008). Glacial bottleneck and postglacial recolonization of a seed parasitic weevil, *Curculio hilgendorfi*, inferred from mitochondrial DNA variation. *Molecular Ecology*, Vol.17, No.14, pp. 3276-3289

Aravalli, R.N., She, Q.X. & Garrett, R.A. (1998). Archaea and the new age of microorganisms. *Trends in Ecology & Evolution*, Vol.13, No.5, pp. 190-194

Avarre, J.C., de Lajudie, P. & Bena, G. (2007). Hybridization of genomic DNA to microarrays: A challenge for the analysis of environmental samples. *Journal of Microbiological Methods*, Vol.69, No.2, pp. 242-248

Avise, J.C. (2009). Phylogeography: retrospect and prospect. *Journal of Biogeography*, Vol.36, No.1, pp. 3-15

Banfield, J.F., Denef, V.J., Kalnejais, L.H., Mueller, R.S., Wilmes, P., Baker, B.J., Thomas, B.C., VerBerkmoes, N.C. & Hettich, R.L. (2010). Proteogenomic basis for ecological divergence of closely related bacteria in natural acidophilic microbial communities. *Proceedings of the National Academy of Sciences of the United States of America*, Vol.107, No.6, pp. 2383-2390

Bartish, I.V., Kadereit, J.W. & Comes, H.P. (2006). Late Quaternary history of *Hippophae rhamnoides* L. (Elaeagnaceae) inferred from chalcone synthase intron (*Chsi*) sequences and chloroplast DNA variation. *Molecular Ecology*, Vol.15, No.13, pp. 4065-4083

Battistuzzi, F.U., Feijao, A. & Hedges, S.B. (2004). A genomic timescale of prokaryote evolution: insights into the origin of methanogenesis, phototrophy, and the colonization of land. *BMC Evolutionary Biology*, Vol.4, pp. 44

Benson, A.K., Kelly, S.A., Legge, R., Ma, F.R., Low, S.J., Kim, J., Zhang, M., Oh, P.L., Nehrenberg, D., Hua, K.J., Kachman, S.D., Moriyama, E.N., Walter, J., Peterson, D.A. & Pomp, D. (2010). Individuality in gut microbiota composition is a complex polygenic trait shaped by multiple environmental and host genetic factors. *Proceedings of the National Academy of Sciences of the United States of America*, Vol.107, No.44, pp. 18933-18938

Biddle, J.F., Fitz-Gibbon, S., Schuster, S.C., Brenchley, J.E. & House, C.H. (2008). Metagenomic signatures of the Peru Margin subseafloor biosphere show a genetically distinct environment. *Proceedings of the National Academy of Sciences of the United States of America*, Vol.105, No.30, pp. 10583-10588

Blank, C.E., Cady, S.L. & Pace, N.R. (2002). Microbial composition of near-boiling silica-depositing thermal springs throughout Yellowstone National Park. *Applied and Environmental Microbiology*, Vol.68, No.10, pp. 5123-5135

Briée, C., Moreira, D. & Lopez-Garcia, P. (2007). Archaeal and bacterial community composition of sediment and plankton from a suboxic freshwater pond. *Research in Microbiology*, Vol.158, No.3, pp. 213-227

Carnaval, A.C. & Bates, J.M. (2007). Amphibian DNA shows marked genetic structure and tracks Pleistocene climate change in northeastern Brazil. *Evolution*, Vol.61, No.12, pp. 2942-2957

Chao, A. (1984). Nonparametric estimation of the number of classes in a population. *Scandinavian Journal of Statistics*, Vol.11, pp. 265-270

Cheng, Y.P., Hwang, S.Y. & Lin, T.P. (2005). Potential refugia in Taiwan revealed by the phylogeographical study of *Castanopsis carlesii* Hayata (Fagaceae). *Molecular Ecology*, Vol.14, No.7, pp. 2075-2085

Chistoserdova, L. (2010). Functional Metagenomics: Recent Advances and Future Challenges. *Biotechnology and Genetic Engineering Reviews*, Vol.26, pp. 335-351

Cohan, F.M. (2006). Towards a conceptual and operational union of bacterial systematics, ecology, and evolution. *Philosophical Transactions of the Royal Society B-Biological Sciences*, Vol.361, No.1475, pp. 1985-1996

Dirnitriu, P.A., Pinkart, H.C., Peyton, B.M. & Mormile, M.R. (2008). Spatial and temporal patterns in the microbial diversity of a meromictic soda lake in Washington State. *Applied and Environmental Microbiology*, Vol.74, No.15, pp. 4877-4888

Doolittle, W.F. & Zhaxybayeva, O. (2009). On the origin of prokaryotic species. *Genome Research*, Vol.19, No.5, pp. 744-756

Drummond, A.J., Rambaut, A., Shapiro, B. & Pybus, O.G. (2005). Bayesian coalescent inference of past population dynamics from molecular sequences. *Molecular Biology and Evolution*, Vol.22, No.5, pp. 1185-1192

Drummond, A.J. & Rambaut, A. (2007). BEAST: Bayesian evolutionary analysis by sampling trees. *BMC Evolutionary Biology*, Vol.7, pp. 214

Ereshefsky, M. (2010). Microbiology and the species problem. *Biology & Philosophy*, Vol.25, No.4, pp. 553-568

Ewens, W.J. (1972). Concepts of substitutional load in finite populations. *Theoretical Population Biology*, Vol.3, No.2, pp. 153-161

Ewens, W.J. (1972). The sampling theory of selectively neutral alleles. *Theoretical Population Biology*, Vol.3, No.1, pp. 87-112

Excoffier, L., Laval, G. & Schneider, S. (2005). Arlequin ver. 3.0: An integrated software package for population genetics data analysis. *Evolutionary Bioinformatics Online*, Vol.1, pp. 47-50

Fang, J.S., Shizuka, A., Kato, C. & Schouten, S. (2006). Microbial diversity of cold-seep sediments in Sagami Bay, Japan, as determined by 16S rRNA gene and lipid analyses. *Fems Microbiology Ecology*, Vol.57, No.3, pp. 429-441

Field, K.G., Gordon, D., Wright, T., Rappe, M., Urbach, E., Vergin, K. & Giovannoni, S.J. (1997). Diversity and depth-specific distribution of SAR11 cluster rRNA genes from marine planktonic bacteria. *Applied and Environmental Microbiology*, Vol.63, No.1, pp. 63-70

Flagstad, O. & Roed, K.H. (2003). Refugial origins of reindeer (*Rangifer tarandus* L.) inferred from mitochondrial DNA sequences. *Evolution*, Vol.57, No.3, pp. 658-670

Foerstner, K.U., von Mering, C., Hooper, S.D. & Bork, P. (2005). Environments shape the nucleotide composition of genomes. *Embo Reports*, Vol.6, No.12, pp. 1208-1213

Franzmann, P.D. (1996). Examination of Antarctic prokaryotic diversity through molecular comparisons. *Biodiversity and Conservation*, Vol.5, No.11, pp. 1295-1305

Fu, Y.X. (1997). Statistical tests of neutrality of mutations against population growth, hitchhiking and background selection. *Genetics*, Vol.147, No.2, pp. 915-925

Fu, Y.X. & Li, W.H. (1999). Coalescing into the 21st century: An overview and prospects of coalescent theory. *Theoretical Population Biology*, Vol.56, No.1, pp. 1-10

Giovannoni, S.J. & Stingl, U. (2005). Molecular diversity and ecology of microbial plankton. *Nature*, Vol.437, No.7057, pp. 343-348

Gray, R.R., Tatem, A.J., Johnson, J.A., Alekseyenko, A.V., Pybus, O.G., Suchard, M.A. & Salemi, M. (2011). Testing spatiotemporal hypothesis of bacterial evolution using Methicillin-Resistant Staphylococcus aureus ST239 genome-wide data within a Bayesian framework. *Molecular Biology and Evolution*, Vol.28, No.5, pp. 1593-1603

Grisi, B., Grace, C., Brookes, P.C., Benedetti, A. & Dell'Abate, M.T. (1998). Temperature effects on organic matter and microbial biomass dynamics in temperate and tropical soils. *Soil Biology & Biochemistry*, Vol.30, No.10-11, pp. 1309-1315

Hall, T.A. (1999). BioEdit: a user-friendly biological sequence alignment editor and analysis program for Windows 95/98/NT. *Nucleic Acids Symposium Series*, Vol.41, pp. 95-98

Hickerson, M.J., Carstens, B.C., Cavender-Bares, J., Crandall, K.A., Graham, C.H., Johnson, J.B., Rissler, L., Victoriano, P.F. & Yoder, A.D. (2010). Phylogeography's past, present, and future: 10 years after Avise, 2000. *Molecular Phylogenetics and Evolution*, Vol.54, No.1, pp. 291-301

Hwang, S.Y., Lin, T.P., Ma, C.S., Lin, C.L., Chung, J.D. & Yang, J.C. (2003). Postglacial population growth of *Cunninghamia konishii* (Cupressaceae) inferred from phylogeographical and mismatch analysis of chloroplast DNA variation. *Molecular Ecology*, Vol.12, No.10, pp. 2689-2695

Jackson, C.R., Langner, H.W., Donahoe-Christiansen, J., Inskeep, W.P. & McDermott, T.R. (2001). Molecular analysis of microbial community structure in an arsenite-oxidizing acidic thermal spring. *Environmental Microbiology*, Vol.3, No.8, pp. 532-542

Johnson, J.A., Dunn, P.O. & Bouzat, J.L. (2007). Effects of recent population bottlenecks on reconstructing the demographic history of prairie-chickens. *Molecular Ecology*, Vol.16, No.11, pp. 2203-2222

Jukes, T.H. & Cantor, C.R. (1969). In: *Mammalian protein metabolism*, H.M. Munro (Ed.^Eds.), 21-132, Academic Press, New York

Kalia, V.C., Lal, S. & Cheema, S. (2008). Phylogeny vs genome reshuffling: horizontal gene transfer. *Indian Journal of Microbiology*, Vol.48, No.2, pp. 228-242

Kelly, D.P., Uchino, Y., Huber, H., Amils, R. & Wood, A.P. (2007). Reassessment of the phylogenetic relationships of *Thiomonas cuprina*. *International Journal of Systematic and Evolutionary Microbiology*, Vol.57, pp. 2720-2724

Kim, K.H., Chang, C.H., Ma, K.F., Chiu, J.M. & Chen, K.C. (2005). Modern seismic observations in the Tatun volcano region of northern Taiwan: Seismic/volcanic hazard adjacent to the Taipei metropolitan area. *Terrestrial Atmospheric and Oceanic Sciences*, Vol.16, No.3, pp. 579-594

Kingman, J.F.C. (1980). *Mathematics of Genetic Diversity*. Society for Industrial and Applied Mathematics, Philadelphia

Kingman, J.F.C. (1982). On the genealogy of large populations. *Journal of Applied Probability*, Vol.19A, pp. 27-43

Kingman, J.F.C. (2000). Origins of the coalescent. 1974-1982. *Genetics*, Vol.156, No.4, pp. 1461-1463

Kitchen, A., Miyamoto, M.M. & Mulligan, C.J. (2008). Utility of DNA viruses for studying human host history: Case study of JC virus. *Molecular Phylogenetics and Evolution*, Vol.46, No.2, pp. 673-682

Klenk, H.P., Auch, A.F., von Jan, M. & Goker, M. (2010). Digital DNA-DNA hybridization for microbial species delineation by means of genome-to-genome sequence comparison. *Standards in Genomic Sciences*, Vol.2, No.1, pp. 117-134

Klepac-Ceraj, V., Bahr, M., Crump, B.C., Teske, A.P., Hobbie, J.E. & Polz, M.F. (2004). High overall diversity and dominance of microdiverse relationships in salt marsh sulphate-reducing bacteria. *Environmental Microbiology*, Vol.6, No.7, pp. 686-698

Konstantinidis, K.T., Ramette, A. & Tiedje, J.M. (2006). The bacterial species definition in the genomic era. *Philosophical Transactions of the Royal Society B-Biological Sciences*, Vol.361, No.1475, pp. 1929-1940

Kraft, N.J.B., Cornwell, W.K., Webb, C.O. & Ackerly, D.D. (2007). Trait evolution, community assembly, and the phylogenetic structure of ecological communities. *American Naturalist*, Vol.170, No.2, pp. 271-283

Kulakov, L.A., Del Casale, A., Flanagan, P.V., Larkin, M.J. & Allen, C.C.R. (2011). Analysis of transduction in wastewater bacterial populations by targeting the phage-derived 16S rRNA gene sequences. *Fems Microbiology Ecology*, Vol.76, No.1, pp. 100-108

Lee, D., Kim, S., Cho, J. & Kim, J. (2008). Microbial population dynamics and temperature changes during fermentation of kimjang kimchi. *Journal of Microbiology*, Vol.46, No.5, pp. 590-593

Lee, S.H., Otawa, K., Onuki, M., Satoh, H. & Mino, T. (2007). Population dynamics of phage-host system of *Microlunatus phosphovorus* indigenous in activated sludge. *Journal of Microbiology and Biotechnology*, Vol.17, No.10, pp. 1704-1707

Liao, P.C., Huang, B.H. & Huang, S. (2007). Microbial community composition of the Danshui river estuary of northern Taiwan and the practicality of the phylogenetic method in microbial barcoding. *Microbial Ecology*, Vol.54, No.3, pp. 497-507

Lin, R.C., Yeung, C.K.L. & Li, S.H. (2008). Drastic post-LGM expansion and lack of historical genetic structure of a subtropical fig-pollinating wasp (*Ceratosolen* sp. 1) of *Ficus septica* in Taiwan. *Molecular Ecology*, Vol.17, No.23, pp. 5008-5022

Llames, M.E. & Vinocur, A. (2007). Phytoplankton structure and dynamics in a volcanic lake in Deception Island (South Shetland Islands, Antarctica). *Polar Biology*, Vol.30, No.7, pp. 849-857

Mikheyev, A.S., Vo, T. & Mueller, U.G. (2008). Phylogeography of post-Pleistocene population expansion in a fungus-gardening ant and its microbial mutualists. *Molecular Ecology*, Vol.17, No.20, pp. 4480-4488

Mills, D.K., Fitzgerald, K., Litchfield, C.D. & Gillevet, P.M. (2003). A comparison of DNA profiling techniques for monitoring nutrient impact on microbial community composition during bioremediation of petroleum-contaminated soils. *Journal of Microbiological Methods*, Vol.54, No.1, pp. 57-74

Milne, I., Wright, F., Rowe, G., Marshall, D.F., Husmeier, D. & McGuire, G. (2004). TOPALi: software for automatic identification of recombinant sequences within DNA multiple alignments. *Bioinformatics*, Vol.20, No.11, pp. 1806-1807

Moon-van der Staay, S.Y., De Wachter, R. & Vaulot, D. (2001). Oceanic 18S rDNA sequences from picoplankton reveal unsuspected eukaryotic diversity. *Nature*, Vol.409, No.6820, pp. 607-610

Moore, L.R., Rocap, G. & Chisholm, S.W. (1998). Physiology and molecular phylogeny of coexisting Prochlorococcus ecotypes. *Nature*, Vol.393, No.6684, pp. 464-467

Morales, S.E. & Holben, W.E. (2011). Linking bacterial identities and ecosystem processes: can 'omic' analyses be more than the sum of their parts? *Fems Microbiology Ecology*, Vol.75, No.1, pp. 2-16

Moreira, D. & Amils, R. (1997). Phylogeny of Thiobacillus cuprinus and other mixotrophic thiobacilli: Proposal for Thiomonas gen nov. *International Journal of Systematic Bacteriology*, Vol.47, No.2, pp. 522-528

Morris, R.M., Rappe, M.S., Connon, S.A., Vergin, K.L., Siebold, W.A., Carlson, C.A. & Giovannoni, S.J. (2002). SAR11 clade dominates ocean surface bacterioplankton communities. *Nature*, Vol.420, No.6917, pp. 806-810

Munster, U., Heikkinen, E. & Knulst, J. (1998). Nutrient composition, microbial biomass and activity at the air-water interface of small boreal forest. *Hydrobiologia*, Vol.363, pp. 261-270

Nagylaki, T. (1989). Gustave Malécot and the transition from classical to modern population genetics. *Genetics*, Vol.122, No.2, pp. 253-268

Nei, M. (1987). *Molecular Evolutionary Genetics*. Columbia Univ. Press, New York

Nesbø, C.L., Dlutek, M. & Doolittle, W.F. (2006). Recombination in thermotoga: Implications for species concepts and biogeography. *Genetics*, Vol.172, No.2, pp. 759-769

Ochman, H. & Wilson, A.C. (1987). Evolution in bacteria: evidence for a universal substitution rate in cellular genomes. *Journal of Molecular Evolution*, Vol.26, No.1-2, pp. 74-86

Ochman, H., Elwyn, S. & Moran, N.A. (1999). Calibrating bacterial evolution. *Proceedings of the National Academy of Sciences of the United States of America*, Vol.96, No.22, pp. 12638-12643

Omoregie, E.O., Mastalerz, V., de Lange, G., Straub, K.L., Kappler, A., Roy, H., Stadnitskaia, A., Foucher, J.P. & Boetius, A. (2008). Biogeochemistry and community composition of iron- and sulfur-precipitating microbial mats at the Chefren mud volcano (Nile Deep Sea fan, Eastern Mediterranean). *Applied and Environmental Microbiology*, Vol.74, No.10, pp. 3198-3215

Papineau, D., Walker, J.J., Mojzsis, S.J. & Pace, N.R. (2005). Composition and structure of microbial communities from stromatolites of Hamelin Pool in Shark Bay, Western Australia. *Applied and Environmental Microbiology*, Vol.71, No.8, pp. 4822-4832

Pavesi, A. (2003). African origin of polyomavirus JC and implications for prehistoric human migrations. *Journal of Molecular Evolution*, Vol.56, No.5, pp. 564-572

Pavesi, A. (2004). Detecting traces of prehistoric human migrations by geographic synthetic maps of polyomavirus JC. *Journal of Molecular Evolution*, Vol.58, No.3, pp. 304-313

Pavesi, A. (2005). Utility of JC polyomavirus in tracing the pattern of human migrations dating to prehistoric times. *Journal of General Virology*, Vol.86, pp. 1315-1326

Perez-Losada, M., Porter, M.L., Tazi, L. & Crandall, K.A. (2007). New methods for inferring population dynamics from microbial sequences. *Infection Genetics and Evolution*, Vol.7, No.1, pp. 24-43

Ptacnik, R., Solimini, A.G., Andersen, T., Tamminen, T., Brettum, P., Lepisto, L., Willen, E. & Rekolainen, S. (2008). Diversity predicts stability and resource use efficiency in natural phytoplankton communities. *Proceedings of the National Academy of Sciences of the United States of America*, Vol.105, No.13, pp. 5134-5138

Rambaut, A. & Drummond, A.J. (2007). Tracer v1.4, Available from http://beast.bio.ed.ac.uk/Tracer pp.

Rocap, G., Larimer, F.W., Lamerdin, J., Malfatti, S., Chain, P., Ahlgren, N.A., Arellano, A., Coleman, M., Hauser, L., Hess, W.R., Johnson, Z.I., Land, M., Lindell, D., Post, A.F., Regala, W., Shah, M., Shaw, S.L., Steglich, C., Sullivan, M.B., Ting, C.S., Tolonen, A., Webb, E.A., Zinser, E.R. & Chisholm, S.W. (2003). Genome divergence in two Prochlorococcus ecotypes reflects oceanic niche differentiation. *Nature*, Vol.424, No.6952, pp. 1042-1047

Rosenberg, N.A. & Nordborg, M. (2002). Genealogical trees, coalescent theory and the analysis of genetic polymorphisms. *Nature Reviews Genetics*, Vol.3, No.5, pp. 380-390

Rosenberg, N.A. (2003). The shapes of neutral gene genealogies in two species: Probabilities of monophyly, paraphyly, and polyphyly in a coalescent model. *Evolution*, Vol.57, No.7, pp. 1465-1477

Rossello-Mora, R. & Amann, R. (2001). The species concept for prokaryotes. *Fems Microbiology Reviews*, Vol.25, No.1, pp. 39-67

Rozas, J., Sanchez-DelBarrio, J.C., Messeguer, X. & Rozas, R. (2003). DnaSP, DNA polymorphism analyses by the coalescent and other methods. *Bioinformatics*, Vol.19, No.18, pp. 2496-2497

Rudi, K. (2009). Environmental shaping of ribosomal RNA nucleotide composition. *Microbial Ecology*, Vol.57, No.3, pp. 469-477

Shanks, O.C., Santo Domingo, J.W., Lamendella, R., Kelty, C.A. & Graham, J.E. (2006). Competitive metagenomic DNA hybridization identifies host-specific microbial genetic markers in cow fecal samples. *Applied and Environmental Microbiology*, Vol.72, No.6, pp. 4054-4060

Shen, C.F. (1996). The biogeography of Taiwan: 1. Background. *Annual of the National Taiwan Museum*, Vol.39, pp. 387-427 (In Chinese)

Sheridan, P.P., Freeman, K.H. & Brenchley, J.E. (2003). Estimated minimal divergence times of the major bacterial and archaeal phyla. *Geomicrobiology Journal*, Vol.20, No.1, pp. 1-14

Skirnisdottir, S., Hreggvidsson, G.O., Hjorleifsdottir, S., Marteinsson, V.T., Petursdottir, S.K., Holst, O. & Kristjansson, J.K. (2000). Influence of sulfide and temperature on species composition and community structure of hot spring microbial mats. *Applied and Environmental Microbiology*, Vol.66, No.7, pp. 2835-2841

Song, S.R., Tsao, S. & Lo, H.J. (2000). Characteristics of the Tatun volcanic eruptions, north Taiwan; implications for a cauldron formation and volcanic evolution. *Journal of the Geological Society of China*, Vol.43, pp. 361-378

Stackebrandt, E. & Ebers, J. (2006). Taxonomic parameters revisited: tarnished gold standards. *Microbiology Today*, Vol.33, pp. 152-155

Staley, J.T. (2006). The bacterial species dilemma and the genomic-phylogenetic species concept. *Philosophical Transactions of the Royal Society B-Biological Sciences*, Vol.361, No.1475, pp. 1899-1909

Tajima, F. (1989). Statistical method for testing the neutral mutation hypothesis by DNA polymorphism. *Genetics*, Vol.123, No.3, pp. 585-595

Tang, B., Sitomer, A. & Jackson, T. (1997). Population dynamics and competition in chemostat models with adaptive nutrient uptake. *Journal of Mathematical Biology*, Vol.35, No.4, pp. 453-479

Temple, K.L. & Colmer, A.R. (1951). The autotrophic oxidation of iron by a new bacterium, *Thiobacillus ferrooxidans*. *Journal of Bacteriology*, Vol.62, pp. 605-611

Thompson, J.D., Gibson, T.J., Plewniak, F., Jeanmougin, F. & Higgins, D.G. (1997). The CLUSTAL_X windows interface: flexible strategies for multiple sequence alignment aided by quality analysis tools. *Nucleic Acids Research*, Vol.25, No.24, pp. 4876-4882

Tsukada, M. (1966). Late Pleistocene vegetatation and climate in Taiwan (Formosa). *Proceedings of the National Academy of Sciences of the United States of America*, Vol.55, pp. 543-548

Vellai, T., Kovacs, A.L., Kovacs, G., Ortutay, C. & Vida, G. (1999). Genome economization and a new approach to the species concept in bacteria. *Proceedings of the Royal Society B-Biological Sciences*, Vol.266, No.1432, pp. 1953-1958

Walker, J.J., Spear, J.R. & Pace, N.R. (2005). Geobiology of a microbial endolithic community in the Yellowstone geothermal environment. *Nature*, Vol.434, No.7036, pp. 1011-1014

Wang, J., Qin, J.J., Li, R.Q., Raes, J., Arumugam, M., Burgdorf, K.S., Manichanh, C., Nielsen, T., Pons, N., Levenez, F., Yamada, T., Mende, D.R., Li, J.H., Xu, J.M., Li, S.C., Li, D.F., Cao, J.J., Wang, B., Liang, H.Q., Zheng, H.S., Xie, Y.L., Tap, J., Lepage, P., Bertalan, M., Batto, J.M., Hansen, T., Le Paslier, D., Linneberg, A., Nielsen, H.B., Pelletier, E., Renault, P., Sicheritz-Ponten, T., Turner, K., Zhu, H.M., Yu, C., Li, S.T., Jian, M., Zhou, Y., Li, Y.R., Zhang, X.Q., Li, S.G., Qin, N., Yang, H.M., Wang, J., Brunak, S., Dore, J., Guarner, F., Kristiansen, K., Pedersen, O., Parkhill, J., Weissenbach, J., Bork, P., Ehrlich, S.D. & Consortium, M. (2010). A human gut microbial gene catalogue established by metagenomic sequencing. *Nature*, Vol.464, No.7285, pp. 59-U70

Ward, D.M. (1998). A natural species concept for prokaryotes. *Current Opinion in Microbiology*, Vol.1, No.3, pp. 271-277

Ward, D.M., Cohan, F.M., Bhaya, D., Heidelberg, J.F., Kuhl, M. & Grossman, A. (2008). Genomics, environmental genomics and the issue of microbial species. *Heredity*, Vol.100, No.2, pp. 207-219

Wilkins, J.S. (2006). The concept and causes of microbial species. *History and Philosophy of the Life Sciences*, Vol.28, No.3, pp. 389-407

Yakimov, M.M., Giuliano, L., Crisafi, E., Chernikova, T.N., Timmis, K.N. & Golyshin, P.N. (2002). Microbial community of a saline mud volcano at San Biagio-Belpasso, Mt. Etna (Italy). *Environmental Microbiology*, Vol.4, No.5, pp. 249-256

Ying, Y.L., Lv, Z.M., Min, H. & Cheng, J. (2008). Dynamic changes of microbial community diversity in a photohydrogen producing reactor monitored by PCR-DGGE. *Journal of Environmental Sciences-China*, Vol.20, No.9, pp. 1118-1125

Zimmer, C. (2008). What is a species? *Scientific American*, Vol.298, No.6, pp. 72-79

6

Microsatellites as Tools for Genetic Diversity Analysis

Andrea Akemi Hoshino[1], Juliana Pereira Bravo[2],
Paula Macedo Nobile[1] and Karina Alessandra Morelli[3]
[1]Universidade Estadual de Campinas, SP,
[2]Universidade Estadual Paulista, SP,
[3]Fundação Oswaldo Cruz, RJ,
Brazil

1. Introduction

Powerful tools for the analysis of genetic biodiversity are molecular markers, which are based on DNA sequence polymorphisms. Indeed, DNA sequences determine the diversity of organisms, and therefore, the techniques used to evaluate DNA polymorphisms directly measure the genetic diversity. Because molecular markers show Mendelian inheritance, it is possible to trace the fingerprint of each organism and determine the evolutionary history of the species by phylogenetic analysis, studies of genetic relationship, population genetic structures and genetic mapping.

According to technical principles, there are three classes of molecular markers: (i) nucleic acid hybridization based on complementary bases, e.g., restriction fragment length polymorphisms (RFLPs), (ii) Polymerase Chain Reaction (PCR) based on DNA amplification, e.g., random amplification of polymorphic DNAs (RAPD), amplified fragment length polymorphisms (AFLP), microsatellites or simple sequence repeats (SSRs) and (iii), single nucleotide polymorphisms (SNPs). The first technique, RFLP, has been decreasingly used due to the difficulties involved in manipulating high throughput sampling and the third technique, SNPs, represents high costs related to large-scale genotyping. However, the cost-effective PCR-based techniques have been largely used.

With the advent of PCR technology in the mid 1980s (Mullis & Faloona, 1987; Saiki et al., 1985), new perspectives have evolved for molecular biology fields that have largely impacted several applied purposes, e.g., diagnostics, plant and animal breeding programs, forensics and others. Microsatellites were detected in eukaryote genomes almost thirty years ago and they are the most promising PCR-based markers. Microsatellites are tandemly repeated motifs of variable lengths that are distributed throughout the eukaryotic nuclear genome in both coding and noncoding regions (Jarne & Lagoda, 1996). They also appear in prokaryotic and eukaryotic organellar genomes, e.g., chloroplast (Powell et al., 1995) and mitochondria (Soranzo et al., 1999).

Due to the high mutation rate of microsatellites, they are potentially the most informative molecular marker with the advantage of easy and low-cost detection by PCR. Moreover, the bottleneck caused by the high cost and laborious approaches to isolate microsatellite loci has been overcome by new sequencing technologies. Large databases of genomic and EST sequences, that could be screened using bioinformatics tools, are now available and many published loci could be transferred from related species. Another great advantage of microsatellites is their co-dominant feature. Unlike RAPD and AFLP, which are dominant markers that detect only the presence or absence of a locus, microsatellite markers detect both homozygote and heterozygote genotypes.

A search using the Web of Science facility (Thomson Reuters, http://webofknowledge.com) confirmed that the microsatellite has been the most used molecular marker to address genetic diversity (Table 1). This marker has been applied for the germplasm conservation, phylogenetic analyses, plant and animal breeding programs, constructing linkage maps, mapping economically important quantitative traits and identifying genes responsible for desired traits.

Science Category	Molecular marker				
	Microsatellite or SSR	SNP	RAPD	AFLP	RFLP
Biochemistry Molecular Biology	1178	185	83	96	38
Evolutionary Biology	989	33	18	78	11
Ecology	989	23	17	49	16
Genetics Heredity	1134	493	124	131	32
Biodiversity Conservation	405	4	3	8	3
Total	4690	1.269	925	668	531

Table 1. A search using the Web of Science facility for the number of articles published in the last five years whose title contains the name of one of the markers: microsatellite or SSR, SNP, RAPD, AFLP or RFLP.

2. Identification and features

Microsatellites are DNA sequences of mono-, di-, tri-, tetra- and pentanucleotide units repeated *in tandem*, which are widely distributed in the genome (Powell et al., 1996). Litt & Luty first used the term "microsatellites" in 1989 when analyzing the abundance and dispersion of $(TG)_n$ in the cardiac actin gene. Microsatellites were originally designed to research degenerative and neurology diseases in humans but showed great applicability for other species.

Many authors classified the markers according to the number of bases, i.e., short repeats (10-30 bases) are microsatellites and longer repeats are minisatellites (between 10-100 bases). Microsatellites have been also been classified according to the type of repeated sequence presented: (i) perfect, when showing only perfect repetitions, e.g., $(AT)_{20}$, (ii) imperfect repeats, when the repeated sequence is interrupted by different nucleotides that are not repeated, e.g., $(AT)_{12}GC(AT)_8$, and (iii) composite, when there are two or more different motifs *in tandem*, e.g., $(AT)_7(GC)_6$. The composite repeats can be perfect or imperfect. The sequences of di-, tri- and tetranucleotide repeats are the most common choices for molecular genetic studies (Selkoe & Toonen, 2006).

In addition to their co-dominant feature, i.e., the identification of all alleles of a given locus, microsatellites can also be amplified using polymerase chain reaction (PCR) in stringent conditions that usually only permit the amplification of single loci, thus facilitating data integration (Bravo et al., 2006). Furthermore, microsatellites are widely distributed throughout the genome, highly polymorphic and transferable between species. These features provide the foundation for their successful application in a wide range of fundamental and applicable fields (Chistiakov et al., 2006).

The presence of SSRs in eukaryotes was verified from diverse genome regions, including 3'-UTRs, 5'-UTRs, exons and introns (Rajendrakumar et al., 2007). Furthermore, their localization could potentially interfere with different aspects of DNA structure, DNA recombination, DNA replication and gene expression as illustrated by Chistiakov et al. (2006). The transposable elements might contain one or more sites that are predisposed to microsatellite formation and enables SSRs dispersion throughout the genome (Bhargava & Fuentes, 2010). Microsatellites are also commonly located in proximity of interspersed repetitive elements, such as short interspersed repeats (SINEs) and long interspersed repeats (LINEs). Kashi et al. (1997) reported that in promoter regions, the presence and length of SSRs could influence transcriptional activity.

The microsatellites can also be present in organellar genomes, such as chloroplast and mitochondria, and nuclear DNA. Powell et al. (1995) provided experimental evidence of length variation in the mononucleotide repeats of the chloroplast genome of angiosperms, and polymorphisms within these regions might be used to study both intraspecific and interspecific variability. Soranzo et al. (1999) was the first to show length variation at a mitochondrial SSR locus in conifers.

Knowledge of the complete genome sequence of many species in the public domain now permits the determination of SSR frequencies at the whole genome level, decreases the economic limitations and accelerates the process of SSR analysis. The accessibility and data analysis of microsatellite content in whole genome sequences would also facilitate comprehensive studies on the direct role of microsatellites in genome organization, recombination, gene regulation, quantitative genetic variation and the evolution of genes (Katti et al., 2001). The density analyses of SSRs in fully sequenced eukaryotic genomes showed a higher density in mammals and the initial analysis of the human genome sequence concluded that approximately 3% of all DNA is represented by SSRs. The human genome is estimated to contain on an average 10-fold more microsatellites than plant genomes (Powell et al., 1996). The analyses of microsatellite distribution in the genomes of many species revealed that compared with *Drosophila, Arabidopsis, Caenorhabditis elegans* and yeast, human chromosomes 21 and 22 are rich in mono- and tetranucleotide repeats. *Drosophila* chromosomes have higher frequencies of di- and trinucleotide repeats and, surprisingly, the *C. elegans* genome contains less SSRs per million base pairs of sequence than the yeast genome (Katti et al., 2001).

3. Isolation and analysis

3.1 Isolation

Since the first studies using microsatellites were performed, the methods of SSR loci isolation have been improved and several protocols were published. There are published reviews concerning this topic (Weising et al., 2005; Zane et al., 2002), but with the recent development of technology and evolution of methodology, new methods and modifications have been proposed.

The published microsatellite isolation protocols can be grouped into three types: (i) the standard method, where a library is screened for repeated sequences; (ii) the automated method, where the SSR sequences are searched in sequence databases and (iii) the sequencing method, where the whole genome or parts of the genome are sequenced using high-throughput technologies. Each of these methods was modified and optimized to many species and conditions, generating a large number of protocols. Here, we will present an overview of the commonly used protocols.

3.1.1 Standard method

This method requires the creation of a library. There are various protocols to create and screen a genomic, cDNA or PCR fragment library [revised by Mittal & Dubey (2009) and Weising et al. (2005)], but the main steps can be summarized as follows:

1. The DNA is fragmented by sonication or enzymatic digestion.
2. The DNA fragments are ligated into a vector and transformed into *Escherichia coli*.
3. The clones are analyzed for the presence of SSR sequences by Southern blot. Then, the positive clones are sequenced.

The number of positive clones obtained by this methodology ranges from 0.04 to 12%, with the lowest yields occurring in birds (Zane et al., 2002). These protocols are efficient; however, the cost of developing a microsatellite marker is high because the use of a total genomic DNA library requires the evaluation of a large number of clones to find those containing repeated sequences. Ito et al. (1992) proposed the use of a biotinylated oligonucleotide to screen the plasmids of a restriction fragment library. The oligonucleotide and plasmid interact to form a triple helix, and the positive clones could be recovered using streptavidin coated magnetic beads. Subsequently, the microsatellite-enriched plasmids are purified and transformed into *E. coli*. However, this technique is limited to sequence motifs that are capable of triple helix formation (such as GA- and GAA-repeats).

Another technique to increase the number of positive clones or enrich the libraries relies on the extension of the library of single-stranded genomic DNA using repeat specific primers. For example, Paetkau (1999) amplified genomic libraries using biotinylated oligonucleotides, which were complementary to the microsatellite sequence, as primers. The single-stranded biotinylated sequences were recovered with streptavidin bound to magnetic particles, made double-stranded and transformed into *E. coli*. In this case, the enrichment efficiency was 100% for the dinucleotide $(CA)_{18}$. However, the enrichment efficiency depends on the size of the genomic library.

The most popular enrichment methods for SSR sequences are based on hybridization selection (Weising et al., 2005). Therefore, the following steps are added after DNA fragmentation:

1. The DNA fragments are ligated to adapters and amplified by PCR.
2. The PCR products are hybridized to microsatellite sequences that are attached to nylon membranes or biotin, and the hybrid sequences are eluted from the membrane or recovered via streptavidin-coated magnetic beads.
3. The selected PCR products are ligated into a vector and transformed into *E. coli*.

Researchers using hybridization selection have reported up to 80% of clones containing a microsatellite. Using two rounds of amplification and hybridization with biotin/streptavidin, Kandpal et al. (1994) generated a high enrichment efficiency of approximately 90% for CA repeats.

Yue et al. (2009) described another method to enrich microsatellite libraries. These authors applied a duplex-specific nuclease to normalize a pool of cDNA prior to cloning and generated 30 times more positive clones as compared with direct sequencing methods. Recently, Santana et al. (2009) and Malausa et al. (2011) applied pyrosequencing to enriched DNA libraries of many species and demonstrated that this methodology is more rapid, effective and economical than others.

3.1.2 Automated method

Microsatellite identification and development is also made possible through the use of public DNA databases to search for repeated sequences. Initially, database searches were performed using unspecific alignment tools, such as BLASTN (Altschul et al., 1990). Subsequently, several computer-based software programs were developed and the SSR search became easier. Mittal & Dubey (2009) reported a list of programs, their applications and references. Because microsatellites located in expressed sequences are more conserved and gene related, many studies have described and applied EST-SSRs [as reviewed by Varshney et al. (2005) for plants].

This automated approach reduces the costs associated with microsatellite marker development but is limited to species with available sequences.

3.1.3 Sequencing method

The new high-throughput sequencing technologies have allowed whole or expressed genome sequencing (Abdelkrim et al., 2009; Mikheyev et al., 2010). These technologies do not require the creation of libraries (total DNA or RNA can be sequenced), produce a huge amount of sequences quickly and because many steps have been skipped, have lower costs than other methods.

Following the isolation of microsatellite sequences, it is necessary to develop PCR primer pairs flanking these sequences to test new loci for robust amplification, genomic copy number and sufficient polymorphism. Arthofer et al. (2011) reviewed published research concerning microsatellite isolation and showed that approximately half of all loci were lost due to inconsistent PCR amplification, multicopy status in the genome or monomorphism, regardless of the isolation strategy used. Moreover, these authors demonstrated the applicability of high-resolution melting (HRM) analyses to screen candidate loci for marker development, reducing the costs of traditional tests.

3.2 Analyses

In microsatellite loci analyses, variations in the amplification product size are related to the number of repeated motifs and would indicate the polymorphism level of that specific locus in a population. There are many protocols to amplify and detect microsatellite loci variation. Weising et al. (2005) described the most frequently used methods.

The protocol choice depends on the availability of equipment and reagents and the desired accuracy of the polymorphism detection. Agarose gels stained with ethidium bromide are easy to handle and are one of the cheapest protocols but do not allow precise fragment size determination. However, one of the most accurate methods requires an automated sequencer and fluorescent-labeled primers. The combined use of multiplex reactions (with primers labeled with different fluorochromes) with capillary DNA sequencers allow high-precision genotyping and high-throughput.

Regardless of the electrophoretic technique chosen to determine the banding pattern of the amplified fragments, the next step is statistical analysis. Molecular markers with known band sizes are usually added to electrophoresis gels to estimate the fragment size.

There are several methods and computer programs that can be used in data analysis, depending on the final application. Excoffier & Heckel (2006), Labate (2000) and Weising et al. (2005) reviewed many of them and summarized their main applications. Several statistical analyses are based on genetic distances, and as a first step, the pairwise similarity is quantified. Most commonly, the similarity index is calculated from band sharing data and the complement to this index is the genetic distance between the samples (Weising et al., 2005). When large number of samples are involved, it is difficult to interpret genetic distances. In these instances, the use of ordination, clustering and dendrograms condenses the differences into fewer characters and permits the visualization of these entries in a multidimensional space (Weising et al., 2005).

Unfortunately, most of the computer programs use a specific data file format, but there are several that can read or write data from, or to, other file formats. It is essential to avoid the limitation of a single program or having to reformat the data manually. Excoffier & Heckel (2006) identified two conversion programs considered as starting points for formatting input data files: Convert (Glaubitz, 2004) and Formatomatic (Manoukis, 2007). These programs can create input files for several other formats.

A critical point in data analyses is that most computer programs conceal the mathematical complexities from the user, but they rely on crucial assumptions that should be taken in account for the correct interpretation of the results (Excoffier & Heckel, 2006).

4. Transferability

Microsatellites are transferable because their flanking regions are highly conserved across taxa, allowing cross-species amplification, i.e., primers developed in one species can be used in others of the same genus or family, especially for vertebrates, such as fishes, reptiles and mammals (Peakall et al., 1998; Rico et al., 1996). The transferability of SSRs derived from EST databases (EST-SSR) is greater than that of SSRs derived from enriched genomic DNA libraries. The EST-SSRs originate from expressed regions, and therefore, they are more conserved across a number of related species than non-coding regions (Varshney et al., 2005).

Many researchers have studied the transferability of SSRs. Zhao et al. (2011) showed the high transferability (86%) of *Brachypodium* SSR markers to *Miscanthus sinensis*. Moreover, 18

(31%) of the transferable markers produced perfect polymorphic and easy-scoring bands; consequently, this study confirms the significance of *Brachypodium* as a model plant for *Miscanthus*. Faria et al. (2010) used *Eucalyptus* EST databases to develop, select and conduct a detailed characterization of a novel set of 20 microsatellite markers that are polymorphic and transferable across 6 of the major *Eucalyptus* plant species. The primers were developed from more conserved transcribed regions; therefore, the transferability and polymorphism of these microsatellites likely extended to the other 300 or more species within the same subgenus *Symphyomyrtus*, further highlighting their applied value for *Eucalyptus* genetics and breeding. Pépin et al. (1995) showed that an estimated 40 per cent of the microsatellites isolated from cattle were useful to study the caprine genome and characterize economically important genetic loci in this species. Moreover, bovine microsatellites were shown to be useful tools for the study of the genetic diversity of Artiodactyla. Dawson et al. (2010) developed primer sets for 33 polymorphic loci that are highly useful in the study of passerine, shorebirds and other non-passerine birds and for genotyping in species belonging to the Passeridae and Fringillidae families.

5. Evolution and mutation models

Microsatellites have a wide variety of applications in life sciences. In addition, these markers are related to several human neurodegenerative diseases and have demonstrated roles in regulating transcription and expression of various genes. Despite the great interest in the functions of these sequences and their applicability as molecular markers, knowledge about the mutational and evolutionary dynamics of microsatellites is still controversial.

The methods used for studying evolution of microsatellites involve pedigree analysis, sequence structure analysis of the alleles within species, sequence comparison of orthologous loci in related species and analysis of microsatellite instability through cloning and maintenance of sequence *in vivo* (Ellegren, 2004). More recently, data from complete genomes coupled with bioinformatics analysis has helped researchers to understand the distribution and variability of microsatellites in genomes.

5.1 Origin of microsatellite

The origin of microsatellites in genomes appears to be nonrandom, with an imbalance between the mechanisms that promote and those that prevent the microsatellites initiation (Bhargava & Fuentes, 2010). Currently, there are two non-mutually exclusive hypotheses to explain the origin of microsatellites:

-   *De novo* microsatellites (Messier et al., 1996) - suggests that the birth of microsatellites was a consequence of the creation of a proto-microsatellite, a short region of as few as 3 or 4 repeated units within cryptically simple sequences, which are defined as a scramble of repetitive motifs lacking a clear tandem arrangement. Proto-microsatellites were originated from base substitutions or indel events; the latter is supported by the observation that insertions tend to copy adjacent bases (Buschiazzo & Gemmell, 2006). Once a 'proto-microsatellite' is initiated, maintenance and multiplication is favored by its propensity to undergo strand slippage during replication and, depending primarily on the repeat motif, its capacity to form unusual DNA conformations and participate in recombination and transposition events. The number of repeat units correlates positively with the mutability of the microsatellite, but the minimum repeat number required for strand slippage or other mechanisms of microsatellite mutation is debatable (Jentzsch et al., 2008).

- Adopted microsatellites (Wilder & Hollocher, 2001) – suggests that microsatellites arise from other genomic regions via transposable elements. The transposable elements might contain one or more sites that are predisposed to microsatellite formation and hence favor the dispersal of microsatellites in genomes. Transposable elements can be divided into two main classes based on their mechanisms of movement: class I (retrovirus-like transposons) and class II (so called cut and paste transposons). Both of these elements can leave traces of their presence and movement during the transposition process across DNA sequences, which resemble microsatellites, especially poly A arrays. A poly A tail is added to the 3′ end of class I retrotransposons after mRNA transcription, which then gets inserted together with the transposed sequence into the new position. Retrotransposons can also contain other microsatellite-like stretches, dinucleotide and tetranucleotide repeats, within their sequences. Class II transposons preferentially insert into certain DNA sequences, which can be either inverted repeats or tandem repeat sequences. This suggests a reciprocal association in which microsatellites act as 'retroposition navigator sequences,' while retrotransposons generate more microsatellites during their dispersion throughout the genome. An example of a retrotransposon-mediated microsatellite genesis in humans is the origin of A/T rich microsatellites with motifs ranging from one to six nucleotides in length from Alu elements (Jentzsch et al., 2008).

5.2 Evolutionary dynamics

Microsatellites are highly mutable as compared with point mutations in coding genes and mutation rates range from 10^{-6} to 10^{-2} events per locus per generation. These rates are highly affected by multiple factors, which influence both the probability of mutations generation and the repair efficiency of these mutations. Mutation mechanisms, DNA repair, structure and characteristics of microsatellite, genomic and individual context and selective biological influences are factors that interact and control the evolutionary dynamics of microsatellites.

5.2.1 Mutation mechanisms and DNA repair

Currently, two mechanisms have been proposed as mutation models in microsatellites: (i) replication slippage and (ii) unequal crossing over during meiosis. The mechanism of DNA replication slippage is most widely observed in microsatellites.

Replication slippage - DNA slippage is a symmetrical process, where the same number of repeats are added and removed. This process inevitably leads to either the loss of microsatellites or the insertion of a high number of repeats (Schlötterer, 2000). The misalignment that gives rise to mutations occurs between a newly synthesized DNA strand and its complementary template strand. The two strands dissociate and reanneal incorrectly, forming a loop, which is stable due to the repetitive nature of the sequence. If the loop is formed on the nascent strand, the resulting mutation will be a repeated expansion, while loops on the template strand result in a reduction of the repeat length (Jentzsch et al., 2008).

Unequal crossing over during meiosis (recombination) - this mechanism is usually associated with the exchange of repeated units between homologous chromosomes, and therefore, plays a limited role in microsatellite mutation. However, this mechanism might be responsible for microsatellite multistep mutations (Grover & Sharma, 2011).

The relative rates of point mutations and slippage might be altered by changes in the efficiency of MMR (mismatch repair) and proofreading. Failure of the MMR system during replication results in a 10^{-3}-fold increase in microsatellite instability (Strand et al., 1993). These proteins govern the balance between enrichment and prevention of microsatellites within genomes. In a given species, MMR proteins play a role in the mutational variability among alleles, loci and individuals, and because they are driven by selective forces, are certainly the cause of differential allele distributions between species. Moreover, the proteins involved in MMR can vary in number and nature among eukaryotes, suggesting variability in their intrinsic efficiency (Buschiazzo & Gemmell, 2006).

5.2.2 Structure characteristics of microsatellite, genomics and biological contexts

Mutation rates might vary greatly among loci and alleles of the same locus depending of the structure of the microsatellite (length and number of repeat units, interruptions within the motif, motif nucleotide composition, motif length and flanking sequences).

Length of microsatellite - mutation increases with the increasing number of repeat units and this is presumable because the more repeating units, the more opportunities for replication slippage to occur. Therefore, loci with a large number of repeats are more polymorphic (Ellegren, 2004).

Interruptions - point mutations and other interruptions within the repeat reduce the mutation rate. Any mutation within the repeated region that causes an interruption will split the original repeat into two shorter units, which would increase locus stability by reducing the substrate for polymer slippage (Bhargava & Fuentes, 2010).

Motif nucleotide composition - repeats with certain motifs have a heightened propensity to form secondary structures and alter DNA structure. Secondary structures, such as hairpins, quadruplex structures, H-DNA or sticky DNA are intermediate DNA hybrid forms that increase the likelihood of strand misalignment and subsequent polymerase slippage. A conformational change in the DNA structure, such as Z-DNA formed by long AC tracts, will affect both polymerases and repair enzymes (Jentzsch et al., 2008).

Motif length - dinucleotide repeats have the highest mutation rate, followed by tri- and tetranucleotide repeats. Shorter microsatellites motifs allow more opportunities for misalignment, whereas motifs longer than three nucleotides require higher dissociation energy and are thus less likely to generate enough single-stranded DNA to form a stable loop. Moreover, motif length can affect MMR efficiency. If the loop is too large, the efficiency of MMR is reduced (Buschiazzo & Gemmell, 2006; Jentzsch et al., 2008).

Flanking sequence - the mutability of microsatellites greatly depends on the genomic constitution of their flanking sequences. Large scale mutation of a sequence that contains or flanks a microsatellite will modify the genomic context of the microsatellite and may change the mutability of the locus (Bhargava & Fuentes, 2010).

The influence of genomic context on the mutation rate of microsatellites becomes clear when the effect of the mutations has a high probability of being disadvantageous and is strongly counteracted by selection, i.e., where the distribution of microsatellites in coding regions is observed (Buschiazzo & Gemmell, 2006; Jentzsch et al., 2008).

The mode of reproduction, metabolic rate and generation time could also influence the mutational dynamics of microsatellites at the species level. It has been reported that the sex and age of some organisms could also influence the mutation rate of microsatellites. For example, men have more cell divisions than women for the production of gametes and it is therefore expected that the microsatellites would undergo more mutations per generation (Buschiazzo & Gemmell, 2006)

5.3 Mutation models

A mutation model of microsatellites evolution is needed for the estimation of population parameters, such as number of migrants, population structure and effective population size. A wide range of models has been proposed to explain the mutational dynamics of microsatellites and some of them are discussed below. For more details see Balloux & Lugon-Molin (2002) and Bhargava & Fuentes (2010).

Infinite Allele Model (IAM) – This model was first described by Kimura & Crow (1964) and assumes that every mutation results in the creation of a new allele. This model does not allow for homoplasy; identical alleles share the same ancestry and are identical-by-descent, which best describes the unusual dynamics of compound/complex microsatellites.

K-Allele model (KAM) – This model was developed by Crow & Kimura (1970) and assumes that there are K possible allelic states and any allele has a constant probability of mutating towards any of the other K-1 allelic states. This model treats all alleles as equivalents with the potential to mutate from one allele to any other allele and allows homoplasy, which is more suitable for data where the mutation pattern is unknown.

Stepwise mutation model (SMM) – This model was developed by Otha & Kimura (1973) and assumes that each mutation creates a novel allele by either adding or deleting a single microsatellite repeated unit. Alleles of very different sizes will be more distantly related than alleles of similar sizes (memory of allele size); this model is often used when estimating relatedness between individuals and population sub-structuring, except when homoplasy is present.

Two-phase-model (TPM) – This model was developed by Di Rienzo et al. (1994) and is an extension of the SMM, which allows for infrequent multistep mutations; one-step mutations are more likely to occur and follow the SMM, whereas the magnitude of multistep mutations follows a truncated geometric distribution.

6. Applications

Due to all of the previously discussed features, microsatellites have been a class of molecular markers chosen for diverse applications. In this review, the SSR applications will be summarized into four categories: (i) genome mapping and marker-assisted selection, (ii) genetic diversity and individual identification, (iii) population and phylogenetic relationships and (iv) bioinvasion and epidemiology.

6.1 Genome mapping and marker-assisted selection

Genome mapping includes genetic, comparative, physical and association mapping. Genetic mapping is one of the major research fields in which microsatellites have been applied because they are highly polymorphic and require a small amount of DNA for each test. Linkage maps are known as recombination maps and define the order and distance of loci along a chromosome on the basis of inheritance in families or mapping populations (Chistiakov et al., 2006). Association mapping links a locus to a phenotypic trait and comparative mapping aligns chromosome fragments of related species based on genetic mapping to trace the history of chromosome rearrangements during the evolution of a species (Wang et al., 2009a). However, in physical mapping, markers anchor large pieces of DNA fragments, such as bacterial artificial clones (BACs), and provide the actual physical distance between the markers (Wang et al., 2009a). Apotikar et al. (2011) constructed a SSR-based skeleton linkage map of two linkage groups of sorghum in a population of 135 recombinant inbred lines derived from a cross between IS18551 (resistant to shoot fly) and 296B (susceptible to shoot fly) varieties. The authors found 14 markers that were mapped to each linkage group and three quantitative trait loci (QTL) governing more than one trait (pleiotropic QTLs). The identification of genomic regions/QTLs that influence resistance can help breeders to introgress them into the breeding lines using the linked molecular markers. Baranski et al. (2010) analyzed the flesh color and growth related traits in salmonids with 128 informative microsatellite loci, distributed across all 29 linkage groups, in individuals from four F2 families. Chromosomes 26 and 4 presented the strongest evidence for significant QTLs that affect flesh color, while chromosomes 10, 5 and 4 presented the strongest evidence for significant QTLs that affect growth traits (length and weight). These potential QTLs provide a starting point for further characterization of the genetic components underlying flesh color and growth in salmonids and are strong candidates for marker-assisted selection.

The use of the markers to indicate the presence of a gene (trait) is the basis for marker-assisted selection (MAS). Therefore, the construction of high-density and high-resolution genetic maps is necessary to select for markers that are tightly linked to the target locus (gene) (Chistiakov et al., 2006). Once a linkage is established between a locus and the gene of interest, the inheritance of the gene can be traced, which could greatly enhance the efficiency of breeding programs (Wang et al., 2009a).

6.2 Genetic diversity and individual identification

Genetic diversity refers to any variation in nucleotides, genes, chromosomes or whole genomes of organisms (Wang et al., 2009a). Genetic diversity can be assessed among different accessions/individuals within same species (intraspecific), among species (interspecific) and between genus and families (Mittal & Dubey, 2009).

Even crops with advanced studies in genomics (e.g., rice, corn, soybean and apple) have been recently evaluated by SSRs to access the genetic diversity. As mentioned previously, large-scale screening requires low-cost technologies. In a recent publication, Ali et al. (2011) evaluated the genetic and agro-morphological diversity of rice (*Oryza sativa*) among subpopulations and their geographic distribution. A selection of 409 Asian landraces and cultivars were chosen from 79 countries representing all of the major rice growing regions of the world. This rice diversity panel with the accompanying genetic and phenotypic information provides a valuable foundation for association mapping and understanding the basis of both genotypic and phenotypic differences within and between subpopulations.

Microsatellite markers have also been used for plants with poor genomic knowledge. For example, in an interspecific analysis, Hoshino et al. (2006) evaluated 76 accessions of 34 species from nine *Arachis* sections and showed that this germplasm bank possessed high variability, even when a species was represented by few accessions. This information was used to maintain *Arachis* genetic diversity during the storage and conservation process. Beatty & Provan (2011) published research utilizing intraspecific analysis through SSR markers. These authors assessed the genetic diversity of glacial and temperate plant species, respectively *Orthilia secunda* (one-sided wintergreen) and *Monotropa hypopitys* (yellow bird's nest). In this case, microsatellites were extremely useful to evaluate biogeographical distributions and the impact of changes in the species ranges on total intraspecific diversity. These authors concluded the following: "given that future species distribution modeling suggests northern range shifts and loss of suitable habitat in the southern parts of the species' current distribution; extinction of genetically diverse rear edge populations could have a significant effect in the range wide intraspecific diversity of both species, but particularly in *M. hypopitys*".

The great variability detected by microsatellites could be used to identify a person, a cultivar or a population. A set of SSR markers could be selected for each species/situation to distinguish one cultivar/genotype from all others. This practice is employed to protect the intellectual property rights of new varieties by commercial companies (Wang et al., 2009a). It is also used in paternity testing, when a progeny inherits one allele from the male parent and another allele from the female parent (Chistiakov et al., 2006). The genotypic profile is highly discriminating, which suggests that a random individual would have a low probability of matching a given genotype and if only a few potential parents are being consider, paternity could be determined by exclusion (Weising et al., 2005).

"Assignment tests" (assignment of an individual to the population) can be used in forensics, conservation biology and molecular ecology. An interesting example is the study of Primmer et al. (2000), which used this approach to identify a case of fishing competition fraud. The assignment of the SSR genotype of the suspect fish to its most likely original population indicated a high level of improbability that the fish originated from Lake Saimaa (where the competition occurred). When this evidence was presented, the offender confessed purchasing the salmon at a local fish shop and criminal charges were laid.

6.3 Population and phylogenetic relationships

Microsatellite markers can be used to determine the population structure within and among populations (Wang et al., 2009a). Evaluations of population differentiation permit the estimation of the migration rate between populations, assuming that these populations are in equilibrium (e.g., no selection, identical mutation rates and generation time) (Weising et al., 2005). In plants, migration rates correspond with the gene flow through seeds and pollen (Weising et al., 2005). Microsatellite markers are a powerful system for revealing inter or intraspecific phylogenetic relationships, even in closely related species (Wang et al., 2009a). Phylogenetic relationships reflect the relatedness of a group of species based on a calculated genetic distance in their evolutionary history. Genomic SSRs, specifically EST-SSR markers, are the best choice for cross-species phylogenetics (Mittal & Dubey, 2009). However, the high incidence of homoplasy increases evolutionary distances and might undermine the confidence of the phylogenetic hypotheses, compromise the accuracy of the analysis and limit the depth of the phylogenetic inference (Jarne & Lagoda, 1996). Another problem with SSR-based phylogenetic inference is that primer transferability might not work well in all taxa and even when it is possible to amplify, the sequences might not be similar enough to permit a confident orthology assessment. Flanking regions of microsatellites have also been used in phylogenetic relationships between species and families because they evolve more slowly than repeated sequences (Chistiakov et al., 2006).

Microsatellites have been used successfully in some phylogenetic cases. Using EST-SSR markers derived from *Medicago*, cowpea and soybean, the genetic diversity of the USDA *Lespedeza* germplasm collection was assessed and its phylogenetic relationship with the genus *Kummerowia* was clarified (Wang et al., 2009b), despite the fact that phylogenetic analysis with morphological reexamination provides a more complete approach to classify accessions in plant germplasm collection and conservation. Orsini et al. (2004) used a set of 48 polymorphic microsatellites derived from *Drosophila virilis* to infer phylogenetic relationships in the *D. virilis* clade and found results consistent with previous studies (*D. virilis* and *D. lummei* were the most basal group of the species). Furthermore, these authors detected differentiations between *D. americana texana*, *D. americana americana* and *D. novamexicana* that were previously supported by FST analyses and a model-based clustering method for multilocus genotype data. Rout et al. (2008) assessed the phylogenetic relationships of Indian goats using 17 microsatellite markers. Breeds were sampled from their natural habitats, covering different agroclimatic zones. Analyses showed that the results of the microsatellite analysis were consistent with mitochondrial DNA data, which classifies Indian goat populations into distinct genetic groups or breeds. The phylogenetic and principal component analysis showed the clustering of goats according to their geographical origin. The authors concluded that although the goat breeding tracts overlapped and spread countrywide, they still maintain genetic distinctions while in their natural habitats.

In the scope of biodiversity conservation and evolutionary genetics, microsatellites have been used to contribute accurate information on issues of population dynamics, demography and ecological/biological factors intrinsic to species and populations. As examples of this approach, we can cite Palstra et al. (2007) and Becquet et al. (2007). Palstra et al. (2007) examined the population structure and connectivity of Atlantic salmon (*Salmo salar*) from Newfoundland and Labrador, which are regions where populations of this species are relatively pristine. Using the genetic variation of 13 microsatellite loci from samples (n=1346) collected from a total of 20 rivers, the connectivity at several regional and temporal scales was verified, and the hypothesis that the predominant direction of the gene flow is from large into small populations was tested. However, this hypothesis was rejected by evidence that the temporal scale in which gene flow is assessed affects the directionality of migration. Whereas large populations tend to function as sources of dispersal over contemporary timescales, such patterns are often changed and even reversed over evolutionary and coalescent-derived timescales. Furthermore, these patterns of population structure vary among different regions and are compatible with demographic and life-history attributes. No evidence for sex-biased dispersal underlying gene flow asymmetry was found. These results are inconsistent with generalizations concerning the directionality of the gene flow in Atlantic salmon and emphasize the necessity of detailed regional study, if such information is to be meaningfully applied in conservation and management of salmonids.

Becquet et al. (2007) used 310 microsatellite markers genotyped in 78 common chimpanzees and six bonobos, allowing a high-resolution genetic analysis of chimpanzee population structure. These chimpanzees have been traditionally classified into three populations: western, central and eastern. While the morphological or behavioral differences are small, genetic studies of mitochondrial DNA and the Y chromosome have supported the geography-based designations. The findings showed that the populations seem to be discontinuous and provided weak evidence for gradients of variation reflecting hybridization among chimpanzee populations. In addition, the results demonstrated that central and eastern chimpanzees are more closely related to each other in time than to western chimpanzees.

6.4 Bioinvasions and epidemiology

The analysis of genetic diversity, population structure and demographic inferences using microsatellite has been useful to elucidate the processes of bioinvasion, understand the epidemiological patterns and aid in controlling and eradicating diseases.

The characterization of the genetic structure of invasive populations is important because genetically variable populations tend to be more successful as invaders than those that are relatively genetically homogeneous, and genetic data might provide an important tool to resource managers concerned with invasion risk assessments and predictions. To examine the invasion genetics of the Eurasian spiny water flea, *Bythotrephes longimanus*, which is a predacious zooplankter with increased range in Europe that is rapidly invading inland waterbodies throughout North America's Great Lakes region, Colautti et al. (2005) employed microsatellite markers. Three populations where *B. longimanus* has been historically present (Switzerland, Italy and Finland), a European-introduced population (the Netherlands) and three North American populations (Lakes Erie, Superior and Shebandowan) were sampled. Consistent with a bottleneck during colonization, the average heterozygosities of the four European populations were higher than the three North American populations. The pairwise FST estimated among North American populations was not significantly different from zero and was much lower than that among European populations. This result is consistent with a scenario of higher gene flow among North American populations. The assignment tests identified several migrant genotypes in all introduced populations (the Netherlands, Erie, Superior and Shebandowan), but rarely in native ones (Switzerland, Italy and Finland). A large number of genotypes from North America were assigned to Italian populations, suggesting a second invasion in the region of northern Italy that was previously unidentified. These results support a bottleneck in the invasion of North American populations that has been largely offset by the gene flow from multiple native sources and among introduced populations.

Microsatellites have also been chosen to evaluate the genetic variability and dynamics of the invasion of *Ambrosia artemisiifolia*, an aggressive North American annual weed, found particularly in sunflower and cornfields. Besides its economic impact on crop yield, this plant represents a major health problem because of strongly allergenic pollen. The results of Genton et al. (2005) suggested that the French invasive populations include plants from a mixture of sources. The reduced diversity in populations distant from the original introduction area indicated that ragweed range expansion probably occurred through sequential bottlenecks from the original populations and not from subsequent new introductions.

Understanding the epidemiology of the disease is related to knowledge about the basic biology of the organisms involved. Population genetic studies can provide information about the taxonomic status of species, the spatial limits of populations and the nature of the gene flow among populations. Examples of the important results in this approach are Pérez de Rosas et al. (2007) and Fitzpatrick et al. (2008). Pérez de Rosas et al. (2007) examined the genetic structure in populations of the Chagas disease vector, *Triatoma infestans*. Levels of genetic variability (assessed by microsatellites) were compared in populations of *T. infestans* from areas with different periods after insecticide treatment and from areas that never received treatment. These authors found that genetic drift and limited gene flow appear to have generated a substantial degree of genetic differentiation among the populations of *T. infestans* and the microgeographical analysis supports the existence of subdivision in *T. infestans* populations. Levels of genetic diversity in the majority of *T. infestans* populations from insecticide-treated localities were similar or higher than those detected in populations from areas without treatment. This study supports the hypothesis of vector population recovery from survivors of the insecticide-treated areas, and therefore highlights the value of population genetic analyses in assessing the effectiveness of the Chagas disease vector control programs. Fitzpatrick et al. (2008) investigated the identity of silvatic *Rhodnius* (vector of Chagas' disease) using sequencing and microsatellites and whether silvatic populations of *Rhodnius* are isolated from domestic populations in Venezuela. Sequencing confirmed the presence of *R. prolixus* in palms and that silvatic bugs can colonize houses. The analyses of microsatellites revealed a lack of genetic structure between silvatic and domestic ecotopes (non-significant FST values), which is indicative of unrestricted gene flow. These results demonstrate that silvatic *R. prolixus* presents an unquestionable threat to the control of Chagas disease in Venezuela.

7. Limitations

Although microsatellites have many advantages over other molecular markers, all data sets might include some errors and genotyping errors remain a subject in population genetics because they might bias the final conclusions (Bonin et al., 2004). Microsatellite genotyping errors result from many variables (reagent quality, *Taq* polymerase error or contamination), as reviewed by Pompanon et al. (2005), and the primary consequence is the misinterpretation of allele banding patterns.

Microsatellite markers are mainly limited by:

1. Null alleles: locus deletion or mutations in the annealing primer site prevent locus amplification and heterozygous identification and lead to erroneous estimations of allele frequencies and segregation rates. Primer redesign might resolve this problem.

2. Homoplasy: alleles identical in state (length) but not by descent are homoplasic alleles (Jarne & Lagoda, 1996). They can be identical in length but not in sequence or identical in length and sequence but with different evolutionary history (Anmarkrud et al., 2008). Because homoplasy is disregarded, the actual divergence between populations is underestimated. Sequencing could be used to identify differences in sequences, but differences in evolutionary history can only be identified by mutations documented in known pedigrees.

3. Linkage disequilibrium: deviations from the random association of alleles in a population, which are primarily caused by population substructuring and high levels of inbreeding (Weising et al., 2005). It is especially problematic for population studies and paternal exclusion. Computer programs or an offspring analysis could detect the problems.

The error impact depends on the data application. In population genetic analyses, homoplasy is not a significant problem (Estoup et al., 2002), except for hypermutable markers that have increased slippage rates (Anmarkrud et al., 2008). However, error rates as low as 0.01 per allele resulted in a rate of false paternity exclusion exceeding 20%, making even modest error rates strongly influential (Hoffman & Amos, 2005). There are a lot of informatics tools that account for genotyping errors (listed by Pompanon et al., 2005). These authors also proposed a protocol for estimating error rates that should be used to attest the reliability of published genotyping studies.

8. Perspectives

The utilization of microsatellites has been demonstrated by a large number of studies applying this marker and by the variety of areas that apply microsatellites for several purposes. Furthermore, novel technologies have enabled the development of markers for previously neglected species through the generation of new sequences and a more refined search in databases.

Nevertheless, there are some bottlenecks that need to be overcome as they hamper the best and widespread use of SSR data, e.g., an exchangeable data format to allow users to access different kinds of analyses and computer programs easily (Excoffier & Heckel, 2006) and the best understanding about microsatellite evolution and mutation mechanisms.

9. Acknowledgment

We are grateful to the Brazilian funding agencies (CAPES, CNPq and FAPESP) for their financial support. PMN is the recipient of a FAPESP postdoctoral fellowship (2010/08238-0).

10. References

Abdelkrim, J.; Robertson, B.C.; Stanton, J.A.L. & Gemmell, N.J. (2009). Fast, cost-effective development of species-specific microsatellite markers by genomic sequencing. *BioTechniques*, Vol.46, No.3, (March 2009), pp. 185-192, ISSN 0736-6205.

Ali, M.L.; McClung, A.M.; Jia, M.H.; Kimball, J.A.; McCouch, S.R. & Eizenga, G.C. (2011). A Rice Diversity Panel Evaluated for Genetic and Agro-Morphological Diversity between Subpopulations and its Geographic Distribution. *Crop Science*, Vol.51, No.5, (September–October 2011), pp. 2021-2035, ISSN 1435-0653.

Altschul, S.F.; Gish, W.; Miller, W.; Myers, E.W. & Lipman, D.J. (1990). Basic local alignment search tool. *Journal of Molecular Biology*, Vol.215, No.3, (October 1990), pp. 403-410, ISSN 0022-2836.

Anmarkrud, J.A.; Kleven, O.; Bachmann, L. & Lifjeld, J.T. (2008). Microsatellite evolution: Mutations, sequence variation, and homoplasy in the avian microsatellite locus HrU10. *BMC Evolutionary Biology*, Vol.8, (May 2008), pp. 138, ISSN 1471-2148.

Apotikar, D.B.; Venkateswarlu, D.; Ghorade, R.B.; Wadaskar, R.M.; Patil, J.V. & Kulwal, P.L. (2011). Mapping of shoot fly tolerance loci in sorghum using SSR markers. *Journal of Genetics*, Vol.90, No.1, (April 2011), pp. 59-66, ISSN 0973-7731.

Arthofer, W.; Steiner, F.M. & Schlick-Steiner, B.C. (2011). Rapid and cost-effective screening of newly identified microsatellite loci by high-resolution melting analysis. *Molecular Genetics and Genomics*,Vol.286, No.3-4, (August 2011), pp. 225-235, ISSN 1617-4623.

Balloux, F. & Lugon-Moulin, N. (2002). The estimation of population differentiation with microsatellite markers. *Molecular Ecology*, Vol.11, No.2, (February 2002), pp. 155–165, ISSN 1365-294X.

Baranski, M.; Moen, T. & Vage D.I. (2010). Mapping of quantitative trait loci for flesh colour and growth traits in Atlantic salmon (*Salmo salar*). *Genetics Selection Evolution*, Vol.42, (June 2010), pp. 17, ISSN 1297-9686.

Beatty, G.E. & Provan, J. (2011). Comparative phylogeography of two related plant species with overlapping ranges in Europe, and the potential effects of climate change on their intraspecific genetic diversity. *BMC Evolutionary Biology*, Vol.11, (January 2011), pp. 29, ISSN 1471-2148.

Becquet, C.; Patterson, N.; Stone, A.C.; Przeworski, M. & Reich, D. (2007). Genetic Structure of Chimpanzee Populations. *PLoS Genetics*, Vol.3, No.4, (April 2007), pp. 617-626, ISSN 1553-7404.

Bhargava, A. & Fuentes, F.F. (2010). Mutational Dynamics of Microsatellites. *Molecular Biotechnology*, Vol.44, No.3, (March 2010), pp. 250–266, ISSN 1559-0305.

Bonin, A.; Bellemain, E.; Bronken Eidesen, P.; Pompanon, F.; Brochmann, C. & Taberlet, P. (2004). How to track and assess genotyping errors in population genetics studies. *Molecular Ecology*, Vol.13, No.11, (November 2004), pp. 3261–3273, ISSN 1365-294X.

Bravo, J.P.; Hoshino, A.A.; Angelici, C.M.L.C.D.; Lopes C.R. & Gimenes, M.A. (2006). Transferability and use of microsatellite markers for the genetic analysis of the germplasm of some Arachis section species of the genus Arachis. *Genetics and Molecular Biology*, Vol.29, No.3, pp. 516-524, ISSN 1415-4757.

Buschiazzo, E. & Gemmell, N.J. (2006). The rise, fall and renaissance of microsatellites in eukaryotic genomes. *Bioessays*, Vol.28, No.10, (September 2006), pp. 1040–1050, ISSN 1521-1878.

Chistiakov, D.A.; Hellemans, B. & Volckaert, F.A.M. (2006). Microsatellites and their genomic distribution, evolution, function and applications: A review with special reference to fish genetics. *Aquaculture*, Vol.255, (May 2006), pp.1-29, ISSN 0044-8486.

Colautti, R.I.; Manca, M. & Viljanen, M. (2005). Invasion genetics of the Eurasian spiny waterflea: evidence for bottlenecks and gene flow using microsattelites. *Molecular Ecology*, Vol.14, No.7, (June 2005), pp. 1869–1879, ISSN 1365-294X.

Crow, J. & Kimura, M. (1970). *An introduction to population genetics theory.* Burgess Publishing Company, ISBN 0-8087-2901-2, Minneapolis, United States of America.

Dawson, D.A.; Horsburgh, G.J.; Küpper, C.; Stewart, I.R.; Ball, A.D.; Durrant, K.L.; Hansson, B.; Bacon, I.; Bird, S.; Klein, A.; Krupa, A.P.; Lee, J.W.; Martín-Gálvez, D.; Simeoni, M.; Smith, G.; Spurgin, L.G. & Burke, T. (2010). New methods to identify conserved microsatellite loci and develop primer sets of high cross-species utility – as demonstrated for birds. *Molecular Ecology Resources*, Vol.10, No.3, (May 2010), pp. 475–494, ISSN 1755-0998.

Di Rienzo, A.; Peterson, A.C.; Garza, J.C.; Valdes, A.M & Slatkin, M. (1994). Mutational processes of simple-sequence repeat loci in human populations. *Proceedings of the National Academy of Sciences of the United States of America*, Vol.91, No.8, (April 1994), pp. 3166–3170, ISSN 0027-8424.

Ellegren, H. (2004). Microsatellites: simple sequences with complex evolution. *Nature Reviews Genetics*, Vol.5, No.6, (June 2004), pp. 435–445, ISSN 1471-005.

Estoup, A.; Jarne, P. & Cornuet, J.M. (2002). Homoplasy and mutation model at microsatellite loci and their consequences for population genetics analysis. *Molecular Ecology*, Vol.11, No.9, (September 2002), pp. 1591–1604, ISSN 1365-294X.

Excoffier, L. & Heckel, G. (2006). Computer programs for population genetics data analysis: a survival guide. *Nature Reviews Genetics*, Vol.7, No.10, (October 2006), pp. 745-758, ISSN 1471-0056.

Faria, A.D.; Mamani, E.M.C.; Pappas, M.R.; Pappas Jr, G.J. & Gratapaglia D. (2010). A Selected Set of EST-Derived Microsatellites, Polymorphic and Transferable across 6 Species of Eucalyptus. *Journal of Heredity*, Vol.101, No.4, (March 2010), pp. 512-520, ISSN 1465-7333.

Fitzpatrick, S.; Feliciangeli, D.; Sanchez-Martin, M., Monteiro, F.A. & Miles, M.A. (2008). Molecular genetics reveal that silvatic Rhodnius prolixus do colonise rural houses. *PLoS Neglected Tropical Diseases*, Vol.2, No.4, (April 2008), pp. 1-16, ISSN 1935-2735.

Genton, B.J.; Shykoff, J.A. & Giraud, T. (2005). High genetic diversity in French invasive populations of common ragweed, Ambrosia artemisiifolia, as a result of multiple sources of introduction. *Molecular Ecology*, Vol.14, No.14, (December 2005), pp. 4275-4285, ISSN 1365-294X.

Glaubitz, J.C. (2004). CONVERT: A user-friendly program to reformat diploid genotypic data for commonly used population genetic software packages. *Molecular Ecology Notes*, Vol.4, No.2, (June 2004), pp. 309-310, ISSN 1755-0998.

Grover, A. & Sharma, P.C. (2011). Is spatial occurrence of microsatellites in the genome a determinant of their function and dynamics contributing to genome evolution? *Current Science*, Vol.100, No.6, (March 2011), pp. 859-869, ISSN 0011-3891.

Hoffman, J.I. & Amos, W. (2005). Microsatellite genotyping errors: detection approaches, common sources and consequences for paternal exclusion. *Molecular Ecology*, Vol.14, No.2, (February 2005), pp. 599-612, ISSN 1365-294X.

Hoshino, A.A.; Bravo, J.P.; Angelici, C.M.L.C.D.; Barbosa, A.V.G.; Lopes, C.R. & Gimenes M.A. (2006). Heterologous microsatellite primer pairs informative for the whole genus Arachis. *Genetics and Molecular Biology*, Vol.29, No.4, pp.665-675, ISSN 1415-4757.

Ito, T.; Smith, C.L. & Cantor, C.R. (1992). Sequence-specific DNA purification by triplex affinity capture. *Proceedings of the National Academy of Sciences*, Vol.89, No.2, (January 1992), pp. 495–498, ISSN 1091-6490.

Jarne, P. & Lagoda, P.J.L. (1996). Microsatellites, from molecules to populations and back. *Trends in Ecology & Evolution*, Vol.11, No.10, (October 1996), pp. 424-429, ISSN 0169-5347.

Jentzsch, V.I.M.; Bagshaw, A.; Buschiazzo, E.; Merkel, A. & Gemmell, J.E. (2008). Evolution of microsatellite DNA. In: *Encyclopedia of Life Sciences (ELS)*, 1-12, John Wiley & Sons Ltd., ISBN 9780470015902, Chichester, United Kingdom.

Kandpal, R.P.; Kandpal, G. & Weissman, S.M. (1994). Construction of libraries enriched for sequence repeats and jumping clones, and hybridization selection for region-specific markers. *Proceedings of the National Academy of Sciences*, Vol.91, No.1, (January 1994), pp. 88–92, ISSN 1091-6490.

Kashi, Y.; King, D. & Soller, M. (1997). Simple sequence repeats as a source of quantitative genetic variation. *Trends in Genetics*, Vol.13, No.2, (February 1997), pp. 74–78, ISSN 0168-9525.

Katti, M.V.; Ranjekar, P.K. & Gupta, V.S. (2001). Differential distribution of simple sequence repeats in eukaryotic genome sequences. *Molecular Biology Evolution*, Vol.18, No.7, (July 2001), pp. 1161–1167, ISSN 1537-1719.

Kimura, M. & Crow, J.F. (1964). The number of alleles that can be maintained in a finite population. *Genetics*, Vol.49, No.4, (April 1964), pp. 725–738, ISSN 1943-2631.

Labate, J.A. (2000). Software for population genetic analyses of molecular marker data. *Crop Science*, Vol.40, No.6, (November 2000), pp. 1521-1528, ISSN 1435-0653.

Litt, M. & Luty, J.A. (1989). A hypervariable microsatellite revealed by in vitro amplification of a dinucleotide repeat within the cardiac muscle actin gene. *American Journal of Human Genetics*, Vol.44, No.3, (March 1989), pp. 397-401, ISSN 0002-9297.

Malausa, T.; Gilles, A.; Meglécz, E.; Blanquart, H.; Duthoy, S.; Costedoat, C.; Dubut, V.; Pech, N.; Castagnone-Sereno, P.; Délye, C.; Feau, N.; Frey, P.; Gauthier, P.; Guillemaud, T.; Hazard, L.; Le Corre, V.; Lung-Escarmant, B.; Malé, P.-J. G.; Ferreira, S. & Martin, J.F. (2011). High-throughput microsatellite isolation through 454 GS-FLX Titanium pyrosequencing of enriched DNA libraries. *Molecular Ecology Resources*, Vol.11, No.4, (July 2011), pp. 638–644, ISSN 1755-0998.

Manoukis, N.C. (2007). FORMATOMATIC: a program for converting diploid allelic data between common formats for population genetic analysis. *Molecular Ecology Notes*, Vol.7, No.4, (July 2007), pp. 592–593, ISSN 1755-0998.

Messier, M.; Li, S.H. & Stewart, C.B. (1996). The birth of microsatellites. *Nature*, Vol.381, (June 1996), pp. 483, ISSN 0028-0836.

Mikheyev, A.S.; Vo, T.; Wee, B.; Singer, M.C. & Parmesan, C. (2010). Rapid Microsatellite Isolation from a Butterfly by De Novo Transcriptome Sequencing: Performance and a Comparison with AFLP-Derived Distances. *PLoS ONE,* Vol.5, No.6, (June 2010), pp. e11212, ISSN 1932-6203.

Mittal, N. & Dubey, A.K. (2009). Microsatellite markers - A new practice of DNA based markers in molecular genetics. *Pharmacognosy Reviews,* Vol.3, No.6, (July-December 2009), pp. 235-246, ISSN 0973-7847.

Mullis, K. & Faloona, F. (1987). Specific synthesis of DNA in vitro via a polymerase catalyzed chain reaction. *Methods in Enzymology,* Vol.155, pp. 335-350, ISSN 0076-6879.

Orsini, L.; Huttunen, S. & Schlotterer, C. (2004). A multilocus microsatellite phylogeny of the Drosophila virilis group. *Heredity,* Vol.93, No.2, (June 2004), pp. 161-165, ISSN 0018-067X.

Otha, T. & Kimura, M. (1973). A model of mutation appropriate to estimate the number of electrophoretically detectable alleles in a finite population. *Genetical Research,* Vol.22, No.2, (October 1973), pp. 201–204, ISSN 0016-6723.

Paetkau, D. (1999). Microsatellites obtained using strand extension: an enriched protocol. *BioTechniques,* Vol.26, No.4, (April 1999), pp. 690-697, ISSN 0736-6205.

Palstra, F.P.; O'Connell, M.F. & Ruzzante, D.E. (2007). Population structure and gene flow reversals in Atlantic salmon (*Salmo salar*) over contemporary and long-term temporal scales: effects of population size and life history. *Molecular Ecology,* Vol.16, No.21, (October 2007), pp. 4504–4522, ISSN 1365-294X.

Peakall, R.; Gilmore, S.; Keys, W.; Morgante, M. & Rafalski, A. (1998). Cross-species amplification of soybean (*Glycine max*) simple sequence repeats (SSRs) within the genus and other legume genera: implications for the transferability of SSRs in plants. *Molecular Biology and Evolution,* Vol.15, No.10, (October 1998), pp. 1275–1287, ISSN 0737-4038.

Pépin, L.; Amigues, Y.; Lépingle, A.; Berthier, J.L.; Bensaid, A. & Vaiman, D. (1995). Sequence conservation of microsatellites between *Bos taurus* (cattle), *Capra hircus* (goat) and related species. Examples of use in parentage testing and phylogeny analysis. *Heredity,* Vol.74, No.1, (January 1995), pp. 53–61, ISSN 0018-067X.

Pérez de Rosas, A.R.; Segura, E.L. & García, B.A. (2007). Microsatellite analysis of genetic structure in natural *Triatoma infestans* (Hemiptera: Reduviidae) populations from Argentina: its implication in assessing the effectiveness of Chagas' disease vector control programmes. *Molecular Ecology,* Vol.16, No.7, (April 2007), pp. 1401-1412, ISSN 1365-294X.

Pompanon, F.; Bonin, A.; Bellemain, E. & Taberlet, P. (2005). Genotyping errors: causes, consequences and solutions. *Nature Reviews Genetics,* Vol.6, No.11, (November 2005), pp. 847-846, ISSN 1471-005.

Powell, W.; Machray, G.C. & Provan, J. (1996). Polymorphism revealed by simple sequence repeats. *Trends in Plant Science,* Vol.1, No.7, (July 1996), pp. 215-222, ISSN 1360-1385.

Powell, W.; Morgante, M.; Andre, C.; McNicol, J.W.; Machray, G.C.; Doyle, J.J.; Tingey, S.V. & Rafalski, J.A. (1995). Hypervariable microsatellites provide a general source of polymorphic DNA markers for the chloroplast genome. *Current Biology,* Vol.5, No.9, (September 1995), pp. 1023-1029, ISSN 0960-9822.

Primmer, C.R.; Koskinen, M.T. & Piironen, J. (2000). The one that did not get away: individual assignment using microsatellite data detects a case of fishing competition fraud. *Proceedings of the Royal Society of London*, Vol.267, (March 2000), pp. 1699–1704, ISSN 09628452.

Rajendrakumar, P.; Biswal, A.K.; Balachandran, S.M.; Srinivasarao, K. & Sundaram, R.M. (2007). Simple sequence repeats in organellar genomes of rice: frequency and distribution in genic and intergenic regions. *Bioinformatics*, Vol.23, No.1, (January 2007), pp. 1–4, ISSN 1460-2059.

Rico, C.; Rico, I. & Hewitt, G. (1996). 470 million years of conservation of microsatellite loci among fish species. *Proceedings Biological Sciences*, Vol. 263, No.1370, (May 1996), pp. 549–557, ISSN 0962-8452.

Rout, P.; Joshi, M.; Mandal, A.; Laloe, D.; Singh, L. & Thangaraj, K. (2008). Microsatellite-based phylogeny of Indian domestic goats. *BMC Genetics*, Vol.9, (January 2008), pp. 11, ISSN 1471-2164.

Saiki, R.K.; Scharf, S.J.; Fallona, F.; Mullis, K.B.; Horn, G.T.; Erlich, H.A. & Arnheim, N. (1985). Enzymatic amplification of beta-globin genomic sequences and restriction site analysis for diagnosis of sickle cell anemia. *Science*, Vol.230, No.4732, (December 1985), pp. 1350-1354, ISSN 1095-9203.

Santana, Q.; Coetzee, M.; Steenkamp, E.; Mlonyeni, O.; Hammond, G.; Wingfield, M. & Wingfield, B. (2009). Microsatellite discovery by deep sequencing of enriched genomic libraries. *Biotechniques*, Vol.46, No.3, (March 2009), pp. 217-223, ISSN 0736-6205.

Schlötterer, C. (2000). Evolutionary dynamics of microsatellite DNA. *Chromosoma*, Vol.109, No.6, (June 2000), pp. 365–371, ISSN 0009-5915.

Selkoe, K.A. & Toonen, R.J. (2006). Microsatellites for ecologists: a practical guide to using and evaluating microsatellite markers. *Ecology Letters*, Vol.9, No.5, (May 2006), pp. 615–629, ISSN 1461-0248.

Soranzo, N.; Provan, J. & Powell, W. (1999). An example of microsatellite length variation in the mitochondrial genome of conifers. *Genome*, Vol.42, No.1, (February 1999), pp. 158–161, ISSN 0831-2796.

Strand, M.; Prolla, T.A.; Liskay, R.M. & Petes, T.D. (1993). Destabilization of tracts of simple repetitive DNA in yeast by mutations affecting DNA mismatch repair. *Nature*, Vol.365, (September 1993), pp. 274–276, ISSN 0028-0836.

Varshney, R.K.; Graner, A. & Sorrels, M.E. (2005). Genetic microsatellite markers in plants: features and applications. *Trends in Biotechnology*, Vol.23, No.1, (January 2005), pp. 48-55, ISSN 0167-7799.

Wang, M.L.; Barkley, N.A. & Jenkins, T.M. (2009a). Microsatellite Markers in Plants and Insects. Part I: Applications of Biotechnology. *Genes, Genomes and Genomics*, Vol.3, No.1, (December 2009), pp. 54-67, ISSN 1749-0383.

Wang, M.L.; Mosjidis, J.A.; Morris, J.B.; Chen, Z.B.; Barkley, N.A. & Pederson, G.A. (2009b). Evaluation of *Lespedeza* germplasm genetic diversity and its phylogenetic relationship with the genus *Kummerowia*. *Conservation Genetics*, Vol.10, No.1, (February 2009), pp. 79-85, ISSN 1566-0621.

Weising, K.; Nybom, H.; Wolff, K. & Kahl, G. (2005). *DNA Fingerprinting in Plants: Principles, Methods and Applications,* CRC Press, ISBN 0-8493-1488-7, Boca Raton, United States.

Wilder, J. & Hollocher, H. (2001). Mobile elements and the genesis of microsatellites in dipterans. *Molecular Biology and Evolution,* Vol.18, No.3, (March 2001), pp.384-392, ISSN 1537-1719.

Yue, G.H.; Zhu, Z.Y.; Wang, C.M. & Xia, J.H. (2009). A simple and efficient method for isolating polymorphic microsatellites from cDNA. *BMC Genomics,* Vol.10, (March 2009), pp. 125, ISSN 1471-2164.

Zane, L.; Bargelloni, L. & Patarnello, T. (2002). Strategies for microsatellite isolation: a review. *Molecular Ecology,* Vol.11, No.1, (January 2002), pp. 1-16, ISSN 1365-294X.

Zhao, H.; Yu, J.; You, F.M.; Luo, M. & Peng, J. (2011). Transferability of Microsatellite Markers from *Brachypodium distachyon* to *Miscanthus sinensis,* a Potential Biomass Crop. *Journal of Integrative Plant Biology,* Vol.53, No.3, (Março 2011), pp. 232–245, ISSN 1744-7909.

7

HIV-1 Diversity and Its Implications in Diagnosis, Transmission, Disease Progression, and Antiretroviral Therapy

Inês Bártolo and Nuno Taveira

Centro de Investigação Interdisciplinar Egas Moniz (CiiEM),
Instituto Superior de Ciências da Saúde Egas Moniz,
Monte de Caparica and Unidade dos Retrovírus e Infecções Associadas,
Centro de Patogénese Molecular,
Faculdade de Farmácia de Lisboa,
Portugal

1. Introduction

Due to an error prone reverse transcriptase, HIV-1 has diversified into multiple variants. Currently, there are four phylogenetic groups named M, N, O and P (Gurtler et al., 1994; Plantier et al., 2009b; Simon et al., 1998). HIV-1 group M, which is responsible for most of the infections in the world, has diversified into nine subtypes named A, B, C, D, F, G, H, J and K, seven sub-subtypes (A1-A5 and F1-F2), multiple circulating recombinant forms (CRFs) and countless unique recombinant forms (UFRs) (Gao et al., 2001; Meloni et al., 2004; Triques et al., 1999; Vidal et al., 2009; Vidal et al., 2006). At present, 49 CRFs were recognized of which 37 are first generation recombinants and 12 are second generation recombinants (Los Alamos Sequence Database, 2011).

Five HIV-1 strains dominate the global epidemic: subtypes A, B, and C, along with CRF01_AE and CRF02_AG, with subtype C accounting for almost 50% of all HIV-1 infections worldwide (Buonaguro, Tornesello, and Buonaguro, 2007; McCutchan, 2006; Santos and Soares, 2010; Taylor and Hammer, 2008). Molecular epidemiological studies show that, with the exception of sub-Saharan Africa where almost all subtypes, CRFs and several URFs have been detected there is a specific geographic distribution pattern for HIV-1 subtypes (Buonaguro, Tornesello, and Buonaguro, 2007; McCutchan, 2006; Santos and Soares, 2010; Taylor and Hammer, 2008). Subtype A is prevalent in Central and Eastern Africa (Kenya, Uganda, Tanzania, and Rwanda) (Harris et al., 2002; Morison et al., 2001; Songok et al., 2004), Iran (Tagliamonte et al., 2007), Eastern Europe (Bobkov et al., 1997; Bobkov et al., 2004), and Central Asia (Beyrer et al., 2009; Eyzaguirre et al., 2007). Subtype B predominates in developed countries, such as United States of America (USA) and Canada (Akouamba et al., 2005; Brennan et al., 2010; Carr et al., 2010a; Jayaraman et al., 2003), in Brazil (Dumans et al., 2004; Monteiro-Cunha et al., 2011; Santos et al., 2006), countries of Western and Central Europe (Abecasis et al., 2011; Castro et al., 2010; De Mendoza et al., 2009; Easterbrook et al., 2010; Galimand et al., 2010; Habekova et al., 2010; Kousiappa, Van

De Vijver, and Kostrikis, 2009; Lai et al., 2010; Parczewski et al., 2010) and Australia (Herring et al., 2003; Ryan et al., 2004), and is also common in several countries of Southeast Asia (Chen et al., 2010; Lau et al., 2010), northern Africa (Annaz et al.) and the Middle East (Sarrami-Forooshani et al., 2006). Subtype C is the overwhelming prevailing strain in Southern Africa (Bartolo et al., 2009b; Gonzalez et al., 2010; Lahuerta et al., 2008; Papathanasopoulos et al., 2010), in India and neighbor countries (Neogi et al., 2009a; Neogi et al., 2009b) and in the southern region of Brazil (Monteiro-Cunha et al., 2011). Subtype D strains are found mainly in East Africa, and to a lesser extent in West Africa (Conroy et al., 2010; Harris et al., 2002; Laukkanen et al., 2000; Songok et al., 2004). Subtype F predominates in Central Africa (Carr et al., 2010b; Soares et al., 2010), South America (Avila et al., 2002; Munerato et al., 2010) and Eastern Europe (Fernandez-Garcia et al., 2009; Paraschiv, Foley, and Otelea, 2011). Subtype G viruses are prevalent in Central and Western Africa (Hawkins et al., 2009; Kalish et al., 2004), as well as in Portugal (Abecasis et al., 2011; Esteves et al., 2003; Esteves et al., 2002; Palma et al., 2007) and Spain (De Mendoza et al., 2009; Trevino et al., 2011). Subtypes H and J were described in Central Africa (Janssens et al., 2000; Mokili et al., 1999; Yamaguchi et al., 2010) and in Angola (Bartolo et al., 2005; Bartolo et al., 2009d). Subtype K was identified in DRC and Cameroon (Triques et al., 2000).

Some CRFs have high impact in local AIDS epidemics, such as CRF01_AE in Southeast Asia (Gao et al., 1996; Magiorkinis et al., 2002) and CRF02_AG in Western and Central Africa (Cornelissen et al., 2000; Fischetti et al., 2004a; Fischetti et al., 2004b). CRF11_cpx was the first second generation CRF described in 2000 in patients from Cameroon (Tscherning-Casper et al., 2000). This CRF circulates in Cameroon, Central African Republic, Gabon, DRC, and Angola although its exact prevalence rate remains to be determined (Djoko et al., 2011). Second generation CRFs are becoming common in complex epidemics with multiple subtypes and recombinant forms. At present they have been detected in Africa, East Asia, Thailand, Malaysia, Estonia and Saudi Arabia (Djoko et al., 2011).

HIV-1 group O seems to be endemic in Cameroon and neighboring countries in West-Central Africa and represents only about 1-5% of HIV-1 positive samples in this region (Peeters et al., 1997; Yamaguchi et al., 2004). Elsewhere in the world, group O viruses have been identified mainly from people with epidemiological links to the referred Central African countries (Lemey et al., 2004). HIV-1 groups N and P circulate exclusively in Cameroon (Brennan et al., 2008; Plantier et al., 2009b; Vallari et al., 2010; Vallari et al., 2011). In some regions of the world, little information is available about HIV diversity, particularly in North Africa, the Middle East, and parts of Central Asia.

Differential characteristics of viral subtypes and their interactions with the human host may influence HIV transmission and disease progression. HIV-1 genetic diversity may also impact the susceptibility and resistance to antiretroviral drug as well as the performance of diagnostic and viral load assays. The aim of this Chapter was to review the current knowledge on HIV-1 diversity and its implications in diagnosis, transmission, disease progression, and antiretroviral therapy and resistance.

2. Impact of HIV-1 diversity in transmission

Although earlier studies found an association between CRF01_AE and heterosexual transmission and between subtype B and intravenous drug use (Gao et al., 1996; Soto-

Ramirez et al., 1996), a more recent longitudinal study performed in Thailand found an increased probability of CRF01_AE transmission among IDUs compared with subtype B (Hudgens et al., 2002). A recent study performed in HIV-discordant couples in Uganda found that subtype A was associated with a significant higher rate of heterosexual transmission than subtype D (P=0.01) (Kiwanuka et al., 2009). The rate of transmission may reflect differences in subtype-specific coreceptor tropism. HIV-1 can use as coreceptors CCR5 (R5 variants), CXCR4 (X4 variants) or both (R5X4 variants or dual tropic) to enter the cells (Schuitemaker, van 't Wout, and Lusso, 2011). R5 variants are largely prevalent during primary infection and seem to be more easily transmitted or established in the newly infected host than X4 strains (Schuitemaker, van 't Wout, and Lusso, 2011). An increased prevalence of X4 variants has been reported for subtype D which may in part explain their reduced heterosexual transmissibility when compared to other genetic forms (Huang et al., 2007; Kaleebu et al., 2007; Kiwanuka et al., 2009; Tscherning et al., 1998).

Several studies have looked for variations between clades in mother-to-child HIV-1 transmission (MTCT) rates. In Kenya, MTCT appeared to be more common among mothers infected with subtype D compared with subtype A (P=0.002) and this association was independent of other risk factors for MTCT, such as maternal HIV viral load, episiotomy or perineal tear, and low birth weight (Yang et al., 2003). On the other hand, in Tanzania HIV-1 subtypes A (odds ratio, 3.8; 95% CI, 0.8-24.7%) and C (odds ratio, 5.1; 95% CI, 1.3-30.8%) were more frequently transmitted from mother-to-child than subtype D (Renjifo et al., 2001). In another study, the risk of MTCT was higher in women infected with subtype C, followed by subtype A, and was lowest in women infected with subtype D (John-Stewart et al., 2005). It was found that pregnant women, infected with subtype C were more likely than those infected with subtype A or D (P=0.006) to shed HIV-1–infected vaginal cells even after adjusting for age, CD4 cell count, and plasma HIV-1 viral load. Another study in Tanzania presented similar results, with preferential in-utero transmission of HIV-1 subtype C compared to HIV-1 subtype A or D (P=0.026) (Renjifo et al., 2004). Other researchers, however, found no association between subtype and rates of MTCT (Martinez et al., 2006; Murray et al., 2000; Tapia et al., 2003). Moreover, in studies where pregnant women received a single-dose nevirapine (sdNVP) prophylaxis no significant differences were observed in the rate of MTCT in women infected with HIV-1 subtype A, D or C (Eshleman et al., 2006; Eshleman et al., 2005a).

Many factors, such as maternal stage of the disease, maternal immunological status, viral load, mode of delivery, duration of breast-feeding, ARV prophylaxis, maternal plasma vitamin A (associated with AIDS progression), and close maternal-child Human Leukocyte Antigen (HLA) matching, can contribute to these differences (Fowler and Rogers, 1996; MacDonald et al., 1998; McGowan and Shah, 2000). Nevertheless, the role of viral determinants in MTCT has yet to be well established (Dickover et al., 2001). Several studies have shown that viral diversity in the mother is generally higher than that present in the infant, suggesting that maternal viruses are selected before transmission (Ahmad, 2005; Zhang et al., 2010b). Several factors like specific viral selection (Wolinsky et al., 1992), neutralization resistance (Dickover et al., 2006; Wu et al., 2006; Zhang et al., 2010a) and enhanced replicative capacity of the transmitted viruses (Kong et al., 2008) have been associated with a bottleneck type transmission. The basis for the MTCT bottleneck is an issue that needs further clarification.

In summary, it remains to be determined whether there is a true association between subtypes and adult or MTCT transmission of HIV-1 or whether the differences in transmission probabilities found in some studies are associated with several other factors that can influence HIV transmission, e.g. behavioral, epidemiological and immunological (Attia et al., 2009). More longitudinal and well controlled studies, preferentially performed in a single area and with a single ethnic group of HIV-1 infected patients, are needed to identify HIV-1 determinants for adult and vertical transmission and to evaluate the potential association of subtype with transmission.

3. Impact of HIV-1 diversity in disease progression

Another important question is whether clade differences result in variable rates of disease progression. There have been several prospective, observational studies of the course of HIV-related disease in cohorts infected with various HIV-1 genetic forms. Although some studies did not find an association between HIV-1 clades and disease progression (Alaeus et al., 1999; Amornkul et al., 1999; Galai et al., 1997; Laurent et al., 2002; Taylor and Hammer, 2008), more recent studies established this association (Easterbrook et al., 2010; Keller et al., 2009; Kiwanuka et al., 2010). A retrospective cohort study (1996-2007) reported that Africans patients infected with HIV-1 non-B subtypes (A, C, F-K, AC, AE, AG, BF and DF) had slower rates of disease progression compared to Haitians (P=0.0001) and Canadians (P=0.02) infected with subtype B viruses (Keller et al., 2009).

Earlier studies found that subtype D was associated with the most rapid disease progression relative to other subtypes (Taylor and Hammer, 2008). A very recent study in patients from Rakai, Uganda, reported that infection with subtype D is associated with significantly faster rates of CD4 T-cell loss than subtype A (P<0.001), which might explain the more rapid disease progression for subtype D compared with subtype A (Kiwanuka et al., 2010). Along the same lines, a study conducted in 2010 in an ethnically diverse population of HIV-1-infected patients in South London showed a faster CD4 cell decline and higher rate of subsequent virological failure with subtype D infection than with subtypes B (P=0.02), A (P=0.004) or C (P=0.01) (Easterbrook et al., 2010).

An important unanswered question is the biological basis for these differences. A possible clue comes from data suggesting that emergence of X4 variants, which in subtype B are associated with increased CD4 depletion and disease progression [302-304], was more common in HIV-1 subtype D compared with subtype A (P=0.040) (Kaleebu et al., 2007; Tscherning et al., 1998). Other study found that subtype D may be dual tropic more frequently than the other subtypes (Huang et al., 2007). The earlier switch to X4 usage in subtype D isolates may explain the faster rate of CD4 decline and disease progression with this subtype (Kaleebu et al., 2007; Tscherning et al., 1998).

One study reported the presence of X4 or R5X4 isolates at early stages of infection, in addition to a decrease in CD4+ counts, in all patients infected with CRF14_BG (Perez-Alvarez et al., 2006). We have shown recently that most CRF14_BG strains (78.9%)

sequenced to date use CXCR4 (Bártolo et al., 2011) and that patients infected with this CRF can progress very quickly to AIDS and death (Bartolo et al., 2009a). Together, these results suggest that, like HIV-1 subtype D, CRF14_BG may be highly pathogenic (Kuritzkes, 2008; Sacktor et al., 2009). The rapid disease progression associated with CRF14_BG may be due to an earlier switch to X4 phenotype driven by the selective pressure of neutralizing antibodies (Bártolo et al., 2011). However, the relationship between higher tendency for X4 use and higher disease progression may not hold for other subtypes. For instance, the percentage of X4 virus appears to be lower in subtype C than in subtype B, even when the viruses are obtained from patients with AIDS (Casper et al., 2002; Cilliers et al., 2003; Ping et al., 1999).

It is important to note that most of these studies of disease progression have confounder factors such as access to medical care, nutritional status, host genetic factors, and mode of viral transmission, which may contribute to the divergent results (Pereyra et al., 2010). More studies are needed to confirm previous conflicting results, and to elucidate the host–viral interactions that may lead to more favorable outcomes in individuals infected with various genetic forms of HIV-1. This kind of studies should be longitudinal, performed with a higher number of patients, preferentially in primary infection, in a single country to better control ethnic and genetic factors of the patients, and with several genetic forms of HIV-1.

4. Impact of HIV-1 diversity in diagnosis and disease management

Acute infection with HIV-1 can be identified and quantified using several virological and immunological markers (Figure 1) (Fiebig et al., 2003). In the first eleven days after infection viral markers are undetectable in the blood (window period). Plasma HIV RNA levels begin to increase at the 10th day, peaking around 20 days after infection. HIV p24 levels typically peak around 20 days after infection. Antibody response starts to be detectable by ELISA assays after 20 days of infection, on average. Serological and molecular assays have been designed to detect and/or quantify one or more of these HIV infection markers. These assays should be able to detect all genetic forms of HIV but the very high genetic and antigenic evolution of this virus along with the continued diversification and global redistribution of HIV groups, subtypes and recombinants may affect their performance.

4.1 Immunoassays

HIV fourth-generation assays detect both HIV antibodies and the p24 antigen. These assays provide an advantage for detection of infection during the window period prior to seroconversion since the diagnostic window may be reduced by an average of 5 days relative to an IgM-sensitive EIA (Fiebig et al., 2003; Weber et al., 1998). However, some fourth-generation assays showed low sensitivity in the detection of p24 antigen from some non-subtype B HIV-1 strains (A, C, F, H, CRF01_AE, O) (Kwon et al., 2006; Ly et al., 2007; Ly et al., 2004; Ly et al., 2001; Weber, 2002). This low sensitivity in antigen detection may be attributed to differences in viral epitopes of the different HIV genetic forms which may not be recognized by the monoclonal antibody used in the assay (Plantier et al., 2009a).

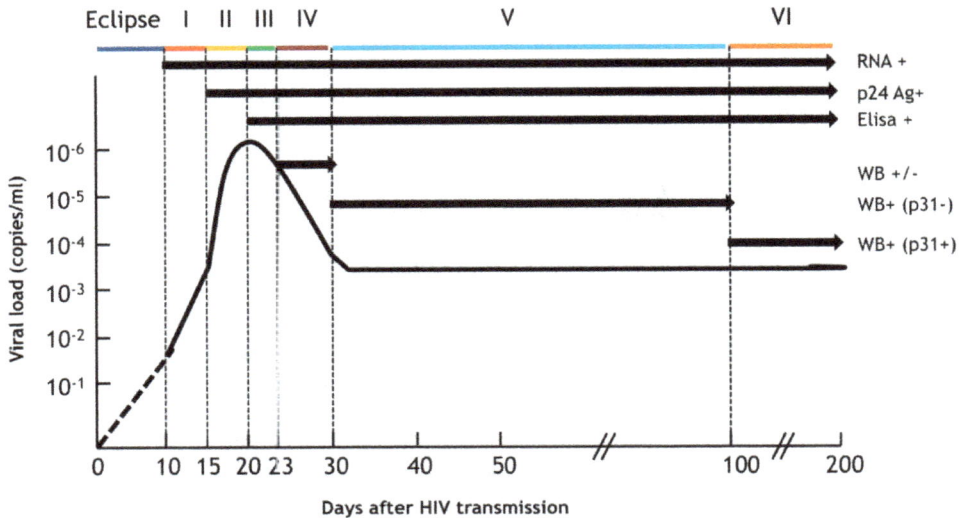

Fig. 1. Progression of HIV-1 markers in acute infection. WB, Western blot; RNA, HIV RNA; LS–Ab, HIV antibody determined by sensitive/less sensitive enzyme immunoassay testing strategy; p24 Ag, HIV p24 antigen, from time of exposure (day 0) through the first 200 days of infection. Eclipse, eclipse period (undetectable viral markers in blood samples); Stage I (definitive HIV RNA viremia), stage II (p24 antigenemia), stage III (HIV EIA antibody reactive), stage IV (I, Western blot indeterminate), stage V (Western blot positive without p31 pol band) and stage VI (P, Western blot positive with p31 pol band). Adapted from (Fiebig et al., 2003).

The major target for HIV-1 antibodies in immunoassays is the *env* gp41 immunodominant region (IDR). Key epitope(s) targeted by these assays might be modified or eliminated by the occurrence of natural polymorphisms within the IDR region associated with the genetic variation of HIV-1, ultimately leading to reduced sensitivity or lack of antibody detection (Brennan et al., 2006; Gaudy et al., 2004). A few cases of false-negative results involving, for example, subtypes B, C, and F, and resulting from major mutations of the IDR epitope have been described (Aghokeng et al., 2009; Gaudy et al., 2004; Ly et al., 2007; Ly et al., 2004; Ly et al., 2001; Zouhair et al., 2006).

Earlier analysis of specimens from patients infected with group O viruses revealed that some commercial immunoassays failed to detect group O infections (Eberle et al., 1997; Loussert-Ajaka et al., 1994; Schable et al., 1994; Simon et al., 1994). This ultimately led to incorporation of group O specific antigens and/or peptides into the assays to improve detection of group O infections (van Binsbergen et al., 1996). Nonetheless, false-negative results continue to be reported for some patients infected with HIV-1 group O (Henquell et al., 2008; Plantier et al., 2009a; Zouhair et al., 2006).

Despite the high genetic divergence between HIV-1 groups M and N, all group N infections studied until now were detected by five commercial HIV immunoassays (Vallari et al., 2010). Group P infections may not be efficiently detected by the current HIV screening tests due to the absence of group P-specific reagents for antibody detection (Vallari et al., 2011). Nevertheless, Plantier et al., in the first report regarding detection of group P infections, found that several HIV-1 screening tests were reactive against this group (Plantier et al., 2009b). Despite the absence of either HIV-1 group N or group P specific antigens in most assays, antibodies targeting some group M specific antigens may cross-react with group N and P antigens allowing for the serologic detection of infections by HIV-1 isolates from these groups.

Serological diagnosis of HIV-1 infection in Sub-Saharan Africa is mostly done with rapid tests (Plate, 2007). This kind of assay is simple, rapid, instrument-free and relatively cheap. However some of these assays have shown problems in detecting HIV-1 subtypes D, F, H, CRF02_AG, group O and HIV-2 (Aghokeng et al., 2009; Beelaert and Fransen, 2010; Chaillet et al., 2010; Holguin et al., 2009; Laforgerie et al., 2010; Pavie et al., 2010). Minor antigenic differences between isolates of different clades and the peptides/recombinant proteins used in these assays could explain the problems in the detection of some HIV genetic forms (Aghokeng et al., 2009; Laforgerie et al., 2010; Makuwa et al., 2002; Pavie et al., 2010). Low sensitivity of some of these tests can also be associated with low level of HIV-specific antibodies due to recent seroconversion, early and stringent control of viral replication by antiretroviral therapy, or immune exhaustion in end-stage AIDS patients (Apetrei et al., 1996; Ferreira Junior et al., 2005; Jurriaans et al., 2004; Laforgerie et al., 2010; Makuwa et al., 2002; Pavie et al., 2010; Spivak et al., 2010).

4.2 Viral load assays

A variety of nucleic acid based diagnostic assays that quantify plasma HIV-1 RNA levels have been developed and used to monitor disease progression and response to antiviral therapy, detect primary infection [plasma HIV RNA levels begin to be detectable about 11 days after infection (Figure 1)], detect HIV infection among perinatally exposed infants and HIV vaccine recipients, and detect HIV infection in the absence of antibodies (Table 1)(Bill & Melinda Gates Foundation, 2009; European AIDS Clinical Society, 2009; Korenromp et al., 2009; Mellors et al., 1997; Thompson et al., 2010). These assays rely on HIV-1 sequence-specific primers and/or probes and use technologies such as reverse transcriptase polymerase chain reaction (RT-PCR) amplification, isothermal nucleic acid sequence-based amplification (NASBA), branched-chain DNA signal amplification (bDNA) and real-time (RT) PCR (Collins et al., 1997; de Mendoza et al., 2005; Dyer et al., 1999; Johanson et al., 2001; Rouet et al., 2005; Stevens et al., 2005; Sun et al., 1998; Yao et al., 2005). The genetic variation of HIV-1 presents challenges to the design of quantitative assays that measure HIV-1 RNA or DNA levels. Reliable quantification can be compromised by natural polymorphisms occurring in primer/probe sequences that have the potential to reduce or abolish hybridization (Christopherson, Sninsky, and Kwok, 1997; Kwok et al., 1990). Genetically divergent variants may go unrecognized since, usually, subtype and target sequence information is not known at the time of testing.

Several comparative studies have shown that the sensitivity and specificity of viral load assays varies depending on HIV-1 group or subtype, especially in non-B subtypes, complex recombinant forms and groups O, N, and P viruses (Bourlet et al., 2011; Church et al., 2011;

Assays	Technology	Probe target	Linear range (RNA copies/ml)	HIV-1 clade recognition
Abbot Real Time HIV-1	RT-PCR	*pol*-INT	40-10,000,000	Group M (subtypes A-H), several CRFs and Groups O and N
Amplicor HIV-1 Monitor Test v1.5	RT-PCR	*gag*-p24	Standard protocol: 400 to >750,000; Ultra-sensitive: 50 - >100,000	HIV-1 Group M (subtypes A-H)
Cobas Amplicor HIV-1 Monitor Test, v1.5	RT-PCR	*gag*	Standard protocol: 400 to >750,000; Ultrasensitive: 50 - >100,000	HIV-1 Group M (subtypes A-H)
Cobas AmpliPrep/Cobas TaqMan HIV-1 Test, v2.0	RT-PCR	*gag*-p41 and 5´LTR	20-10,000,000	Group M, several CRFs, and Group O
Versant HIV-1 RNA 1.0 Assay (kPCR)	RT-PCR	*gag*-p24	37-11,000,000	HIV-1 Groups M and O
Versant HIV-1 RNA 3.0 Assay	bDNA	*pol*-INT	50-500,000	HIV-1 Group M
NucliSens EasyQ HIV-1 v2.0	NASBA	*gag*-p24	10- 10,000,000	HIV-1 Group M (subtypes A-J), CRF01_AE, and CRF02_AG

Adapted from (World Health Organization, June 2010).

Table 1. Viral load assays approved by FDA and recommended by WHO

Geelen et al., 2003; Gottesman et al., 2006; Holguin et al., 2008; Katsoulidou et al., 2011; Plantier et al., 2009b; Rouet et al., 2010; Scott et al., 2009; Swanson et al., 2005; Swanson et al., 2006; Swanson et al., 2007; Tang et al., 2007; Wirden et al., 2009; Xu et al., 2008). However, the newer quantitative real-time PCR (qRT-PCR) methods (i.e., m2000rt Abbot Real Time HIV-1 Assay or Cobas AmpliPrep/COBAS TaqMan) showed a higher performance on HIV viral load testing of patients with subtype B as well as patients with non-B subtype infections (Bourlet et al., 2011; Church et al., 2011; Katsoulidou et al., 2011; Swanson et al., 2007; Tang et al., 2007). Abbot Real Time HIV-1 Assay seems to be the only assay prepared to detect all HIV-1 subtypes, several CRFs, as well as group N, and O viruses (Church et al., 2011; Swanson et al., 2007; Tang et al., 2007). This is probably related with the high level of genetic conservation of the integrase gene that this test amplifies (Young et al., 2011). In contrast, bDNA (Versant v3.0) and NASBA (EasyQ) assays are considerably less reliable for accurate viral load measurements across HIV clades (Bourlet et al., 2011; Church et al., 2011; Katsoulidou et al., 2011; Swanson et al., 2007; Tang et al., 2007). In summary, available data indicates that HIV-1 assays targeting the highly conserved *pol* integrase region of the HIV-1 genome may be subject to less variability than assays targeting the *gag* gene (Geelen et al.,

2003; Swanson et al., 2005; Swanson et al., 2006; Swanson et al., 2007). As HIV genetic diversity evolves, evaluations of all commercially licensed HIV-1 viral load assays should be performed regularly in populations with patients infected with all viral subtypes.

Failed detection or unreliable quantification of HIV infection can have significant consequences in early detection of MTCT (Creek et al., 2007; Geelen et al., 2003). Early diagnosis in infants can only be achieved with tests that detect HIV-1 DNA or RNA, since maternal HIV antibodies can persist in the infant until month 18 (Read, 2007) thereby precluding the use of antibody detection tests. It has been recommended that diagnostic testing with HIV-1 DNA or RNA assays be performed within the first 14-21 days of life, at 1-2 months and at 4-6 months of age (AIDSinfo, May 24, 2010). Additionally, if any of these test results are positive, repeat testing on a second sample has to be done to confirm the diagnosis of HIV-1 infection. A diagnosis of HIV-1 infection can be made on the basis of 2 separate positive HIV-1 DNA or RNA assay results (New York State Department of Health AIDS Institute, 2010).

Viral load assay	Extraction method	Lower detection limit (log10 HIV-1 RNA copies/ml)	Reference
COBAS TaqMan RT-PCR Assay	Nuclisens MiniMAG	3.0 (96.4% detected)	(Andreotti et al., 2010)
COBAS TaqMan RT-PCR Assay	Primagen	3.0 (96 % detected)	(Waters et al., 2007)
Nuclisens EasyQ HIV-1 v2.0	Nuclisens EasyMAG	2.9 (95% detected)	(van Deursen et al., 2010)
Nuclisens EasyQ HIV-1 v2.0	Manual Nuclisens	3.5 (100% detected)	(Johannessen et al., 2009)
Nuclisens EasyQ HIV-1	Nuclisens MiniMAG	2.9 (100% detected)	(Kane et al., 2008)
Amplicor HIV-1 Monitor Test v1.5	In-house method	3.0 (100% detected)	(Ikomey et al., 2009)
Abbot Real Time HIV-1	m2000 RT system	3.7 (100% detected)	(Garrido et al., 2009)
Abbot Real Time HIV-1	m2000 RT system	2.6 (99% detected)	(Lofgren et al., 2009)
Abbot Real Time HIV-1	m2000 RT system	3.0 (100% detected)	(Mbida et al., 2009)

Adapted from (Johannessen, 2010).

Table 2. Recent studies comparing HIV-1 viral load assays in DBS and plasma

Amplicor HIV-1 DNA PCR test version 1.0 (Roche Diagnostic), the first commercial HIV-1 qualitative DNA PCR assay, lacked optimal sensitivity to detect non-B HIV-1 subtypes (Bogh et al., 2001; Kline, Schwarzwald, and Kline, 2002; Obaro et al., 2005). In May 2005,

Roche Diagnostics replaced Amplicor HIV-1 DNA PCR version 1.0 by version 1.5, which has been shown to have excellent sensitivity and specificity in testing adult venous blood samples and infant DBS (Germer et al., 2006; Patton et al., 2007). This test is highly accurate in detecting the multiple HIV-1 subtypes circulating in Africa (Stevens et al., 2008), is standardized and supported for use in Africa, and has been used by researchers and infant diagnosis pilot programs in several countries (Creek et al., 2007). However, Amplicor HIV-1 DNA PCR 1.5 uses primers for the relatively variable *gag* gene and was developed to amplify HIV-1 group M strains (Roche Diagnostics). So, it is likely that sensitivity problems arise with groups O, N and P. New DNA amplification assays that cover all HIV-1 genetic forms are needed

5. Impact of HIV-1 diversity in response to antiretroviral therapy

Differences in amino acid composition between HIV-1 clades can lead to differences in susceptibility to ARV drugs. This is best illustrated by HIV-1 group O and HIV-2 isolates that show high-level of innate resistance to NNRTIs and T20 (Descamps et al., 1997; Poveda et al., 2004; Smith et al., 2009). This innate resistance is due to resistance mutations that are present as natural polymorphisms. For instance, HIV-1 group O isolates naturally present a cysteine at RT position 181 (Y181C) which is considered a major drug resistance mutation (DRM) to NNRTIs; the secondary NNRTI DRM A98G is also a natural polymorphism in group O (Descamps et al., 1997; Poveda et al., 2004) (Table 3).

Susceptibility of non-B subtypes to ARV drugs has been less well studied than subtype B mainly because of the predominance of subtype B in developed countries where ARVs first became available, coupled with the availability of genotypic and phenotypic ARV drug resistance testing (Brenner, 2007). Some studies of sdNVP for prevention of MTCT have demonstrated a statistically significant disparity in the overall drug resistance among subtypes, with frequencies of 69-87%, 55.3-36%, 19-42%, and 21% resistance against NVP in women with subtypes C, D, A, and CRF02_AG infections, respectively (Eshleman et al., 2005b; Flys et al., 2006; Johnson et al., 2005; Toni et al., 2005). There were no significant differences in the pre-NVP frequency of NVP resistance mutations or the pre-NVP levels of K103N-containing variants in women with subtypes A, C, and D that could explain the subtype-based differences in mutations after sdNVP exposure (Flys et al., 2006). However, there are other factors that may be associated with NVP resistance in women after the administration of sdNVP and which include: higher viral load and lower CD4+ T cell count prior to NVP exposure, increased pharmacokinetic exposure to NVP (e.g., longer half-life and decreased oral clearance of NVP), and the timing of sample collection (Eshleman et al., 2005b). Additional studies are needed that take in account all these factors to better understand the biological causes of these subtype differences in sdNVP resistance.

In the Pediatric European Network for Treatment of AIDS (PENTA) 5 trial, where 128 children were enrolled in a randomised trial to evaluate the antiviral effect of NRTI combinations (3TC+Abacavir, 3TC+ZDV, Abacavir+ZDV) and the tolerability of adding NFV, there was no significant difference according to HIV-1 subtype in the virologic response to treatment or in the frequency of development of resistance among children (Pillay et al., 2002). A French cohort study of 416 adult patients, 24% of whom carried non-B subtypes, showed that at 3, 6 and 12 months after initiation of ARV therapy (first line regimens, subtype B: 65% PI-based regimens, 25% NNRTI-based regimens, 10% NRTI only; non-B subtype: 65% PI-based regimens, 30% NNRTI-based regimens, 5% NRTI only) HIV-1 subtype did not affect clinical progression, CD4 cell count, or viral load in response to treatment (Bocket et al., 2005). Frater *et al.* studied patients of African origin who were infected with a non-B subtype of HIV-1 and were living in London, and found no significant difference in the response to therapy (first line regimens, 50% PI-based regimens, 50% NNRTI-based regimens) among patients infected with subtypes A, C and D (Frater et al., 2001). Geretti and collaborators reported that patients infected with subtypes A, C, D and CRF02_AG were as likely to achieve viral load suppression (NRTI backbone: AZT+3TC or TDF+FTC or TDF+3TC or 3TC+d4T or d4T+ddI or ABC+3TC; third drug: EFV or NVP or RTV boosted PI) as those infected with subtype B and showed comparable rates of CD4 cell count recovery (Geretti et al., 2009). Other studies have analyzed virologic and immunologic responses to antiretroviral therapy according to the HIV-1 subtype and also did not find any differences (Alexander et al., 2002; Atlas et al., 2005; Bannister et al., 2006; De Wit et al., 2004; Nicastri et al., 2004). Thus, overall, it appears that HIV-1 subtypes do not have major differences in the response to ARV therapy. However, further studies should be designed, firstly to assess the efficacy of specific drug regimens in patients with non-B subtypes and secondly to evaluate the efficacy of these regimens in patients infected with particular non-B subtype species including the highly divergent H, J and K subtypes, complex CRFs and URFs which are common in African countries as wells as in patients infected with group O, N and P viruses. These studies should be performed within a single country in order to control for the many variables that might influence response to therapy, namely, adherence, ethnicity, psychosocial support and drug regimens.

Polymorphisms	Prevalence in subtype B	Prevalence in non-B subtypes
A98S	5%	70% G and 98% O
K103R	2.7%	98% O
V179E	0.4%	98% O
V179I	3.2%	50% A
Y181C	0%	100% O

Adapted from (Wainberg and Brenner, 2010).

Table 3. Polymorphisms in the RT that may impact HIV-1 resistance to NNRTIs

6. Impact of HIV-1 diversity in drug resistance

In the absence of any drug exposure, RT and PR sequences from B and non-B subtypes are polymorphic in about 40% of the first 240 RT amino acids and 30% of the 99 PR amino acids (Bartolo et al., 2009b; Bartolo et al., 2009c; Kantor and Katzenstein, 2004). Polymorphisms in the RT of non-B subtype viruses normally do not occur in known sites of resistance to NRTIs (Kantor and Katzenstein, 2003); in contrast, the PR from drug naive patients may contain amino acid substitutions associated with secondary resistance to some PIs in subtype B (ex. K20R, M36I, H69KQ) (Table 4) (Grossman et al., 2001; Holguin et al., 2004). However, these genotypic changes by themselves do not consistently confer decreased susceptibility to PIs when viral strains are subject to phenotypic testing (Descamps et al., 2005; Grossman et al., 2004; Ly et al., 2005; Maljkovic et al., 2003; Nkengafac et al., 2007; Palma et al., 2007; Paraskevis et al., 2005; Roudinskii et al., 2004; Tee, Kamarulzaman, and Ng, 2006; Vazquez de Parga et al., 2005; Wensing et al., 2005). Consistent with this, most observational studies performed *in vitro* and *in vivo* suggest that the currently available PR and RT inhibitors are as active against non-B subtype viruses as they are against subtype B viruses (Santos and Soares, 2010).

Different HIV genetic forms carry in their genomes genetic signatures and polymorphisms that could alter the structure of viral proteins which are targeted by drugs, thus impairing ARV drug binding and efficacy (Tables 3 and 4). A single nucleotide substitution from the wild-type codon found in subtype C can generate the mutation V106M, which is associated with NNRTIs resistance, while at least two substitutions are needed for the wild-type subtype B codon (Brenner et al., 2003; Loemba et al., 2002). This suggested that subtype C could have a lower genetic barrier to resistance to NNRTIs than subtype B, and that this V106M mutation could be more frequent in subtype C infected patients failing therapy, than in subtype B infected patients. Indeed, the clinical importance of the V106M mutation in non-B subtypes has been confirmed in several studies showing that V106M is more frequently seen in subtype C (and CRF01_AE) after therapy with EFV or NVP (Deshpande et al., 2007; Hsu et al., 2005; Marconi et al., 2008; Rajesh et al., 2009). The G190A mutation was also relatively more frequent in subtype C Indian and Israeli patients failling NNRTI-based regimens than in subtype B (Deshpande et al., 2007; Grossman et al., 2004).

In vitro, the emergence of the K65R mutation after therapy with TDF is faster in subtype C (15 weeks after TDF) than in subtype B (34-74 weeks) (Brenner et al., 2006; Coutsinos et al., 2010; Coutsinos et al., 2009). In contrast, K65R may be less frequent in subtype A than in all other subtypes (Gupta et al., 2005). Several studies suggest that there is a higher risk of development of K65R in subtype C infected patients failing ddI and d4T-containing regimens (Brenner and Coutsinos, 2009; Deshpande et al., 2010; Doualla-Bell et al., 2006; Hosseinipour et al., 2009; Orrell et al., 2009). A study from Israel reported a high frequency of K65R in subtype C viruses from Ethiopian immigrants in ARV therapy (Turner et al., 2009). However, K65R did not appear to emerge frequently in subtype C patients who participated in large clinical trials in which they received either TDF or TDF/FTC as part of a triple therapy regimen (Miller et al., 2007). In Malawi, in patients with subtype C viruses, differences observed in the emergence of the K65R mutation were significantly related to treatment regimen and disease stage (Hosseinipour et al., 2009). In addition, development of K65R in subtype C and CRF01_AE has been associated with the Y181C NVP mutation within the viral backbone (Brenner and Coutsinos, 2009; Zolfo et al., 2010 Set 22 [Epub ahead of print].). The presence of higher rates of the K65R mutation in subtype C in some studies (Doualla-Bell et al., 2006; Hosseinipour et al., 2009; Orrell et al., 2009) suggests that these viruses may have a particular predisposition toward acquiring this mutation. It has been proposed that a RNA template mechanism could explain the higher rates of K65R in subtype C viruses than in other subtypes. In this subtype, there is an intrinsic difficulty in synthesizing pol-A homopolymeric sequences that leads to template pausing at codon 65, facilitating the acquisition of K65R under selective drug pressure (Coutsinos et al., 2010; Coutsinos et al., 2009). The natural polymorphisms found in the RT of treatment-naive patients (10% in 726 patients) infected with HIV-1 non-B subtypes had no significant impact on susceptibility to ETR (Cotte et al., 2009; Derache et al., 2008; Maiga et al., 2010).

In PR, polymorphisms do not impair drug susceptibility but may affect the genetic pathway of resistance as soon as the virus generates a major resistant mutation (Martinez-Cajas et al., 2008). The rare minor V11I mutation, which is associated with DRV resistance, is a natural polymorphism in all CRF37_cpx isolates and some subtype A isolates (Bartolo et al., 2009b; Bartolo et al., 2009c; de Meyer et al., 2008; Poveda et al., 2007; Powell et al., 2007a), suggesting that these viruses may have a lower genetic barrier to DRV resistance. The V82I natural polymorphism in subtype G led to the emergence of I82M/T/S with treatment failure to IDV (Camacho et al., 2005). A study suggested that polymorphisms at position 36 in PR may play important roles in determining the emergence of specific patterns of resistance mutations among viruses of different subtypes (Lisovsky et al., 2010). Gonzáles *et al.* (Gonzalez et al., 2003) compared clinical isolates of C subtype with and without the I93L polymorphism, finding that hipersusceptibility to LPV in subtype C is strongly associated with the presence of that mutation.

Minor mutations	ARV	Prevalence in subtype B	Prevalence in non-B viruses
V11I	DRV	1%	100% CRF37_cpx and 4% in subtype A
I13V	TPV	13%	90%–98% in subtypes A, G and CRF02_AG, 4%–78% in other non-B subtypes
K20I	ATV	2%	93%–98% in subtypes G and CRF02_AG, 1%–3.5% in subtypes A, F and CRF01_AE
M36I	ATV, IDV, NFV and TPV	13%	81%–99% in several non-B subtypes
H69K	TPV	2%	96%–97% in subtypes A, C and G, CRF01_AE and CRF02_AG, 2% in subtype F
V82I	ATV	2%	87% in subtype G, 1%–6% in several non-B subtypes
I93L	ATV	33%	94% in subtype C, 5%–40% in several non-B subtypes

DRV, darunavir; TPV, tripranavir; ATV, atazanavir; IDV, indinavir; NFV, nelfinavir. Adapted from (Santos and Soares, 2010).

Table 4. Polymorphisms on the PR of HIV-1 non-B subtypes associated with resistance to PIs

The D30N mutation was not observed in CRF02_AG and CRF02_AE isolates from patients failing NFV therapy; rather, the N88S mutation emerged after NFV use in CRF01_AE and after IDV use in subtype B (Ariyoshi et al., 2003; Chaix et al., 2005). The M89I/V mutations have been observed in C, F and G subtypes in PI experienced patients (NFV, APV, IDV, LPV, ATV) but not in other subtypes (Abecasis et al., 2005). The L90M mutation, that confers resistance to NFV and SQV, is rare in subtype F but common in subtype B in patients from Brazil (Calazans et al., 2005). D30N has a stronger negative impact in the replicative capacity of C subtype than in B subtype (Gonzalez et al., 2004), which could explain the low frequency of this mutation observed in subtype C infected individuals failing NFV-containing regimens. A recent study in Portuguese patients, reported that mutation I54V/L was selected by NFV in subtype G isolates, a mutation not previously described for this drug in subtype B (Santos et al., 2009).

The frequency of polymorphisms in gp41 among different HIV-1 clades from T20 drug-naive patients is higher in non-B subtypes and recombinants than in subtype B viruses (P<0.001) especially at positions Q32L/T/N/K, R46K/Q, N43H, I37L, and V69L that are associated with resistance to T20 (Carmona et al., 2005). The N42S polymorphism, associated with increased susceptibility to T20, is detected more frequently in non-B subtypes than in B subtype (13% in B, 73% in A and 90% in G) (Carmona et al., 2005). A30V (in subtype G and CRF06_cpx) and Q56K/R (in subtypes A and J, CRF04_cpx, CRF09_cpx, CRF11_cpx, and CRF13_cpx), Q56R and S138A in group O, and S138A in group N are natural polymorphisms associated with T20 resistance (Holguin, De Arellano, and Soriano, 2007).

Integrase inhibitor-associated mutations (primary and secondary) are normally absent from HIV-1 subtype B isolates from patients receiving ARV regimens without raltegravir (RAL) and from untreated patients (Ceccherini-Silberstein et al., 2010a; Ceccherini-Silberstein et al., 2010b). A study found integrase gene polymorphisms present in more than 10% of the 97 analyzed sequences (subtype B and non-B) from patients treated with RAL but these polymorphisms showed no impact on virological outcome either at week 24 or at week 48 (Charpentier et al., 2010). N155H, Q148H/R/K with G140S/A, and Y143R/C are the described mutational patterns that confer resistance to RAL, with or without secondary mutations (Garrido et al., 2010). A study on natural polymorphisms and mutations associated with resistance to RAL in drug-naive and ARV-treated patients (all RAL naive) found that CRF02_AG and subtype C isolates could have a higher genetic barrier to the development of G140C or G140S compared to subtype B (Brenner et al., 2011). On group O viruses, natural presence of the E157Q mutation or E157E/Q mixture seems to confer resistance to RAL (Leoz et al., 2008).

These observations in non-B subtype viruses suggest that differences in drug resistance pathways between HIV-1 subtypes do exist. However, the accumulated evidence is insufficient to adequately assess the contribution of the innate genetic diversity of HIV-1 to resistance. Larger and more rigorous prospective studies in drug naive and treated patients are required to validate these hypotheses and it will be necessary to evaluate these mutations by the analysis of site-directed mutants in phenotypic resistance assays.

7. Impact of HIV-1 diversity in the performance of genotypic and phenotypic drug resistance assays

Drug resistance testing is extremely important for the management of ART therapy failure in HIV patients (Grant and Zolopa, 2009; Shafer, 2002; Taylor, Jayasuriya, and Smit, 2009). Genotypic and phenotypic assays are both used to detect resistance to ARV drugs that could compromise response to treatment (Vandamme et al., 2011). All current clinically used genotypic assays involve sequencing the genes whose proteins are targeted by the different antiretroviral drugs [pol (RT, PR and IN) and env], to detect mutations that are known to confer phenotypic drug resistance. There are two approved genotyping resistance assays commercially available, the ViroSeq HIV-1 genotyping system, version 2.0 (Eshleman et al., 2004) and the Trugene HIV-1 genotyping kit for drug resistance (Grant et al., 2003). Phenotypic assays measure the ability of an HIV-1 isolate to grow *in vitro* in the presence of an inhibitor, in comparison with a known susceptible strain.

The European HIV Drug Resistance Guidelines Panel recommends genotyping in most situations using updated and clinically evaluated interpretation systems (Vandamme et al., 2011). Genotypic assays are faster and cheaper than phenotypic assays (Vandamme et al., 2004). Nonetheless, the commercial genotypic tests are too expensive to be used in low-income countries. In-house methods for genotyping drug resistance mutations are recommended by WHO for surveillance of primary and secondary drug resistance (Bennett et al., 2008). The reported rate of success in amplification and sequencing with these methods in low-income countries ranges from 41 up to 100 % with non-B isolates (Bartolo et al., 2009b; Bartolo et al., 2009c; Bennett et al., 2008; Oliveira et al., 2011).

Several studies analyzed the performance of commercially available genotypic resistance assays and in-house methods in B and non-B strains (Aghokeng et al., 2011; Beddows et al., 2003; Fontaine et al., 2001; Jagodzinski et al., 2003; Maes et al., 2004). In commercial kits a greater degree of success was obtained when sequencing subtype B isolates compared to non-B isolates, and some studies report that alternative amplification/sequencing primers had to be used for some samples belonging to non-B subtypes. A Belgian study analyzed the performance of the ViroSeq HIV-1 Genotyping System in 383 samples comprising 12 different subtypes (Maes et al., 2004). Amplification failed in 8.4% of the samples and there was a lower performance in the amplification of non-B subtypes. The sequencing performance on the different subtypes showed a significant decrease of positive results for subtypes A, G and recombinant strains. As a result of sequencing problems, 18.5% of the samples had to be processed with in-house procedures. In Cameroon, where all groups of HIV-1 circulate, the sequencing efficiency of the ViroSeq assay was also evaluated (Aghokeng et al., 2011). The sequencing failures involved mainly the 5´end of the PR and the 3` end of the RT genes because of the high failure rate of primers A, D, F, and H. There was a high degree of polymorphism in non-B isolates in the areas for which these primers are designed. One study compared the two commercially available sequencing kits with a in-house genotyping system in HIV-1 samples from treated and untreated patients belonging to subtypes A through J (Fontaine et al., 2001). All the samples could be amplified and sequenced by the three systems; however, for all systems, alternative amplification/sequencing primers had to be used for some samples belonging to non-B subtypes.

Several studies have evaluated the use of DBS for HIV-1 genotypic resistance testing (reviewed in (Johannessen, 2010)). Nucleotide similarity between the two sample types ranged from 98.1 to 99.9%. Drug-resistant mutations found in plasma were detected in 82–100% of the corresponding DBS specimens. In all, these findings indicate that the performance of amplification and sequencing primers must be improved to allow good sequencing results and consequently fast and reliable resistance testing for all HIV-1 genetic forms. Validated in-house methods with primers designed on the basis of the local HIV genetic diversity are needed for low–resources settings.

Drug resistance interpretation algorithms are user friendly and helpful in the clinical setting to follow up HIV-infected patients. These algorithms have been developed to interpret complex patterns of resistance mutations in HIV-1 subtype B. The most frequently used clinically available systems are listed in Table 5. There are two types of systems, geno2pheno and *Virtual*Phenotype which try to predict viral phenotype under the assumption that phenotype predicts treatment response, whereas all others algorithms try primarily to predict treatment response based on information extracted from databases of genotypic and correlated phenotypic or treatment response data (Vandamme et al., 2011). Several studies have compared these algorithms in drug naive and treated patients infected with non-B subtypes to examine the influence of pre-existing polymorphisms on predictions of drug susceptibilities and the subsequent choice of therapy (Champenois et al., 2008; Depatureaux et al., 2011; Snoeck et al., 2006; Vergne et al., 2006a; Vergne et al., 2006b; Yebra et al., 2009). Most of these studies found some discordance between algorithms, which was related to the presence of naturally occurring polymorphisms in non-B subtypes (Champenois et al., 2008; Depatureaux et al., 2011; Snoeck et al., 2006; Vergne et al., 2006a; Vergne et al., 2006b; Yebra et al., 2009). A study showed that, according to available resistance algorithms, both B and non-B subtypes from drug naive patients were considered fully susceptible to PIs, except for TPV/RTV for which the ANRS algorithm scored non-B subtypes as naturally resistant (Champenois et al., 2008). The discordant results for TPV/RTV were due to differences in the mutations that are considered by the algorithms in the analysis. The ANRS algorithm takes in account TPV/RTV mutations that are considered natural polymorphisms in non-B subtypes (e.g. M36I, H69K and L89M). In another study, 68 drug naive and 9 highly ARV-experienced HIV-1 group O infected patients were analyzed (Depatureaux et al., 2011). Twelve minor resistance mutations, present in more than 75% of the PR sequences, led to the different algorithms giving discrepant results for NFV and SQV susceptibility.

A large study (5030 patients infected with different HIV-1 clades) found that the four algorithms analyzed agreed well on the level of resistance scored and that the discordances could be attributed to specific (subtype-dependent) combinations of mutations (Snoeck et al., 2006). In a comparison of five algorithms in HIV-1 sequences from drug naive patients, discordances were significantly higher in non-B vs. B variants for ddI, NVP, TPV, and fAPV, and were attributed to natural patterns of mutations in non-B subtypes (Yebra et al., 2009). Several other studies demonstrated that there was a lack of concordance between algorithms that predict treatment response based on phenotype and genotype (Holguin, Hertogs, and Soriano, 2003; Ross et al., 2005; Santos et al., 2009). These discrepancies indicate that the patterns of drug resistance mutations have not yet been completely clarified in non-B subtype variants. The use of certain algorithms could lead to an overestimation of the resistance in the analysis of specific non-B subtypes because of the lack of consensus in the resistance mutations considered although with increasing knowledge such discrepancies tend to diminish.

Tropism testing is recommended before the use of a CCR5 antagonist drug (Vandekerckhove et al., 2011). In general, the enhanced sensitivity Trofile (ESTA) assay (phenotypic assay) and V3 population genotyping are the recommended methods. A multicenter prospective study evaluated the performance of genotypic algorithms for prediction of HIV-1 coreceptor usage in comparison with a phenotypic assay for the determination of coreceptor usage (Recordon-Pinson et al., 2010). Researchers reported important differences between 13 algorithms in the sensitivity of detection of X4 isolates.

The most sensitive were PSSM and Geno2pheno, with sensitivities of about 60%; on the other hand the specificity was high for most algorithms. In other studies, higher sensitivities could be found for the same genotypic algorithms (Chueca et al., 2009; Raymond et al., 2008). Geno2pheno presented sensitivities of 88-93.7% and specificity of 87%, and PSSM with sensitivities of 77% and specificity of 94%. Overall, these studies validate genotypic algorithms for prediction of HIV-1 coreceptor use in antiretroviral-experienced patients infected with subtype B. Few studies have evaluated the performance of genotypic algorithms for prediction of HIV-1 coreceptor use in non-B subtype viruses. An initial report showed a poor performance of genotypic tools for non-B subtypes (A-J, CRF01_AE, CRF02_AG, CRF11, CRF12_BF, CRF14_BG, URFs, and U samples), where they particularly failed to detect X4 strains (Garrido et al., 2008). Other studies found that main genotypic algorithms perform well when applied to CRF02_AG (Raymond et al., 2009) and subtype C viruses (Raymond et al., 2010). Additional studies are needed to evaluate the performance of these genotypic tools to predict coreceptor use in non-B subtypes.

System	Levels of resistance	Web Site
HIV DB Stanford	S, PL, LL, IR, HR	http://hivdb.stanford.edu/
REGA	S, I, R	http://www.kuleuven.ac.be/rega/cev/links/
ANRS	S, I, R	http://www.hivfrenchresistance.org/index.html
GenoSure	S, RP, R	http://www.monogramhiv.com
ResRis	S, I, R	http://www.retic-ris.net
HIVGrade	S, I, LS, R	http://www.hiv-grade.de
AntiRetroScan	100/75/50/25/0#	http://www.hivarca.net/includeGenpub/AntiRetroScan.html
HIV-TRePS	Quantitative*	http://www.eurist.org
EuResist Network	Quantitative*	http://www.eurist.org
Geno2pheno	Quantitative, S, I, R	http://www.geno2pheno.org
Virco	Quantitative+	http://www.vircolab.com
ViroSeq	S, P, R	http://www.abbotmolecular.com
TruGene	S, I, R	http://www.labnews.com

S, susceptible; PL, possible low-level resistance; LL, low-level resistance; IR or I, intermediate resistance; HR, high level resistance; R, resistance; RP or P, resistance possible; LS, low susceptibility; #100/75/50/25/0 in %activity with drug-GSS weighting factor; *probability for short-term response with specific drug combinations; +lower clinical cut-off at 20% of loss of response, upper to 80%. Adapted from (Vandamme et al., 2011).

Table 5. Drug resistance interpretation algorithms

8. References

Abecasis, A. B., Deforche, K., Snoeck, J., Bacheler, L. T., McKenna, P., Carvalho, A. P., Gomes, P., Camacho, R. J., and Vandamme, A. M. (2005). Protease mutation M89I/V is linked to therapy failure in patients infected with the HIV-1 non-B subtypes C, F or G. *AIDS* 19(16), 1799-806.

Abecasis, A. B., Martins, A.. Costa, I., Carvalho, A. P., Diogo, I., Gomes, P., and Camacho, R. J. (2011). Molecular epidemiological analysis of paired pol/env sequences from Portuguese HIV type 1 patients. *AIDS Res Hum Retroviruses* 27(7), 803-5.

Adojaan, M., Kivisild, T., Mannik, A., Krispin, T., Ustina, V., Zilmer, K., Liebert, E., Jaroslavtsev, N., Priimagi, L., Tefanova, V., Schmidt, J., Krohn, K., Villems, R., Salminen, M., and Ustav, M. (2005). Predominance of a rare type of HIV-1 in Estonia. *J Acquir Immune Defic Syndr* 39(5), 598-605.

Aghokeng, A. F., Mpoudi-Ngole, E., Chia, J. E., Edoul, E. M., Delaporte, E., and Peeters, M. (2011). High failure rate of the ViroSeq HIV-1 genotyping system for drug resistance testing in Cameroon, a country with broad HIV-1 genetic diversity. *J Clin Microbiol* 49(4), 1635-41.

Aghokeng, A. F., Mpoudi-Ngole, E., Dimodi, H., Atem-Tambe, A., Tongo, M., Butel, C., Delaporte, E., and Peeters, M. (2009). Inaccurate diagnosis of HIV-1 group M and O is a key challenge for ongoing universal access to antiretroviral treatment and HIV prevention in Cameroon. *PLoS One* 4(11), e7702.

Ahmad, N. (2005). The vertical transmission of human immunodeficiency virus type 1: molecular and biological properties of the virus. *Crit Rev Clin Lab Sci* 42(1), 1-34.

AIDSInfo (May 24, 2010). Panel on Treatment of HIV-Infected Pregnant Women and Prevention of Perinatal Transmission. Recommendations for Use of Antiretroviral Drugs in Pregnant HIV-1-Infected Women for Maternal Health and Interventions to Reduce Perinatal HIV Transmission in the United States. http://aidsinfo.nih.gov/ContentFiles/PerinatalGL.pdfAIDSinfo.

Akouamba, B. S., Viel, J., Charest, H., Merindol, N., Samson, J., Lapointe, N., Brenner, B. G., Lalonde, R., Harrigan, P. R., Boucher, M., and Soudeyns, H. (2005). HIV-1 genetic diversity in antenatal cohort, Canada. *Emerg Infect Dis* 11(8), 1230-4.

Alaeus, A., Lidman, K., Bjorkman, A., Giesecke, J., and Albert, J. (1999). Similar rate of disease progression among individuals infected with HIV-1 genetic subtypes A-D. *AIDS* 13(8), 901-7.

Alexander, C. S., Montessori, V., Wynhoven, B., Dong, W., Chan, K., O'Shaughnessy, M. V., Mo, T., Piaseczny, M., Montaner, J. S., and Harrigan, P. R. (2002). Prevalence and response to antiretroviral therapy of non-B subtypes of HIV in antiretroviral-naive individuals in British Columbia. *Antivir Ther* 7(1), 31-5.

Amornkul, P. N., Tansuphasawadikul, S., Limpakarnjanarat, K., Likanonsakul, S., Young, N., Eampokalap, B., Kaewkungwal, J., Naiwatanakul, T., Von Bargen, J., Hu, D. J., and Mastro, T. D. (1999). Clinical disease associated with HIV-1 subtype B' and E infection among 2104 patients in Thailand. *AIDS* 13(14), 1963-9.

Andreotti, M., Pirillo, M., Guidotti, G., Ceffa, S., Paturzo, G., Germano, P., Luhanga, R., Chimwaza, D., Mancini, M. G., Marazzi, M. C., Vella, S., Palombi, L., and Giuliano, M. (2010). Correlation between HIV-1 viral load quantification in plasma, dried blood spots, and dried plasma spots using the Roche COBAS Taqman assay. *J Clin Virol* 47(1), 4-7.

Annaz, H., Recordon-Pinson, P., Baba, N., Sedrati, O., Mrani, S., and Fleury, H. (2011). Presence of Drug Resistance Mutations Among Drug-Naive Patients in Morocco. AIDS Res Hum Retroviruses.27 (8): 917-920.

Apetrei, C., Loussert-Ajaka, I., Descamps, D., Damond, F., Saragosti, S., Brun-Vezinet, F., and Simon, F. (1996). Lack of screening test sensitivity during HIV-1 non-subtype B seroconversions. *AIDS* 10(14), F57-60.

Ariyoshi, K., Matsuda, M., Miura, H., Tateishi, S., Yamada, K., and Sugiura, W. (2003). Patterns of point mutations associated with antiretroviral drug treatment failure in CRF01_AE (subtype E) infection differ from subtype B infection. *J Acquir Immune Defic Syndr* 33(3), 336-42.

Atlas, A., Granath, F., Lindstrom, A., Lidman, K., Lindback, S., and Alaeus, A. (2005). Impact of HIV type 1 genetic subtype on the outcome of antiretroviral therapy. *AIDS Res Hum Retroviruses* 21(3), 221-7.

Attia, S., Egger, M., Muller, M., Zwahlen, M., and Low, N. (2009). Sexual transmission of HIV according to viral load and antiretroviral therapy: systematic review and meta-analysis. *AIDS* 23(11), 1397-404.

Avila, M. M., Pando, M. A., Carrion, G., Peralta, L. M., Salomon, H., Carrillo, M. G., Sanchez, J., Maulen, S., Hierholzer, J., Marinello, M., Negrete, M., Russell, K. L., and Carr, J. K. (2002). Two HIV-1 epidemics in Argentina: different genetic subtypes associated with different risk groups. *J Acquir Immune Defic Syndr* 29(4), 422-6.

Bannister, W. P., Ruiz, L., Loveday, C., Vella, S., Zilmer, K., Kjaer, J., Knysz, B., Phillips, A. N., and Mocroft, A. (2006). HIV-1 subtypes and response to combination antiretroviral therapy in Europe. *Antivir Ther* 11(6), 707-15.

Bártolo, I., Abecasis, A. B., Borrego, P., Barroso, H., McCutchan, F., Gomes, P., Camacho, R., and Taveira, N. (2011). Origin and Epidemiological History of HIV-1 CRF14_BG. *PLoS ONE, Accepted*

Bartolo, I., Camacho, R., Barroso, H., Bezerra, V., and Taveira, N. (2009a). Rapid clinical progression to AIDS and death in a persistently seronegative HIV-1 infected heterosexual young man. *AIDS* 23(17), 2359-62.

Bartolo, I., Casanovas, J., Bastos, R., Rocha, C., Abecasis, A. B., Folgosa, E., Mondlane, J., Manuel, R., and Taveira, N. (2009b). HIV-1 genetic diversity and transmitted drug resistance in health care settings in Maputo, Mozambique. *J Acquir Immune Defic Syndr* 51(3), 323-31.

Bartolo, I., Epalanga, M., Bartolomeu, J., Fonseca, M., Mendes, A., Gama, A., and Taveira, N. (2005). High genetic diversity of human immunodeficiency virus type 1 in Angola. *AIDS Res Hum Retroviruses* 21(4), 306-10.

Bartolo, I., Rocha, C., Bartolomeu, J., Gama, A., Fonseca, M., Mendes, A., Cristina, F., Thamm, S., Epalanga, M., Silva, P. C., and Taveira, N. (2009c). Antiretroviral drug resistance surveillance among treatment-naive human immunodeficiency virus type 1-infected individuals in Angola: evidence for low level of transmitted drug resistance. *Antimicrob Agents Chemother* 53(7), 3156-8.

Bartolo, I., Rocha, C., Bartolomeu, J., Gama, A., Marcelino, R., Fonseca, M., Mendes, A., Epalanga, M., Silva, P. C., and Taveira, N. (2009d). Highly divergent subtypes and new recombinant forms prevail in the HIV/AIDS epidemic in Angola: New insights into the origins of the AIDS pandemic. *Infect Genet Evol* 9, 672-682. .

Beddows, S., Galpin, S., Kazmi, S. H., Ashraf, A., Johargy, A., Frater, A. J., White, N., Braganza, R., Clarke, J., McClure, M., and Weber, J. N. (2003). Performance of two commercially available sequence-based HIV-1 genotyping systems for the detection of drug resistance against HIV type 1 group M subtypes. *J Med Virol* 70(3), 337-42.

Beelaert, G., and Fransen, K. (2010). Evaluation of a rapid and simple fourth-generation HIV screening assay for qualitative detection of HIV p24 antigen and/or antibodies to HIV-1 and HIV-2. *J Virol Methods* 168(1-2), 218-22.

Bennett, D. E., Myatt, M., Bertagnolio, S., Sutherland, D., and Gilks, C. F. (2008). Recommendations for surveillance of transmitted HIV drug resistance in countries scaling up antiretroviral treatment. *Antivir Ther* 13 Suppl 2, 25-36.

Beyrer, C., Patel, Z., Stachowiak, J. A., Tishkova, F. K., Stibich, M. A., Eyzaguirre, L. M., Carr, J. K., Mogilnii, V., Peryshkina, A., Latypov, A., and Strathdee, S. A. (2009). Characterization of the emerging HIV type 1 and HCV epidemics among injecting drug users in Dushanbe, Tajikistan. *AIDS Res Hum Retroviruses* 25(9), 853-60.

(2009). Assays to Estimate HIV Incidence and Detect Acute HIV Infection. Bill & Melinda Gates Foundation.

Bobkov, A., Cheingsong-Popov, R., Selimova, L., Ladnaya, N., Kazennova, E., Kravchenko, A., Fedotov, E., Saukhat, S., Zverev, S., Pokrovsky, V., and Weber, J. (1997). An HIV type 1 epidemic among injecting drug users in the former Soviet Union caused by a homogeneous subtype A strain. *AIDS Res Hum Retroviruses* 13(14), 1195-201.

Bobkov, A. F., Kazennova, E. V., Sukhanova, A. L., Bobkova, M. R., Pokrovsky, V. V., Zeman, V. V., Kovtunenko, N. G., and Erasilova, I. B. (2004). An HIV type 1 subtype A outbreak among injecting drug users in Kazakhstan. *AIDS Res Hum Retroviruses* 20(10), 1134-6.

Bocket, L., Cheret, A., Deuffic-Burban, S., Choisy, P., Gerard, Y., de la Tribonniere, X., Viget, N., Ajana, F., Goffard, A., Barin, F., Mouton, Y., and Yazdanpanah, Y. (2005). Impact of human immunodeficiency virus type 1 subtype on first-line antiretroviral therapy effectiveness. *Antivir Ther* 10(2), 247-54.

Bogh, M., Machuca, R., Gerstoft, J., Pedersen, C., Obel, N., Kvinesdal, B., Nielsen, H., and Nielsen, C. (2001). Subtype-specific problems with qualitative Amplicor HIV-1 DNA PCR test. *J Clin Virol* 20(3), 149-53.

Bourlet, T., Signori-Schmuck, A., Roche, L., Icard, V., Saoudin, H., Trabaud, M. A., Tardy, J. C., Morand, P., Pozzetto, B., Ecochard, R., and Andre, P. (2011). HIV-1 load comparison using four commercial real-time assays. *J Clin Microbiol* 49(1), 292-7.

Brennan, C. A., Bodelle, P., Coffey, R., Devare, S. G., Golden, A., Hackett, J., Jr., Harris, B., Holzmayer, V., Luk, K. C., Schochetman, G., Swanson, P., Yamaguchi, J., Vallari, A., Ndembi, N., Ngansop, C., Makamche, F., Mbanya, D., Gurtler, L. G., Zekeng, L., and Kaptue, L. (2008). The prevalence of diverse HIV-1 strains was stable in Cameroonian blood donors from 1996 to 2004. *J Acquir Immune Defic Syndr* 49(4), 432-9.

Brennan, C. A., Bodelle, P., Coffey, R., Harris, B., Holzmayer, V., Luk, K. C., Swanson, P., Yamaguchi, J., Vallari, A., Devare, S. G., Schochetman, G., and Hackett, J., Jr. (2006). HIV global surveillance: foundation for retroviral discovery and assay development. *J Med Virol* 78 Suppl 1, S24-9.

Brennan, C. A., Yamaguchi, J., Devare, S. G., Foster, G. A., and Stramer, S. L. (2010). Expanded evaluation of blood donors in the United States for human immunodeficiency virus type 1 non-B subtypes and antiretroviral drug-resistant strains: 2005 through 2007. *Transfusion* 50(12), 2707-12.

Brenner, B., Oliveira, M., Doualla-Bell, F., Moisi, D., Ntemgwa, M., Frankel, F., Essex, M., and Wainberg, M. (2006). HIV-1 subtype C viruses rapidly develop K65R resistance to tenofovir in cell culture. *AIDS* 20(9), F9-13.

Brenner, B., Turner, D., Oliveira, M., Moisi, D., Detorio, M., Carobene, M., Marlink, R. G., Schapiro, J., Roger, M., and Wainberg, M. A. (2003). A V106M mutation in HIV-1 clade C viruses exposed to efavirenz confers cross-resistance to non-nucleoside reverse transcriptase inhibitors. *AIDS* 17(1), F1-5.

Brenner, B. G. (2007). Resistance and viral subtypes: how important are the differences and why do they occur? *Curr Opin HIV AIDS* 2(2), 94-102.

Brenner, B. G., and Coutsinos, D. (2009). The K65R mutation in HIV-1 reverse transcriptase: genetic barriers, resistance profile and clinical implications. *HIV Ther* 3(6), 583-594.

Brenner, B. G., Lowe, M., Moisi, D., Hardy, I., Gagnon, S., Charest, H., Baril, J. G., Wainberg, M. A., and Roger, M. (2011). Subtype diversity associated with the development of HIV-1 resistance to integrase inhibitors. *J Med Virol* 83(5), 751-9.

Buonaguro, L., Tornesello, M. L., and Buonaguro, F. M. (2007). Human immunodeficiency virus type 1 subtype distribution in the worldwide epidemic: pathogenetic and therapeutic implications. *J Virol* 81, 10209-10219.

Calazans, A., Brindeiro, R., Brindeiro, P., Verli, H., Arruda, M. B., Gonzalez, L. M., Guimaraes, J. A., Diaz, R. S., Antunes, O. A., and Tanuri, A. (2005). Low accumulation of L90M in protease from subtype F HIV-1 with resistance to protease inhibitors is caused by the L89M polymorphism. *J Infect Dis* 191(11), 1961-70.

Camacho, R., Godinho, A., Gomes, P., Abecasis, A., Vandamme, A.-M., Palma, C., Carvalho, A., Cabanas, J., and Gonçalves, J. (2005). Different substitutions under drug pressure at protease codon 82 in HIV-1 subtype G compared to subtype B infected individuals including a novel I82M resistance mutation. *Antivir Ther.* 10, Suppl 1.

Carmona, R., Perez-Alvarez, L., Munoz, M., Casado, G., Delgado, E., Sierra, M., Thomson, M., Vega, Y., Vazquez de Parga, E., Contreras, G., Medrano, L., and Najera, R. (2005). Natural resistance-associated mutations to Enfuvirtide (T20) and polymorphisms in the gp41 region of different HIV-1 genetic forms from T20 naive patients. *J Clin Virol* 32(3), 248-53.

Carr, J. K., Osinusi, A., Flynn, C. P., Gilliam, B. L., Maheshwari, V., and Zhao, R. Y. (2010a). Two independent epidemics of HIV in Maryland. *J Acquir Immune Defic Syndr* 54(3), 297-303.

Carr, J. K., Wolfe, N. D., Torimiro, J. N., Tamoufe, U., Mpoudi-Ngole, E., Eyzaguirre, L., Birx, D. L., McCutchan, F. E., and Burke, D. S. (2010b). HIV-1 recombinants with multiple parental strains in low-prevalence, remote regions of Cameroon: evolutionary relics? *Retrovirology* 7, 39.

Casper, C., Naver, L., Clevestig, P., Belfrage, E., Leitner, T., Albert, J., Lindgren, S., Ottenblad, C., Bohlin, A. B., Fenyo, E. M., and Ehrnst, A. (2002). Coreceptor change appears after immune deficiency is established in children infected with different HIV-1 subtypes. *AIDS Res Hum Retroviruses* 18(5), 343-52.

Castro, E., Khonkarly, M., Ciuffreda, D., Burgisser, P., Cavassini, M., Yerly, S., Pantaleo, G., and Bart, P. A. (2010). HIV-1 drug resistance transmission networks in southwest Switzerland. *AIDS Res Hum Retroviruses* 26(11), 1233-8.

Ceccherini-Silberstein, F., Malet, I., Fabeni, L., Dimonte, S., Svicher, V., D'Arrigo, R., Artese, A., Costa, G., Bono, S., Alcaro, S., Monforte, A., Katlama, C., Calvez, V., Antinori, A., Marcelin, A. G., and Perno, C. F. (2010a). Specific HIV-1 integrase polymorphisms change their prevalence in untreated versus antiretroviral-treated HIV-1-infected patients, all naive to integrase inhibitors. *J Antimicrob Chemother* 65(11), 2305-18.

Ceccherini-Silberstein, F., Van Baelen, K., Armenia, D., Trignetti, M., Rondelez, E., Fabeni, L., Scopelliti, F., Pollicita, M., Van Wesenbeeck, L., Van Eygen, V., Dori, L., Sarmati, L., Aquaro, S., Palamara, G., Andreoni, M., Stuyver, L. J., and Perno, C. F. (2010b). Secondary integrase resistance mutations found in HIV-1 minority quasispecies in integrase therapy-naive patients have little or no effect on susceptibility to integrase inhibitors. *Antimicrob Agents Chemother* 54(9), 3938-48.

Chaillet, P., Tayler-Smith, K., Zachariah, R., Duclos, N., Moctar, D., Beelaert, G., and Fransen, K. (2010). Evaluation of four rapid tests for diagnosis and differentiation of HIV-1 and HIV-2 infections in Guinea-Conakry, West Africa. *Trans R Soc Trop Med Hyg* 104(9), 571-6.

Chaix, M. L., Rouet, F., Kouakoussui, K. A., Laguide, R., Fassinou, P., Montcho, C., Blanche, S., Rouzioux, C., and Msellati, P. (2005). Genotypic human immunodeficiency virus type 1 drug resistance in highly active antiretroviral therapy-treated children in Abidjan, Cote d'Ivoire. *Pediatr Infect Dis J* 24(12), 1072-6.

Champenois, K., Bocket, L., Deuffic-Burban, S., Cotte, L., Andre, P., Choisy, P., and Yazdanpanah, Y. (2008). Expected response to protease inhibitors of HIV-1 non-B subtype viruses according to resistance algorithms. *AIDS* 22(9), 1087-9.

Charpentier, C., Roquebert, B., Colin, C., Taburet, A. M., Fagard, C., Katlama, C., Molina, J. M., Jacomet, C., Brun-Vezinet, F., Chene, G., Yazdanpanah, Y., and Descamps, D. (2010). Resistance analyses in highly experienced patients failing raltegravir, etravirine and darunavir/ritonavir regimen. *AIDS* 24(17), 2651-6.

Chen, J. H., Wong, K. H., Chen, Z., Chan, K., Lam, H. Y., To, S. W., Cheng, V. C., Yuen, K. Y., and Yam, W. C. (2010). Increased genetic diversity of HIV-1 circulating in Hong Kong. *PLoS One* 5(8), e12198.

Christopherson, C., Sninsky, J., and Kwok, S. (1997). The effects of internal primer-template mismatches on RT-PCR: HIV-1 model studies. *Nucleic Acids Res* 25(3), 654-8.

Chueca, N., Garrido, C., Alvarez, M., Poveda, E., de Dios Luna, J., Zahonero, N., Hernandez-Quero, J., Soriano, V., Maroto, C., de Mendoza, C., and Garcia, F. (2009). Improvement in the determination of HIV-1 tropism using the V3 gene sequence and a combination of bioinformatic tools. *J Med Virol* 81(5), 763-7.

Church, D., Gregson, D., Lloyd, T., Klein, M., Beckthold, B., Laupland, K., and Gill, J. (2011). HIV-1 Viral Load Multi-Assay Comparison of the RealTime HIV-1, COBAS TaqMan 48 v 1.0, Easy Q v1.2 and Versant v3.0 assays in a Cohort of Canadian Patients with Diverse HIV Subtype Infections. *J Clin Microbiol* 49(1), 118-124.

Cilliers, T., Nhlapo, J., Coetzer, M., Orlovic, D., Ketas, T., Olson, W. C., Moore, J. P., Trkola, A., and Morris, L. (2003). The CCR5 and CXCR4 coreceptors are both used by human immunodeficiency virus type 1 primary isolates from subtype C. *J Virol* 77(7), 4449-56.

Collins, M. L., Irvine, B., Tyner, D., Fine, E., Zayati, C., Chang, C., Horn, T., Ahle, D., Detmer, J., Shen, L. P., Kolberg, J., Bushnell, S., Urdea, M. S., and Ho, D. D. (1997). A branched DNA signal amplification assay for quantification of nucleic acid targets below 100 molecules/ml. *Nucleic Acids Res* 25(15), 2979-84.

Conroy, S. A., Laeyendecker, O., Redd, A. D., Collinson-Streng, A., Kong, X., Makumbi, F., Lutalo, T., Sewankambo, N., Kiwanuka, N., Gray, R. H., Wawer, M. J., Serwadda, D., and Quinn, T. C. (2010). Changes in the distribution of HIV type 1 subtypes D and A in Rakai District, Uganda between 1994 and 2002. *AIDS Res Hum Retroviruses* 26(10), 1087-91.

Cornelissen, M., van Den Burg, R., Zorgdrager, F., and Goudsmit, J. (2000). Spread of distinct human immunodeficiency virus type 1 AG recombinant lineages in Africa. *J Gen Virol* 81(Pt 2), 515-23.

Cotte, L., Trabaud, M. A., Tardy, J. C., Brochier, C., Gilibert, R. P., Miailhes, P., Trepo, C., and Andre, P. (2009). Prediction of the virological response to etravirine in clinical practice: Comparison of three genotype algorithms. *J Med Virol* 81(4), 672-7.

Coutsinos, D., Invernizzi, C. F., Xu, H., Brenner, B. G., and Wainberg, M. A. (2010). Factors affecting template usage in the development of K65R resistance in subtype C variants of HIV type-1. *Antivir Chem Chemother* 20(3), 117-31.

Coutsinos, D., Invernizzi, C. F., Xu, H., Moisi, D., Oliveira, M., Brenner, B. G., and Wainberg, M. A. (2009). Template usage is responsible for the preferential acquisition of the K65R reverse transcriptase mutation in subtype C variants of human immunodeficiency virus type 1. *J Virol* 83(4), 2029-33.

Creek, T. L., Sherman, G. G., Nkengasong, J., Lu, L., Finkbeiner, T., Fowler, M. G., Rivadeneira, E., and Shaffer, N. (2007). Infant human immunodeficiency virus diagnosis in resource-limited settings: issues, technologies, and country experiences. *Am J Obstet Gynecol* 197(3 Suppl), S64-71.

De Mendoza, C., Garrido, C., Poveda, E., Corral, A., Zahonero, N., Trevino, A., Anta, L., and Soriano, V. (2009). Changes in drug resistance patterns following the introduction of HIV type 1 non-B subtypes in Spain. *AIDS Res Hum Retroviruses* 25(10), 967-72.

de Mendoza, C., Koppelman, M., Montes, B., Ferre, V., Soriano, V., Cuypers, H., Segondy, M., and Oosterlaken, T. (2005). Multicenter evaluation of the NucliSens EasyQ HIV-1 v1.1 assay for the quantitative detection of HIV-1 RNA in plasma. *J Virol Methods* 127(1), 54-9.

de Meyer, S., Vangeneugden, T., van Baelen, B., de Paepe, E., van Marck, H., Picchio, G., Lefebvre, E., and de Bethune, M. P. (2008). Resistance profile of darunavir: combined 24-week results from the POWER trials. *AIDS Res Hum Retroviruses* 24(3), 379-88.

De Wit, S., Boulme, R., Poll, B., Schmit, J. C., and Clumeck, N. (2004). Viral load and CD4 cell response to protease inhibitor-containing regimens in subtype B versus non-B treatment-naive HIV-1 patients. *AIDS* 18(17), 2330-1.

Depatureaux, A., Charpentier, C., Leoz, M., Unal, G., Damond, F., Kfutwah, A., Vessiere, A., Simon, F., and Plantier, J. C. (2011). Impact of HIV-1 Group O Genetic Diversity on Genotypic Resistance Interpretation by Algorithms Designed for HIV-1 Group M. *J Acquir Immune Defic Syndr* 56(2), 139-145.

Derache, A., Maiga, A. I., Traore, O., Akonde, A., Cisse, M., Jarrousse, B., Koita, V., Diarra, B., Carcelain, G., Barin, F., Pizzocolo, C., Pizarro, L., Katlama, C., Calvez, V., and Marcelin, A. G. (2008). Evolution of genetic diversity and drug resistance mutations in HIV-1 among untreated patients from Mali between 2005 and 2006. *J Antimicrob Chemother* 62(3), 456-63.

Descamps, D., Chaix, M. L., Andre, P., Brodard, V., Cottalorda, J., Deveau, C., Harzic, M., Ingrand, D., Izopet, J., Kohli, E., Masquelier, B., Mouajjah, S., Palmer, P., Pellegrin, I., Plantier, J. C., Poggi, C., Rogez, S., Ruffault, A., Schneider, V., Signori-Schmuck, A., Tamalet, C., Wirden, M., Rouzioux, C., Brun-Vezinet, F., Meyer, L., and Costagliola, D. (2005). French national sentinel survey of antiretroviral drug resistance in patients with HIV-1 primary infection and in antiretroviral-naive chronically infected patients in 2001-2002. *J Acquir Immune Defic Syndr* 38(5), 545-52.

Descamps, D., Collin, G., Letourneur, F., Apetrei, C., Damond, F., Loussert-Ajaka, I., Simon, F., Saragosti, S., and Brun-Vezinet, F. (1997). Susceptibility of human immunodeficiency virus type 1 group O isolates to antiretroviral agents: in vitro phenotypic and genotypic analyses. *J Virol* 71(11), 8893-8.

Deshpande, A., Jauvin, V., Magnin, N., Pinson, P., Faure, M., Masquelier, B., Aurillac-Lavignolle, V., and Fleury, H. J. (2007). Resistance mutations in subtype C HIV type 1 isolates from Indian patients of Mumbai receiving NRTIs plus NNRTIs and experiencing a treatment failure: resistance to AR. *AIDS Res Hum Retroviruses* 23(2), 335-40.

Deshpande, A., Jeannot, A. C., Schrive, M. H., Wittkop, L., Pinson, P., and Fleury, H. J. (2010). Analysis of RT sequences of subtype C HIV-type 1 isolates from indian patients at failure of a first-line treatment according to clinical and/or immunological WHO guidelines. *AIDS Res Hum Retroviruses* 26(3), 343-50.

Dickover, R., Garratty, E., Yusim, K., Miller, C., Korber, B., and Bryson, Y. (2006). Role of maternal autologous neutralizing antibody in selective perinatal transmission of human immunodeficiency virus type 1 escape variants. *J Virol* 80(13), 6525-33.

Dickover, R. E., Garratty, E. M., Plaeger, S., and Bryson, Y. J. (2001). Perinatal transmission of major, minor, and multiple maternal human immunodeficiency virus type 1 variants in utero and intrapartum. *J Virol* 75(5), 2194-203.

Djoko, C. F., Rimoin, A. W., Vidal, N., Tamoufe, U., Wolfe, N. D., Butel, C., Lebreton, M., Tshala, F. M., Kayembe, P. K., Muyembe, J. J., Edidi-Basepeo, S., Pike, B. L., Fair, J. N., Mbacham, W. F., Saylors, K. E., Mpoudi-Ngole, E., Delaporte, E., Grillo, M., and Peeters, M. (2011). High HIV Type 1 Group M pol Diversity and Low Rate of Antiretroviral Resistance Mutations Among the 13 Uniformed Services in Kinshasa, DRC. AIDS Res Hum Retroviruses. 27 (3): 323-329Doualla-Bell, F., Avalos, A., Brenner, B., Gaolathe, T., Mine, M., Gaseitsiwe, S., Oliveira, M., Moisi, D., Ndwapi, N., Moffat, H., Essex, M., and Wainberg, M. A. (2006). High prevalence of the K65R mutation in human immunodeficiency virus type 1 subtype C isolates from infected patients in Botswana treated with didanosine-based regimens. *Antimicrob Agents Chemother* 50(12), 4182-5.

Dumans, A. T., Soares, M. A., Machado, E. S., Hue, S., Brindeiro, R. M., Pillay, D., and Tanuri, A. (2004). Synonymous genetic polymorphisms within Brazilian human immunodeficiency virus Type 1 subtypes may influence mutational routes to drug resistance. *J Infect Dis* 189(7), 1232-8.

Dyer, J. R., Pilcher, C. D., Shepard, R., Schock, J., Eron, J. J., and Fiscus, S. A. (1999). Comparison of NucliSens and Roche Monitor assays for quantitation of levels of human immunodeficiency virus type 1 RNA in plasma. *J Clin Microbiol* 37(2), 447-9.

Easterbrook, P. J., Smith, M., Mullen, J., O'Shea, S., Chrystie, I., de Ruiter, A., Tatt, I. D., Geretti, A. M., and Zuckerman, M. (2010). Impact of HIV-1 viral subtype on disease progression and response to antiretroviral therapy. *J Int AIDS Soc* 13, 4.

Eberle, J., Loussert-Ajaka, I., Brust, S., Zekeng, L., Hauser, P. H., Kaptue, L., Knapp, S., Damond, F., Saragosti, S., Simon, F., and Gurtler, L. G. (1997). Diversity of the immunodominant epitope of gp41 of HIV-1 subtype O and its validity for antibody detection. *J Virol Methods* 67(1), 85-91.

Eshleman, S. H., Church, J. D., Chen, S., Guay, L. A., Mwatha, A., Fiscus, S. A., Mmiro, F., Musoke, P., Kumwenda, N., Jackson, J. B., Taha, T. E., and Hoover, D. R. (2006). Comparison of HIV-1 mother-to-child transmission after single-dose nevirapine prophylaxis among African women with subtypes A, C, and D. *J Acquir Immune Defic Syndr* 42(4), 518-21.

Eshleman, S. H., Guay, L. A., Mwatha, A., Brown, E., Musoke, P., Mmiro, F., and Jackson, J. B. (2005a). Comparison of mother-to-child transmission rates in Ugandan women with subtype A versus D HIV-1 who received single-dose nevirapine prophylaxis: HIV Network For Prevention Trials 012. *J Acquir Immune Defic Syndr* 39(5), 593-7.

Eshleman, S. H., Hackett, J., Jr., Swanson, P., Cunningham, S. P., Drews, B., Brennan, C., Devare, S. G., Zekeng, L., Kaptue, L., and Marlowe, N. (2004). Performance of the Celera Diagnostics ViroSeq HIV-1 Genotyping System for sequence-based analysis of diverse human immunodeficiency virus type 1 strains. *J Clin Microbiol* 42(6), 2711-7.

Eshleman, S. H., Hoover, D. R., Chen, S., Hudelson, S. E., Guay, L. A., Mwatha, A., Fiscus, S. A., Mmiro, F., Musoke, P., Jackson, J. B., Kumwenda, N., and Taha, T. (2005b). Nevirapine (NVP) resistance in women with HIV-1 subtype C, compared with subtypes A and D, after the administration of single-dose NVP. *J Infect Dis* 192(1), 30-6.

Esteves, A., Parreira, R., Piedade, J., Venenno, T., Franco, M., Germano de Sousa, J., Patricio, L., Brum, P., Costa, A., and Canas-Ferreira, W. F. (2003). Spreading of HIV-1 subtype G and envB/gagG recombinant strains among injecting drug users in Lisbon, Portugal. *AIDS Res Hum Retroviruses* 19(6), 511-7.

Esteves, A., Parreira, R., Venenno, T., Franco, M., Piedade, J., Germano De Sousa, J., and Canas-Ferreira, W. F. (2002). Molecular epidemiology of HIV type 1 infection in Portugal: high prevalence of non-B subtypes. *AIDS Res Hum Retroviruses* 18(5), 313-25.

European AIDS Clinical Society (2009). Guidelines for the clinical management and treatment of HIV-infected adults in Europe. European AIDS Clinical Society.

Eyzaguirre, L. M., Erasilova, I. B., Nadai, Y., Saad, M. D., Kovtunenko, N. G., Gomatos, P. J., Zeman, V. V., Botros, B. A., Sanchez, J. L., Birx, D. L., Earhart, K. C., and Carr, J. K. (2007). Genetic characterization of HIV-1 strains circulating in Kazakhstan. *J Acquir Immune Defic Syndr* 46(1), 19-23.

Fernandez-Garcia, A., Cuevas, M. T., Vinogradova, A., Rakhmanova, A., Perez-Alvarez, L., de Castro, R. O., Osmanov, S., and Thomson, M. M. (2009). Near full-length genome characterization of a newly identified HIV type 1 subtype F variant circulating in St. Petersburg, Russia. *AIDS Res Hum Retroviruses* 25(11), 1187-91.

Ferreira Junior, O. C., Ferreira, C., Riedel, M., Widolin, M. R., and Barbosa-Junior, A. (2005). Evaluation of rapid tests for anti-HIV detection in Brazil. *AIDS* 19 Suppl 4, S70-5.

Fiebig, E. W., Wright, D. J., Rawal, B. D., Garrett, P. E., Schumacher, R. T., Peddada, L., Heldebrant, C., Smith, R., Conrad, A., Kleinman, S. H., and Busch, M. P. (2003). Dynamics of HIV viremia and antibody seroconversion in plasma donors: implications for diagnosis and staging of primary HIV infection. *AIDS* 17(13), 1871-9.

Fischetti, L., Opare-Sem, O., Candotti, D., Lee, H., and Allain, J. P. (2004a). Higher viral load may explain the dominance of CRF02_AG in the molecular epidemiology of HIV in Ghana. *AIDS* 18(8), 1208-10.

Fischetti, L., Opare-Sem, O., Candotti, D., Sarkodie, F., Lee, H., and Allain, J. P. (2004b). Molecular epidemiology of HIV in Ghana: dominance of CRF02_AG. *J Med Virol* 73(2), 158-66.

Flys, T. S., Chen, S., Jones, D. C., Hoover, D. R., Church, J. D., Fiscus, S. A., Mwatha, A., Guay, L. A., Mmiro, F., Musoke, P., Kumwenda, N., Taha, T. E., Jackson, J. B., and Eshleman, S. H. (2006). Quantitative analysis of HIV-1 variants with the K103N resistance mutation after single-dose nevirapine in women with HIV-1 subtypes A, C, and D. *J Acquir Immune Defic Syndr* 42(5), 610-3.

Fontaine, E., Riva, C., Peeters, M., Schmit, J. C., Delaporte, E., Van Laethem, K., Van Vaerenbergh, K., Snoeck, J., Van Wijngaerden, E., De Clercq, E., Van Ranst, M., and Vandamme, A. M. (2001). Evaluation of two commercial kits for the detection of genotypic drug resistance on a panel of HIV type 1 subtypes A through J. *J Acquir Immune Defic Syndr* 28(3), 254-8.

Fowler, M. G., and Rogers, M. F. (1996). Overview of perinatal HIV infection. *J Nutr* 126(10 Suppl), 2602S-2607S.

Frater, A. J., Beardall, A., Ariyoshi, K., Churchill, D., Galpin, S., Clarke, J. R., Weber, J. N., and McClure, M. O. (2001). Impact of baseline polymorphisms in RT and protease on outcome of highly active antiretroviral therapy in HIV-1-infected African patients. *AIDS* 15(12), 1493-502.

Galai, N., Kalinkovich, A., Burstein, R., Vlahov, D., and Bentwich, Z. (1997). African HIV-1 subtype C and rate of progression among Ethiopian immigrants in Israel. *Lancet* 349(9046), 180-1.

Galimand, J., Frange, P., Rouzioux, C., Deveau, C., Avettand-Fenoel, V., Ghosn, J., Lascoux, C., Goujard, C., Meyer, L., and Chaix, M. L. (2010). Short communication: evidence of HIV type 1 complex and second generation recombinant strains among patients infected in 1997-2007 in France: ANRS CO06 PRIMO Cohort. *AIDS Res Hum Retroviruses* 26(6), 645-51.

Gao, F., Robertson, D. L., Morrison, S. G., Hui, H., Craig, S., Decker, J., Fultz, P. N., Girard, M., Shaw, G. M., Hahn, B. H., and Sharp, P. M. (1996). The heterosexual human immunodeficiency virus type 1 epidemic in Thailand is caused by an intersubtype (A/E) recombinant of African origin. *J Virol* 70(10), 7013-29.

Gao, F., Vidal, N., Li, Y., Trask, S. A., Chen, Y., Kostrikis, L. G., Ho, D. D., Kim, J., Oh, M. D., Choe, K., Salminen, M., Robertson, D. L., Shaw, G. M., Hahn, B. H., and Peeters, M. (2001). Evidence of two distinct subsubtypes within the HIV-1 subtype A radiation. *AIDS Res Hum Retroviruses* 17(8), 675-88.

Garrido, C., Geretti, A. M., Zahonero, N., Booth, C., Strang, A., Soriano, V., and De Mendoza, C. (2010). Integrase variability and susceptibility to HIV integrase inhibitors: impact of subtypes, antiretroviral experience and duration of HIV infection. *J Antimicrob Chemother* 65(2), 320-6.

Garrido, C., Roulet, V., Chueca, N., Poveda, E., Aguilera, A., Skrabal, K., Zahonero, N., Carlos, S., Garcia, F., Faudon, J. L., Soriano, V., and de Mendoza, C. (2008). Evaluation of eight different bioinformatics tools to predict viral tropism in different human immunodeficiency virus type 1 subtypes. *J Clin Microbiol* 46(3), 887-91.

Garrido, C., Zahonero, N., Corral, A., Arredondo, M., Soriano, V., and de Mendoza, C. (2009). Correlation between human immunodeficiency virus type 1 (HIV-1) RNA measurements obtained with dried blood spots and those obtained with plasma by use of Nuclisens EasyQ HIV-1 and Abbott RealTime HIV load tests. *J Clin Microbiol* 47(4), 1031-6.

Gaudy, C., Moreau, A., Brunet, S., Descamps, J. M., Deleplanque, P., Brand, D., and Barin, F. (2004). Subtype B human immunodeficiency virus (HIV) type 1 mutant that escapes detection in a fourth-generation immunoassay for HIV infection. *J Clin Microbiol* 42(6), 2847-9.

Geelen, S., Lange, J., Borleffs, J., Wolfs, T., Weersink, A., and Schuurman, R. (2003). Failure to detect a non-B HIV-1 subtype by the HIV-1 Amplicor Monitor test, version 1.5: a case of unexpected vertical transmission. *AIDS* 17(5), 781-2.

Geretti, A. M., Harrison, L., Green, H., Sabin, C., Hill, T., Fearnhill, E., Pillay, D., and Dunn, D. (2009). Effect of HIV-1 subtype on virologic and immunologic response to starting highly active antiretroviral therapy. *Clin Infect Dis* 48(9), 1296-305.

Germer, J. J., Gerads, T. M., Mandrekar, J. N., Mitchell, P. S., and Yao, J. D. (2006). Detection of HIV-1 proviral DNA with the AMPLICOR HIV-1 DNA Test, version 1.5, following sample processing by the MagNA Pure LC instrument. *J Clin Virol* 37(3), 195-8.

Gonzalez, L. M., Brindeiro, R. M., Aguiar, R. S., Pereira, H. S., Abreu, C. M., Soares, M. A., and Tanuri, A. (2004). Impact of nelfinavir resistance mutations on in vitro phenotype, fitness, and replication capacity of human immunodeficiency virus type 1 with subtype B and C proteases. *Antimicrob Agents Chemother* 48(9), 3552-5.

Gonzalez, L. M., Brindeiro, R. M., Tarin, M., Calazans, A., Soares, M. A., Cassol, S., and Tanuri, A. (2003). In vitro hypersusceptibility of human immunodeficiency virus type 1 subtype C protease to lopinavir. *Antimicrob Agents Chemother* 47(9), 2817-22.

Gonzalez, S., Gondwe, C., Tully, D. C., Minhas, V., Shea, D., Kankasa, C., M'Soka, T., and Wood, C. (2010). Short communication: antiretroviral therapy resistance mutations present in the HIV type 1 subtype C pol and env regions from therapy-naive patients in Zambia. *AIDS Res Hum Retroviruses* 26(7), 795-803.

Gottesman, B. S., Grossman, Z., Lorber, M., Levi, I., Shitrit, P., Katzir, M., Shahar, E., Gottesman, G., and Chowers, M. (2006). Comparative performance of the Amplicor HIV-1 Monitor Assay versus NucliSens EasyQ in HIV subtype C-infected patients. *J Med Virol* 78(7), 883-7.

Grant, P. M., and Zolopa, A. R. (2009). The use of resistance testing in the management of HIV-1-infected patients. *Curr Opin HIV AIDS* 4(6), 474-80.

Grant, R. M., Kuritzkes, D. R., Johnson, V. A., Mellors, J. W., Sullivan, J. L., Swanstrom, R., D'Aquila, R. T., Van Gorder, M., Holodniy, M., Lloyd Jr, R. M., Jr., Reid, C., Morgan, G. F., and Winslow, D. L. (2003). Accuracy of the TRUGENE HIV-1 genotyping kit. *J Clin Microbiol* 41(4), 1586-93.

Grossman, Z., Istomin, V., Averbuch, D., Lorber, M., Risenberg, K., Levi, I., Chowers, M., Burke, M., Bar Yaacov, N., and Schapiro, J. M. (2004). Genetic variation at NNRTI resistance-associated positions in patients infected with HIV-1 subtype C. *AIDS* 18(6), 909-15.

Grossman, Z., Vardinon, N., Chemtob, D., Alkan, M. L., Bentwich, Z., Burke, M., Gottesman, G., Istomin, V., Levi, I., Maayan, S., Shahar, E., and Schapiro, J. M. (2001). Genotypic variation of HIV-1 reverse transcriptase and protease: comparative analysis of clade C and clade B. *AIDS* 15(12), 1453-60.

Gupta, R. K., Chrystie, I. L., O'Shea, S., Mullen, J. E., Kulasegaram, R., and Tong, C. Y. (2005). K65R and Y181C are less prevalent in HAART-experienced HIV-1 subtype A patients. *AIDS* 19(16), 1916-9.

Gurtler, L. G., Hauser, P. H., Eberle, J., von Brunn, A., Knapp, S., Zekeng, L., Tsague, J. M., and Kaptue, L. (1994). A new subtype of human immunodeficiency virus type 1 (MVP-5180) from Cameroon. *J Virol* 68(3), 1581-5.

Habekova, M., Takacova, M., Lysy, J., Mokras, M., Camacho, R., Truska, P., and Stanekova, D. (2010). Genetic subtypes of HIV type 1 circulating in Slovakia. *AIDS Res Hum Retroviruses* 26(10), 1103-7.

Harris, M. E., Serwadda, D., Sewankambo, N., Kim, B., Kigozi, G., Kiwanuka, N., Phillips, J. B., Wabwire, F., Meehen, M., Lutalo, T., Lane, J. R., Merling, R., Gray, R., Wawer, M., Birx, D. L., Robb, M. L., and McCutchan, F. E. (2002). Among 46 near full length HIV type 1 genome sequences from Rakai District, Uganda, subtype D and AD recombinants predominate. *AIDS Res Hum Retroviruses* 18(17), 1281-90.

Hawkins, C. A., Chaplin, B., Idoko, J., Ekong, E., Adewole, I., Gashau, W., Murphy, R. L., and Kanki, P. (2009). Clinical and genotypic findings in HIV-infected patients with the K65R mutation failing first-line antiretroviral therapy in Nigeria. *J Acquir Immune Defic Syndr* 52(2), 228-34.

Henquell, C., Jacomet, C., Antoniotti, O., Chaib, A., Regagnon, C., Brunet, S., Peigue-Lafeuille, H., and Barin, F. (2008). Difficulties in diagnosing group o human immunodeficiency virus type 1 acute primary infection. *J Clin Microbiol* 46(7), 2453-6.

Herring, B. L., Ge, Y. C., Wang, B., Ratnamohan, M., Zheng, F., Cunningham, A. L., Saksena, N. K., and Dwyer, D. E. (2003). Segregation of human immunodeficiency virus type 1 subtypes by risk factor in Australia. *J Clin Microbiol* 41(10), 4600-4.

Holguin, A., De Arellano, E. R., and Soriano, V. (2007). Amino acid conservation in the gp41 transmembrane protein and natural polymorphisms associated with enfuvirtide resistance across HIV-1 variants. *AIDS Res Hum Retroviruses* 23(9), 1067-74.

Holguin, A., Gutierrez, M., Portocarrero, N., Rivas, P., and Baquero, M. (2009). Performance of OraQuick Advance Rapid HIV-1/2 Antibody Test for detection of antibodies in oral fluid and serum/plasma in HIV-1+ subjects carrying different HIV-1 subtypes and recombinant variants. *J Clin Virol* 45(2), 150-2.

Holguin, A., Hertogs, K., and Soriano, V. (2003). Performance of drug resistance assays in testing HIV-1 non-B subtypes. *Clin Microbiol Infect* 9(4), 323-6.

Holguin, A., Lopez, M., Molinero, M., and Soriano, V. (2008). Performance of three commercial viral load assays, Versant human immunodeficiency virus type 1 (HIV-1) RNA bDNA v3.0, Cobas AmpliPrep/Cobas TaqMan HIV-1, and NucliSens HIV-1 EasyQ v1.2, testing HIV-1 non-B subtypes and recombinant variants. *J Clin Microbiol* 46(9), 2918-23.

Holguin, A., Paxinos, E., Hertogs, K., Womac, C., and Soriano, V. (2004). Impact of frequent natural polymorphisms at the protease gene on the in vitro susceptibility to protease inhibitors in HIV-1 non-B subtypes. *J Clin Virol* 31(3), 215-20.

Hosseinipour, M. C., van Oosterhout, J. J., Weigel, R., Phiri, S., Kamwendo, D., Parkin, N., Fiscus, S. A., Nelson, J. A., Eron, J. J., and Kumwenda, J. (2009). The public health approach to identify antiretroviral therapy failure: high-level nucleoside reverse transcriptase inhibitor resistance among Malawians failing first-line antiretroviral therapy. *AIDS* 23(9), 1127-34.

Hsu, L. Y., Subramaniam, R., Bacheler, L., and Paton, N. I. (2005). Characterization of mutations in CRF01_AE virus isolates from antiretroviral treatment-naive and -experienced patients in Singapore. *J Acquir Immune Defic Syndr* 38(1), 5-13.

Huang, W., Eshleman, S. H., Toma, J., Fransen, S., Stawiski, E., Paxinos, E. E., Whitcomb, J. M., Young, A. M., Donnell, D., Mmiro, F., Musoke, P., Guay, L. A., Jackson, J. B., Parkin, N. T., and Petropoulos, C. J. (2007). Coreceptor tropism in human immunodeficiency virus type 1 subtype D: high prevalence of CXCR4 tropism and heterogeneous composition of viral populations. *J Virol* 81(15), 7885-93.

Hudgens, M. G., Longini, I. M., Jr., Vanichseni, S., Hu, D. J., Kitayaporn, D., Mock, P. A., Halloran, M. E., Satten, G. A., Choopanya, K., and Mastro, T. D. (2002). Subtype-specific transmission probabilities for human immunodeficiency virus type 1 among injecting drug users in Bangkok, Thailand. *Am J Epidemiol* 155(2), 159-68.

Ikomey, G. M., Atashili, J., Okomo-Assoumou, M. C., Mesembe, M., and Ndumbe, P. M. (2009). Dried blood spots versus plasma for the quantification of HIV-1 RNA using the manual (PCR-ELISA) amplicor monitor HIV-1 version 1.5 assay in Yaounde, Cameroon. *J Int Assoc Physicians AIDS Care (Chic)* 8(3), 181-4.

Jagodzinski, L. L., Cooley, J. D., Weber, M., and Michael, N. L. (2003). Performance characteristics of human immunodeficiency virus type 1 (HIV-1) genotyping systems in sequence-based analysis of subtypes other than HIV-1 subtype B. *J Clin Microbiol* 41(3), 998-1003.

Janssens, W., Laukkanen, T., Salminen, M. O., Carr, J. K., Van der Auwera, G., Heyndrickx, L., van der Groen, G., and McCutchan, F. E. (2000). HIV-1 subtype H near-full length genome reference strains and analysis of subtype-H-containing inter-subtype recombinants. *AIDS* 14(11), 1533-43.

Jayaraman, G. C., Gleeson, T., Rekart, M. L., Cook, D., Preiksaitis, J., Sidaway, F., Harmen, S., Dawood, M., Wood, M., Ratnam, S., Sandstrom, P., and Archibald, C. (2003). Prevalence and determinants of HIV-1 subtypes in Canada: enhancing routinely collected information through the Canadian HIV Strain and Drug Resistance Surveillance Program. *Can Commun Dis Rep.* 29(4), 29-36.

Johannessen, A. (2010). Dried blood spots in HIV monitoring: applications in resource-limited settings. *Bioanalysis* 2(11), 1893-908.

Johannessen, A., Garrido, C., Zahonero, N., Sandvik, L., Naman, E., Kivuyo, S. L., Kasubi, M. J., Gundersen, S. G., Bruun, J. N., and de Mendoza, C. (2009). Dried blood spots perform well in viral load monitoring of patients who receive antiretroviral treatment in rural Tanzania. *Clin Infect Dis* 49(6), 976-81.

Johanson, J., Abravaya, K., Caminiti, W., Erickson, D., Flanders, R., Leckie, G., Marshall, E., Mullen, C., Ohhashi, Y., Perry, R., Ricci, J., Salituro, J., Smith, A., Tang, N., Vi, M., and Robinson, J. (2001). A new ultrasensitive assay for quantitation of HIV-1 RNA in plasma. *J Virol Methods* 95(1-2), 81-92.

John-Stewart, G. C., Nduati, R. W., Rousseau, C. M., Mbori-Ngacha, D. A., Richardson, B. A., Rainwater, S., Panteleeff, D. D., and Overbaugh, J. (2005). Subtype C Is associated with increased vaginal shedding of HIV-1. *J Infect Dis* 192(3), 492-6.

Johnson, J. A., Li, J. F., Morris, L., Martinson, N., Gray, G., McIntyre, J., and Heneine, W. (2005). Emergence of drug-resistant HIV-1 after intrapartum administration of single-dose nevirapine is substantially underestimated. *J Infect Dis* 192(1), 16-23.

Jurriaans, S., Sankatsing, S. U., Prins, J. M., Schuitemaker, H., Lange, J., Van Der Kuyl, A. C., and Cornelissen, M. (2004). HIV-1 seroreversion in an HIV-1-seropositive patient treated during acute infection with highly active antiretroviral therapy and mycophenolate mofetil. *AIDS* 18(11), 1607-8.

Kaleebu, P., Nankya, I. L., Yirrell, D. L., Shafer, L. A., Kyosiimire-Lugemwa, J., Lule, D. B., Morgan, D., Beddows, S., Weber, J., and Whitworth, J. A. (2007). Relation between chemokine receptor use, disease stage, and HIV-1 subtypes A and D: results from a rural Ugandan cohort. *J Acquir Immune Defic Syndr* 45(1), 28-33.

Kalish, M. L., Robbins, K. E., Pieniazek, D., Schaefer, A., Nzilambi, N., Quinn, T. C., St Louis, M. E., Youngpairoj, A. S., Phillips, J., Jaffe, H. W., and Folks, T. M. (2004). Recombinant viruses and early global HIV-1 epidemic. *Emerg Infect Dis* 10(7), 1227-34.

Kane, C. T., Ndiaye, H. D., Diallo, S., Ndiaye, I., Wade, A. S., Diaw, P. A., Gaye-Diallo, A., and Mboup, S. (2008). Quantitation of HIV-1 RNA in dried blood spots by the real-time NucliSENS EasyQ HIV-1 assay in Senegal. *J Virol Methods* 148(1-2), 291-5.

Kantor, R., and Katzenstein, D. (2003). Polymorphism in HIV-1 non-subtype B protease and reverse transcriptase and its potential impact on drug susceptibility and drug resistance evolution. *AIDS Rev* 5(1), 25-35.

Kantor, R., and Katzenstein, D. (2004). Drug resistance in non-subtype B HIV-1. *J Clin Virol* 29(3), 152-9.

Katsoulidou, A., Rokka, C., Issaris, C., Haida, C., Tzannis, K., Sypsa, V., Detsika, M., Paraskevis, D., and Hatzakis, A. (2011). Comparative evaluation of the performance of the Abbott RealTime HIV-1 assay for measurement of HIV-1 plasma viral load on genetically diverse samples from Greece. *Virol J* 8, 10.

Keller, M., Lu, Y., Lalonde, R. G., and Klein, M. B. (2009). Impact of HIV-1 viral subtype on CD4+ T-cell decline and clinical outcomes in antiretroviral naive patients receiving universal healthcare. *AIDS* 23(6), 731-7.

Kiwanuka, N., Laeyendecker, O., Quinn, T. C., Wawer, M. J., Shepherd, J., Robb, M., Kigozi, G., Kagaayi, J., Serwadda, D., Makumbi, F. E., Reynolds, S. J., and Gray, R. H. (2009). HIV-1 subtypes and differences in heterosexual HIV transmission among HIV-discordant couples in Rakai, Uganda. *AIDS* 23(18), 2479-84.

Kiwanuka, N., Robb, M., Laeyendecker, O., Kigozi, G., Wabwire-Mangen, F., Makumbi, F. E., Nalugoda, F., Kagaayi, J., Eller, M., Eller, L. A., Serwadda, D., Sewankambo, N. K., Reynolds, S. J., Quinn, T. C., Gray, R. H., Wawer, M. J., and Whalen, C. C. (2010). HIV-1 viral subtype differences in the rate of CD4+ T-cell decline among HIV seroincident antiretroviral naive persons in Rakai district, Uganda. *J Acquir Immune Defic Syndr* 54(2), 180-4.

Kline, N. E., Schwarzwald, H., and Kline, M. W. (2002). False negative DNA polymerase chain reaction in an infant with subtype C human immunodeficiency virus 1 infection. *Pediatr Infect Dis J* 21(9), 885-6.

Kong, X., West, J. T., Zhang, H., Shea, D. M., M'Soka T, J., and Wood, C. (2008). The human immunodeficiency virus type 1 envelope confers higher rates of replicative fitness to perinatally transmitted viruses than to nontransmitted viruses. *J Virol* 82(23), 11609-18.

Korenromp, E. L., Williams, B. G., Schmid, G. P., and Dye, C. (2009). Clinical prognostic value of RNA viral load and CD4 cell counts during untreated HIV-1 infection--a quantitative review. *PLoS One* 4(6), e5950.

Kousiappa, I., Van De Vijver, D. A., and Kostrikis, L. G. (2009). Near full-length genetic analysis of HIV sequences derived from Cyprus: evidence of a highly polyphyletic and evolving infection. *AIDS Res Hum Retroviruses* 25(8), 727-40.

Kuritzkes, D. R. (2008). HIV-1 subtype as a determinant of disease progression. *J Infect Dis* 197(5), 638-9.

Kwok, S., Kellogg, D. E., McKinney, N., Spasic, D., Goda, L., Levenson, C., and Sninsky, J. J. (1990). Effects of primer-template mismatches on the polymerase chain reaction: human immunodeficiency virus type 1 model studies. *Nucleic Acids Res* 18(4), 999-1005.

Kwon, J. A., Yoon, S. Y., Lee, C. K., Lim, C. S., Lee, K. N., Sung, H. J., Brennan, C. A., and Devare, S. G. (2006). Performance evaluation of three automated human immunodeficiency virus antigen-antibody combination immunoassays. *J Virol Methods* 133(1), 20-6.

Laforgerie, E., Boucher, B., Ly, T. D., Maisoneuve, L., Izopet, J., Delaugerre, C., and Simon, F. (2010). Sensitivity of 8 CE (European Community)-approved rapid disposable tests for anti-HIV antibody detection during and after seroconversion. *J Virol Methods* 165(1), 105-7.

Lahuerta, M., Aparicio, E., Bardaji, A., Marco, S., Sacarlal, J., Mandomando, I., Alonso, P., Martinez, M. A., Menendez, C., and Naniche, D. (2008). Rapid spread and genetic diversification of HIV type 1 subtype C in a rural area of southern Mozambique. *AIDS Res Hum Retroviruses* 24(2), 327-35.

Lai, A., Riva, C., Marconi, A., Balestrieri, M., Razzolini, F., Meini, G., Vicenti, I., Rosi, A., Saladini, F., Caramma, I., Franzetti, M., Rossini, V., Galli, A., Galli, M., Violin, M.,

Zazzi, M., and Balotta, C. (2010). Changing patterns in HIV-1 non-B clade prevalence and diversity in Italy over three decades. *HIV Med* 11(9), 593-602.

Lau, K. A., Wang, B., Miranda-Saksena, M., Boadle, R., Kamarulzaman, A., Ng, K. P., and Saksena, N. K. (2010). Evidence for possible biological advantages of the newly emerging HIV-1 circulating recombinant form from Malaysia - CRF33_01B in comparison to its progenitors - CRF01_AE and subtype B. *Curr HIV Res* 8(3), 259-71.

Laukkanen, T., Carr, J. K., Janssens, W., Liitsola, K., Gotte, D., McCutchan, F. E., Op de Coul, E., Cornelissen, M., Heyndrickx, L., van der Groen, G., and Salminen, M. O. (2000). Virtually full-length subtype F and F/D recombinant HIV-1 from Africa and South America. *Virology* 269(1), 95-104.

Laurent, C., Bourgeois, A., Faye, M. A., Mougnutou, R., Seydi, M., Gueye, M., Liegeois, F., Kane, C. T., Butel, C., Mbuagbaw, J., Zekeng, L., Mboup, S., Mpoudi-Ngole, E., Peeters, M., and Delaporte, E. (2002). No difference in clinical progression between patients infected with the predominant human immunodeficiency virus type 1 circulating recombinant form (CRF) 02_AG strain and patients not infected with CRF02_AG, in Western and West-Central Africa: a four-year prospective multicenter study. *J Infect Dis* 186(4), 486-92.

Lemey, P., Pybus, O. G., Rambaut, A., Drummond, A. J., Robertson, D. L., Roques, P., Worobey, M., and Vandamme, A. M. (2004). The molecular population genetics of HIV-1 group O. *Genetics* 167(3), 1059-68.

Leoz, M., Depatureaux, A., Vessiere, A., Roquebert, B., Damond, F., Rousset, D., Roques, P., Simon, F., and Plantier, J. C. (2008). Integrase polymorphism and HIV-1 group O diversity. *AIDS* 22(10), 1239-43.

Li, Y., Tee, K. K., Liao, H., Hase, S., Uenishi, R., Li, X. J., Tsuchiura, T., Yang, R., Govindasamy, S., Yong, Y. K., Tan, H. Y., Pybus, O. G., Kamarulzaman, A., and Takebe, Y. (2010). Identification of a novel second-generation circulating recombinant form (CRF48_01B) in Malaysia: a descendant of the previously identified CRF33_01B. *J Acquir Immune Defic Syndr* 54(2), 129-36.

Lisovsky, I., Schader, S. M., Martinez-Cajas, J. L., Oliveira, M., Moisi, D., and Wainberg, M. A. (2010). HIV-1 protease codon 36 polymorphisms and differential development of resistance to nelfinavir, lopinavir, and atazanavir in different HIV-1 subtypes. *Antimicrob Agents Chemother* 54(7), 2878-85.

Loemba, H., Brenner, B., Parniak, M. A., Ma'ayan, S., Spira, B., Moisi, D., Oliveira, M., Detorio, M., and Wainberg, M. A. (2002). Genetic divergence of human immunodeficiency virus type 1 Ethiopian clade C reverse transcriptase (RT) and rapid development of resistance against nonnucleoside inhibitors of RT. *Antimicrob Agents Chemother* 46(7), 2087-94.

Lofgren, S. M., Morrissey, A. B., Chevallier, C. C., Malabeja, A. I., Edmonds, S., Amos, B., Sifuna, D. J., von Seidlein, L., Schimana, W., Stevens, W. S., Bartlett, J. A., and Crump, J. A. (2009). Evaluation of a dried blood spot HIV-1 RNA program for early infant diagnosis and viral load monitoring at rural and remote healthcare facilities. *AIDS* 23(18), 2459-66.

Los Alamos Sequence Database (2011). Los Alamos, New Mexico: Los Alamos National Laboratory http://www.hiv.lanl.gov/

Loussert-Ajaka, I., Ly, T. D., Chaix, M. L., Ingrand, D., Saragosti, S., Courouce, A. M., Brun-Vezinet, F., and Simon, F. (1994). HIV-1/HIV-2 seronegativity in HIV-1 subtype O infected patients. *Lancet* 343(8910), 1393-4.

Ly, N., Recordon-Pinson, P., Phoung, V., Srey, C., Kruy, L. S., Koum, K., Chhum, V., Glaziou, P., Fleury, H. J., and Reynes, J. M. (2005). Characterization of mutations in HIV type 1 isolates from 144 Cambodian recently infected patients and pregnant women naive to antiretroviral drugs. *AIDS Res Hum Retroviruses* 21(11), 971-6.

Ly, T. D., Ebel, A., Faucher, V., Fihman, V., and Laperche, S. (2007). Could the new HIV combined p24 antigen and antibody assays replace p24 antigen specific assays? *J Virol Methods* 143(1), 86-94.

Ly, T. D., Laperche, S., Brennan, C., Vallari, A., Ebel, A., Hunt, J., Martin, L., Daghfal, D., Schochetman, G., and Devare, S. (2004). Evaluation of the sensitivity and specificity of six HIV combined p24 antigen and antibody assays. *J Virol Methods* 122(2), 185-94.

Ly, T. D., Martin, L., Daghfal, D., Sandridge, A., West, D., Bristow, R., Chalouas, L., Qiu, X., Lou, S. C., Hunt, J. C., Schochetman, G., and Devare, S. G. (2001). Seven human immunodeficiency virus (HIV) antigen-antibody combination assays: evaluation of HIV seroconversion sensitivity and subtype detection. *J Clin Microbiol* 39(9), 3122-8.

MacDonald, K. S., Embree, J., Njenga, S., Nagelkerke, N. J., Ngatia, I., Mohammed, Z., Barber, B. H., Ndinya-Achola, J., Bwayo, J., and Plummer, F. A. (1998). Mother-child class I HLA concordance increases perinatal human immunodeficiency virus type 1 transmission. *J Infect Dis* 177(3), 551-6.

Maes, B., Schrooten, Y., Snoeck, J., Derdelinckx, I., Van Ranst, M., Vandamme, A. M., and Van Laethem, K. (2004). Performance of ViroSeq HIV-1 Genotyping System in routine practice at a Belgian clinical laboratory. *J Virol Methods* 119(1), 45-9.

Magiorkinis, G., Paraskevis, D., Magiorkinis, E., Vandamme, A. M., and Hatzakis, A. (2002). Reanalysis of the HIV-1 circulating recombinant form A/E (CRF01_AE): evidence of A/E/G recombination. *J Acquir Immune Defic Syndr* 30(1), 124-9.

Maiga, A. I., Descamps, D., Morand-Joubert, L., Malet, I., Derache, A., Cisse, M., Koita, V., Akonde, A., Diarra, B., Wirden, M., Tounkara, A., Verlinden, Y., Katlama, C., Costagliola, D., Masquelier, B., Calvez, V., and Marcelin, A. G. (2010). Resistance-associated mutations to etravirine (TMC-125) in antiretroviral-naive patients infected with non-B HIV-1 subtypes. *Antimicrob Agents Chemother* 54(2), 728-33.

Makuwa, M., Souquiere, S., Niangui, M. T., Rouquet, P., Apetrei, C., Roques, P., and Simon, F. (2002). Reliability of rapid diagnostic tests for HIV variant infection. *J Virol Methods* 103(2), 183-90.

Maljkovic, I., Wilbe, K., Solver, E., Alaeus, A., and Leitner, T. (2003). Limited transmission of drug-resistant HIV type 1 in 100 Swedish newly detected and drug-naive patients infected with subtypes A, B, C, D, G, U, and CRF01_AE. *AIDS Res Hum Retroviruses* 19(11), 989-97.

Mamadou, S., Vidal, N., Montavon, C., Ben, A., Djibo, A., Rabiou, S., Soga, G., Delaporte, E., Mboup, S., and Peeters, M. (2003). Emergence of complex and diverse CRF02-AG/CRF06-cpx recombinant HIV type 1 strains in Niger, West Africa. *AIDS Res Hum Retroviruses* 19(1), 77-82.

Marconi, V. C., Sunpath, H., Lu, Z., Gordon, M., Koranteng-Apeagyei, K., Hampton, J., Carpenter, S., Giddy, J., Ross, D., Holst, H., Losina, E., Walker, B. D., and Kuritzkes, D. R. (2008). Prevalence of HIV-1 drug resistance after failure of a first highly active antiretroviral therapy regimen in KwaZulu Natal, South Africa. *Clin Infect Dis* 46(10), 1589-97.

Marechal, V., Jauvin, V., Selekon, B., Leal, J., Pelembi, P., Fikouma, V., Gabrie, P., Heredeibona, L. S., Goumba, C., Serdouma, E., Ayouba, A., and Fleury, H. (2006). Increasing HIV type 1 polymorphic diversity but no resistance to antiretroviral drugs in untreated patients from Central African Republic: a 2005 study. *AIDS Res Hum Retroviruses* 22(10), 1036-44.

Martinez-Cajas, J. L., Pant-Pai, N., Klein, M. B., and Wainberg, M. A. (2008). Role of genetic diversity amongst HIV-1 non-B subtypes in drug resistance: a systematic review of virologic and biochemical evidence. *AIDS Rev* 10(4), 212-23.

Martinez, A. M., Hora, V. P., Santos, A. L., Mendoza-Sassi, R., Von Groll, A., Soares, E. A., D'Avila, N., Silveira, J., Leal, R. G., Tanuri, A., and Soares, M. A. (2006). Determinants of HIV-1 mother-to-child transmission in Southern Brazil. *An Acad Bras Cienc* 78(1), 113-21.

Mbida, A. D., Sosso, S., Flori, P., Saoudin, H., Lawrence, P., Monny-Lobe, M., Oyono, Y., Ndzi, E., Cappelli, G., Lucht, F., Pozzetto, B., Oukem-Boyer, O. O., and Bourlet, T. (2009). Measure of viral load by using the Abbott Real-Time HIV-1 assay on dried blood and plasma spot specimens collected in 2 rural dispensaries in Cameroon. *J Acquir Immune Defic Syndr* 52(1), 9-16.

McCutchan, F. E. (2006). Global epidemiology of HIV. *J Med Virol* 78 Suppl 1, S7-S12.

McGowan, J. P., and Shah, S. S. (2000). Prevention of perinatal HIV transmission during pregnancy. *J Antimicrob Chemother* 46(5), 657-68.

Mellors, J. W., Munoz, A., Giorgi, J. V., Margolick, J. B., Tassoni, C. J., Gupta, P., Kingsley, L. A., Todd, J. A., Saah, A. J., Detels, R., Phair, J. P., and Rinaldo, C. R., Jr. (1997). Plasma viral load and CD4+ lymphocytes as prognostic markers of HIV-1 infection. *Ann Intern Med* 126(12), 946-54.

Meloni, S. T., Kim, B., Sankale, J. L., Hamel, D. J., Tovanabutra, S., Mboup, S., McCutchan, F. E., and Kanki, P. J. (2004). Distinct human immunodeficiency virus type 1 subtype A virus circulating in West Africa: sub-subtype A3. *J Virol* 78(22), 12438-45.

Miller, M. D., Margot, N., McColl, D., and Cheng, A. K. (2007). K65R development among subtype C HIV-1-infected patients in tenofovir DF clinical trials. *AIDS* 21(2), 265-6.

Mintsa-Ndong, A., Caron, M., Plantier, J. C., Makuwa, M., Le Hello, S., Courgnaud, V., Roques, P., and Kazanji, M. (2009). High HIV Type 1 prevalence and wide genetic diversity with dominance of recombinant strains but low level of antiretroviral drug-resistance mutations in untreated patients in northeast Gabon, Central Africa. *AIDS Res Hum Retroviruses* 25(4), 411-8.

Mokili, J. L., Wade, C. M., Burns, S. M., Cutting, W. A., Bopopi, J. M., Green, S. D., Peutherer, J. F., and Simmonds, P. (1999). Genetic heterogeneity of HIV type 1 subtypes in Kimpese, rural Democratic Republic of Congo. *AIDS Res Hum Retroviruses* 15(7), 655-64.

Montavon, C., Vergne, L., Bourgeois, A., Mpoudi-Ngole, E., Malonga-Mouellet, G., Butel, C., Toure-Kane, C., Delaporte, E., and Peeters, M. (2002). Identification of a new circulating recombinant form of HIV type 1, CRF11-cpx, involving subtypes A, G, J, and CRF01-AE, in Central Africa. *AIDS Res Hum Retroviruses* 18(3), 231-6.

Monteiro-Cunha, J. P., Araujo, A. F., Santos, E., Galvao-Castro, B., and Alcantara, L. C. (2011). Lack of high-level resistance mutations in HIV type 1 BF recombinant strains circulating in northeast Brazil. *AIDS Res Hum Retroviruses* 27(6), 623-31.

Morison, L., Buve, A., Zekeng, L., Heyndrickx, L., Anagonou, S., Musonda, R., Kahindo, M., Weiss, H. A., Hayes, R. J., Laga, M., Janssens, W., and van der Groen, G. (2001). HIV-1 subtypes and the HIV epidemics in four cities in sub-Saharan Africa. *AIDS* 15 Suppl 4, S109-16.

Munerato, P., Sucupira, M. C., Oliveros, M. P., Janini, L. M., de Souza, D. F., Pereira, A. A., Inocencio, L. A., and Diaz, R. S. (2010). HIV type 1 antiretroviral resistance mutations in subtypes B, C, and F in the City of Sao Paulo, Brazil. *AIDS Res Hum Retroviruses* 26(3), 265-73.

Murray, M. C., Embree, J. E., Ramdahin, S. G., Anzala, A. O., Njenga, S., and Plummer, F. A. (2000). Effect of human immunodeficiency virus (HIV) type 1 viral genotype on mother-to-child transmission of HIV-1. *J Infect Dis* 181(2), 746-9.

Neogi, U., Sood, V., Banerjee, S., Ghosh, N., Verma, S., Samrat, S., Sharma, Y., Saxena, A., Husain, S., Ramachandran, V. G., Das, S., Sreedhar, K. V., Goel, N., Wanchu, A., and Banerjea, A. C. (2009a). Global HIV-1 molecular epidemiology with special reference to genetic analysis of HIV-1 subtypes circulating in North India: functional and pathogenic implications of genetic variation. *Indian J Exp Biol* 47(6), 424-31.

Neogi, U., Sood, V., Chowdhury, A., Das, S., Ramachandran, V. G., Sreedhar, V. K., Wanchu, A., Ghosh, N., and Banerjea, A. C. (2009b). Genetic analysis of HIV-1 Circulating Recombinant Form 02_AG, B and C subtype-specific envelope sequences from Northern India and their predicted co-receptor usage. *AIDS Res Ther* 6, 28.

(2010). DIAGNOSIS OF PEDIATRIC HIV INFECTION IN HIV-EXPOSED INFANTS. www.hivguidelines.org New York State Department of Health AIDS Institute.

Nicastri, E., Sarmati, L., d'Ettorre, G., Parisi, S. G., Palmisano, L., Montano, M., Buonomini, A. R., Galluzzo, C., Vullo, V., Concia, E., Vella, S., and Andreoni, M. (2004). Non-B HIV type 1 subtypes: replicative capacity and response to antiretroviral therapy. *AIDS Res Hum Retroviruses* 20(8), 816-8.

Nkengafac, A., Tina, S., Sua, F., Mason, T., Auyuketta, N., and Oben, S. (2007). *XVI Drug Resistance Workshop: Basic Principles and Clinical Implications*.

Obaro, S. K., Losikoff, P., Harwell, J., and Pugatch, D. (2005). Failure of serial human immunodeficiency virus type 1 DNA polymerase chain reactions to identify human immunodeficiency virus type 1 clade A/G. *Pediatr Infect Dis J* 24(2), 183-4.

Oliveira, V., Bártolo, I., Borrego, P., Rocha, C., Valadas, E., Barreto, J., Almeida, E., Antunes, F., and Taveira, N. (2011). Genetic diversity and drug resistance profiles in HIV-1 and HIV-2 infected patients from Cape Verde Islands. *AIDS Research & Human Retroviruses, accepted.*

Orrell, C., Walensky, R. P., Losina, E., Pitt, J., Freedberg, K. A., and Wood, R. (2009). HIV type-1 clade C resistance genotypes in treatment-naive patients and after first virological failure in a large community antiretroviral therapy programme. *Antivir Ther* 14(4), 523-31.

Palma, A. C., Araujo, F., Duque, V., Borges, F., Paixao, M. T., and Camacho, R. (2007). Molecular epidemiology and prevalence of drug resistance-associated mutations in newly diagnosed HIV-1 patients in Portugal. *Infect Genet Evol* 7(3), 391-8.

Papathanasopoulos, M. A., Vardas, E., Wallis, C., Glashoff, R., Butto, S., Poli, G., Malnati, M., Clerici, M., and Ensoli, B. (2010). Characterization of HIV type 1 genetic diversity among South African participants enrolled in the AIDS Vaccine Integrated Project (AVIP) study. *AIDS Res Hum Retroviruses* 26(6), 705-9.

Paraschiv, S., Foley, B., and Otelea, D. (2011). Diversity of HIV-1 subtype C strains isolated in Romania. *Infect Genet Evol* 11(2), 270-5.

Paraskevis, D., Magiorkinis, E., Katsoulidou, A., Hatzitheodorou, E., Antoniadou, A., Papadopoulos, A., Poulakou, G., Paparizos, V., Botsi, C., Stavrianeas, N., Lelekis, M., Chini, M., Gargalianos, P., Magafas, N., Lazanas, M., Chryssos, G., Petrikkos, G., Panos, G., Kordossis, T., Theodoridou, M., Sypsa, V., and Hatzakis, A. (2005).

Prevalence of resistance-associated mutations in newly diagnosed HIV-1 patients in Greece. *Virus Res* 112(1-2), 115-22.

Paraskevis, D., Magiorkinis, M., Paparizos, V., Pavlakis, G. N., and Hatzakis, A. (2000). Molecular characterization of a recombinant HIV type 1 isolate (A/G/E/?): unidentified regions may be derived from parental subtype E sequences. *AIDS Res Hum Retroviruses* 16(9), 845-55.

Parczewski, M., Leszczyszyn-Pynka, M., Bander, D., Urbanska, A., Stanczak, G., and Boron-Kaczmarska, A. (2010). Characteristics of HIV-1 non-B subtype infections in Northwest Poland. *J Med Virol* 82(8), 1306-13.

Patton, J. C., Akkers, E., Coovadia, A. H., Meyers, T. M., Stevens, W. S., and Sherman, G. G. (2007). Evaluation of dried whole blood spots obtained by heel or finger stick as an alternative to venous blood for diagnosis of human immunodeficiency virus type 1 infection in vertically exposed infants in the routine diagnostic laboratory. *Clin Vaccine Immunol* 14(2), 201-3.

Pavie, J., Rachline, A., Loze, B., Niedbalski, L., Delaugerre, C., Laforgerie, E., Plantier, J. C., Rozenbaum, W., Chevret, S., Molina, J. M., and Simon, F. (2010). Sensitivity of five rapid HIV tests on oral fluid or finger-stick whole blood: a real-time comparison in a healthcare setting. *PLoS One* 5(7), e11581.

Peeters, M., Gueye, A., Mboup, S., Bibollet-Ruche, F., Ekaza, E., Mulanga, C., Ouedrago, R., Gandji, R., Mpele, P., Dibanga, G., Koumare, B., Saidou, M., Esu-Williams, E., Lombart, J. P., Badombena, W., Luo, N., Vanden Haesevelde, M., and Delaporte, E. (1997). Geographical distribution of HIV-1 group O viruses in Africa. *AIDS* 11(4), 493-8.

Pereyra, F., Jia, X., McLaren, P. J., Telenti, A., de Bakker, P. I., Walker, B. D., Ripke, S., Brumme, C. J., Pulit, S. L., Carrington, M., Kadie, C. M., Carlson, J. M., Heckerman, D., Graham, R. R., Plenge, R. M., Deeks, S. G., Gianniny, L., Crawford, G., Sullivan, J., Gonzalez, E., Davies, L., Camargo, A., Moore, J. M., Beattie, N., Gupta, S., Crenshaw, A., Burtt, N. P., Guiducci, C., Gupta, N., Gao, X., Qi, Y., Yuki, Y., Piechocka-Trocha, A., Cutrell, E., Rosenberg, R., Moss, K. L., Lemay, P., O'Leary, J., Schaefer, T., Verma, P., Toth, I., Block, B., Baker, B., Rothchild, A., Lian, J., Proudfoot, J., Alvino, D. M., Vine, S., Addo, M. M., Allen, T. M., Altfeld, M., Henn, M. R., Le Gall, S., Streeck, H., Haas, D. W., Kuritzkes, D. R., Robbins, G. K., Shafer, R. W., Gulick, R. M., Shikuma, C. M., Haubrich, R., Riddler, S., Sax, P. E., Daar, E. S., Ribaudo, H. J., Agan, B., Agarwal, S., Ahern, R. L., Allen, B. L., Altidor, S., Altschuler, E. L., Ambardar, S., Anastos, K., Anderson, B., Anderson, V., Andrady, U., Antoniskis, D., Bangsberg, D., Barbaro, D., Barrie, W., Bartczak, J., Barton, S., Basden, P., Basgoz, N., Bazner, S., Bellos, N. C., Benson, A. M., Berger, J., Bernard, N. F., Bernard, A. M., Birch, C., Bodner, S. J., Bolan, R. K., Boudreaux, E. T., Bradley, M., Braun, J. F., Brndjar, J. E., Brown, S. J., Brown, K., Brown, S. T., Burack, J., Bush, L. M., Cafaro, V., Campbell, O., Campbell, J., Carlson, R. H., Carmichael, J. K., Casey, K. K., Cavacuiti, C., Celestin, G., Chambers, S. T., Chez, N., Chirch, L. M., Cimoch, P. J., Cohen, D., Cohn, L. E., Conway, B., Cooper, D. A., Cornelson, B., Cox, D. T., Cristofano, M. V., Cuchural, G., Jr., Czartoski, J. L., Dahman, J. M., Daly, J. S., Davis, B. T., Davis, K., Davod, S. M., DeJesus, E., Dietz, C. A., Dunham, E., Dunn, M. E., Ellerin, T. B., Eron, J. J., Fangman, J. J., Farel, C. E., Ferlazzo, H., Fidler, S., Fleenor-Ford, A., Frankel, R., Freedberg, K. A., French, N. K., Fuchs, J. D., Fuller, J. D., Gaberman, J., Gallant, J. E., Gandhi, R. T., Garcia, E., Garmon, D., Gathe, J. C., Jr., Gaultier, C. R., Gebre, W., Gilman, F. D., Gilson, I., Goepfert, P. A., Gottlieb, M. S., Goulston, C., Groger, R. K., Gurley, T. D., Haber, S., Hardwicke, R., Hardy, W.

D., Harrigan, P. R., Hawkins, T. N., Heath, S., Hecht, F. M., Henry, W. K., Hladek, M., Hoffman, R. P., Horton, J. M., Hsu, R. K., Huhn, G. D., Hunt, P., Hupert, M. J., Illeman, M. L., Jaeger, H., Jellinger, R. M., John, M., Johnson, J. A., Johnson, K. L., Johnson, H., Johnson, K., Joly, J., Jordan, W. C., Kauffman, C. A., Khanlou, H., Killian, R. K., Kim, A. Y., Kim, D. D., Kinder, C. A., Kirchner, J. T., Kogelman, L., Kojic, E. M., Korthuis, P. T., Kurisu, W., Kwon, D. S., LaMar, M., Lampiris, H., Lanzafame, M., Lederman, M. M., Lee, D. M., Lee, J. M., Lee, M. J., Lee, E. T., Lemoine, J., Levy, J. A., Llibre, J. M., Liguori, M. A., Little, S. J., Liu, A. Y., Lopez, A. J., Loutfy, M. R., Loy, D., Mohammed, D. Y., Man, A., Mansour, M. K., Marconi, V. C., Markowitz, M., Marques, R., Martin, J. N., Martin, H. L., Jr., Mayer, K. H., McElrath, M. J., McGhee, T. A., McGovern, B. H., McGowan, K., McIntyre, D., McLeod, G. X., Menezes, P., Mesa, G., Metroka, C. E., Meyer-Olson, D., Miller, A. O., Montgomery, K., Mounzer, K. C., Nagami, E. H., Nagin, I., Nahass, R. G., Nelson, M. O., Nielsen, C., Norene, D. L., O'Connor, D. H., Ojikutu, B. O., Okulicz, J., Oladehin, O. O., Oldfield, E. C., 3rd, Olender, S. A., Ostrowski, M., Owen, W. F., Jr., Pae, E., Parsonnet, J., Pavlatos, A. M., Perlmutter, A. M., Pierce, M. N., Pincus, J. M., Pisani, L., Price, L. J., Proia, L., Prokesch, R. C., Pujet, H. C., Ramgopal, M., Rathod, A., Rausch, M., Ravishankar, J., Rhame, F. S., Richards, C. S., Richman, D. D., Rodes, B., Rodriguez, M., Rose, R. C., 3rd, Rosenberg, E. S., Rosenthal, D., Ross, P. E., Rubin, D. S., Rumbaugh, E., Saenz, L., Salvaggio, M. R., Sanchez, W. C., Sanjana, V. M., Santiago, S., Schmidt, W., Schuitemaker, H., Sestak, P. M., Shalit, P., Shay, W., Shirvani, V. N., Silebi, V. I., Sizemore, J. M., Jr., Skolnik, P. R., Sokol-Anderson, M., Sosman, J. M., Stabile, P., Stapleton, J. T., Starrett, S., Stein, F., Stellbrink, H. J., Sterman, F. L., Stone, V. E., Stone, D. R., Tambussi, G., Taplitz, R. A., Tedaldi, E. M., Theisen, W., Torres, R., Tosiello, L., Tremblay, C., Tribble, M. A., Trinh, P. D., Tsao, A., Ueda, P., Vaccaro, A., Valadas, E., Vanig, T. J., Vecino, I., Vega, V. M., Veikley, W., Wade, B. H., Walworth, C., Wanidworanun, C., Ward, D. J., Warner, D. A., Weber, R. D., Webster, D., Weis, S., Wheeler, D. A., White, D. J., Wilkins, E., Winston, A., Wlodaver, C. G., van't Wout, A., Wright, D. P., Yang, O. O., Yurdin, D. L., Zabukovic, B. W., Zachary, K. C., Zeeman, B., and Zhao, M. (2010). The major genetic determinants of HIV-1 control affect HLA class I peptide presentation. *Science* 330(6010), 1551-7.

Perez-Alvarez, L., Munoz, M., Delgado, E., Miralles, C., Ocampo, A., Garcia, V., Thomson, M., Contreras, G., and Najera, R. (2006). Isolation and biological characterization of HIV-1 BG intersubtype recombinants and other genetic forms circulating in Galicia, Spain. *J Med Virol* 78(12), 1520-8.

Pillay, D., Walker, A. S., Gibb, D. M., de Rossi, A., Kaye, S., Ait-Khaled, M., Munoz-Fernandez, M., and Babiker, A. (2002). Impact of human immunodeficiency virus type 1 subtypes on virologic response and emergence of drug resistance among children in the Paediatric European Network for Treatment of AIDS (PENTA) 5 trial. *J Infect Dis* 186(5), 617-25.

Ping, L. H., Nelson, J. A., Hoffman, I. F., Schock, J., Lamers, S. L., Goodman, M., Vernazza, P., Kazembe, P., Maida, M., Zimba, D., Goodenow, M. M., Eron, J. J., Jr., Fiscus, S. A., Cohen, M. S., and Swanstrom, R. (1999). Characterization of V3 sequence heterogeneity in subtype C human immunodeficiency virus type 1 isolates from Malawi: underrepresentation of X4 variants. *J Virol* 73(8), 6271-81.

Plantier, J. C., Djemai, M., Lemee, V., Reggiani, A., Leoz, M., Burc, L., Vessiere, A., Rousset, D., Poveda, J. D., Henquell, C., Gautheret-Dejean, A., and Barin, F. (2009a). Census and analysis of persistent false-negative results in serological diagnosis of human immunodeficiency virus type 1 group O infections. *J Clin Microbiol* 47(9), 2906-11.

Plantier, J. C., Leoz, M., Dickerson, J. E., De Oliveira, F., Cordonnier, F., Lemee, V., Damond, F., Robertson, D. L., and Simon, F. (2009b). A new human immunodeficiency virus derived from gorillas. *Nat Med* 15(8), 871-2.

Plate, D. K. (2007). Evaluation and implementation of rapid HIV tests: the experience in 11 African countries. *AIDS Res Hum Retroviruses* 23(12), 1491-8.

Poveda, E., de Mendoza, C., Martin-Carbonero, L., Corral, A., Briz, V., Gonzalez-Lahoz, J., and Soriano, V. (2007). Prevalence of darunavir resistance mutations in HIV-1-infected patients failing other protease inhibitors. *J Antimicrob Chemother* 60(4), 885-8.

Poveda, E., Rodes, B., Toro, C., and Soriano, V. (2004). Are fusion inhibitors active against all HIV variants? *AIDS Res Hum Retroviruses* 20(3), 347-8.

Powell, R. L., Zhao, J., Konings, F. A., Tang, S., Ewane, L., Burda, S., Urbanski, M. M., Saa, D. R., Hewlett, I., and Nyambi, P. N. (2007a). Circulating recombinant form (CRF) 37_cpx: an old strain in Cameroon composed of diverse, genetically distant lineages of subtypes A and G. *AIDS Res Hum Retroviruses* 23(7), 923-33.

Powell, R. L., Zhao, J., Konings, F. A., Tang, S., Nanfack, A., Burda, S., Urbanski, M. M., Saa, D. R., Hewlett, I., and Nyambi, P. N. (2007b). Identification of a novel circulating recombinant form (CRF) 36_cpx in Cameroon that combines two CRFs (01_AE and 02_AG) with ancestral lineages of subtypes A and G. *AIDS Res Hum Retroviruses* 23(8), 1008-19.

Rajesh, L., Karunaianantham, R., Narayanan, P. R., and Swaminathan, S. (2009). Antiretroviral drug-resistant mutations at baseline and at time of failure of antiretroviral therapy in HIV type 1-coinfected TB patients. *AIDS Res Hum Retroviruses* 25(11), 1179-85.

Raymond, S., Delobel, P., Mavigner, M., Cazabat, M., Souyris, C., Encinas, S., Sandres-Saune, K., Pasquier, C., Marchou, B., Massip, P., and Izopet, J. (2009). Genotypic prediction of human immunodeficiency virus type 1 CRF02-AG tropism. *J Clin Microbiol* 47(7), 2292-4.

Raymond, S., Delobel, P., Mavigner, M., Cazabat, M., Souyris, C., Sandres-Saune, K., Cuzin, L., Marchou, B., Massip, P., and Izopet, J. (2008). Correlation between genotypic predictions based on V3 sequences and phenotypic determination of HIV-1 tropism. *AIDS* 22(14), F11-6.

Raymond, S., Delobel, P., Mavigner, M., Ferradini, L., Cazabat, M., Souyris, C., Sandres-Saune, K., Pasquier, C., Marchou, B., Massip, P., and Izopet, J. (2010). Prediction of HIV type 1 subtype C tropism by genotypic algorithms built from subtype B viruses. *J Acquir Immune Defic Syndr* 53(2), 167-75.

Read, J. S. (2007). Diagnosis of HIV-1 infection in children younger than 18 months in the United States. *Pediatrics* 120(6), e1547-62.

Recordon-Pinson, P., Soulie, C., Flandre, P., Descamps, D., Lazrek, M., Charpentier, C., Montes, B., Trabaud, M. A., Cottalorda, J., Schneider, V., Morand-Joubert, L., Tamalet, C., Desbois, D., Mace, M., Ferre, V., Vabret, A., Ruffault, A., Pallier, C., Raymond, S., Izopet, J., Reynes, J., Marcelin, A. G., and Masquelier, B. (2010). Evaluation of the genotypic prediction of HIV-1 coreceptor use versus a phenotypic assay and correlation with the virological response to maraviroc: the ANRS GenoTropism study. *Antimicrob Agents Chemother* 54(8), 3335-40.

Renjifo, B., Fawzi, W., Mwakagile, D., Hunter, D., Msamanga, G., Spiegelman, D., Garland, M., Kagoma, C., Kim, A., Chaplin, B., Hertzmark, E., and Essex, M. (2001).

Differences in perinatal transmission among human immunodeficiency virus type 1 genotypes. *J Hum Virol* 4(1), 16-25.

Renjifo, B., Gilbert, P., Chaplin, B., Msamanga, G., Mwakagile, D., Fawzi, W., and Essex, M. (2004). Preferential in-utero transmission of HIV-1 subtype C as compared to HIV-1 subtype A or D. *AIDS* 18(12), 1629-36.

Roche Diagnostics Instruction manual, Amplicor HIV-1 DNA test, version 1.5. Singapore.

Ross, L., Boulme, R., Fisher, R., Hernandez, J., Florance, A., Schmit, J. C., and Williams, V. (2005). A direct comparison of drug susceptibility to HIV type 1 from antiretroviral experienced subjects as assessed by the antivirogram and PhenoSense assays and by seven resistance algorithms. *AIDS Res Hum Retroviruses* 21(11), 933-9.

Roudinskii, N. I., Sukhanova, A. L., Kazennova, E. V., Weber, J. N., Pokrovsky, V. V., Mikhailovich, V. M., and Bobkov, A. F. (2004). Diversity of human immunodeficiency virus type 1 subtype A and CRF03_AB protease in Eastern Europe: selection of the V77I variant and its rapid spread in injecting drug user populations. *J Virol* 78(20), 11276-87.

Rouet, F., Ekouevi, D. K., Chaix, M. L., Burgard, M., Inwoley, A., Tony, T. D., Danel, C., Anglaret, X., Leroy, V., Msellati, P., Dabis, F., and Rouzioux, C. (2005). Transfer and evaluation of an automated, low-cost real-time reverse transcription-PCR test for diagnosis and monitoring of human immunodeficiency virus type 1 infection in a West African resource-limited setting. *J Clin Microbiol* 43(6), 2709-17.

Rouet, F., Foulongne, V., Viljoen, J., Steegen, K., Becquart, P., Valea, D., Danaviah, S., Segondy, M., Verhofstede, C., and Van de Perre, P. (2010). Comparison of the Generic HIV Viral Load assay with the Amplicor HIV-1 monitor v1.5 and Nuclisens HIV-1 EasyQ v1.2 techniques for plasma HIV-1 RNA quantitation of non-B subtypes: the Kesho Bora preparatory study. *J Virol Methods* 163(2), 253-7.

Ryan, C. E., Elliott, J. H., Middleton, T., Mijch, A. M., Street, A. C., Hellard, M., Crofts, N., Crowe, S. M., and Oelrichs, R. B. (2004). The molecular epidemiology of HIV type 1 among Vietnamese Australian injecting drug users in Melbourne, Australia. *AIDS Res Hum Retroviruses* 20(12), 1364-7.

Sacktor, N., Nakasujja, N., Skolasky, R. L., Rezapour, M., Robertson, K., Musisi, S., Katabira, E., Ronald, A., Clifford, D. B., Laeyendecker, O., and Quinn, T. C. (2009). HIV subtype D is associated with dementia, compared with subtype A, in immunosuppressed individuals at risk of cognitive impairment in Kampala, Uganda. *Clin Infect Dis* 49(5), 780-6.

Santos, A. F., Abecasis, A. B., Vandamme, A. M., Camacho, R. J., and Soares, M. A. (2009). Discordant genotypic interpretation and phenotypic role of protease mutations in HIV-1 subtypes B and G. *J Antimicrob Chemother* 63(3), 593-9.

Santos, A. F., and Soares, M. A. (2010). HIV Genetic Diversity and Drug Resistance *Viruses* 2(2), 503-531.

Santos, A. F., Sousa, T. M., Soares, E. A., Sanabani, S., Martinez, A. M., Sprinz, E., Silveira, J., Sabino, E. C., Tanuri, A., and Soares, M. A. (2006). Characterization of a new circulating recombinant form comprising HIV-1 subtypes C and B in southern Brazil. *AIDS* 20(16), 2011-9.

Sarrami-Forooshani, R., Das, S. R., Sabahi, F., Adeli, A., Esmaeili, R., Wahren, B., Mohraz, M., Haji-Abdolbaghi, M., Rasoolinejad, M., Jameel, S., and Mahboudi, F. (2006). Molecular analysis and phylogenetic characterization of HIV in Iran. *J Med Virol* 78(7), 853-63.

Schable, C., Zekeng, L., Pau, C. P., Hu, D., Kaptue, L., Gurtler, L., Dondero, T., Tsague, J. M., Schochetman, G., Jaffe, H., and et al. (1994). Sensitivity of United States HIV antibody tests for detection of HIV-1 group O infections. *Lancet* 344(8933), 1333-4.

Schuitemaker, H., van 't Wout, A. B., and Lusso, P. (2011). Clinical significance of HIV-1 coreceptor usage. *J Transl Med* 9 Suppl 1, S5.

Scott, L. E., Noble, L. D., Moloi, J., Erasmus, L., Venter, W. D., and Stevens, W. (2009). Evaluation of the Abbott m2000 RealTime human immunodeficiency virus type 1 (HIV-1) assay for HIV load monitoring in South Africa compared to the Roche Cobas AmpliPrep-Cobas Amplicor, Roche Cobas AmpliPrep-Cobas TaqMan HIV-1, and BioMerieux NucliSENS EasyQ HIV-1 assays. *J Clin Microbiol* 47(7), 2209-17.

Shafer, R. W. (2002). Genotypic testing for human immunodeficiency virus type 1 drug resistance. *Clin Microbiol Rev* 15(2), 247-77.

Simon, F., Ly, T. D., Baillou-Beaufils, A., Fauveau, V., De Saint-Martin, J., Loussert-Ajaka, I., Chaix, M. L., Saragosti, S., Courouce, A. M., Ingrand, D., and et al. (1994). Sensitivity of screening kits for anti-HIV-1 subtype O antibodies. *AIDS* 8(11), 1628-9.

Simon, F., Mauclere, P., Roques, P., Loussert-Ajaka, I., Muller-Trutwin, M. C., Saragosti, S., Georges-Courbot, M. C., Barre-Sinoussi, F., and Brun-Vezinet, F. (1998). Identification of a new human immunodeficiency virus type 1 distinct from group M and group O. *Nat Med* 4(9), 1032-7.

Smith, R. A., Anderson, D. J., Pyrak, C. L., Preston, B. D., and Gottlieb, G. S. (2009). Antiretroviral drug resistance in HIV-2: three amino acid changes are sufficient for classwide nucleoside analogue resistance. *J Infect Dis* 199(9), 1323-6.

Snoeck, J., Kantor, R., Shafer, R. W., Van Laethem, K., Deforche, K., Carvalho, A. P., Wynhoven, B., Soares, M. A., Cane, P., Clarke, J., Pillay, C., Sirivichayakul, S., Ariyoshi, K., Holguin, A., Rudich, H., Rodrigues, R., Bouzas, M. B., Brun-Vezinet, F., Reid, C., Cahn, P., Brigido, L. F., Grossman, Z., Soriano, V., Sugiura, W., Phanuphak, P., Morris, L., Weber, J., Pillay, D., Tanuri, A., Harrigan, R. P., Camacho, R., Schapiro, J. M., Katzenstein, D., and Vandamme, A. M. (2006). Discordances between interpretation algorithms for genotypic resistance to protease and reverse transcriptase inhibitors of human immunodeficiency virus are subtype dependent. *Antimicrob Agents Chemother* 50(2), 694-701.

Soares, E., Makamche, M., Siqueira, J., Lumngwena, E., Mbuagbaw, J., Kaptue, L., Asonganyi, T., Seuánez, H., Soares, M., and Alemnji, G. (2010). Molecular diversity and polymerase gene genotypes of HIV-1 among treatment-naïve Cameroonian subjects with advanced disease. *J. Clin Virol.* 48(3), 173-179.

Songok, E. M., Lwembe, R. M., Kibaya, R., Kobayashi, K., Ndembi, N., Kita, K., Vulule, J., Oishi, I., Okoth, F., Kageyama, S., and Ichimura, H. (2004). Active generation and selection for HIV intersubtype A/D recombinant forms in a coinfected patient in Kenya. *AIDS Res Hum Retroviruses* 20(2), 255-8.

Soto-Ramirez, L., Renjifo, B., McLane, M., Marlink, R., O'Hara, C., Sutthent, R., Wasi, C., Vithayasai, P., Vithayasai, V., Apichartpiyakul, C., Auewarakul, P., Peña-Cruz, V., Chui, D., Osathanondh, R., Mayer, K., Lee, T., and Essex, M. (1996). HIV-1 Langerhans' cell tropism associated with heterosexual transmission of HIV. *Science* 271(5253), 1291-3.

Spivak, A. M., Sydnor, E. R., Blankson, J. N., and Gallant, J. E. (2010). Seronegative HIV-1 infection: a review of the literature. *AIDS* 24(10), 1407-14.

Stevens, W., Sherman, G., Downing, R., Parsons, L. M., Ou, C. Y., Crowley, S., Gershy-Damet, G. M., Fransen, K., Bulterys, M., Lu, L., Homsy, J., Finkbeiner, T., and

Nkengasong, J. N. (2008). Role of the laboratory in ensuring global access to ARV treatment for HIV-infected children: consensus statement on the performance of laboratory assays for early infant diagnosis. *Open AIDS J* 2, 17-25.

Stevens, W., Wiggill, T., Horsfield, P., Coetzee, L., and Scott, L. E. (2005). Evaluation of the NucliSens EasyQ assay in HIV-1-infected individuals in South Africa. *J Virol Methods* 124(1-2), 105-10.

Sun, R., Ku, J., Jayakar, H., Kuo, J. C., Brambilla, D., Herman, S., Rosenstraus, M., and Spadoro, J. (1998). Ultrasensitive reverse transcription-PCR assay for quantitation of human immunodeficiency virus type 1 RNA in plasma. *J Clin Microbiol* 36(10), 2964-9.

Swanson, P., de Mendoza, C., Joshi, Y., Golden, A., Hodinka, R. L., Soriano, V., Devare, S. G., and Hackett, J., Jr. (2005). Impact of human immunodeficiency virus type 1 (HIV-1) genetic diversity on performance of four commercial viral load assays: LCx HIV RNA Quantitative, AMPLICOR HIV-1 MONITOR v1.5, VERSANT HIV-1 RNA 3.0, and NucliSens HIV-1 QT. *J Clin Microbiol* 43(8), 3860-8.

Swanson, P., Holzmayer, V., Huang, S., Hay, P., Adebiyi, A., Rice, P., Abravaya, K., Thamm, S., Devare, S. G., and Hackett, J., Jr. (2006). Performance of the automated Abbott RealTime HIV-1 assay on a genetically diverse panel of specimens from London: comparison to VERSANT HIV-1 RNA 3.0, AMPLICOR HIV-1 MONITOR v1.5, and LCx HIV RNA Quantitative assays. *J Virol Methods* 137(2), 184-92.

Swanson, P., Huang, S., Abravaya, K., de Mendoza, C., Soriano, V., Devare, S. G., and Hackett, J., Jr. (2007). Evaluation of performance across the dynamic range of the Abbott RealTime HIV-1 assay as compared to VERSANT HIV-1 RNA 3.0 and AMPLICOR HIV-1 MONITOR v1.5 using serial dilutions of 39 group M and O viruses. *J Virol Methods* 141(1), 49-57.

Tagliamonte, M., Naderi, H. R., Tornesello, M. L., Farid, R., Buonaguro, F. M., and Buonaguro, L. (2007). HIV type 1 subtype A epidemic in injecting drug user (IDU) communities in Iran. *AIDS Res Hum Retroviruses* 23(12), 1569-74.

Tang, N., Huang, S., Salituro, J., Mak, W. B., Cloherty, G., Johanson, J., Li, Y. H., Schneider, G., Robinson, J., Hackett, J., Jr., Swanson, P., and Abravaya, K. (2007). A RealTime HIV-1 viral load assay for automated quantitation of HIV-1 RNA in genetically diverse group M subtypes A-H, group O and group N samples. *J Virol Methods* 146(1-2), 236-45.

Tapia, N., Franco, S., Puig-Basagoiti, F., Menendez, C., Alonso, P. L., Mshinda, H., Clotet, B., Saiz, J. C., and Martinez, M. A. (2003). Influence of human immunodeficiency virus type 1 subtype on mother-to-child transmission. *J Gen Virol* 84(Pt 3), 607-13.

Taylor, B. S., and Hammer, S. M. (2008). The challenge of HIV-1 subtype diversity. *N Engl J Med* 359(18), 1965-6.

Taylor, S., Jayasuriya, A., and Smit, E. (2009). Using HIV resistance tests in clinical practice. *J Antimicrob Chemother* 64(2), 218-22.

Tee, K. K., Kamarulzaman, A., and Ng, K. P. (2006). Short communication: low prevalence of genotypic drug resistance mutations among antiretroviral-naive HIV type 1 patients in Malaysia. *AIDS Res Hum Retroviruses* 22(2), 121-4.

Tee, K. K., Li, X. J., Nohtomi, K., Ng, K. P., Kamarulzaman, A., and Takebe, Y. (2006). Identification of a novel circulating recombinant form (CRF33_01B) disseminating widely among various risk populations in Kuala Lumpur, Malaysia. *J Acquir Immune Defic Syndr* 43(5), 523-9.

Thompson, M. A., Aberg, J. A., Cahn, P., Montaner, J. S., Rizzardini, G., Telenti, A., Gatell, J. M., Gunthard, H. F., Hammer, S. M., Hirsch, M. S., Jacobsen, D. M., Reiss, P.,

Richman, D. D., Volberding, P. A., Yeni, P., and Schooley, R. T. (2010). Antiretroviral treatment of adult HIV infection: 2010 recommendations of the International AIDS Society-USA panel. *JAMA* 304(3), 321-33.

Toni, T. D., Masquelier, B., Lazaro, E., Dore-Mbami, M., Ba-Gomis, F. O., Tea-Diop, Y., Kouakou, K., Diby, J., Sia, E., Soppi, S., Essien, S., Schrive, M. H., Pinson, P., Chenal, H., and Fleury, H. J. (2005). Characterization of nevirapine (NVP) resistance mutations and HIV type 1 subtype in women from Abidjan (Cote d'Ivoire) after NVP single-dose prophylaxis of HIV type 1 mother-to-child transmission. *AIDS Res Hum Retroviruses* 21(12), 1031-4.

Torimiro, J. N., D'Arrigo, R., Takou, D., Nanfack, A., Pizzi, D., Ngong, I., Carr, J. K., Joseph, F. P., Perno, C. F., and Cappelli, G. (2009). Human immunodeficiency virus type 1 intersubtype recombinants predominate in the AIDS epidemic in Cameroon. *New Microbiol* 32(4), 325-31.

Tovanabutra, S., Kijak, G. H., Beyrer, C., Gammon-Richardson, C., Sakkhachornphop, S., Vongchak, T., Jittiwutikarn, J., Razak, M. H., Sanders-Buell, E., Robb, M. L., Suriyanon, V., Birx, D. L., Michael, N. L., Celentano, D. D., and McCutchan, F. E. (2007). Identification of CRF34_01B, a second circulating recombinant form unrelated to and more complex than CRF15_01B, among injecting drug users in northern Thailand. *AIDS Res Hum Retroviruses* 23(6), 829-33.

Tovanabutra, S., Watanaveeradej, V., Viputtikul, K., De Souza, M., Razak, M. H., Suriyanon, V., Jittiwutikarn, J., Sriplienchan, S., Nitayaphan, S., Benenson, M. W., Sirisopana, N., Renzullo, P. O., Brown, A. E., Robb, M. L., Beyrer, C., Celentano, D. D., McNeil, J. G., Birx, D. L., Carr, J. K., and McCutchan, F. E. (2003). A new circulating recombinant form, CRF15_01B, reinforces the linkage between IDU and heterosexual epidemics in Thailand. *AIDS Res Hum Retroviruses* 19(7), 561-7.

Trevino, A., Soriano, V., Rodriguez, C., Arredondo, M., Rivas, P., Herrero-Mendoza, D., Parra, P., Del Romero, J., Anta, L., Puente, S., and de Mendoza, C. (2011). Changing Rate of Non-B Subtypes and Coinfection with Hepatitis B/C Viruses in Newly Diagnosed HIV Type 1 Individuals in Spain. *AIDS Res Hum Retroviruses* 27(6), 633-638.

Triques, K., Bourgeois, A., Saragosti, S., Vidal, N., Mpoudi-Ngole, E., Nzilambi, N., Apetrei, C., Ekwalanga, M., Delaporte, E., and Peeters, M. (1999). High diversity of HIV-1 subtype F strains in Central Africa. *Virology* 259(1), 99-109.

Triques, K., Bourgeois, A., Vidal, N., Mpoudi-Ngole, E., Mulanga-Kabeya, C., Nzilambi, N., Torimiro, N., Saman, E., Delaporte, E., and Peeters, M. (2000). Near-full-length genome sequencing of divergent African HIV type 1 subtype F viruses leads to the identification of a new HIV type 1 subtype designated K. *AIDS Res Hum Retroviruses* 16(2), 139-51.

Tscherning-Casper, C., Dolcini, G., Mauclere, P., Fenyo, E. M., Barre-Sinoussi, F., Albert, J., and Menu, E. (2000). Evidence of the existence of a new circulating recombinant form of HIV type 1 subtype A/J in Cameroon. The European Network on the Study of In Utero Transmission of HIV-1. *AIDS Res Hum Retroviruses* 16(13), 1313-8.

Tscherning, C., Alaeus, A., Fredriksson, R., Bjorndal, A., Deng, H., Littman, D. R., Fenyo, E. M., and Albert, J. (1998). Differences in chemokine coreceptor usage between genetic subtypes of HIV-1. *Virology* 241(2), 181-8.

Turner, D., Shahar, E., Katchman, E., Kedem, E., Matus, N., Katzir, M., Hassoun, G., Pollack, S., Kessner, R., Wainberg, M. A., and Avidor, B. (2009). Prevalence of the K65R resistance reverse transcriptase mutation in different HIV-1 subtypes in Israel. *J Med Virol* 81(9), 1509-12.

Vallari, A., Bodelle, P., Ngansop, C., Makamche, F., Ndembi, N., Mbanya, D., Kaptue, L., Gurtler, L. G., McArthur, C. P., Devare, S. G., and Brennan, C. A. (2010). Four new HIV-1 group N isolates from Cameroon: Prevalence continues to be low. *AIDS Res Hum Retroviruses* 26(1), 109-15.

Vallari, A., Holzmayer, V., Harris, B., Yamaguchi, J., Ngansop, C., Makamche, F., Mbanya, D., Kaptue, L., Ndembi, N., Gurtler, L., Devare, S., and Brennan, C. A. (2011). Confirmation of Putative HIV-1 Group P in Cameroon. *J Virol* 85, 1403-1407.

van Binsbergen, J., de Rijk, D., Peels, H., Dries, C., Scherders, J., Koolen, M., Zekeng, L., and Gurtler, L. G. (1996). Evaluation of a new third generation anti-HIV-1/anti-HIV-2 assay with increased sensitivity for HIV-1 group O. *J Virol Methods* 60(2), 131-7.

van Deursen, P., Oosterlaken, T., Andre, P., Verhoeven, A., Bertens, L., Trabaud, M. A., Ligeon, V., and de Jong, J. (2010). Measuring human immunodeficiency virus type 1 RNA loads in dried blood spot specimens using NucliSENS EasyQ HIV-1 v2.0. *J Clin Virol* 47(2), 120-5.

Vandamme, A. M., Camacho, R. J., Ceccherini-Silberstein, F., De Luca, A., Palmisano, L., Paraskevis, D., Paredes, R., Poljak, M., Schmit, J. C., Soriano, V., Walter, H., and Sonnerborg, A. (2011). European Recommendations for the Clinical Use of HIV Drug Resistance Testing: 2011 Update. *AIDS Rev* 13(2), 77-108.

Vandamme, A. M., Sonnerborg, A., Ait-Khaled, M., Albert, J., Asjo, B., Bacheler, L., Banhegyi, D., Boucher, C., Brun-Vezinet, F., Camacho, R., Clevenbergh, P., Clumeck, N., Dedes, N., De Luca, A., Doerr, H. W., Faudon, J. L., Gatti, G., Gerstoft, J., Hall, W. W., Hatzakis, A., Hellmann, N., Horban, A., Lundgren, J. D., Kempf, D., Miller, M., Miller, V., Myers, T. W., Nielsen, C., Opravil, M., Palmisano, L., Perno, C. F., Phillips, A., Pillay, D., Pumarola, T., Ruiz, L., Salminen, M., Schapiro, J., Schmidt, B., Schmit, J. C., Schuurman, R., Shulse, E., Soriano, V., Staszewski, S., Vella, S., Youle, M., Ziermann, R., and Perrin, L. (2004). Updated European recommendations for the clinical use of HIV drug resistance testing. *Antivir Ther* 9(6), 829-48.

Vandekerckhove, L., Wensing, A., Kaiser, R., Brun-Vezinet, F., Clotet, B., De Luca, A., Dressler, S., Garcia, F., Geretti, A., Klimkait, T., Korn, K., Masquelier, B., Perno, C., Schapiro, J., Soriano, V., Sonnerborg, A., Vandamme, A. M., Verhofstede, C., Walter, H., Zazzi, M., and Boucher, C. (2011). European guidelines on the clinical management of HIV-1 tropism testing. *Lancet Infect Dis* 11(5), 394-407.

Vazquez de Parga, E., Rakhmanova, A., Perez-Alvarez, L., Vinogradova, A., Delgado, E., Thomson, M. M., Casado, G., Sierra, M., Munoz, M., Carmona, R., Vega, Y., Contreras, G., Medrano, L., Osmanov, S., and Najera, R. (2005). Analysis of drug resistance-associated mutations in treatment-naive individuals infected with different genetic forms of HIV-1 circulating in countries of the former Soviet Union. *J Med Virol* 77(3), 337-44.

Vergne, L., Snoeck, J., Aghokeng, A., Maes, B., Valea, D., Delaporte, E., Vandamme, A. M., Peeters, M., and Van Laethem, K. (2006a). Genotypic drug resistance interpretation algorithms display high levels of discordance when applied to non-B strains from HIV-1 naive and treated patients. *FEMS Immunol Med Microbiol* 46(1), 53-62.

Vergne, L., Stuyver, L., Van Houtte, M., Butel, C., Delaporte, E., and Peeters, M. (2006b). Natural polymorphism in protease and reverse transcriptase genes and in vitro antiretroviral drug susceptibilities of non-B HIV-1 strains from treatment-naive patients. *J Clin Virol* 36(1), 43-9.

Vidal, N., Bazepeo, S. E., Mulanga, C., Delaporte, E., and Peeters, M. (2009). Genetic characterization of eight full-length HIV type 1 genomes from the Democratic

Republic of Congo (DRC) reveal a new subsubtype, A5, in the A radiation that predominates in the recombinant structure of CRF26_A5U. *AIDS Res Hum Retroviruses* 25(8), 823-32.

Vidal, N., Mulanga, C., Bazepeo, S. E., Lepira, F., Delaporte, E., and Peeters, M. (2006). Identification and molecular characterization of subsubtype A4 in central Africa. *AIDS Res Hum Retroviruses* 22(2), 182-7.

Wainberg, M. A., and Brenner, B. G. (2010). Role of HIV Subtype Diversity in the Development of Resistance to Antiviral Drugs *Viruses* 2, 2493-2508.

Waters, L., Kambugu, A., Tibenderana, H., Meya, D., John, L., Mandalia, S., Nabankema, M., Namugga, I., Quinn, T. C., Gazzard, B., Reynolds, S. J., and Nelson, M. (2007). Evaluation of filter paper transfer of whole-blood and plasma samples for quantifying HIV RNA in subjects on antiretroviral therapy in Uganda. *J Acquir Immune Defic Syndr* 46(5), 590-3.

Weber, B. (2002). Human immunodeficiency virus (HIV) antigen-antibody combination assays: evaluation of HIV seroconversion sensitivity and subtype detection. *J Clin Microbiol* 40(11), 4402-3; author reply 4403-4.

Weber, B., Fall, E. H., Berger, A., and Doerr, H. W. (1998). Reduction of diagnostic window by new fourth-generation human immunodeficiency virus screening assays. *J Clin Microbiol* 36(8), 2235-9.

Wensing, A. M., van de Vijver, D. A., Angarano, G., Asjo, B., Balotta, C., Boeri, E., Camacho, R., Chaix, M. L., Costagliola, D., De Luca, A., Derdelinckx, I., Grossman, Z., Hamouda, O., Hatzakis, A., Hemmer, R., Hoepelman, A., Horban, A., Korn, K., Kucherer, C., Leitner, T., Loveday, C., MacRae, E., Maljkovic, I., de Mendoza, C., Meyer, L., Nielsen, C., Op de Coul, E. L., Ormaasen, V., Paraskevis, D., Perrin, L., Puchhammer-Stockl, E., Ruiz, L., Salminen, M., Schmit, J. C., Schneider, F., Schuurman, R., Soriano, V., Stanczak, G., Stanojevic, M., Vandamme, A. M., Van Laethem, K., Violin, M., Wilbe, K., Yerly, S., Zazzi, M., and Boucher, C. A. (2005). Prevalence of drug-resistant HIV-1 variants in untreated individuals in Europe: implications for clinical management. *J Infect Dis* 192(6), 958-66.

Wilbe, K., Casper, C., Albert, J., and Leitner, T. (2002). Identification of two CRF11-cpx genomes and two preliminary representatives of a new circulating recombinant form (CRF13-cpx) of HIV type 1 in Cameroon. *AIDS Res Hum Retroviruses* 18(12), 849-56.

Wirden, M., Tubiana, R., Marguet, F., Leroy, I., Simon, A., Bonmarchand, M., Ait-Arkoub, Z., Murphy, R., Marcelin, A. G., Katlama, C., and Calvez, V. (2009). Impact of discrepancies between the Abbott realtime and cobas TaqMan assays for quantification of human immunodeficiency virus type 1 group M non-B subtypes. *J Clin Microbiol* 47(5), 1543-5.

Wolinsky, S. M., Wike, C. M., Korber, B. T., Hutto, C., Parks, W. P., Rosenblum, L. L., Kunstman, K. J., Furtado, M. R., and Munoz, J. L. (1992). Selective transmission of human immunodeficiency virus type-1 variants from mothers to infants. *Science* 255(5048), 1134-7.

(June 2010). Technical Brief on HIV Viral Load Technologies World Health Organization.

Wu, X., Parast, A. B., Richardson, B. A., Nduati, R., John-Stewart, G., Mbori-Ngacha, D., Rainwater, S. M., and Overbaugh, J. (2006). Neutralization escape variants of human immunodeficiency virus type 1 are transmitted from mother to infant. *J Virol* 80(2), 835-44.

Xu, S., Song, A., Li, X., Li, J., Bao, Z., Mao, P., Zhao, Q., and Wang, Y. (2008). Performance of the Abbott RealTime HIV-1 assay for quantification of HIV-1 clades prevalent in China. *J Clin Virol* 41(4), 305-9.

Yamaguchi, J., Badreddine, S., Swanson, P., Bodelle, P., Devare, S. G., and Brennan, C. A. (2008). Identification of new CRF43_02G and CRF25_cpx in Saudi Arabia based on full genome sequence analysis of six HIV type 1 isolates. *AIDS Res Hum Retroviruses* 24(10), 1327-35.

Yamaguchi, J., Bodelle, P., Vallari, A. S., Coffey, R., McArthur, C. P., Schochetman, G., Devare, S. G., and Brennan, C. A. (2004). HIV infections in northwestern Cameroon: identification of HIV type 1 group O and dual HIV type 1 group M and group O infections. *AIDS Res Hum Retroviruses* 20(9), 944-57.

Yamaguchi, J., Vallari, A., Ngansop, C., Makamche, F., Ndembi, N., Mbanya, D., Kaptue, L., Gurtler, L. G., Devare, S. G., and Brennan, C. A. (2010). Near full-length sequence of HIV type 1 subtype J strain 04CMU11421 from Cameroon. *AIDS Res Hum Retroviruses* 26(6), 693-7.

Yang, C., Li, M., Newman, R. D., Shi, Y. P., Ayisi, J., van Eijk, A. M., Otieno, J., Misore, A. O., Steketee, R. W., Nahlen, B. L., and Lal, R. B. (2003). Genetic diversity of HIV-1 in western Kenya: subtype-specific differences in mother-to-child transmission. *AIDS* 17(11), 1667-74.

Yao, J., Liu, Z., Ko, L. S., Pan, G., and Jiang, Y. (2005). Quantitative detection of HIV-1 RNA using NucliSens EasyQ HIV-1 assay. *J Virol Methods* 129(1), 40-6.

Yebra, G., de Mulder, M., del Romero, J., Rodriguez, C., and Holguin, A. HIV-1 non-B subtypes: High transmitted NNRTI-resistance in Spain and impaired genotypic resistance interpretation due to variability. *Antiviral Res* 85(2), 409-17.

Young, T. P., Cloherty, G., Fransen, S., Napolitano, L., Swanson, P., Herman, C., Parkin, N. T., and Hackett, J., Jr. (2011). Performance of the Abbott RealTime HIV-1 viral load assay is not impacted by integrase inhibitor resistance-associated mutations. *J Clin Microbiol* 49(4), 1631-4.

Zhang, H., Rola, M., West, J. T., Tully, D. C., Kubis, P., He, J., Kankasa, C., and Wood, C. (2010a). Functional properties of the HIV-1 subtype C envelope glycoprotein associated with mother-to-child transmission. *Virology* 400(2), 164-74.

Zhang, H., Tully, D. C., Hoffmann, F. G., He, J., Kankasa, C., and Wood, C. (2010b). Restricted genetic diversity of HIV-1 subtype C envelope glycoprotein from perinatally infected Zambian infants. *PLoS One* 5(2), e9294.

Zhao, J., Tang, S., Ragupathy, V., Carr, J. K., Wolfe, N. D., Awazi, B., and Hewlett, I. (2010). Identification and genetic characterization of a novel CRF22_01A1 recombinant form of HIV type 1 in Cameroon. *AIDS Res Hum Retroviruses* 26(9), 1033-45.

Zolfo, M., Schapiro, J. M., Phan, V., Koole, O., Thai, S., Vekemans, M., Fransen, K., and Lynen, L. (2010 Set 22 [Epub ahead of print].). Genotypic Impact of Prolonged Detectable HIV Type 1 RNA Viral Load after HAART Failure in a CRF01_AE-Infected Cohort. *AIDS Res Hum Retroviruses*.

Zouhair, S., Roussin-Bretagne, S., Moreau, A., Brunet, S., Laperche, S., Maniez, M., Barin, F., and Harzic, M. (2006). Group o human immunodeficiency virus type 1 infection that escaped detection in two immmunoassays. *J Clin Microbiol* 44(2), 662-5.

8

Using Retroviral Iterative Genetic Algorithm to Solve Constraint Global Optimization Problems

Renato Simões Moreira[1,3], Otávio Noura Teixeira[1,2,3],
Walter Avelino da Luz Lobato[1,3], Hitoshi Seki Yanaguibashi[1,3]
and Roberto Célio Limão de Oliveira[1]
[1]*Programa de Pós-Graduação em Engenharia Elétrica, Universidade Federal do Pará*
[2]*Laboratório de Computação Natural, Centro Universitário do Estado do Pará*
[3]*Movimento Evolucionário e Cooperativo para a Construçao do Artificial - MEC²A*
Brazil

1. Introduction

Over the years the viruses have been treated as villains in the destruction of organic structures, resulting from the disappearance of entire species and even causing long and fatal epidemics (such as HIV). However, its effectiveness to perpetuate themselves is really impressive, although their acceptance as life forms are still under debate. Retroviruses are the nature's swiftest forms (Carter & Saunders, 2007) and this retroviral feature could not be discarded to develop some computational structure, specifically in the field of evolutionary computation, that can use them as inspiration.

This chapter aims to present a new hybrid nature-inspired metaheuristic developed based on viral structures of family Retroviridae (retroviruses) called as Retroviral Iterative Genetic Algorithm (RIGA). The source of the name comes from the junction of its features: Genetic Algorithm for behaving like a GA, Retroviral for having retroviruses structures and Iterative because occurs every single generation. Also it is made a comparison with another approach based on the junction of Genetic Algorithms and Game Theory called Social Interaction Genetic Algorithm (SIGA).

Both approaches are compared considering the biological versus social approaches applied in the context of Genetic Algorithms and also these two metaheuristics are applied in solving four widespread engineering problems find in the literature, i. e., (1) Welded Beam Design; (2) Design of a Pressure Vessel; (3) Minimization of the Weight of a Tension/Compression String; (4) Speed Reducer Design. In this way, this chapter presents the necessary fundamentals for the conception of RIGA's Algorithm, its structure and the results obtained in the simulations.

2. Biological basement: from viruses to retroviruses

Viruses are compulsory intracellular parasites with a very simple structure. Their acceptance as life forms is very controversial, since they are very different from the most simple bacteria and they have unique features, like the absence cell membrane. They don't have any known organelles and their size is several smaller, thus, the only possible way to see them is by electronic microscopy. They are also metabolically inert unless they are inside a host cell. It is important to notice also that they cannot contain simultaneously DNA and RNA molecules (Hogg, 2005).

The viruses are formed basically by two components: the capsid, consisting of viral proteins, and the core, which contains their genetic information; the combination of these two structures is known as nucleocapsid. The main objective of viruses is to replicate themselves. To achieve this, they need to penetrate a host cell, make copies of themselves and put those copies out of the host cell.

2.1 Retroviruses

Retroviruses are the only known entities that are able to convert RNA into DNA under normal circumstances. After the adsorption and the injection of their genetic material into the host cell, the process of retro transcription takes place in the cytoplasm of the infected cell, using the viral reverse transcriptase enzyme. This process will convert a single stranded molecule of RNA (ssRNA) into a doublestranded DNA (dsDNA) molecule that is larger than the original RNA and has a high error rate, creating DNAs sequences different of which should be (Agut, 2009).

The retroviruses replication process can be described basically in follow steps (Carter & Saunders, 2007):

1. Viral recognition by the receptors present in the host cell surface
2. Penetration into the host cell
3. Reverse transcription (RNA to DNA)
4. Viral integration to the hosts genome, where it will replicate
5. Viral DNA translation (produces viral mRNA that will translated in viral proteins)
6. Viral assembling
7. Viral shedding, when the new viruses leave the host cell

One of the proteins of the virus is the integrase, which is still associated with provirus. This enzyme cuts the chromosomal DNA of the host cell and inserts the viral converted DNA, integrating the provirus into the host cell chromosome as in Fig.1. The next time this infected cell divides, the provirus will be replicated to the daughter cells (Agut, 2009). After the viral genome is integrated in the host cell genome, the virus will be totally dependent of the cellular metabolism to continue its process of transcription, translation, genome replication, viral assembling and shedding.

The concepts of Charles Darwin about reproduction and Natural Selection applied to organics forms are also applied to viruses. Even though their acceptance as an organic form is very questionable, the viruses have genes that are striving to perpetuate the species. The main mechanisms used for viral evolution are mutation, recombination, reassortment and acquisition of cellular genes (Carter & Saunders, 2007).

3. Genetic algorithms

Genetic algorithms are part of probabilistic techniques that try to find different solutions in different executions with the same parameters, and sometimes with the same population

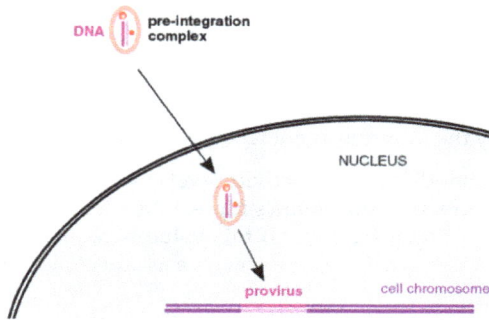

Fig. 1. Provirus creation and reverse transcription

(Goldberg, 1989). According to Haupt & Haupt (2000) some of the main advantages of GA are:

1. Optimization of discrete and continues values;

2. Simultaneous search;

3. Possibility to work with many variables;

4. Provide a set of optimum variables, instead of only specific solution.

The fundamental principle is to explore a population of chromosomes inside a search space, whose evolution depends on mutation and crossover operations, as in the natural evolutionary process. In Fig. 2 is possible to see its structure.

```
1. Begin;
2. Generate Initial Population;
3. Evaluate each indivvual of Initial Population;
4. Repeat until m generations
// Reproduction Phase
    4.1. Repeat until the number of descendents equal to desired quantity (crossover %):
        4.1.1. Select two individuals based on a selection method;
        4.1.2. Make crossover operation in the selected individuals;
        4.1.3. Make mutation operation in the descendent generated in the last step;
        4.1.4. Evaluate all individuals generated;
    4.2. Substitute old individuals by new individuals generated in reproduction;
5. The best individual of population is the problem's solution;
6. End.
```

Fig. 2. Basic structure of Genetic Algorithm

3.1 Genetic algorithm with viral infection (related works)

The viral infection is not an innovation in GA. It was discussed other times like in VEGA (Kubota et al., 1996) and GAVI (Guedes et al., 2005). In both methods is used another population composed by virus, called viral population and infection of chromosomes called transcription. In VEGA the viral population is a subset of chromosomes, created from initial hosts.

During the process of infection, a virus is selected by the same process of rank. However, as the virus has no fitness value, because it doesn't have a complete solution to be evaluated, it is used a parameter of infection, called fitvirus (Kubota et al., 1996). This is an indicator of how well the virus has acted. After selecting the virus that will perform the infection (in a particular chromosome), the transcription is made in this moment, which consists in the modification of infected chromosome by the viral information that contains a section similar to the one represented by the infecting agent Guedes et al. (2005); Kubota et al. (1996).

When a virus is created or modified, its infectivity level is set at a fixed initial value. If the virus infects a chromosome (increase chromosome's fitness) their level of infectivity is increased by 1 (one), otherwise (if turn down the fitness) this value is reduced by 1 (one). If the virus infectivity value reaches 0 (zero) it discards its own parts and copies a part of chromosome for itself (Guedes et al., 2005).

The main difference between VEGA and GAVI is because GAVI uses viral infection as operator, ignoring the operator of mutation, and VEGA is a complete GA (Guedes et al., 2005; Kubota et al., 1996).

4. Retroviral Iterative Genetic Algorithm

The main reason for the use of viral structure in the algorithm is the fact that these viruses are associated with a source of genetic innovation, which is influenced by the rapid rate of replication and changing (Villarreal, 2009).

For biological inspiration of RIGA, the family of retroviruses was chosen. These viruses do not possess correction mechanisms to undo possible genetic mutations that occur naturally during viral multiplication, which causes a high mutation rate, arising genetically modified individuals at each generation, what is considered an important characteristic during the processing time of GA (Haupt & Haupt, 2000; Mitchell, 1999). The acquisition of cellular genes was chosen as a method to viral evolution, since it is quite common in retroviruses (Mitchell, 1999).

There are many differences between RIGA, GAVI and VEGA, some of them:

1. RIGA doesn't change any GA component, GAVI remove the mutation operation;
2. In the GAVI the worst viruses has them genetic material changed, in the RIGA they are completely changed, thereby, the viral population is constantly remade, increase the possibility of infection in chromosome population;
3. The biological basement of RIGA is very specific for the use of retroviral structure;
4. VEGA creates virus only from host chromosomes, RIGA creates virus from host chromosomes too, but, uses the main concept of a retroviruses: high mutation rate;
5. VEGA and GAVI handle a virus as a sequenced subset of chromosome, in the other hand, RIGA handle virus with dispersed information;
6. The virus lifecycle in RIGA is well-defined.

4.1 Viruses

Viruses in RIGA are structures that have the same size of a chromosome, however, with random empty spaces, because the idea is to share genetic material and avoid other population of chromosomes working in parallel. The amount of empty spaces and its positions are determined randomly. Thus for a problem that requires a binary representation of eight positions, some viruses have information as can be seen in Fig. 3 .

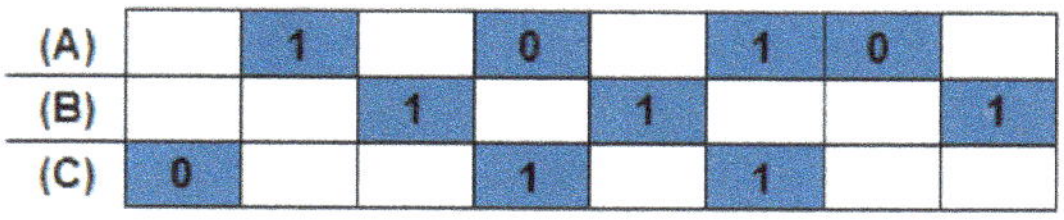

(A)		1		0		1	0	
(B)			1		1			1
(C)	0			1		1		

Fig. 3. Possible viruses for a binary chromosome with eight positions

4.2 Viral population and creation of new viruses

For creation of new viruses that will compose the viral population, RIGA was inspired by the natural process common in retroviruses called reverse transcription. The process consists basically of the steps showed in Fig. 4 .

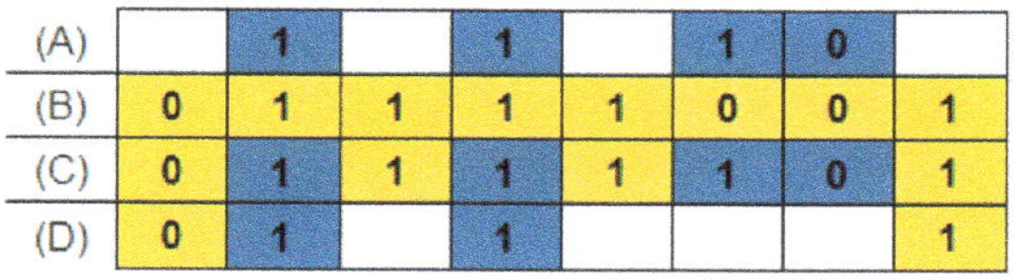

(A)		1		1		1	0	
(B)	0	1	1	1	1	0	0	1
(C)	0	1	1	1	1	1	0	1
(D)	0	1		1				1

Fig. 4. Creating a new virus process (A)Random virus (B)Chromosome from population (C)Auxiliary chromosome contained the mix of genetic material (D)New virus contained genetic material from virus and host

4.3 Infection

Infection is the process of inclusion of the viral genetic material into the host chromosome, which is required a virus and a chromosome. The target chromosome will have changed their genetic material in the same positions where the genes are arranged on the virus, so all the viral genetic information, will be copied to the target chromosome, excepting the empty spaces, which will be filled by the host chromosome. The RIGA infection is represented in Fig. 5 .

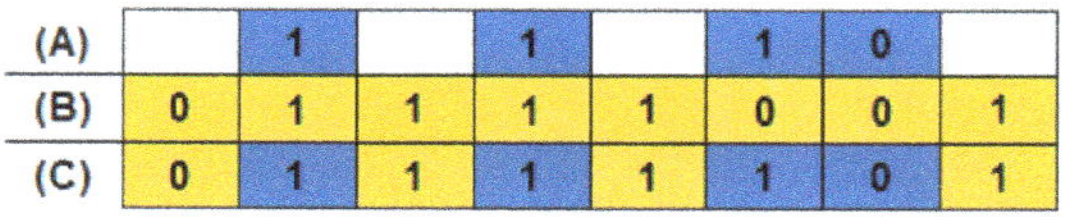

(A)		1		1		1	0	
(B)	0	1	1	1	1	0	0	1
(C)	0	1	1	1	1	1	0	1

Fig. 5. Chromosome infection (A)Virus (B)Chromosome (C)Infected Chromosome

The general view about the creation and infection process is represented in Fig. 6 . In the same figure is possible verify the exchange between genetic material of all structures involved. It's possible verify in Fig. 6 as well, that the new virus (D) is made from an chromosome (B) infected by a virus (A) of viral population. The auxiliary chromosome (C) is made to be template to virus (D).

The infection process depends exclusively on one single factor: increasing chromosomal fitness. The infection is successful when an infected chromosome has an increase on its fitness and unsuccessful when it has a decrease on its fitness. This is an important factor because it determines which viruses will infect the next generation. The viruses with less infection rates will be extinct. For RIGA it is important to restrict the infection only when there is an increase in the chromosomal fitness, because, if any infection was considered, good chromosomes could turn into bad ones. Thus, only the successful infections are important to RIGA.

(A)		1		1		1	0	
(B)	0	1	1	1	1	0	0	1
(C)	0	1	1	1	1	1	0	1
(D)	0	1		1				1
(E)	1	0	0	0	1	1	1	0
(F)	0	1	0	1	1	1	1	1

Fig. 6. Creating a new virus and infection process (A) Random Virus (B) Chromosome from population (C) Auxiliary chromosome contained the mix of genetic material (D) New virus contained genetic material from virus and host (E) Target chromosome (F) Infected chromosome

4.4 Parameters

The RIGA uses the same parameters from classic GA (number of individuals, rate of mutation, crossover and elitism and the type of selection and crossover). However, to apply the concepts of viral infection by retroviruses, some other paramenters are necessary and they are:

1. Infection Population rate: the rate of chromosomes that will be infected;
2. Viral Elitism rate: the rate of viruses that will be kept in the next viral population;
3. Number of Viruses: the number of viruses of viral population;
4. Weakest Infection: this parameter forces the infection of weakest chromosome;
5. Single Infection: this parameter forces a unique infection per chromosome;
6. Internal Infection rate: this parameter indicates the maximum percentage of genetic material from any chromosome that will form a new virus.

4.5 The RIGA algorithm

The algorithm of RIGA method uses the basic structure of Genetic Algorithm presented in Fig. 2 and just extend it as shown in Fig. 7 in the way to include some new steps described as *Virus Application*. In next section the Social Interaction Genetic Algorihm is presented and then compared with RIGA.

5. Social Interaction Genetic Algorithm

The Social Interaction Genetic Algorithm is based on the fundaments of Genetic Algorithms and Game Theory where an individual can take your fitness value changed during the evolutionary process, in other words during the execution of GA. With this option, individuals can now increase your chances to survive and produce offspring, through the struggle of the games in order to maximize their individual gains.

5.1 Basic structure

In order to enable individuals to increase their fitness throughout the implementation process, was inserted a new step called *Social Interaction Phase* in the basic structure of a Genetic Algorithm shown in Figure 2. This new phase was placed before the *Reproductive Phase* and the individuals in it are exposed to an environment, which is nothing more than a strategic

```
1. Begin;
2. Generate Initial Population;
3. Evaluate each indivivual of Initial Population;
4. Repeat until m generations
// Reproduction Phase
    4.1. Repeat until the number of descendents equal to desired quantity (crossover %):
        4.1.1. Select two individuals based on a selection method;
        4.1.2. Make crossover operation in the selected individuals;
        4.1.3. Make mutation operation in the descendent generated in the last step;
        // Viral Application
        4.1.4. If it's the first time, generate a random viral population;
        4.1.5. Generate new virus based on population;
        4.1.6. Infect individuals:
            a. Infect each chosen individual with each existent virus;
            b. For each successful infection, increment 1 to viral fitness otherwise
               decrement 1;
            c. Check the virus with the highest level infection rate and keep
               according to viral elitism rate;
        4.1.7. Evaluate all individuals generated;
    4.2. Substitute old individuals by new individuals generated in reproduction;
5. The best individual of population is the problem's solution;
6. End.
```

Fig. 7. RIGA Algorithm

game, where they have the opportunity to fight for their existence for some period of time. This new approach can be seen in Fig. 8 and the new phase can be expressed by three steps: (1) randomly select two individuals; (2) obtain the behavior of each individual, from their behavioral strategies; (3) change their adaptability (fitness value) based on the behavior adopted and the paytable of the game. Furthermore, this structure allows the use of any type of game, according to their paytable, and also any selection methods, such as roulette and tournament.

```
1. Begin;
2. Generate Initial Population;
3. Repeat until m generations
    // Social Interaction Phase
    3.1. Repeat until d contests
        3.1.1. Random select two individuals;
        3.1.2. Repeat until r rounds
            a. Obtain the behavior of each indivivuals based on their behavior strategy;
            b. Alter their fitness based on their behaviors and the payoff function of
               the game;
    3.2. Evaluate each individual of population;
    // Reproduction Phase
    4.2. Repeat until the descendents number is equal to the desired quantity (crossover %)
        4.2.1. Select two individuals based on a selection method;
        4.2.2. Make crossover operation in the selected individuals;
        4.2.3. Make mutation operation in the descendent generated in the last step;
        4.2.4. Evaluate all individuals generated;
    4.3. Substitute old individuals by new individuals generated in reproduction;
5. The best individual of population is the problem's solution;
6. End.
```

Fig. 8. Structure of SIGA Algorithm

5.2 Implementation process

Based on the structure shown in Figure 8, the aspects related to implementation of SIGA that differ from the GA are: (1) individuals and behavioral strategies, and (2) calculating the value of fitness.

In the classical approach of the GA, individuals are represented by only one chromosome, which contains information about the problem to be solved. In the case of SIGA, there is the need to give individuals the ability to behave strategically. Thus, each now has more than one chromosome, which is responsible for the genetic code related to its strategy. This is randomly generated in the step of generating the initial population and is transmitted to offspring, where each child receives directly the strategy of each parent, and later, this mutated into one of two of their genes.

Thus, the behavioral strategies have been genetically encoded, as can be seen in the Table 1, through a chromosome from two positions and an alphabet composed of a ternary system, containing the values 0, 1 and 2. This system is sufficient to encode all four strategies. They are: *ALL-D*, *ALL-C*, *TFT* (Tit-For-Tat) and *Random*, with a distribution equal to 3:3:2:1, respectively, in the initial population.

Genotype	Chromossome	Phenotype
$B^h B^h$	00	ALL-D
$B^h B^d$	01	TFT
$B^h b$	02	ALL-D
$B^d B^h$	10	TFT
$B^d B^d$	11	ALL-C
$B^d b$	12	ALL-C
$B^h b$	20	ALL-D
$B^d b$	21	ALL-C
b b	22	Random

Table 1. Genetic Coding of Behavioral Strategies

This distribution can be observed by the amount of genetic codings for each strategy. Besides that the notation for the genetic gene is represented by the letter B (Behavior) and the dominance relationship between alleles is given as follows $B^h = B^d > b$.

The calculation of fitness in the context of the classical theory of Genetic Algorithms is based on the solutiont's genotype of an individual ($fitness_{Solution}$) and considering in the case of SIGA that an individual has two chromossomes, there is a need to change the way to calculate it.

Thus, besides having a fitness value related to the solution of the problem, individuals are given a fitness value based on the sum of the gains achieved in the stage of social interaction ($fitness_{Strategy}$), according to the Eq. 1.

$$fitness_{Total} = \alpha(fitness_{Solution}) + \beta(fitness_{Strategy}) \tag{1}$$

The total fitness of an individual ($fitness_{Total}$) is calculated from the weighted sum of fitness values of the solution($fitness_{Solution}$) and the gains made in the disputes of the games, through strategic behavior ($fitness_{Strategy}$).

It is worth mentioning that for the category of problems used here, Constraint Global Optimization Problems, in addition to calculating the fitness, the number of violations of restrictions on each individual is taken into consideration in the selection process. Thus, there is the need to minimize priority violations and subsequently the value of $fitness_{Total}$. Moreover, the approach of SIGA does not take into account the implementation of the fee value of fitness, due to the number of violated constraints, commonly found in other approaches available in the literature, such as in (Coello, 2002).

5.3 Simulation of social interaction phase

In order to illustrate the functioning of the social interaction phase a simulation is shown in Figure 3. The population is composed of four individuals (id, fitness value and strategy behavior), the values of T, R, P and S are respectively 30, 25, 15 and 10 and, at each iteration are held four rounds of disputes the Iterative Prisoner's Dilemma game.

It is interesting to note that in the first interation, for example, the individuals 2 and 3 were selected. In each round in the dispute, the second individual always betrays while the third individual cooperates in the first round and then acts the same way as his opponent in the previous round, ie, he also betrayed only. In each round are assigned values of gain and therefore this change influence the outcome of a tournament selection operation.

Moreover, it is important to note that first is performed the social interation phase and after the tournament selection method. This selection method was used for testing. Thus, two individuals are selected randomly and checked their fitness values and the number of constraintt's violations. Then there are the following:

1. if both do not violate any constraint, then the lowest fitness wins the tournament;
2. if one did not have and the other did have violations, then the individual without violations wins;
3. if both violate constraints wins the one with the least amount of violations;
4. if they violate the same amount of constraints, then the smaller fitness gains.

At the bottom of Fig. 9, it is possible to see the change on the graphic configuration of winners in each one of the games through the partial results of the tournaments. Each individual is represented by a different texture. It is possibile to observe the influence of social interaction phase, which allows less fit individuals to evolve and be selected for crossover operation, as is the case of individual #1 who was less able than individual #4 and became more fit.

6. Evaluation of RIGA and SIGA

In the way to evaluate RIGA and SIGA approaches, they were applied to four widespread engineering problems in literature and they are also well-known as *Constraint Global Optimization Problems*.

6.1 Problem 1 : Welded beam design

According to Rao (1996) this problem aims to minimize the production cost of a Welded Beam subject to some constraints on shear stress, bending stress in the beam, buckling load on the bar, end deflection of the beam and side constraints. There are four design variables: (1) thickness of the weld (h); (2) width of the beam (t); (3) thickness of the beam (b); and, (4) length of welded joint (l), as shown in Fig. 11.

State of population in iteration #0

1	100	ALL-C
2	60	ALL-D
3	130	TFT
4	110	Random

Rounds — 1st game

r.1	Defect x Cooperate	⟹	30 x 10
r.2	Defect x Defect	⟹	15 x 15
r.3	Defect x Defect	⟹	15 x 15
r.4	Defect x Defect	⟹	15 x 15
	Comportamentos		Ganhos

Tournament — Mating Pool

4 110 vs 3 185	⟹	3 185
4 110 vs 1 100	⟹	4 110
3 185 vs 2 135	⟹	3 185
1 100 vs 2 135	⟹	2 135

State of population in iteration #1

1	100	ALL-C
2	135	ALL-D
3	185	TFT
4	110	Random

Rounds — 2nd game

r.1	Defect x Defect	⟹	15 x 15
r.2	Defect x Cooperate	⟹	30 x 10
r.3	Defect x Cooperate	⟹	30 x 10
r.4	Defect x Cooperate	⟹	30 x 10
	Comportamentos		Ganhos

Tournament — Mating Pool

4 155 vs 3 185	⟹	3 185
4 155 vs 1	⟹	4 155
3 185 vs 2 100/240	⟹	2 240
1 100 vs 2 240	⟹	2 240

State of population in iteration #2

1	100	ALL-C
2	240	ALL-D
3	185	TFT
4	155	Random

Rounds — 3rd game

r.1	Coop. x Coop.	⟹	25 x 25
r.2	Coop. x Coop.	⟹	25 x 25
r.3	Coop x Coop.	⟹	25 x 25
r.4	Coop x Coop.	⟹	25 x 25
	Comportamentos		Ganhos

Tournament — Mating Pool

4 155 vs 3 285	⟹	3 285
4 155 vs 1 200	⟹	1 200
3 285 vs 2 240	⟹	3 285
1 200 vs 2 240	⟹	2 240

State of population in iteration #3

1	200	ALL-C
2	240	ALL-D
3	285	TFT
4	155	Random

Rounds — 4th game

r.1	Defect x Defect	⟹	15 x 15
r.2	Defect x Defect	⟹	15 x 15
r.3	Defect x Defect	⟹	15 x 15
r.4	Defect x Defect	⟹	15 x 15
	Comportamentos		Ganhos

Tournament — Mating Pool

4 155 vs 3 345	⟹	3 345
4 155 vs 1 200	⟹	1 200
3 345 vs 2 300	⟹	3 345
1 200 vs 2 300	⟹	2 300

State of population after S.I.

1	200	ALL-C
2	300	ALL-D
3	345	TFT
4	155	Random

T.0	T.1 (G.1)	T.2 (G.2)	T.3 (G.3)	T.4 (G.4)

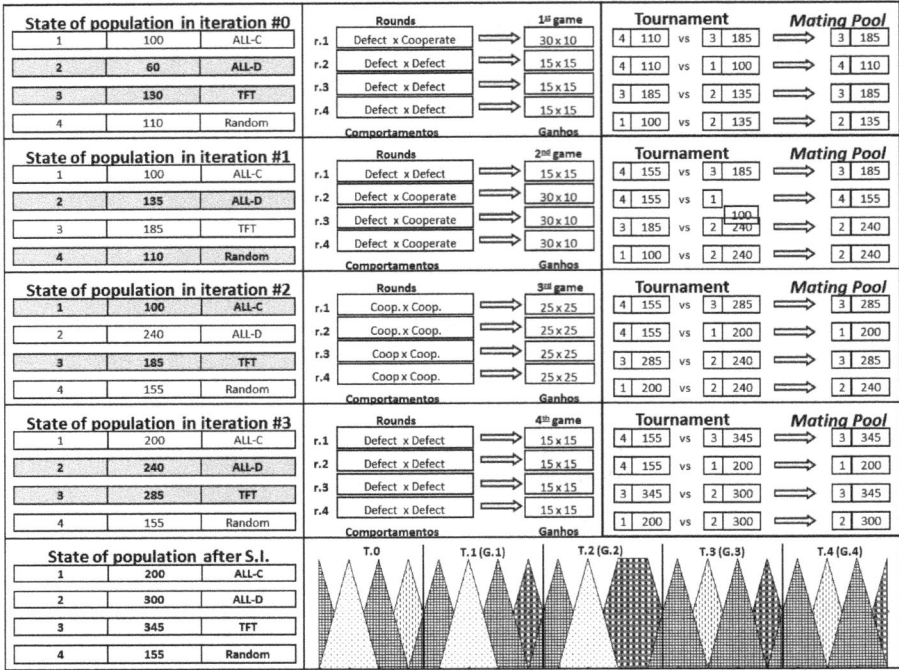

Fig. 9. Simulation of Social Interaction Phase

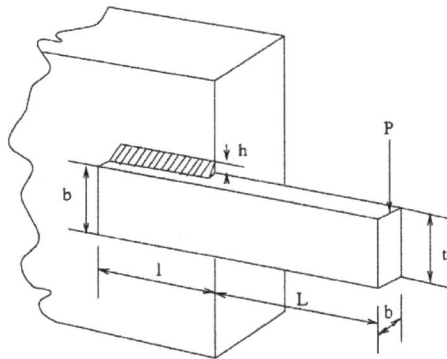

Fig. 10. Welded Beam Design (Alfares & Esat, 2007)

This problem can be stated as follows:

$$Minimize\ f(X) = 1.10471x_1^2 x_2 + 0.04811x_3 x_4 (14.0 + x_2) \tag{2}$$

Subject to:

$$g_1 = \tau - \tau_{max} \le 0 \tag{3}$$

$$g_2 = \sigma - \sigma_{max} \le 0 \tag{4}$$

$$g_3 = h - b \leq 0 \tag{5}$$

$$g_4 = 0.10471h^2 + 0.04811tb(14.0 + l) - 5.0 \leq 0 \tag{6}$$

$$g_5 = 0.125 - h \leq 0 \tag{7}$$

$$g_6 = \delta - \delta_{max} \leq 0 \tag{8}$$

$$g_7 = P - P_c \leq 0 \tag{9}$$

Where:

$$\tau = \sqrt{\left(\tau'\right)^2 + 2\tau'\tau''\frac{1}{2R} + \left(\tau''\right)^2} \tag{10}$$

$$\tau' = \frac{P}{\sqrt{2}hl} \tag{11}$$

$$\tau'' = \frac{MR}{J} \tag{12}$$

$$M = P\left(L + \frac{1}{2}\right) \tag{13}$$

$$R = \sqrt{\frac{t^2}{4} + \left(\frac{h+t}{2}\right)^2} \tag{14}$$

$$J = 2\left\{\sqrt{2}hl\left[\frac{t^2}{12} + \left(\frac{h+t}{2}\right)^2\right]\right\} \tag{15}$$

$$\sigma = \frac{6PL}{bt^2} \tag{16}$$

$$\delta = \frac{4PL^3}{Eb^3} \tag{17}$$

$$P_c = \frac{4.013E\sqrt{\frac{t^2b^6}{36}}}{L^2} + \left(1 - \frac{t}{2L}\sqrt{\frac{E}{4G}}\right) \tag{18}$$

Besides that, the following values were considered: $P = 6000lb$, $L = 14in$, $E = 30x10^6psi$, $G = 12x10^6psi$, $\tau_{max} = 13600psi$, $\sigma_{max} = 30000psi$, $\delta_{max} = 0.25in$ and variable ranges: $0.1 \leq h \leq 2.0$; $0.1 \leq l \leq 10.0$; $0.1 \leq t \leq 10.0$; $0.1 \leq b \leq s\, 2.0$.

Both algorithms, RIGA and SIGA, were executed 30 times for different parameters configurations and in this way it was established a set of parameters with the best results. For SIGA was applied the following values for parameters: {selection method = tourament; tournament size = 2; number of generations = 2000; population size = 200; crossover tax = 85%; mutation tax = 10%; alpha = 1.0; beta = 1.0; games = 100; rounds = 10; R = 5; T = 3; P = 1; S = 0}. And for RIGA was applied the following values for parameters: {selection method = tourament; tournament size = 2; number of generations = 2000; population size = 100; crossover tax = 85%; mutation tax = 10%; Infection Population Rate: 50%; Number of Viruses: 100}.

The best result out of ten plays, one for each of the algorithms, with the parameter set can be seen in Table 2.

Comparing the results of RIGA and SIGA with others results from literature, it is possible to see that they got the best two results, but SIGA was just a little better than RIGA.

Variables	RIGA	SIGA	He & Wang (2007)	Coello & Montes (2002)	Coello (2000)	Deb (1991)
$x1(h)$	0.171937	0.171937	0.202369	0.205986	0.208800	0.248900
$x2(l)$	4.131627	4.122129	3.544214	3.471328	3.420500	6.173000
$x3(t)$	9.587429	9.587429	9.048210	9.020224	8.997500	8.178900
$x4(b)$	0.183010	0.183010	0.205723	0.206480	0.210000	0.253300
$g1(x)$	-32.388778	-8.067400	-12.839796	-0.074092	-0.337812	-5758.603777
$g2(x)$	-39.336830	-39.336800	-1.247467	-0.266227	-353.902604	-255.576901
$g3(x)$	-0.011073	-0.011070	-0.001498	-0.000495	-0.001200	-0.004400
$g4(x)$	-3.466349	-3.467150	-3.429347	-3.430043	-3.411865	-2.982866
$g5(x)$	-0.236389	-0.236390	-0.079381	-0.080986	-0.083800	-0.123900
$g6(x)$	-16.024295	-16.024300	-0.235536	-0.235514	-0.235649	-0.234160
$g7(x)$	-0.046937	-0.046940	-11.681355	-58.666440	-363.232384	-4465.270928
$f(x)$	1.665485	1.664373	1.728024	1.728226	1.748309	2.433116

Table 2. Comparative results for Welded Beam Design

6.2 Problem 2: Design of a pressure vessel

The aim of the problem is to minimize the total cost of designing a pressure vessel, including the cost of the material, forming and welding. A cylindrical vessel is capped at both ends by hemispherical heads as shown in Fig. 11. There are four design variables: (1) thickness of the shell (T_s, x_1); (2) thickness of the head (T_h, x_2), inner radius (R, x_3) and the length of the cylindrical section of the vessel (L, x_4), not including the head. Among the four variables, T_s e T_h are integer multiples of 0.0625in, that are the available thickness of rolled steel plates and R and L are continuous variables.

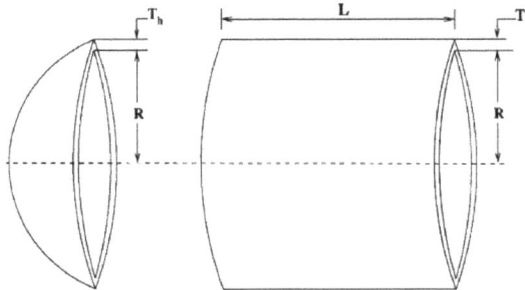

Fig. 11. Design of a Pressure Vessel (He & Wang, 2007)

This problem can be stated as follows:

$$Minimize\ f(X) = 0.6224x_1x_3x_4 + 1.7781x_2x_3^2 + 3.1661x_1^2x_4 + 19.84x_1^2x_3 \tag{19}$$

Subject to:

$$g_1 = -x_1 + 0.0193x_3 \le 0 \tag{20}$$

$$g_2 = -x_2 + 0.00954x_3 \le 0 \tag{21}$$

$$g_3 = -\pi x_3^2 x_4 - \frac{4}{3}\pi x_3^3 + 1296000 \le 0 \tag{22}$$

$$g_4 = x_4 - 240 \le 0 \tag{23}$$

Variables	RIGA	SIGA	He & Wang (2007)	Coello & Montes (2002)	Deb (1997)	Kannan & Kramer (1994)
$x1(Ts)$	0.812500	0.812500	0.812500	0.812500	0.937500	1.125000
$x2(Th)$	0.437500	0.437500	0.437500	0.437500	0.500000	0.625000
$x3(R)$	41.960321	42.092732	42.091266	42.097398	48.329000	58.291000
$x4(L)$	178.379290	176.947780	176.746500	176.654050	112.679000	43.690000
$g1(x)$	-0.002666	-0.000110	-0.000139	-0.000020	-0.004750	0.000016
$g2(x)$	-0.037199	-0.035935	-0.035949	-0.035891	-0.038941	-0.068904
$g3(x)$	-130.284087	-1337.994634	-116.382700	-27.886075	-3652.876838	-21.220104
$g4(x)$	-61.620710	-63.052220	-63.253500	-63.345953	-127.321000	-196.310000
$f(x)$	6077.156337	6066.029360	6061.077700	6059.946300	6410.381100	7.198.042800

Table 3. Comparative Results for Design of a Pressure Vessel

Besides that, the following variable ranges were considered: 1x0.0625 ≤ x_1 ≤ 99x0.0625; 1x0.0625 ≤ x_2 ≤ 99x0.0625; 10 ≤ x_3 ≤ 200; 10 ≤ x_4 ≤ 200.

Both algorithms, RIGA and SIGA, were executed 50 times for different parameters configurations and in this way it was established a set of parameters with the best results. For SIGA was applied the following values for parameters: {selection method = tourament; tournament size = 2; number of generations = 2000; population size = 200; crossover tax = 85%; mutation tax = 20%; alpha = 1.0; beta = 1.0; games = 100; rounds = 10; R = 5; T = 3; P = 1; S = 0}. And for RIGA was applied the following values for parameters: {selection method = tourament; tournament size = 2; number of generations = 2000; population size = 100; crossover tax = 85%; mutation tax = 20%; Infection Population Rate: 50%; Number of Viruses: 100}.

The best result out of ten plays, one for each of the algorithms, with the parameter set can be seen in Table 3.

Comparing the results of RIGA and SIGA with others results from literature, it is possible to see that they got just the 4th and 3rd best results, but SIGA again was just a little better than RIGA.

6.3 Problem 3: Minimization of the weight of a tension/compression string

This problem aims to minimize the weight of a tension/compression string subject to constraints on minimum deflection, shear stress, surge frequency, limits on outside diameter and on design variables, according to (Arora, 1989) e (Belegundu, 1982). The design variables are: (1) the wire diameter (d, x_1); (2) the mean coil diameter (D, x_2); and, (3) the number of active coils (P, x_3), as shown in Fig. 12.

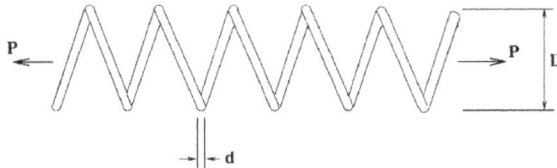

Fig. 12. Minimization of the Weight of a Tension/Compression String (He & Wang, 2007)

Variables	RIGA	SIGA	He & Wang (2007)	Coello & Montes (2002)	Arora (1989)	Belegundu (1982)
$x1(d)$	0.050021	0.050180	0.051728	0.051989	0.053396	0.050000
$x2(D)$	0.282322	0.279604	0.357644	0.363965	0.399180	0.315900
$x3(P)$	2.000075	2.087959	11.244543	10.890522	9.185400	14.250000
$g1(x)$	-0.000290	-0.002840	-0.000845	-0.000013	0.000019	-0.000014
$g2(x)$	-0.235797	-0.249450	-0.0000126	-0.000021	-0.000018	-0.003782
$g3(x)$	-43.106825	-42.176000	-4.051300	-4.061338	-4.123832	-3.938302
$g4(x)$	-0.778519	-0.780140	-0.727090	-0.722698	-0.698283	-0.756067
$f(x)$	0.002824	0.002878	0.0126747	0.0126810	0.0127303	0.0128334

Table 4. Comparative results for Minimization of the Weight of a Tension/Compression String

This problem can be stated as follows:

$$Minimize\ f\ (X) = (x_3 + 2)\ x_2 x_1^2 \tag{24}$$

Subject to:

$$g_1 = 1 - \frac{x_2^3 x_3}{71785 x_1^4} \leq 0 \tag{25}$$

$$g_2 = \frac{4x_2^2 - x_1 x_2}{12566\ (x_2 x_1^3 - x_1^4)} + \frac{1}{5108 x_1^2} - 1 \leq 0 \tag{26}$$

$$g_3 = 1 - \frac{140.45 x_1}{x_2^2 x_3} \leq 0 \tag{27}$$

$$g_4 = \frac{x_1 - x_2}{1.5} - 1 \leq 0 \tag{28}$$

Both algorithms, RIGA and SIGA, were executed 20 times for different parameters configurations and in this way it was established a set of parameters with the best results. For SIGA was applied the following values for parameters: {selection method = tourament; tournament size = 2; number of generations = 2000; population size = 200; crossover tax = 85%; mutation tax = 20%; alpha = 1.0; beta = 1.0; games = 100; rounds = 10; R = 5; T = 3; P = 1; S = 0}. And for RIGA was applied the following values for parameters: {selection method = tourament; tournament size = 2; number of generations = 2000; population size = 100; crossover tax = 85%; mutation tax = 20%; Infection Population Rate: 50%; Number of Viruses: 100}.

The best result out of ten plays, one for each of the algorithms, with the parameter set can be seen in Table 4.

Comparing the results of RIGA and SIGA with others results from literature, it is possible to see that they got the two best results, but this time RIGA was better than SIGA, just a little bit. Also, they results are much better than the others.

6.4 Problem 4: Speed reducer design

In Fig. 13 is possible to see the design of a speed reducer where its weight has to be minimized subject to constraints on bending stress of the gear teeth, surface stress, transverse deflections of the shafts and stresses in the shaft. The design variables are: (1) face width (x_1); (2) module of teeth (x_2); (3) number of teeth on pinion (x_3); (4) length of the first shaft between bearings (x_4); (5) length of the second shaft between bearings (x_5); (6) diameter of the first shaft (x_6); and, (7) diameter of the second shaft (x_7).

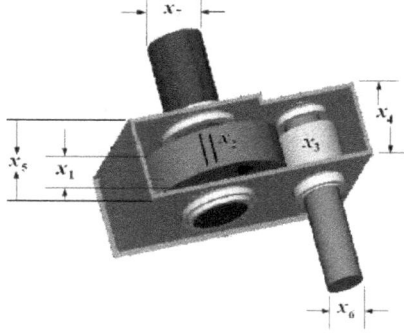

Fig. 13. Speed Reducer Design (Brajevic et al., 2010)

This problem can be stated as follows:

$$Minimizar\ f(X) = 0.7854 x_1 x_2^2 \left(3.3333 x_3^2 + 14.9334 x_3 - 43.0934\right) \tag{29}$$
$$-1.508 x_1 \left(x_6^2 + x_7^2\right) + 7.4777 \left(x_6^3 + x_7^3\right)$$
$$+0.78054 \left(x_4 x_6^2 + x_5 x_7^2\right)$$

Subject to:

$$g_1 = \frac{27}{x_1 x_2^2 x_3} - 1 \leq 0 \tag{30}$$

$$g_2 = \frac{397.5}{x_1 x_2^2 x_3} - 1 \leq 0 \tag{31}$$

$$g_3 = \frac{1.93 x_4^3}{x_2 x_3 x_6^3} - 1 \leq 0 \tag{32}$$

$$g_4 = \frac{1.93 x_5^3}{x_2 x_3 x_7^3} - 1 \leq 0 \tag{33}$$

$$g_5 = \frac{1.0}{110 x_6^3} \sqrt{\left(\frac{750.0 x_4}{x_2 x_3}\right)^2 + 16.9 \times 10^6} - 1 \leq 0 \tag{34}$$

$$g_6 = \frac{1.0}{85 x_7^3} \sqrt{\left(\frac{750.0 x_5}{x_2 x_3}\right)^2 + 157.5 \times 10^6} - 1 \leq 0 \tag{35}$$

$$g_7 = \frac{x_2 x_3}{40} - 1 \leq 0 \tag{36}$$

$$g_8 = \frac{5 x_2}{x_1} - 1 \leq 0 \tag{37}$$

$$g_9 = \frac{x_1}{12 x_2} - 1 \leq 0 \tag{38}$$

Variables	RIGA	SIGA	Brajevic et al. (2010)	Canigna & Esquivel (2008)
$x1$	3.500098	3.500459	3.500000	3.500000
$x2$	0.700000	0.700020	0.700000	0.700000
$x3$	17.000337	17.005030	17.000000	17.000000
$x4$	7.300075	7.300251	7.300000	7.300000
$x5$	7.800014	7.800195	7.800000	7.800000
$x6$	2.900055	2.900041	3.350215	3.350214
$x7$	5.286690	5.286863	5.286683	5.286683
$g1(x)$	-0.073960	-0.074364	-0.073915	-0.073915
$g2(x)$	-0.198053	-0.198624	-0.197996	-0.197998
$g3(x)$	-0.108012	-0.108202	-0.499172	-0.499172
$g4(x)$	-0.901474	-0.901443	-0.901471	-0.901471
$g5(x)$	-1.000000	-1.000000	-2.220E-16	0.000000
$g6(x)$	-0.000004	-0.000102	-3.331E-16	-5.000E-16
$g7(x)$	-0.702494	-0.702403	-0.702500	-0.702500
$g8(x)$	-0.000028	-0.000103	0.000000	-1.000E-16
$g9(x)$	-0.795828	-0.795801	-0.583333	-0.583333
$g10(x)$	-0.143833	-0.143857	-0.051326	-0.051325
$g11(x)$	-0.010853	-0.011074	-0.010852	-0.010852
$f(x)$	2896.372448	2897.531422	2996.348165	2996.348165

Table 5. Comparative results for Speed Reducer Design

$$g_{10} = \frac{1.5x_6 + 1.9}{x_4} - 1 \leq 0 \tag{39}$$

$$g_{11} = \frac{1.1x_7 + 1.9}{x_5} - 1 \leq 0 \tag{40}$$

Besides that, the following variable ranges were considered: $2.6 \leq x_1 \leq 3.6; 0.7 \leq x_2 \leq 0.8; 17 \leq x_3 \leq 28; 7.3 \leq x_4 \leq 8.3; 7.3 \leq x_5 \leq 8.3; 2.9 \leq x_6 \leq 5.0; 7.3 \leq x_7 \leq 5.5$.

Both algorithms, RIGA and SIGA, were executed 50 times for different parameters configurations and in this way it was established a set of parameters with the best results. For SIGA was applied the following values for parameters: {selection method = tourament; tournament size = 2; number of generations = 2000; population size = 200; crossover tax = 85%; mutation tax = 20%; alpha = 1.0; beta = 1.0; games = 100; rounds = 10; R = 5; T = 3; P = 1; S = 0}. And for RIGA was applied the following values for parameters: {selection method = tourament; tournament size = 2; number of generations = 2000; population size = 100; crossover tax = 85%; mutation tax = 20%; Infection Population Rate: 50%; Number of Viruses: 100}.

The best result out of ten plays, one for each of the algorithms, with the parameter set can be seen in Table 5.

Comparing the results of RIGA and SIGA with others results from literature, it is possible to see that they got the two best results, and RIGA again was better than SIGA.

7. Conclusion

This chapter presents a new hybrid nature-inspired metaheuristic called Retroviral Iterative Genetic Algorithm (RIGA) applied in the resolution of four well-known engineering problems in literature. For its design were based on the study of Retroviruses because they are very effective to perpetuate themselves. To establish a comparison at different levels of inspiration in nature, it has been presented the Social Interaction Genetic Algorithm (SIGA), also applied to the same four problems. Besides that some other methods results were used for comparison too. The pratical results show that RIGA is a promising approach and has to be considered in the resolution of Constraint Global Optimization Problems.

8. Acknowledgments

This work has the academic support of Natural Computation Laboratory of the University Center of Para State (LCN-CESUPA) and the Technological Institute of Federal University of Para (ITEC-UFPA).

9. References

Agut, A. (2009). Um sistema estratégio de reprodução, *Scientific American Brasil* pp. 14–19.

Alfares, F. S. & Esat, I. I. (2007). Real-coded quantum inspired evolution algorithm applied to engineering optimization problems, *Proceedings of the IEEE Second International Symposium on Leveraging Applications and Formal Methods, Verification and Validation,* pp. 169–176.

Arora, J. S. (1989). *Introduction to Optimum Design*, McGraw-Hill, New York.

Belegundu, A. D. (1982). A study of mathematical programming methods for structural optimization, *Technical report*, Department of Civil and Environmental Engineering. University of Iowa, Iowa, USA.

Brajevic, I., Tuba, M. & Subotic, M. (2010). Improved artificial bee colony algorithm for constrained problems, *Proceedings of the 11th WSEAS International Conference on Neural Networks, Fuzzy Systems and Evolutionary Computing*, pp. 185–190.

Canigna, L. C. & Esquivel, S. C. (2008). Solving engineering optimization problems with the simple constrained particle swarm optimizer, *Informatica* 32: 319 – 326.

Carter, J. B. & Saunders, V. A. (2007). *Virology: Principles and Applications*, John wiley & Sons Ltd.

Coello, C. A. C. (2000). Use of a self-adaptive penalty approach for engineering optimization problems, *Computers in Industry* 41: 113–127.

Coello, C. A. C. (2002). Theoretical and numerical constraint-handling techniques used with evolutionary algorithms: a survey of the state of the art, *Computer Methods in Applied Mechanics and Engineering* 191(11-12): 1245–1287.

Coello, C. A. C. & Montes, E. M. (2002). Constraint-handling in genetic algorithms through the use of dominance-based tournament selection, *Advanced Engineering Informatics* 16(11-12): 193–203.

Deb, K. (1991). Optimal design of a welded beam via genetic algorithms, *AIAA Journal* 29 (11): 2013–2015.

Deb, K. (1997). Geneas: a robust optimal design technique for mechanical component design, *in* D. Dasgupta & Z. Michalewicz (eds), *Evolutionary Algorithms in Engineering Applications*, Springer, pp. 497–514.

Goldberg, D. (1989). *Genetic Algorithms in Search, Optimization and Machine Learning*, Addison Wesley Longman, Inc.

Guedes, A., Leite, J. & Aloise, D. (2005). Um algoritmo genético com infecção viral para o problema do caixeiro viajante.

Haupt, R. L. & Haupt, S. E. (2000). *Pratical Genetic Algorithms*, John Wiley & Sons, Inc.

He, Q. & Wang, L. (2007). An effective co-evolutionary particle swarm optimization for constrained engineering design problem, *Engineering Applications of Artificial Intelligence, Elsevier* 20 (1): 89–99.

Hogg, S. (2005). *Essential Microbiology*, John Wiley & Sons, Inc.

Kannan, B. K. & Kramer, S. N. (1994). An augmented lagrange multiplier based method for mixed integer discrete continuous optimization and its applications to mechanical design, *Transactions of the ASME Journal of Mechanical Design* 116: 318–320.

Kubota, N., Fukuda, T. & Shimojima, K. (1996). Virus-evolutionary genetic algorithm for a self-organizing manufacturing system, *Computers & Industrial Engineering* 30: 1015–1026.

Mitchell, M. (1999). *An Introduction to Genetic Algorithms*, MIT Press.

Rao, S. S. (1996). *Engineering Optimization*, Wiley, New York, USA.

Villarreal, L. (2009). Vírus são seres vivos?, *Scientific American Brasil* 28: 21–24.

Part 2

Phylogenetics

9

Issues Associated with Genetic Diversity Studies of the Liver Fluke, Fasciola Heptica (Platyhelminthes, Digenea, Fasciolidae)

Denitsa Teofanova, Peter Hristov,
Aneliya Yoveva and Georgi Radoslavov
Institute of Biodiversity and Ecosystem Research,
Bulgarian Academy of Sciences,
Bulgaria

1. Introduction

Parasitic diseases are huge problem for human and veterinary medicine and for economy, agriculture and wildlife management. One of these diseases is fasciolosis, which is caused by two trematode species, *Fasciola hepatica* (liver fluke) and *Fasciola gigantica*. Only *F. hepatica* is a concern in Europe and Americas but the distribution of both species overlaps in many areas of Africa and Asia (Mas-Coma et al., 2005). *F. hepatica* occurs mostly in cattle, sheep and wild ruminants. Recently, worldwide losses in animal productivity due to fasciolosis were estimated at over US$ 3.2 billion per annum (Spithill et al., 1999). The infection of humans has been regarded as accidental for many years. However, fasciolosis is now recognized as an emerging human disease: the World Health Organization (WHO, 2006) has estimated that 2.4 million people are infected with *F. hepatica* and a further 180 million are at risk of infection (Mas-Coma et al., 1999).

Considering all these facts, the increasing of the knowledge of population structure and genetic diversity of liver fluke is necessary. The most often used methods for taxonomical and population studies have been morphological but nowadays they are incomplete and imperfect. Various molecular and genetic techniques are recently utilized, in addition to the classical methods (Ai et al., 2011). Part of the genetic methods is based on molecular markers for identification of genotype variability or genetic differentiation of geographic isolates.

A variety of molecular methods have been applied for genetic structuring of parasitic populations and in particular of the liver fluke. Examples of such methods are: PCR-RFLP of multiple genes (ribosomal, mitochondrial, etc.), RAPD variability, usage of microsatellite markers, etc. Most of these methods are time- and/or labour-consuming, expensive, hard for interpretations and sometimes repetitively impossible. To avoid these negative features, recent developments include single nucleotide polymorphism (SNP) assays after direct sequencing, which has been proved to be the most reliable method used in genetic diversity studies of *F. hepatica*. The information about those variations could increase the knowledge of species belonging, differentiation and diversity of closely related species and intraspecific relationships.

2. PCR based molecular approaches for genetic detection, identification and characterisation – Pros and cons

Most of the molecular techniques utilized for taxonomic, phylogenetic and evolutionary investigation of Fasciola species are based on conventional polymerase chain reaction (PCR) method. PCR is widely used in genomic analysis. One of its main applications is in the DNA markers determination and gene mapping, which is useful in breeding, taxonomy and evolution. Several PCR based methods are available varying in complexity, reliability and information generating capacity. These include random amplified polymorphic DNA (RAPD), simple sequence repeat polymorphism (SSR), restriction fragment length polymorphism (RFLP), the novel sequence-related amplified polymorphism (SRAP) method, etc. Each system has its own advantages and disadvantages.

2.1 RAPD (Random Amplified Polymorphic DNA)

RAPD markers are DNA fragments from PCR amplification of random segments of genomic DNA with single primer of arbitrary nucleotide sequence. Unlike traditional PCR analysis, RAPD does not require any specific knowledge of the DNA sequence of the target organism Different primer sequences will produce different band patterns and possibly will allow a more specific recognition of individual strains. RAPD is a simple method to fingerprint genomic DNA, but poor consistency and low multiplexing output limit its use. That technique is often utilized for genetic variation level establishment among the populations and for species phylogeny determination. RAPD has its limitations as a method. For example nearly all RAPD markers are dominant, i.e. it is not possible to distinguish whether a DNA segment is amplified from a locus that is heterozygous (1 copy) or homozygous (2 copies). Co-dominant RAPD markers, observed as different-sized DNA segments amplified from the same locus, are detected only rarely. PCR is an enzymatic reaction, therefore the quality and concentration of template DNA, concentrations of PCR components, and the PCR cycling conditions may greatly influence the outcome. Thus, the RAPD technique is notoriously laboratory dependent and needs carefully developed laboratory protocols to be reproducible. Mismatches between the primer and the template may result in the total absence of PCR product as well as in a merely decreased amount of the product. Thus, the RAPD results can be difficult for interpretation.

RAPD is a technique for the identification and differentiation of *F. hepatica* and *F. gigantica*. Using that method, only a low level of genetic variation have been detected among *F. gigantica* populations from different hosts (cattle, buffalo, and goat) (Gunasekar et al., 2008). RAPD variability and genetic diversity for *F. hepatica* in particular have been examined from Semyenova et al. (2003) in cattle populations of liver fluke from Ukraine and Armenia. Established results for genotypic diversity have showed higher variability within than between examined populations. Obtained data suggested multiple genetically different parasites but did not provide adequate information about genetic diversity, distribution and population structure of studied liver fluke isolates. Specific RAPD assay have been developed for differentiation of fasciolid species in UK, Peru, Ghana and Sudan (McGarry et al., 2007). That technique enabled distinguishment of *F. hepatica* and *F. gigantica* from cattle and sheep hosts from countries mentioned above. All liver flukes have been correctly identified to species level. In general all these investigations do not give detailed and complete information for genetic structure and diversity of liver fluke intrapopulations and do not established intra- and interpopulation relationships in different geographic regions.

2.2 SRAP (Sequence-Related Amplified Polymorphism)

SRAP is a simple marker technique aimed for the amplification of open reading frames (ORFs) of genomes from related organisms and it is based on two-primer amplification (Li & Quiros, 2001). SRAP technique has been found to be useful for revealing genetic variability within and between *F. hepatica*, *F. gigantica* and evidential for the existence of the "intermediate Fasciola" from different host species and geographical locations in mainland China (Li et al., 2009). The same technique has been utilized for evaluation of genetic diversity of *F. hepatica* populations from different host species and 16 geographical locations in Spain (Alasaad et al., 2008). SRAP polymorphic banding patterns have indicated presence of genetic variability in the coding regions of the genomes within the examined *F. hepatica* representatives even in low degree. The authors have defined four clusters that were not related to particular host species and/or geographical origins of the samples. As RAPD technique the SRAP method also do not provide explanation of obtained genetic structure of Spanish liver fluke populations and what it is due to.

2.3 SSR (Simple Sequence Repeat polymorphism or microsatellite markers)

Microsatellites (sometimes referred to as simple sequence repeats) are short segments of DNA that have a repeated sequence (e.g. CA_n) and they tend to occur in non-coding DNA. In some microsatellites, the repeated unit may occur two to four times, in others it may be seven or thirty. In diploid organisms there will be two copies of any particular microsatellite segment. In comparison with other DNA regions microsatellites have high level of mutations which could be caused by many events (e.g. recombination, "proofreading" mistakes, etc.). Microsatellite markers have many different applications as in forensics, disease diagnostics and identification, conservation biology and population studies. By looking at the variation of microsatellites in populations, inferences can be made about population structure and diversity, genetic drift and even for last common ancestor of the examined population. Microsatellites can be used also to detect sudden changes in population, effects of population fragmentation and interaction of different populations. Microsatellites are useful in identification of new and incipient populations. That method has been used for human population identification studies as well (Pemberton et al., 2009) but recently it is considered that it has a lot of disadvantages. For example it is necessary to have previous information about studied organism genomes so microsatellite analysis could be performed only for well known species. That assay requires huge upfront work and there are a lot of problems associated with PCR of microsatellites. With regards to liver fluke there are only a few investigations reported. Isolation and characterization of microsatellite markers in *F. hepatica* have been performed from Hurtrez-Bousses et al. (2004). In total six microsatellite markers have been isolated from *F. hepatica* and for representatives from Bolivian Altiplano only five were polymorphic. The demonstrated genetic variability does not provide precise information applicable for adequate conclusions about genetic structure of *F. hepatica* population.

2.4 PCR-RFLP (Restriction Fragment Length Polymorphism)

PCR-RFLP method is based on amplification of particular DNA region that contains polymorphic site and subsequent restriction enzyme digestion of target amplicon. The mentioned polymorphism creates or removes specific restriction site for the used endonuclease and different bands could be observed. RFLP detection relies on the possibility of comparing band profiles generated after restriction enzyme digestion of target DNA. There are multiple applications of that technique concerning genetic diversity, genetic relationships between species, origin and evolution of species, genetic drift and selection, whole genome and comparative mapping, etc. PCR-RFLP assay has its advantages as a highly robust methodology with good transferability between laboratories. Its simplicity gives the availability the technique to be readily applied to any organism and through the codominant inheritance heterozygosity can be estimated. To utilize that analysis sequence information is not mandatory required but because it is based on sequence homology PCR-RFLP is recommended for phylogenetic analysis between related species. It also has a discriminatory power and can be applied at the species and/or population levels or individual level. Nevertheless PCR-RFLP has also disadvantages. For example usually large amount of amplified DNA is required and automation is not possible. That technique is not applicable when in some species there is no or very low level of polymorphism and only a few loci can be detected per assay. It is also time consuming comparatively expensive and hard for interpretation method and sometimes different enzyme combinations may be needed as well. With respect to *F. hepatica* researches PCR-RFLP has been often utilized in different cases and for various genes. Marcilla et al. (2002) have developed that assay targeting 28S ribosomal DNA (rDNA) region for *F. hepatica* and *F. gigantica* distinction in Spain. Similar analysis has been used for differentiation between *F. hepatica, F. gigantica* and the "intermediate Fasciola" or identification of *Fasciola* spp. targeting the ITS 2 (internal transcribed spacer 2) of rDNA (Huang et al., 2003), ITS 1 (internal transcribed spacer 1) of rDNA (Ichikawa & Itagaki, 2010) or cox1 (cytochrome c oxidase I) mitochondrial gene (Hashimoto et al., 1997) as genetic markers. Through PCR-RFLP analysis Walker et al. (2007) have defined for *F. hepatica* three mitochondrial DNA (mtDNA) regions with highest polymorphism frequency. Authors also through PCR-RFLP established 52 different complex haplotypes specific for particular geographic isolates and/or hosts. In total PCR-RFLP method is simple and easy to perform but for more detailed and complete interpretation of the results consequent sequence analysis of target DNA is recommended.

2.5 PCR-SSCP (Single-Strand Conformation Polymorphism)

SSCP is the electrophoretic separation of single-stranded nucleic acids often based on single nucleotide variations, which result in a different secondary structure and a measurable difference in mobility through a gel. It offers an inexpensive, convenient, and sensitive method for determining genetic variations (Sunnucks et al., 2000). Like RFLPs, SSCPs are allelic variants of inherited, genetic traits that can be used as genetic markers. However SSCP analysis can detect multiple DNA polymorphisms and larger mutations in DNA (Orita et al., 1989). Nevertheless as a mutation scanning technique, it is often used to analyze the polymorphisms at single loci (Sunnucks et al., 2000). Most experiments involving SSCP are designed to evaluate polymorphisms at single loci and compare the results from different individuals. Recently, more convenient molecular techniques have been developed, although it is a simple method to amplify the double strand and then denature it into single strands instead of searching suitable primers if the targeted sequence is unknown. That is because of some disadvantages and limitations of PCR-SSCP procedures. For example the single-stranded DNA (ssDNA) mobility is temperature dependent and for best results, gel electrophoresis must be run in a constant temperature. Sensitivity of SSCP is also affected by pH. Double-stranded DNA fragments are usually denatured by exposure to basic conditions and lowering of the electrophoresis buffer pH is necessary for increasing the sensitivity and clearer data. Fragment length also influences SSCP analysis. For optimal results, DNA fragment size should fall within the range of 150 to 300 bp (Wagner, 2003).

Specific PCR-SSCP assays have been developed for the accurate identification of *Fasciola* spp. (Alasaad et al., 2011). That study has established a fluorescence-based PCR-linked SSCP (F-PCR-SSCP) assay for the identification of *Fasciola* spp. based on the sequences of ITS 2 of the rDNA and with specific primers labelled by fluorescence dyes. That method has displayed three different SSCP profiles that allowed the identification of *F. hepatica, F. gigantica* and the "intermediate" *Fasciola* from China, Spain, Nigeria, and Egypt. Under optimal conditions, approximately 80 to 90% of the potential base exchanges are detectable by SSCP (Wagner, 2003). If the specific nucleotide responsible for the mobility difference is needed, a technique of SNP may be applied.

2.6 LAMP (Loop-Mediated Isothermal Amplification)

Loop-mediated isothermal amplification (LAMP) is a novel technique for the amplification of DNA which uses single temperature incubation and two or three sets of primers (Notomi et al., 2000). It allows amplification of target nucleic acids under isothermal conditions with high rapidity and precision and has applications for pathogens detection. Detection of the product can be only by photometry. LAMP could be used as a simple, low cost screening method for infectious disease and different pathogens diagnosis. LAMP can also be quantitative. Ai et al. (2010) developed a LAMP assay for the sensitive and rapid detection and discrimination of *F. hepatica* and *F. gigantica* using intergenic spacer (IGS) region of rDNA as a target for PCR amplification.

As it concerns *Fasciola* spp. many other examples for molecular approaches based on PCR technique for species differentiation and genetic characterisation could be presented. Conventional PCR and its modifications like multiplex PCR (Magalhaes et al., 2008) and TaqMan real-time PCR (Alasaad et al., 2011) have been used for amplification of different target DNA regions. The most often utilized are mitochondrial polymorphic genes like nad1 (NADH dehydrogenase 1) and cox1 or ribosomal variable regions like ITS 1 and ITS 2, aiming detailed studies of genetic diversity and/or the origin, evolutionary relationships of fasciolides etc. Although many researches includes above mentioned techniques they are often combined with sequencing and followed by SNP assay. That is because of already pointed disadvantages of these methods and SNP analysis' higher informativeness, precision, detailness and easier results interpretation.

3. Single Nucleotide Polymorphism (SNP) assay

SNP assay is based on analysis of single base pair variations at a specific locus inside and/or outside the coding DNA regions. Single polymorphisms have two alleles per marker; they are very common in the genome (1 per 1000 bp); the necessary PCR product could be either very small (markers will work even with degraded DNA samples) or if needed, larger. Other advantages are associated with complete automation of sample processing and possibility of multiplex SNP assay on chips. Because SNPs are conserved during evolution, they have been used as markers for quantitative trait loci analysis and in other studies in place of mentioned in Section 2 methods. SNPs can also provide a fingerprint for use of species identification, population genetic structuring, molecular taxonomy, phylogeny, evolution and genetic diversity investigations. There are multiple methods for SNP genotyping: hybridization-based methods, enzyme-based methods and other post-amplification methods based on physical properties of DNA. Examples of such methods are: Reverse Dot Blots (based on position of spots blotted with specific allele probes);

HPLC Genotyping (HPLC technique of separating components based on time of moving through the column and resulting on three profiles – two for homozygous and one for heterozygous forms);

TaqMan Assay (uses fluorescent allele specific probes);

Fluorescence Polarization (differentiation of amplicons through four different coloured ddNTPs);

Microchips (uses silicon chips with attached allele specific oligonucleotide probes and binding of fluorescent labelled PCR products);

Pyrosequencing (the way of DNA sequencing detected through releasing of pyrophosphate and light during the dNTPs incorporation);

SNaPshot (e.g. mini-sequencing, uses extension primers up to SNP and again adds ddNTPs labelled with different coloured dye, assay can be multiplexed);

Direct Sequencing, etc.

3.1 SNP genotyping through direct sequencing

The most often utilized and the simplest method for search of unknown SNP is by direct sequencing (Figure 1). It is based on sequencing the region of interest using Sanger dideoxy method. The fragment with the suspected SNP is compared to a known wild type fragment or to another sequenced one. A SNP may or may not be present in the fragment to compare. Correct reading of the gene region sequence is of a high importance for interpretation of obtained data. Many software products facilitate the processing and comparing multiple sequences.

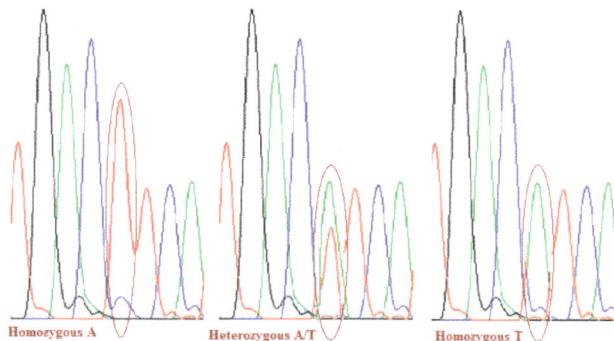

Fig. 1. Example of SNP assay after direct sequencing.

That way of DNA sequences microanalysis allows the detection of polymorphic sites through which it is possible to make inter- and intraspecific differentiation. As these single variations are conservative in particular population and from an evolutionary point of view, they are a reliable marker for genotype mapping. SNP assay after a direct sequencing has been proved to be the most reliable method used in genetic diversity studies of *F. hepatica*. As genetic markers for SNP analyses, multiple genes have been used: ribosomal, mitochondrial and protein coding (somatic). These genes are predominantly highly conservative, with a small number of nucleotide variations. The information about those polymorphisms could increase the knowledge of species identification, differentiation and diversity of closely related species as well as of interspecific relationships.

3.1.1 Ribosomal DNA regions used for SNP assay

Ribosomal DNA (rDNA) has been widely used for taxonomy status establishment and/or inter- and intraspecific differences. It consists of alternating conservative and variable regions (Figure 2).

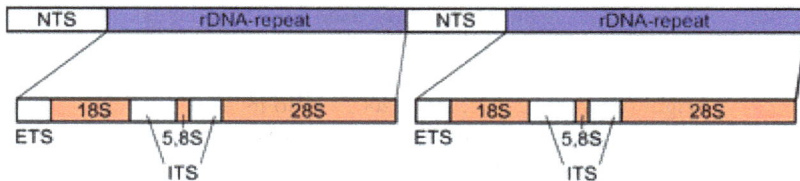

Fig. 2. Structure of eukaryote rDNA genes (18S, 5,8S, 28S) in tandem clusters with internal (ITS 1 and ITS2) and external (ETS) transcribed spacers and nontranscribed spacers (NTS). (http://www.absoluteastronomy.com/topics/Ribosomal_DNA).

Coding rDNA regions provide information for systematics and phylogeny of the species (Hillis & Dixon, 1991). Conservative genes (18S, 28S) are used for revealing the relationships between different or closely related species. For clarifying intraspecific genetic structure of populations only a variable ribosomal regions could be reliable for analysing. Most commonly utilized variable regions are ITS 1, ITS 2 and D-domains (divergent regions in 28S and 18S genes). As it concerns *F. hepatica* many research teams have applied above mentioned rDNA genes in multiple investigations (Lee et al., 2007; Olson et al., 2003; Lotfy et al., 2008; Vara-Del Río et al., 2007; Marcilla et al., 2002, etc.). D1-region from 28S rDNA has been used for genetic structuring of the populations of species from phylum Platyhelmintes (Lee et al., 2007). Differences in nucleotide sequences have been observed even in the same class but nevertheless high phylogenetic similarity has been found in classes and orders. Similar investigations have been obtained with 18S rDNA and D1-D3 regions of 28S rDNA for class Trematoda (Olson et al., 2003). The results from that investigation have enabled clarifying of taxonomy, phylogeny and evolutionary relationships in the class. As it concerns the evolution and origin of fasciolid species and liver fluke in particular, Lotfy et al. (2008) utilized 28S rDNA, ITS 1 and ITS 2 spacers to establish monophyleticity of that group. Using the nucleotide sequences and phylogenetic relationships of mentioned ribosomal regions the authors described evolutionary processes that occurred with fasciolid species and defined their origin. A fragment from 28S rDNA has been used to demonstrate the genetic heterogeneity of *F. hepatica* isolates in Spain (Vara-Del Río et al., 2007). SNP assay of that fragment sequence revealed intraspecific genetic variability. Sequence analysis again of 28S rDNA region in combination with PCR-RFLP assay also allowed differentiation between *F. hepatica* and *F. gigantica* (Marcilla et al., 2002). Although intraspecific polymorphisms have not been proved and only a few iterspecific variation have been found. With the same fragment genetic diversity studies have been performed (Teofanova et al., 2011). 28S rDNA has not been proved as proper molecule marker for that kind of population research but better for species identification. Usually for *Fasciola* spp. identification the researches have been performed with ITS 1 and ITS 2 spacers because of their higher variability (Semyenova et al., 2005; Erensoy et al., 2009; Morgan & Blair, 1995; Prasad et al., 2008; Huang et al., 2004; Amer et al., 2010). That type of investigations have been conducted for different geographic locations – Russia, Ukraine, Armenia, etc. (Semyenova et al., 2005), Turkey (Erensoy et al., 2009), China (Huang et al., 2004) and other countries. These spacers have been commonly used for population structuring and genetic diversity explanations.

3.1.2 Mitochondrial DNA regions used for SNP assay

Mitochondrial DNA (mtDNA) is also frequently used in molecular taxonomy and genetic diversity investigations. That is due to its specific characteristics. For example maternal inheritance gives the opportunity to trace out the genealogy that defines population relationships and origin (Le et al., 2000a, 2001, 2002). Also its evolution, faster than nuclear DNA evolution, assumes more frequent mutation events and often gene variations or rearrangements are specific for particular taxa. Mitochondrial haplotype defining enables the genetic diversity determination among and within the populations (including liver fluke populations). mtDNA genes sequence analysis assists sometimes extremely hard differentiation of two species and/or subspecies closely related in accordance with their morphological and physiological characteristics. There is a detailed mtDNA data for a number of organisms and in particular full nucleotide sequence of *F. hepatica* mitochondrion have been described from Le et al., (2001). As a conservative molecule not all of the gene regions of liver fluke mtDNA are "plastic" and therefore informative enough for population structuring, phylogenetic and genetic diversity studies. Three areas of it have been defined as potentially most polymorphic (FhmtDNACOX3/ND4, FhmtDNAATP6/ND1 и FhmtDNACOX1/l–rRNA) (Figure 3) (Walker et al., 2007).

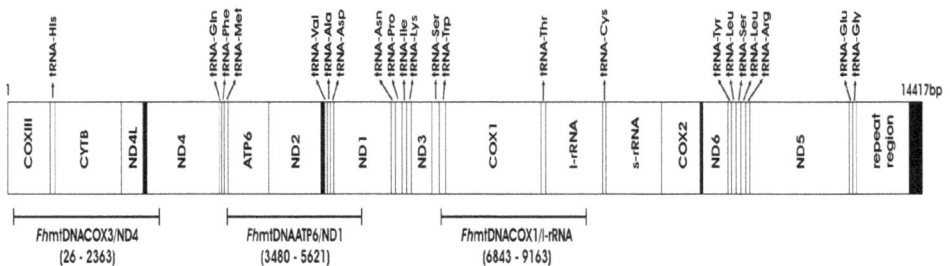

Fig. 3. Localization of the three most polymorphic regions in mtDNA of *F.hepatica* (Walker et al., 2007).

Most often used is variation analysis of the nad1 and cox1 genes (Hashimoto et al., 1997; Semyenova et al., 2006; Lee et al., 2007; Amer et al., 2010). The comprehensive information about them helps definition of polymorphic sites as molecule markers for explanation of liver fluke populations' genetic diversity. Sequence analysis of cox1 gene has been performed for establishment of taxonomical belonging of Japanese *Fasciola* sp. (Hashimoto et al., 1997). *F. hepatica* differs with 25-28 nucleotides from Japanese *Fasciola* sp. but *F. gigantica* differs only with 4-5 nucleotides. Respectively amino acid variations are 4 for *F. hepatica* and only one for *F. gigantica*. This data in combination with PCR-RFLP profiles of full mtDNA enables the confirmation by the authors of taxonomical identity of Japanese fasciolides and *F. gigantica*. cox1 gene has been also used for revealing the philogenetic relationships of the species from Phylum Platyhelmintes (Lee et al., 2007). It has been shown that even in one and the same class there are differences not only in PCR products length and GC content but also in nucleotide sequence and the presence of gaps which are 9 and 5 for trematodes and cestodes respectively. nad1 и cox1 gene sequences have been analysed also for differentiation between Eastern European and Western Asian liver fluke populations (Semyenova et al., 2006). 13 nad1 and 10 cox1 haplotypes have been identified in that research. They have been dispersed within two main lineages (I and II) with 1.07% nucleotide difference and unequal distribution in studied geographic areas. Obtained data has been interpreted by the authors as different maternal origin from two genetically different ancestral populations. It has been hypothesized that lineage I has Asian origin but both of them present in European population of *F. hepatica*. Nucleotide sequence of nad1 gene has been used also aiming the search of evolutionary history and origin of fasciolide species by Lotfy et al. (2008). Noncoding regions of mtDNA (long (LNR) and short (SNR)) have been studied (Korchagina et al., 2009) and it has been assumed that the polymorphisms and structural differences among them are associated with different haplotype groups of *F. hepatica*. In comparison with all other researchers, Teofanova et al. (2011), Walker et al. (2011) and Kantzoura et al.(2011), performed the investigations not with commonly used nad1 and cox1 genes from respectively second and third "plastic" regions of mtDNA (Figure 3) but with part from the first one which found to be most variable. Teofanova et al. (2011) have defined the population structure of liver fluke in Eastern Europe and Walker et al. (2011) have studied population dynamics of *F. hepatica* and genetic diversity of its populations in the Netherlands. Kantzoura et al. (2011) have established the correlation between defined South-Eastern European genetic variants and physicogeographic characteristics of those regions and prognosticated these haplogroups distribution through mathematical modelling.

3.1.3 Protein coding genes used for SNP assay

Population genetics researches could apply sequence analysis of protein coding genes as well. For that purpose it is important the selection of highly conservative genes. With regard to parasitic helminths commonly used are genes encoding target proteins for variety potential antihelmintic drugs. Different proteins and protein coding genes have been studied for genetic structuring of populations. These proteins are for example cathepsin L-like enzymes, cathepsin L protease family, β-tubulin 3, etc. (Robinson et al., 2001, 2002; Irving et al., 2003; Ryan et al., 2008). For fasciolides exhaustive sequence and evolutionary analysis of 18 different cathepsin L-like enzymes have been obtained (Irving et al., 2003). Respectively amino acid assay have been performed by the authors aiming the definition of specific sites in polypeptide chain which are presumable targets of positive adaptive evolution. The built phylogenetic structure of these fasciolide enzymes have been used for divergence timing between closely related species *F. hepatica* and *F. gigantica*.

Target proteins for antihelmintic drugs as β-tubulin 3 are not well studied. It has been established that *F. hepatica* express different isotypes of tubulin proteins and many of them are associated with antihelmintic resistance and because of that they are subject of recent investigations (Robinson et al., 2001, 2002; Ryan et al., 2008). The β-tubulin 3 gene have been utilized for genetic diversity research of liver fluke populations but it has been found to be not suitable as a molecule marker because of the multiple polymorphism without structuring properties (Teofanova et al., 2011).

4. Issues associated with genetic diversity studies of Fasciola heptica

Despite of the high quality of SNP analyses, data and results interpretations bring up many questions. They concern the presence of heteroplasmy within the mitochondrial haplotypes, the presence of recombination events in ribosomal genes and/or protein coding genes and their indicative value for the presence of cross-fertilization in populations of liver flukes. A substantial problem is also to find the correct correlation between SNP analyses data and the species origin and genetic diversity of geographic isolates. Despite the fact that *F. hepatica* has been an object for many years of research, its origin, biogeography, evolution and genetic diversity are still very intensive and not entirely complete which is also a problem to resolve.

4.1 Choosing a gene of an interest to study

The choice of gene regions to study is determined by the search and identification of the most proper molecule marker for population and genetic diversity studies. These regions could be with different type of inheritance and genetic status (e.g. organelle or nuclear localisation). Detailed analysis of chosen gene sequence would provide information about genetic drift, phylogeny of the species, evolution, genetic diversity and its relation to geographic localization and/or a host. Data comparison of variety of investigation would contribute to obtain complete view of genetic structure of *F. hepatica* populations. Nevertheless more of them do not point out any particular gene to be the most suitable marker and what are the advantages or disadvantages of that gene region usage. Comparative analyse of three gene fragments with different types of inheritance has been performed only in the research of Teofanova et al. (2011). It has been established that with respect to population structuring, phylogenetic and genetic diversity studies mtDNA genes are most adequate molecule markers. rDNA only allows clarification of taxonomic belonging of the organism and only a partial explanation of genetic diversity and evolution. Somatic protein coding gene used in that investigation, regardless its insignificant role for establishment of inter- and intrapopulational relationships, have been proved to be important for confirmation of cross-fertilization between liver flukes.

4.1.1 28S rDNA gene regions analysis

SNP assay of 28S rDNA gene region has been used for taxonomic belonging confirmation to *F. hepatica* species (Teofanova et al., 2011) which is not surprising taking the fact that in

Europe it is the only representative of *Fasciola* spp. (Mas-Coma et al., 1999; WHO, 2006). Investigations for species differentiation (Marcilla et al., 2002) and genetic heterogeneity (Vara-Del Rio et al., 2007) also have been performed (Section 3.1.1) but only the study of Teofanova et al. (2011) gives the opportunity for lineage definition (e.g. b105A, b105G and b105C/T) aiming determination of genetic diversity of liver fluke populations. It also has contributed to establishment of relations between the lineages and geographic location and for the first time the reason of such distribution of these lineages has been shown (e.g. polymorphism at 105 bp) (Figure 4). Some of the lineages have been found specific for the country of origin (e.g. b105C/T for Poland). The interpretation of that small group is dabatable concidering the low frequency of presence. Although the study of Vara-Del Rio et al. (2007) shows variation at the same nucleotide position it has not been concretely defined and only commented as associated with sensitive and/or resistant to antihelmithic drugs liver fluke isolates in Spain but not in a view of population structuring.

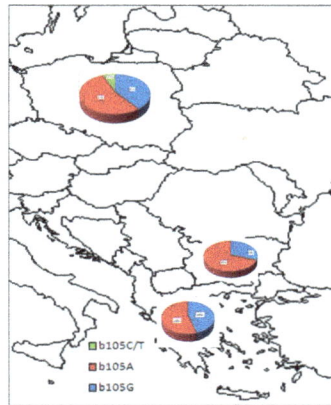

Fig. 4. Geographic distribution of 28s rDNA lineages for *F. hepatica* populations (Teofanova et al., 2011).

In total all the investigations on 28S rDNA are related to a single geographic isolates of *F. hepatica* but not at the populational level which is one of the main issues using such a conservative genes. It is a common casus finding of additional and not depended on the distribution of the basic lineages genotype groups (e.g. S28f and S28e; Teofanova et al., 2011) which can not be previously predicted and more hardly to explain with a higher level of speculation.

4.1.2 mtDNA gene regions analysis

mtDNA is the most common used for interpopulational structuring, intra- and interspecific definition studies. That is due to its characteristics such as high conservativeness, specific inhetitance (maternal and very rarely parental) and absence of recombinant events (haploid genome). The most valuable and most common mitochondrial genes were presented in Section 3.1.2. Only three studies (Teofanova et al., 2011; Walker et al., 2011; Kantzoura et al., 2011) concerned different but found to be most polymorphic fragments from mtDNA of *F. hepatica* (consisted from partial cox3 (cytochrome c oxidase 3) gene, tRNA-His, partial cob (cytochrome B) gene). Again predominance of the researches refers to single geographic isolates of *F. hepatica* but not forming a worldwide picture of populational groups' distribution. Only in some cases analysis of nucleotide variations of mitochondrion genes was performed for species identification (Section 3.1.2). Choosing "plastic"fragments of liver fluke mitochondrion means choosing the most informative and adequate markers for obtaining a complete view on genetic diversity of *F. hepatica* in the world. Nevertheless only three available in bibliographic data bases studies could give such a picture (Semyenova et al., 2006; Teofanova et al., 2011; Walker et al., 2011). One of them is genetic diversity investigation of liver fluke populations from Eastern Europe (Teofanova et al., 2011). Obtained data appears to be very informative and allows defining of genetic intrapopulational structure, evolution history and migration processes of *F. hepatica*. It also enables the differentiation of two main lineages (CtCmt1 and CtCmt2) and divergent CtCmt2 mytotypes from first and second level, which are specific for Southerneastern or Northernestern Europe populations (Figure 5). These lineages and haplogroups are well described according to the corresponding single nucleotide polymorphisms and their presence frequency. That frequency gives the opportunity for phylogenetic ageing of each group having in mind that higher frequency correlates to evolutionaty earliar origin.

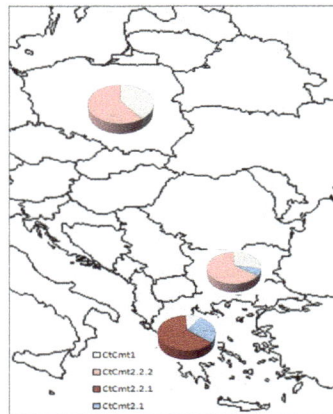

Fig. 5. Geographic distribution of mtDNA lineages for *F. hepatica* populations (Teofanova et al., 2011).

It is also possible that revealed after a detailed aminoacid sequence analysis substitutions in the studied mitochondrial proteins to be with physiological influence with direct adaptive evolutionary role for liver fluke individuals. This data could be compared only with mentioned above two other studies – about the genetic differentiation in Eastern European and Western Asian populations of *F. hepatica* (Semyenova et al., 2006) and about population dynamics of *F. hepatica* and genetic diversity of its populations in the Netherlands (Walker et al., 2011). Althogh the investigation of Semyenova et al. (2006) has been performed using nad1 and cox 1 gene both lineage I and II were described through the characteristic nucleotide variations as well as in investigation of Teofanova et al. (2011) (Figure 6). Mediated by the full mitochondrion sequence of *F. hepatica* (AF216697; Le et al., 2000b) it is possible to claim the identity of lineage I and CtCmt1. By analogy to lineage I the origin of

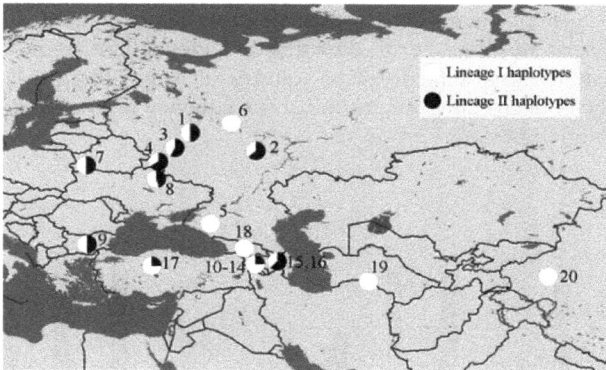

Fig. 6. Geographic distribution of two mitochondrial lineages for *F. hepatica* populations
(Semyenova et al., 2006).

CtCmt1 was assumed to be Asian. That is also due to the fact that unlike CtCmt2 lineage
CtCmt1 does not divergate to additional haplogroups and it is almost equally distributed in
Eastern Europe. Probably there is also a correlation between lineage II and CtCmt2 but there
are no direct evidences about that. Some of the mentioned above explainations are an
example of solving the problem of finding the correct correlation between SNP assay data
and the species origin and genetic diversity of variety of geographic isolates. It also reveals
the properties of well described and interpreted nucleotide variations data for polymorphic
mtDNA gene regions. Unfortunately such analysis is rarely observed in published
researches and that decrease the opportunity for establishment of whole and complete
picture of worldwide liver fluke genetic diversity. Another debatable case could be found
after elaboration of Walker et al. (2011) investigation about the Dutch population of *F.
hepatica* and after a search for correlation between available in GenBank™ sequences from
the Netherlands and those typical for CtCmt1/CtCmt2 lineages and the divergent
mytotypes. Dutch liver fluke representatives have been differentiated in three detached
clades but the belonging of GenBank™ sequences to them has not been described by the
authors. Sequence comparative assay enables their assignation to CtCmt1, CtCmt2.1.3 and
CtCmt2.2.2 haplogroups from the study of Teofanova et al. (2011). That confirms that
CtCmt2.2.2 is characteristic mytotype for Northern populations and gives the opportunity to
presume that CtCmt2.1.3 haplotype is typical for Western European populations but not for
Greek ones as it has been mentioned by Teofanova et al. (2011). That once again shows that a
correct and complete genetic diversity analysis is very difficult and associated with multiple
issues but realizable as it concerns mtDNA genes as molecule markers for SNP assay.

4.1.3 β-tubulin 3 (somatic) gene analysis

It is not so common to use somatic genes for phylogenetic and populational genetics
analysis. The reason is the higher variability and the presence of heterozygosity in
consequence of recombination events. They are also different than genes described in
Sections 4.1.1. and 4.1.2 with respect to the inheritance type. β-tubulin 3 in particular has
never been used for populational research except for study of Teofanova et al. (2011). Even
in that investigation that gene analysis has not been proved to be adequate for definition of

explicit lineages because of the multiple polymorphism, heterozygosity and recombination. That problem has been resolved by its directing to independent frequency evaluation of polymorphisms in sequences with different country of origin (Figure 7). There has been observed a tendency of decreasing polymorphisms frequency in liver fluke populations in a South/North direction in Eastern Europe.

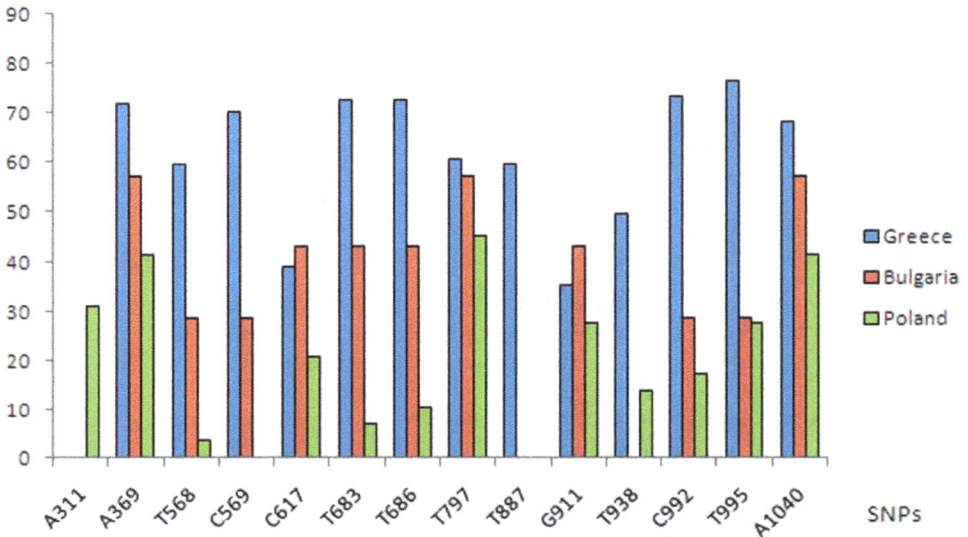

Fig. 7. Geographic distribution of β-tubulin 3 gene polymorphisms for *F. hepatica* populations (Teofanova et al., 2011).

Although that somatic gene has not been found suitable as a molecule marker for populational structuring and phylogenetic studies, it supports the analysis data for ribosomal and mitochondrial gene regions in that particular study for discrimination of Southern and Northern *F. hepatica* populations from Eastern Europe.

4.2 Heterozygosity and heteroplasmy

A diploid organism is heterozygous at a gene locus when its cells contain two different alleles of a gene.

Heteroplasmy could be explained by the inheritance of mixed organelle populations (e.g. mitochondrial) (Doublet et al., 2008; Reiner et al., 2010) and is found only for haploid genomes.

As it concerns somatic genes, the heterozygosity in them is a commonly observed as a multiple alleles resulting from high level of recombination events which are usual for outbreeding systems. In the case of β-tubulin 3 gene (Section 4.1.3) the heterozygosity could be used as an evidence for cross-fertilization processes together with hermaphroditic sexual reproduction which was considered to be the only way of reproduction of *F. hepatica* in past.

Now some investigations are attempting to prove the presence of mating between different liver fluke individuals as well (Hurtrez-Bousses et al., 2001; Fletcher et al., 2004; Amer et al., 2010; Teofanova et al., 2011).

More strange and unusual is observed heterozygosity in rDNA (Teofanova et al., 2011) which could be defined as a phenomenon and is not well studied. The heterozygosity in 105th nucleotide could have different explanations. One of them is the possibility that is highly mutable position. On the other side it could be due to recombination events by analogy with somatic genes. There are no much published data about this event and more of it concerns yeasts (Petes, 1980) so the first presumption seems more likely. Heterozygosity has been found also in additional S28f and S28e genotype groups of rDNA (Teofanova et al., 2011). In that case heterozygosity has been with high rate and could be an attestation for their evolutionary history defining them as evolutionary young groups. Analogically the presence of only one polymorphic nucleotide site (105 bp) with low level of heterozygosity would be determinative for evolutionary old mutation which is proved by almost equal distribution of basic lineages (b105A and b105G). Of course all of that is quite speculative and could not be used as a definitive evidence for the reason of observed hetezygosity. Only in one other case was mentioned about the heterozygosity in 28S rDNA and that is in single Spanish isolate (Vara-Del Río et al., 2007) but with no reasoning explanation.

It is possible to say that heteroplasmy is "heterozygosity of haploid orgnaisms" and represents a mixture of more than one type of an organellar genome (mitochondrial DNA or plastid DNA) within a cell or individual. The presence of heteroplasmic individuals is probably due to recent evolutionary processes. It is quite uncommon commenting that effect of SNP assay of mitochondrial sequences. Only a few authors mention its observation but more information about it could give the opportunity for correct and complete data interpretation (Semyenova et al., 2006; Korchagina et al., 2009; Teofanova et al., 2011). In the investigation of Semyenova et al. (2006) for population structure of liver flukes with respect to nad1 and cox1 mtDNA genes heteroplasmy has been defined as a source of the population haplotype diversity. Korchagina et al. (2009) discuss more widely the heteroplasmy in noncoding mtDNA regions from *F. hepatica* but pointed that the right mechanism that cause and sustain heteroplasmy is unknown. Changed copies are thought to result from polymerase slippage during DNA replication (Levinson & Gutman, 1987), mispairing (Buroker et al., 1990), and, in some cases, DNA recombination (Lunt & Hyman, 1997). In the investigation of Teofanova et al. (2011) heteroplasmy has been observed with high frequency in small haplogroups (C3mt(193-206) and CBmt(710/711)) and it could be presumably suggested that it is due to a recent evolutionary prosseses and these groups are young and still forming their appearance. It is obvious that if the heteroplasmy presence problem would be resolved the obtained data could be more detailed and precise.

4.3 Cladogram visualisation issues

When working with sequence analysis a variety of software products could facilitate the processing (e.g. ChromasPro 1.5 (Copyright © 2003-2009 Technelysium Pty Ltd), ClustalW2 (Larkin et al., 2007), BLAST (Basic Local Alignment Search Tool), MEGA 4.0 (Kumar et al., 2008)). That always means that many mistakes are possible for example due to statistical problems or not taking into account the scientific background. For example only single nucleotide variations or polymorphisms with very low and insignificant frequencies must not be considered and are detected as sequencing mistakes. Nevertheless some programs do include such variations when processing the analysis. Moreover most of the programs do not evaluate mutation occurance frequencies of any studied SNPs and could not detect their phylogenetic and evolutionary significance. Another problem concerns cladogram visualisation. Sometimes it shows hard for interpretation and contradictory data and in that case it is not suitable for SNP assays. For example some programs when building Neighbor Joining Trees could put the heterozygous or heteroplasmic representatives in a separate group which is not correct. Also two different programs could build two different cladograms independently of used tool for that. The process was checked with sequences from investigation of Teofanova et al. (2011) which are available in GenBank™ as populational sets and with BLAST tool (at http://www.ncbi.nlm.nih.gov), with ClustalW2 and MEGA 4.0 programs and built cladograms appeared to be different between each other and not coinciding with data obtaind from polymorphisms description.

4.4 Evolutionary and phylogenetic reasoning for genetic diversity

Data for the origin, biogeography, evolution and their association with genetic diversity of *F. hepatica* is not entirely complete. It is a problem that not many investigations discuss and usually there are a lot of speculations and presumptions with only an indirect evidencing. Still there are some researchers trying to solve these isssues (Irving et al., 2003; Semyenova et al., 2006; Lotfy et al., 2008; Teofanova et al., 2011). All these studies performed their hypothesis based on obtained results. Modern theories for the evolution and origin of fasciolid species are connected to the coevolution of definitive (mammals) and intermediate (freshwater snails) hosts during the Neozoan age. It has been assumed that fasciolides'originated in Africa about 50 MYA (million years ago) during the Eocene Epoch followed by their dispersion in Eurasia (Lotfy et al., 2008). Authors have accepted the hypothesis of coevolution between fasciolides and their hosts and because of that they have suggested that definitive hosts of the ancient fasciolid species were predecessors of the elephants. As it concerns intermediate hosts according to Lotfy et al. (2008) that are freshwater planorbid (Planorbidae) and/or lymnaeid (Lymnaeidae) snails. It has been presumed that the switch between planorbids (specific for ancient fasciolides from Africa) to lymnaeids had happened in Eurasia which favoured emergence of Fasciolinae. Because of that it has been hypothesized that *F. hepatica* has Eurasian origin and colonization of Africa occurred secondary. The evidence for that is parasite preference to the lymnaeid snail *Galba truncatula*. Subsequently according to Irving et al. (2003) the divergence between *F. hepatica* and the closely related to it *F. gigantica* has occurred about 28 - 16 MYA during the Miocene Epoch. Hypothesises about the following evolution of the liver flukes are connected with the evolution of cloven-footed animals (mainly sheep, goats, and cows). Because of that there are many searches for correlation between particular definitive hosts and specific genetic lineages or clades of *F. hepatica* (Hiendleder et al., 2002; Semyenova et al., 2006) but there are no direct evidences for that. As it was mentioned in Section 4.1.2 there is a lineage with Asian origin (Semyenova et al., 2006) subsequently spread in Europe. Geological timing of that recolonization has not been defined but it is very probably to be associated with the end of the last ice age and/or with human migration from Asia to Europe at the end of Pleistocene Epoch. These evolutionary and phylogenetic hypothesises are combined and visualised on Figure 8.

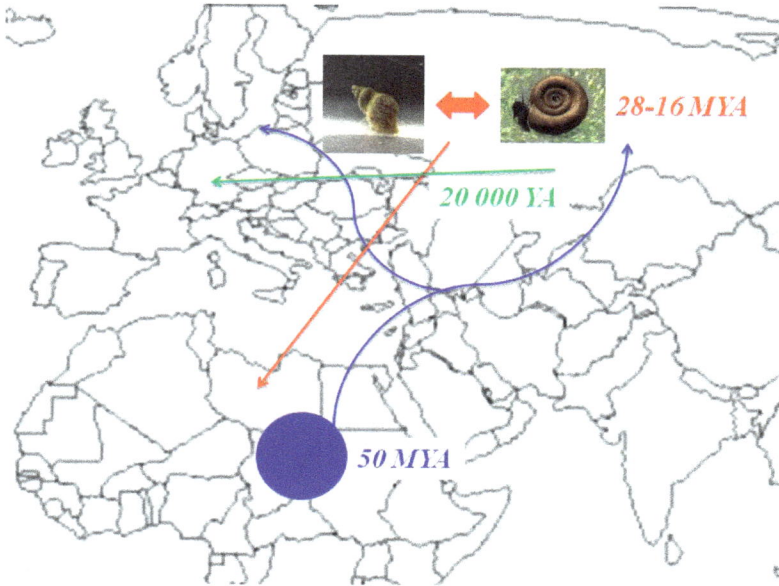

Fig. 8. Evolutionary and phylogenetic hypothesises for *F. hepatica*. Blue arrows – African origin of fasciolid species and distribution in Eurasia (Lotfy et al., 2008); Red arrows – Eurasian origin of *F. hepatica* as a result of switch between planorbids and lymnaeids and recolonization of Africa (Irving et al., 2003); Green arrow – Mitochondrial lineage with Asian origin and colonization of Europe (Semyenova et al., 2006).

The investigation of Teofanova et al. (2011) proposed some more additional hypothesises. First one has resulted from determination of haplogroup belonging of Australian (AF216697), (Le et al., 2000b) and Tanzanian sequence (EU282862), (Walker et al., 2008) to CtCmt1 and CtCmt2.2.2 respectively. Because mentioned haplogroups have been defined as specific for Northern European liver fluke populations the presence of *F. hepatica* in Australia and Tanzania has been assumed to be result of those geographic areas colonization from Northern European liver fluke populations (mainly from UK). Second theory in that study has been based on territorial and populational isolation of chosen geographic areas in past due to the continental drift. During the Miocene about 19 MYA Africa collided with Eurasia (Seyfert & Sirkin, 1979). During that period present-days' part of Italia and Greece (part of Western Balkans) have not been connected to Eurasian continent. Although it is quite speculative, that could be an explanation of observed differences between the liver fluke populations in Central Europe (Poland) and Southern Eastern Europe (Greece) regions. It is possible either Eurasian liver fluke populations or populations from newly joined to Eurasia Greek areas to perform divergation and colonization.

Third hypothesis from the investigation of Teofanova et al. (2011) has been associated with *F. hepatica* rerecolonization after last ice age during the Pleistocene Epoch about 20 000 years (Figure 9).

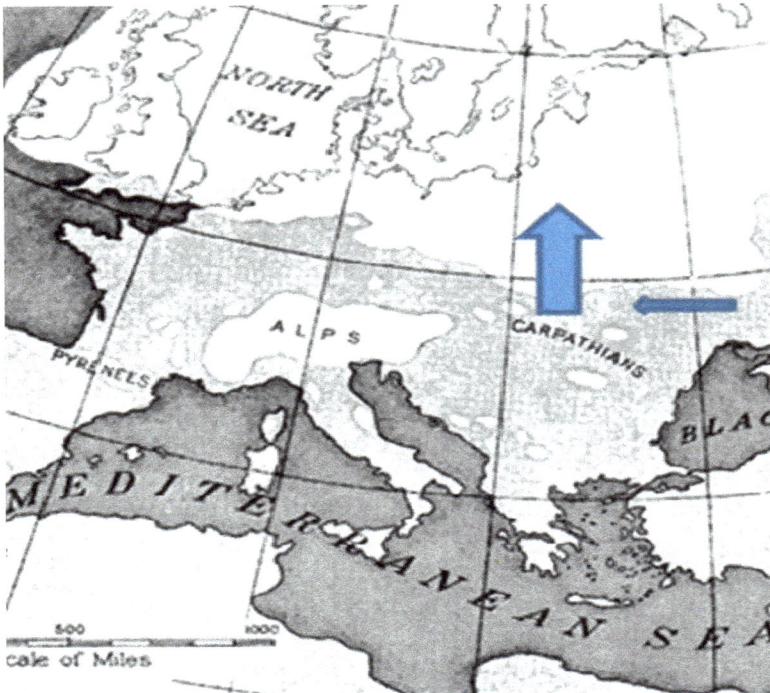

Fig. 9. Spreading of glacial sheets during the last ice age during Pleistocene Epoch and colonization possibilities of newly released geographic regions (Blue arrows).

During the last ice age glacial sheets did not cover the territories of Bulgaria and Greece and reach only Northern Carpathian Mountains (Seyfert & Sirkin, 1979). According to the study of Teofanova et al. (2011), re-colonization of Northern areas after ice age originated not from far South (Greece) but from boundary regions of the glacial sheets. This hypothesis is confirmed by the presented data for the b-tubulin 3 gene and mitochondrial profile of Bulgarian population. The parallel genetic drift could be also realized from East to West (from Asia to Europe). That could be also related to human migration and the transfer of domestic animals. The evidence was found in the mitochondrial lineage I with Asian origin from the investigation of Semyenova et al. (2006) and identical to CtCmt1 lineage (Teofanova et al., 2011).

In total it is obvious that the results of any population structuring or genetic diversity investigation are associated with variety of issues and problems but their resolution is not impossible and could contributed to a better understanding of the genetic structure and heterogeneity of liver fluke intrapopulations worlwide.

5. Conclusion

The increasing of the knowledge of population structure and genetic diversity of liver fluke is necessary. A variety of molecular methods have been applied for genetic structuring of parasitic populations and in particular of the liver fluke. Each one of them has its advantages and disadvantages. To avoid their negative features, recent developments include single nucleotide polymorphism assays after direct sequencing, which has been proved to be the most reliable method used in genetic diversity studies of *Fasciola hepatica*. Despite of the high quality of that assay it is pursueted of a lot of issues and problems to resolve (heteroplasmy within the mitochondrial genes, recombination events in ribosomal genes and/or somatic genes, etc.). Their proper and complete analysis and evaluation in combination with correct data interpretation could solve a substantial problem to find the correlation between nucleotide variations and the species origin and genetic diversity of geographic isolates. In view of using data of genetic polymorphism for solving practical issues (studying differences in the pathogenic effects of various haplotypes, infectivity monitoring of pastures and wetlands, etc.), it is of a significant importance to clarify the unresolved issues.

6. Acknowledgment

This study was supported by the grant WETLANET funded by FP7 EC (FP7 CSA – SUPPORT ACTION, GA 229802) and by the European Union funded DELIVER Project (Contract No.: FOOD-CT-2004-023025).

7. References

Ai, L., Chen, M.-X., Alasaad, S., Elsheikha, H.M., Li, J., Li, H.-L., Lin, R.-Q., Zou, F.-C., Zhu, X.-Q. & Chen, J.-X. (2011). Genetic characterization, species differentiation and detection of *Fasciola* spp. by molecular approaches, *Parasites & Vectors*, Vol.4, No.101, pp. 1-6, ISSN 1756-3305

Ai, L., Li, C., Elsheikha, H.M., Hong, S.J., Chen, J.X., Chen, S.H., Li, X., Cai, X.Q., Chen, M.X. & Zhu, X.Q. (2010). Rapid identification and differentiation of *Fasciola hepatica* and *Fasciola gigantica* by a loop-mediated isothermal amplification (LAMP) assay, *Veterinary parasitology*, Vol.174, No.3-4, pp. 228–233, ISSN 0304-4017

Alasaad, S., Li, Q.Y., Lin, R.Q., Martín-Atance, P., Granados, J.E., Díez-Baños, P., Pérez, J.M. & Zhu, X.Q. (2008). Genetic variability among Fasciola hepatica samples from different host species and geographical localities in Spain revealed by the novel SRAP marker, *Parasitology research*, Vol.103, No.1, pp. 181–186, ISSN 0932-0113

Alasaad, S., Soriguer, R.C., Abu-Madi, M., El Behairy, A., Baños, P.D., Píriz, A., Fickel, J. & Zhu, X.Q. (2011). A fluorescence-based polymerase chain reaction-linked singlestrandconformation polymorphism (F-PCR-SSCP) assay for the identification of *Fasciola spp.*, *Parasitology research*, Vol.108, No.6, pp. 1513–1517, ISSN 0932-0113

Alasaad, S., Soriguer, R.C., Abu-Madi, M., El Behairy, A., Baños, P.D., Píriz, A., Fickel, J. & Zhu, X.Q. (2011). A TaqMan real-time PCR-based assay for the identification of *Fasciola spp.*, *Veterinary parasitology*, Vol.179, No.1-3, pp. 266–271, ISSN 0304-4017

Amer, S., Dar, Y., Ichikawa, M., Fukuda, Y., Tada, C., Itagaki, T. & Nakai, Y. (2010). Identification of Fasciola species isolated from Egypt based on sequence analysis of genomic (ITS1 and ITS2) and mitochondrial (NDI and COI) gene markers, *Parasitology International*, Vol.60, No.1, pp. 5-12, ISSN 1383-5769

Buroker, N.E., Brown, J.R., Gilbert, T.A., O'Hara, P.J., Beckenbach, A.T., Thomas, W.K. & Smith, M.J. (1990). Length heteroplasmy of sturgeon mitochondrial DNA: An illegitimate elongation model, *Genetics*, Vol.124, No.1, pp. 157–163, ISSN 0016-6731

Doublet, V., Souty-Grosset, C., Bouchon, D., Cordaux, R. & Marcade, I. (2008). A Thirty Million Year-Old Inherited Heteroplasmy, *PLoS ONE*, Vol.3, No.8, e2938, ISSN 1932-6203

Erensoy, A., Kuk, S. & Ozden, M. (2009). Genetic identification of *Fasciola hepatica* by ITS-2 sequence of nuclear ribosomal DNA in Turkey, *Parasitology Research*, Vol.105, No.2, pp. 407–412, ISSN 0932-0113

Fletcher, H.L., Hoey, E.M., Orr, N., Trudgett, A., Fairweather, I. & Robinson, M.W. (2004). The occurrence and significance of triploidy in the liver fluke, *Fasciola hepatica*, *Parasitology*, Vol.128, No.1, pp. 69–72, ISSN 0031-1820

Gunasekar, K.R., Tewari, A.K., Sreekumar, C., Gupta, S.C. & Rao, J.R. (2008) Elucidation ofgenetic variability among different isolates of *Fasciola gigantica* (giant liverfluke) using random-amplified polymorphic DNA polymerase chain reaction, *Parasitology research*, Vol.103, No.5, pp. 1075–1081, ISSN 0932-0113

Hashimoto, K., Watanobe, T., Liu, C.X., Init, I., Blair, D., Ohnishi, S. & Agatsuma, T. (1997). Mitochondrial DNA and nuclear DNA indicate that the Japanese Fasciola species is *F. gigantica*, *Parasitology research*, Vol.83, No.3, pp. 220–225, ISSN 0932-0113

Hiendleder, S., Kaupe, B., Wassmuth, R. & Janke, A. (2002). Molecular analysis of wild and domestic sheep questions current nomenclature and provides evidence for domestication from two different subspecies, *Proceedings of the Royal Society of London. Series B. Biological Science*, Vol.269, No.1494, pp. 893-904, ISBN 3-900051-07-0

Hillis, D.M. & Dixon, M.T. (1991). Ribosomal DNA: molecular evolution and phylogenetic inference, *The Quarterly review of biology*, Vol.66, No.4, pp. 411-453, ISSN 0033-5770

Huang, W.Y., He, B., Wang, C.R. & Zhu, X.Q. (2004). Characterisation of Fasciola species from mainland China by ITS-2 ribosomal DNA sequence, *Veterinary parasitology*, Vol.120, No.1-2, pp. 75–83, ISSN 0304-4017

Hurtrez-Bousses, S., Durand, P., Jabbour-Zahab, R., Guegan, J.-F., Meunier, C., Bargues, M.-D., Mas-Coma, S., & Renaud, F. (2004). Isolation and characterization of microsatellite markers in the liver fluke (*Fasciola hepatica*), *Molecular Ecology Notes*, Vol.4, No.4, pp. 689-690, ISSN 1471-8278

Hurtrez-Bousses, S., Meunier, C., Durand, P. & Renaud, F. (2001). Dynamics of host–parasite interactions: the example of population biology of the liver fluke (*Fasciola hepatica*), *Microbes and Infection*, Vol.3, No.10, pp. 841-849, ISSN 1286-4579

Ichikawa, M. & Itagaki, T. (2010). Discrimination of the ITS1 types of *Fasciola spp.* basedon a PCR-RFLP method, *Parasitology research*, Vol.106, No.3, pp. 757–761, ISSN: 0932-0113

Irving, J.A., Spithill, T.W., Pike, R.N., Whisstock, J.C. & Smooker, P.M. (2003). The Evolution of Enzyme Specificity in *Fasciola spp.*, *Journal of Molecular Evolution*, Vol.57, No.1, pp. 1-15, ISSN 0022-2844

Kantzoura, V., Kouam, M.K., Feidas, H., Teofanova, D. & Theodoropoulos, G. (2011). Geographic distribution modelling for ruminant liver flukes (*Fasciola hepatica*) in South-eastern Europe, *International Journal for Parasitology*, Vol.41, No.7, pp. 747-753, ISSN 0020-7519

Kumar, S., Dudley, J., Nei, M. & Tamura, K. (2008). MEGA: A biologist-centric software for evolutionary analysis of DNA and protein sequences, *Briefings in Bioinformatics*, Vol.9, No.4, pp. 299-306, ISSN 1467-5463

Larkin, M.A., Blackshields, G., Brown, N.P., Chenna, R., McGettigan, P.A., McWilliam, H., Valentin, F., Wallace, I.M., Wilm, A., Lopez, R., Thompson, J.D., Gibson, T.J. & Higgins, D.G. (2007). ClustalW and ClustalX version 2, *Bioinformatics*, Vol.23, No.21, pp. 2947-2948, ISSN 1367-4803

Le, T.H., Blair, D. & McManus, D.P. (2000a). Mitochondrial genomes of human helminths and their use as markers in population genetics and phylogeny, *Acta Tropica*, Vol.77, No.3, pp. 243-256, ISSN 0001-706X

Le, T.H., Blair, D., Agatsuma, T., Humair, P.F., Campbell, N.J., Iwagami, M., Littlewood, D.T., Peacock, B., Johnston, D.A., Bartley, J., Rollinson, D., Herniou, E.A., Zarlenga, D.S. & McManus, D.P. (2000b). Phylogenies inferred from mitochondrial gene orders-a cautionary tale from the parasitic flatworms, *Molecular Biology and Evolution*, Vol.17, No.7, pp. 1123-1125, ISSN 0737-4038

Le, T.H., Blair, D. & McManus, D.P. (2001). Complete DNA sequence and gene organization of the mitochondrial genome of the liver fuke, *Fasciola hepatica* L. (Platyhelminthes; Trematoda), *Parasitology*, Vol.123, No.6, pp. 609-621, ISSN 0031-1820

Le, T.H., Blair, D. & McManus, D.P. (2002). Mitochondrial genomes of parasitic flatworms, *Trends in Parasitology*, Vol.18, No.5, pp. 206-213, ISSN 1471-4922

Lee, S.-U., Chun, H.-C. & Huh, S. (2007). Molecular phylogeny of parasitic Platyhelminthes based on sequences of partial 28S rDNA D1 and mitochondrial cytochrome c oxidase subunit I, *The Korean Journal of Parasitology*, Vol.45, No.3, pp. 181-189, ISSN 0023-4001

Levinson, G. & Gutman, G.A. (1987). Slipped-strand mispairing: A major mechanism for DNA sequence evolution, *Molecular Biology and Evolution*, Vol.4, No.3, pp. 203–221, ISSN 0737-4038

Li, G. & Quiros, C.F. (2001). Sequence-related amplified polymorphism (SRAP), a new marker system based on a simple PCR reaction: its application to mapping and gene tagging in Brassica, *Theoretical and applied genetics*, Vol.103, No.2-3, pp. 455-461, ISSN 0040-5752

Li, Q.Y., Dong, S.J., Zhang, W.Y., Lin, R.Q., Wang, C.R., Qian, D.X., Lun, Z.R., Song, H.Q. & Zhu, X.Q. (2009). Sequence-related amplified polymorphism, an effective molecularapproach for studying genetic variation in *Fasciola spp.* of human andanimal health significance, *Electrophoresis*, Vol.30, No.2, pp. 403–409, ISSN 0173-0835

Lotfy, W.M., Brant, S.V., DeJong, R.J., Le, T.H., Demiaszkiewicz, A., Jayanthe, R.R.P.V., Perera, V.B.V.P., Laursen, J.R. & Loker, E.S. (2008). Evolutionary Origins, Diversification and Biogeography of Liver Flukes (Digenea, Fasciolidae), *The American Journal of Tropical Medicine and Hygiene*, Vol.79, No.2, pp. 248-255, ISSN 0002-9637

Lunt, D.H. & Hyman, B.C. (1997). Animal mitochondrial DNA recombination, *Nature*, Vol.387, No.6630, pp. 247, ISSN 0028-0836

Magalhaes, K.G., Jannotti-Passos, L.K., Caldeira, R.L., Berne, M.E., Muller, G., Carvalho, O.S. & Lenzi, H.L. (2008). Isolation and detection of *Fasciola hepatica* DNA in *Lymnaea viatrix* from formalin-fixed and paraffin-embedded tissues through multiplex-PCR, *Veterinary parasitology*, Vol.152, No.3-4, pp. 333–338, ISSN 0304-4017

Marcilla, A., Bargues, M.D. & Mas-Coma, S. (2002). A PCR-RFLP assay for the distinction between *Fasciola hepatica* and *Fasciola gigantica, Molecular and Cellular Probes*, Vol.16, No.5, pp.327-333, ISSN 0890-8508

Mas-Coma, S., Esteban, J.G. & Bargues M.D. (1999). Epidemiology of human fascioliasis: a review and proposed new classification, *Bulletin of the World Health Organization*, Vol.77, No.4, pp. 340-346, ISSN 0042-9686

Mas-Coma, S., Bargues, M.D. & Valero, M.A. (2005). Fascioliasis and other plant-borne trematode zoonoses. International Journal for Parasitology, Vol.35, No.11-12, pp. 1255-1278, ISSN 0020-7519

McGarry, J.W., Ortiz, P.L., Hodgkinson, J.E., Goreish, I. & Williams D.J.L. (2007). PCR-based differentiation of *Fasciola* species (Trematoda: Fasciolidae), using primers based on RAPD-derived sequences, *Annals of Tropical Medicine and Parasitology*, Vol.101, No.5, pp. 415-421, ISSN 0003-4983

Morgan, J.A.T. & Blair, D. (1995). Nuclear rDNA ITS sequence variation in the trematode genus Echinostoma: an aid to establishing relationships within the 37-collar spine group, *Parasitology*, Vol.111, No.5, pp.609-615, ISSN 0031-1820

Notomi, T., Okayama, H., Masubuchi, H., Yonekawa, T., Watanabe, K., Amino, N. & Hase, T. (2000). Loop-mediated isothermal amplification of DNA, *Nucleic Acids Research*, Vol.28, No.12, e63, ISSN 0305-1048

Olson, P.D., Cribb, T.H., Tkach, V.V., Bray, R.A. & Littlewood, D.T.J. (2003). Phylogeny and classification of the Digenea (Platyhelminthes: Trematoda), *International Journal for Parasitology*, Vol.33, No.7, pp. 733-755, ISSN 0020-7519

Orita, M., Iwahana, H., Kanazawa, H., Hayashi, K. & Sekiya, T. (1989).Detection of polymorphisms of human DNA by gel electrophoresis as single-strand conformation polymorphisms (mobility shift of separated strands/point mutation/riction franment length polymorphism), *Proceedings of the National Academy of Sciences of the United States of America, Genetics*, Vol.86, No.8, pp. 2766-2770, ISSN 0027-8424

Pemberton, T.J., Sandefur,C.I., Jakobsson,M. & Rosenberg, N.A. (2009). Sequence determinants of human microsatellite variability, *BMC Genomics*, Vol.10, No.1, pp. 612-631, ISSN 1471-2164

Petes, T.D. (1980). Unequal meiotic recombination within tandem arrays of yeast ribosomal DNA genes, *Cell*, Vol.19, No.3, pp. 765-774, ISSN 0092-8674

Prasad, P.K., Tandon, V., Biswal, D.K., Goswami, L.M. & Chatterjee, A. (2008). Molecular identification of the Indian liver fluke, *Fasciola* (Trematoda: Fasciolidae) based on the ribosomal internal transcribed spacer regions, *Parasitology Research*, Vol.103, No.6, pp. 1247-1255, ISSN 0932-0113

Reiner, J.E., Kishore, R.B., Levin, B.C., Albanetti, T. & Boire, N. (2010). Detection of Heteroplasmic Mitochondrial DNA in Single Mitochondria, *PLoS ONE*, Vol.5, No.12, e14359, ISSN 1932-6203

Robinson, M.W., Hoey, E.M., Fairweather, I., Dalton, J.P., McGonigle, S. & Trudgett, A. (2001). Characterization of a β-tubulin gene from the liver fuke, *Fasciola hepatica, International Journal for Parasitology* Vol.31, No.11, pp. 1264-1268, ISSN 0020-7519

Robinson, M.W., Trudgett, A., Hoey, E.M. & Fairweather, I. (2002). Triclabendazole-resistant
    Fasciola hepatica: β-tubulin and response to in vitro treatment with triclabendazole,
    Parasitology, Vol.124, No.3, pp. 325-338, ISSN 0031-1820

Rokas,A., Ladoukakis, E. & Zouros, E. (2003). Animal mitochondrial DNA recombination
    revisited, *Trends in Ecology and Evolution*, Vol.18, No.8, pp. 411-417, ISSN 0169-5347

Ryan, L.A., Hoey, E., Trudgett, A., Fairweather, I., Fuchs, M., Robinson, M.W., Chambers, E.,
    Timson, D.J., Ryana, E., Feltwell, T., Ivens, A., Bentley, G. & Johnston, D. (2008).
    Fasciola hepatica expresses multiple α- and β-tubulin isotypes, *Molecular and
    Biochemical Parasitology*, Vol.159, No.1, pp. 73-78, ISSN 0166-6851

Semyenova, S.K., Morozova, E.V., Chrisanfova, G.G., Asatrian, A.M., Movsessian, S.O. &
    Ryskov, A.P. (2003). RAPD variability and genetic diversity in two populations of
    liver fluke, *Fasciola hepatica*, *Acta Parasitologica*, Vol.48, No.2, pp. 1230-2821, ISSN
    1230-2821

Semyenova, S.K., Morozova, E.V., Vasilyev, V.A., Gorokhov, V.V., Moskvin, S., Movsessyan,
    S.O. & Ryskov, A.P. (2005). Polymorphism of internal transcribed spacer 2 (ITS-2)
    sequences and genetic relationships between *Fasciola hepatica* and *Fasciola gigantica*,
    Acta Parasitologica, Vol.50, No.3, pp. 240-243, ISSN 1230-2821

Semyenova, S.K., Morozova, E.V., Chrisanfova, G.G., Gorokhov, V.V., Arkhipov, I.A.,
    Moskvin, A.S., Movsessyan, S.O. & Ryskov, A.P. (2006). Genetic differentiation in
    Eastern European and Western Asian populations of the liver fluke, *Fasciola
    hepatica*, as revealed by mitochondrial *nad1* and *cox1* genes, *The Journal of
    Parasitology*, Vol.92, No.3, pp. 525-530, ISSN 0022-3395

Seyfert, C.K. & Sirkin, L.D. (1979). *Earth history and Plate Tectonics*, Harper and Row
    Publishers, pp. 600, ISBN 0060459190, New York

Spithill, T.W., Smooker, P.M. & Copeman, D.B. (1999). *Fasciola gigantica*: epidemiology,
    control, immunology and molecular biology, In: *Fasciolosis*, J.P. Dalton, (Ed.), pp.
    465-525, ISBN 0-85199-260-9, Wallingford, Oxon, UK: CABI Pub.

Sunnucks, P., Wilson, A.C., Beheregaray, L.B., Zenger, K., French, J. & Taylor, A.C. (2000).
    SSCP is not so difficult: the application and utility of single-stranded conformation
    polymorphism in evolutionary biology and molecular ecology, *Molecular Ecology*,
    Vol.9, No.11, pp. 1699-1710, ISSN 0962-1083

Teofanova, D., Kantzoura, V, Walker, S., Radoslavov, G., Hristov, P., Theodoropoulos, G.,
    Bankov, I. & Trudgett, A. (2011). Genetic diversity of liver flukes (*Fasciola hepatica*)
    from Eastern Europe, *Infection, Genetics and Evolution*, Vol.11, No.1, pp. 109-115,
    ISSN 1567-1348

Vara-Del Río, M.P., Villa, H., Martinez-Valladares, M. & Rojo-Vázquez, F.A. (2007). Genetic
    heterogeneity of *Fasciola hepatica* isolates in the Northwest of Spain, *Parasitology
    Research*, Vol.101, No.4, pp. 1003-1006, ISSN 0932-0113

Wagner, J. (February 2003). Screening Methods for Detection of Unknown Point Mutations,
    Available from
    http://www.users.med.cornell.edu/~jawagne/screening_for_mutations.html#Sin
    gle-Strand.Conformational.Polymorphism

Walker, S.M., Prodöhl, P.A., Fletcher, H.L., Hanna, R.E.B., Kantzoura, V., Hoey, E.M. &
    Trudgett, A. (2007). Evidence for multiple mitochondrial lineages of *Fasciola hepatica*
    (liver fluke) within infrapopulations from cattle and sheep, *Parasitology Research*,
    Vol.101, No.1, pp. 117-125, ISSN 0932-0113

Walker, S.M., Johnston, C., Hoey, E.M., Fairweather, I., Borgsteede, F., Gaasenbeek, C., Prodöhl, P.A. & Trudgett, A. (2011). Population dynamics of the liver fluke, *Fasciola hepatica*: the effect of time and spatial separation on the genetic diversity of fluke populations in the Netherlands, *Parasitology*, Vol.138, No.2, pp. 215-223, ISSN 0031-1820

World Health Organization Headquarters. (2006). Report of the WHO Informal Meeting on use of triclabendazole in fascioliasis control, ISBN 978-92-4-1564090, Geneva, Switzerland

10

Genetic Diversity of Brazilian Cyanobacteria Revealed by Phylogenetic Analysis

Maria do Carmo Bittencourt-Oliveira[1,2] and Viviane Piccin-Santos[1,2]
[1]Universidade de São Paulo
[2]Universidade Estadual Paulista
Brazil

1. Introduction

Cyanobacteria are prokaryotic microorganisms with a long evolutionary history. They are usually aquatic, perform oxigenic photosynthesis and have been responsible for initial rise of atmospheric O_2. Cyanobacteria are predominant in the phytoplankton of continental waters, reaching an ample diversity of shapes due to their morphological, biochemical and physiological adaptabilities acquired all along their evolutionary history. Some cyanobacteria, like the *Microcystis, Cylindrospermopsis, Anabaena, Aphanizomenon* and *Planktothrix*, could give rise to blooms with the liberation of a wide range of toxins.

Cyanobacteria constitute a monophyletic group within the Bacteria domain (Castenholz, 2001). However, the group taxonomy is traditionally based on morphological characters, according to botanic criteria (Anagnostidis & Komárek 1985, 1988, 1990; Komárek & Anagnostidis 1986, 1989, 1999, 2005), and added to ecological data.

The high morphological variability and the low number of phenotypic characters used in the cyanobacterial taxonomy leads to serious identification problems. Several studies using molecular techniques have questioned the use of morphological characteristics for identification at the species, as well as, at the genus level. These studies showed the importance of using molecular tools other than morphology to help their identification and phylogenetic studies (Bittencourt-Oliveira et al., 2007a, 2009b; Bolch et al., 1996, 1999; Margheri et al., 2003; Tomitani et al., 2006; Valério et al., 2009; Wood et al., 2002).

Several techniques based on DNA polymorphism detection have been developed and applied on cyanobacterial phylogenetic studies as, RFLP (*Restriction Fragment Length Polymorphism*) (Bittencourt-Oliveira et al., 2009a; Bolch et al., 1996; Iteman et al., 2002; Lu et al., 1997; Lyra et al., 2001), RAPD (*Random Amplified Polymorphic DNA*) (Casamatta et al., 2003; Nishihara et al., 1997; Prabina et al., 2005; Shalini et al., 2007), STRR (*Short Tandemly Repeated Repetitive*) (Chonudomkul et al., 2004; Valério et al., 2009; Wilson et al., 2000) and HIP1 (*Highly Iterated Palindrome*) sequences (Bittencourt-Oliveira et al., 2007a, 2007b; Neilan et al., 2003; Orcutt et al., 2002; Pomati et al., 2004; Wilson et al., 2005; Zheng et al., 2002).

However, the most important breakthrough in phylogenetical studies took place with the advent of direct sequencing of molecular markers. For cyanobacteria, the phycocyanin operon are among the most used molecular markers. The c-phycocyanin genes (*cpc*B and *cpc*A) and the intervening intergenic spacer (IGS) show variations in their sequences which are capable of differentiating genotypes below the generic level. Besides, they are relatively large-sized in comparison with other genes encoding for photosynthetic pigments (~700-800bp), belonging to all cyanobacteria and they are almost totally restricted to this group of organism when in freshwater ecosystems (Barker et al., 2000; Bittencourt-Oliveira et al., 2001, 2009b; Bolch et al., 1996; Dyble et al., 2002; Haverkamp et al., 2008, 2009; Neilan et al., 1995; Tan et al., 2010; Wu et al., 2010).

Among cyanobacteria, several taxa need to be studied in order to clarify their taxonomic position as the *Geitlerinema amphibium* (Agardh ex Gomont) Anagnostidis, *G. unigranulatum* (R.N. Singh) Komárek & Azevedo and *Microcystis panniformis* Komárek et al.

Species of *Geitlerinema* are cosmopolitan and frequently found in different kinds of habitats, as well as in freshwater or marine ecosystems (Kirkwood & Henley, 2006; Margheri et al., 2003; Rippka et al., 1979, Romo et al., 1993; Silva et al., 1996), and they can also form blooms in reservoirs (Torgan & Paula, 1994).

Geitlerinema amphibium and *G. unigranulatum* are morphologically similar to each other. According to Romo et al. (1993) and Komárek & Anagnostidis (2005), they could be differentiated only by their dimensions and the number of cyanophycin granules close to the cross-walls. However, Bittencourt-Oliveira et al. (2009b) used transmission electronic and optical microscopy to study strains of these two morphospecies showing that there is a large overlap between them, in both cell dimensions and the number of granules per cell. Therefore, the authors concluded that it was not possible to distinguish *G. amphibium* from *G unigranulatum* by means of morphological data.

Cyanobacterial blooms of the genus *Microcystis* (Chroococcales, Cyanobacteria) are of serious ecological and public health concern due to their ability to dominate the planktonic environment and produce toxins. These toxins can affect aquatic and terrestrial organisms and humans. *M. aeruginosa* (Kützing) Kützing, *M. ichthyoblabe* Kützing, *M. novacekii* (Komárek) Compère, *M. flos-aquae* (Wittrock) Kirchner ex Forti and *M. viridis* (A. Braun in Rabenhorst) Lemmermann are commonly reported species causing hepatotoxicity and odor problems in lakes and water supply systems (Carmichael, 1996; Codd et al., 1999). However, in tropical regions such as in Brazil, *M. panniformis* is also a potential microcystin-producing morphospecies. It is morphologically characterized by flattened irregular colonies and it is closely related to *M. aeruginosa* morphospecies (Bittencourt-Oliveira et al., 2005).

In the same way that happens to other species of the genus, *M. panniformis* colonies can have morphologically different stages during their life cycle, which make difficult to define and to establish taxonomic limits for its identification (Bittencourt-Oliveira, 2000; Otsuka et al., 2000, 2001). Furthermore, it has been observed that in *Microcystis* populations a genotype could represent more than one morphotype , or that distinct morphotypes could represent a single genotype (Bittencourt-Oliveira et al., 2001; Hannde et al., 2007).

HIP1 sequences is a powerful tool to study genetic diversity of cyanobacteria strains or closely related taxa, and it was used by Bittencourt-Oliveira et al. (2007a, 2007b) to investigate *M. panniformis*, *G. amphibium* and *G. unigranulatum*. However, it is recommended that studies on molecular phylogeny do confirm, by the use of DNA sequences, previous findings which have used fingerprinting techniques like HIP1.

The goal of this study was the investigation of *Geitlerinema amphibium*, *G. unigranulatum* and *Microcystis panniformis* taxonomic position, using the phycocyanin gene partial sequencing for the build up of phylogenetic trees.

2. Material and methods

2.1 Field sampling, isolation and growth conditions

Sequences from 14 clonal and non-axenic strains of *Geitlerinema* and 17 of *Microcystis* from the Brazilian Cyanobacteria Collection of the University of São Paulo (BCCUSP; previously named FCLA), were used in this study (Figure 1, Table 1). These strains were isolated from aquatic habitats situated in localities of Brazil. One strain of *Geitlerinema amphibium* (BCCUSP31) was donated by Dr. Romo from University of Valencia, Spain. For isolation purpose one individual colony or thricome was removed by micromanipulation techniques with Pasteur pipettes at magnifications of 100 – 400X. Each isolate was washed, by transferring it through several consecutive drops of water until all other microorganisms be removed, and subsequently transferred to glass tubes containing 10 ml of BG-11 medium (Rippka et al., 1979). All strains were maintained in incubators at 21°C ± 1°C and 30 ± 5 µmol photons · m^{-2} · s^{-1} (photometer Li-Cor mod. 250), under a 14:10 hour light:dark photoperiod. The cultures are maintained at the Brazilian Cyanobacteria Collection of the University of São Paulo, Brazil (BCCUSP). The Spanish strain (BCCUSP 31) was acclimatized for three months in the BCCUSP at the same conditions as for those before the beginning of the experiment.

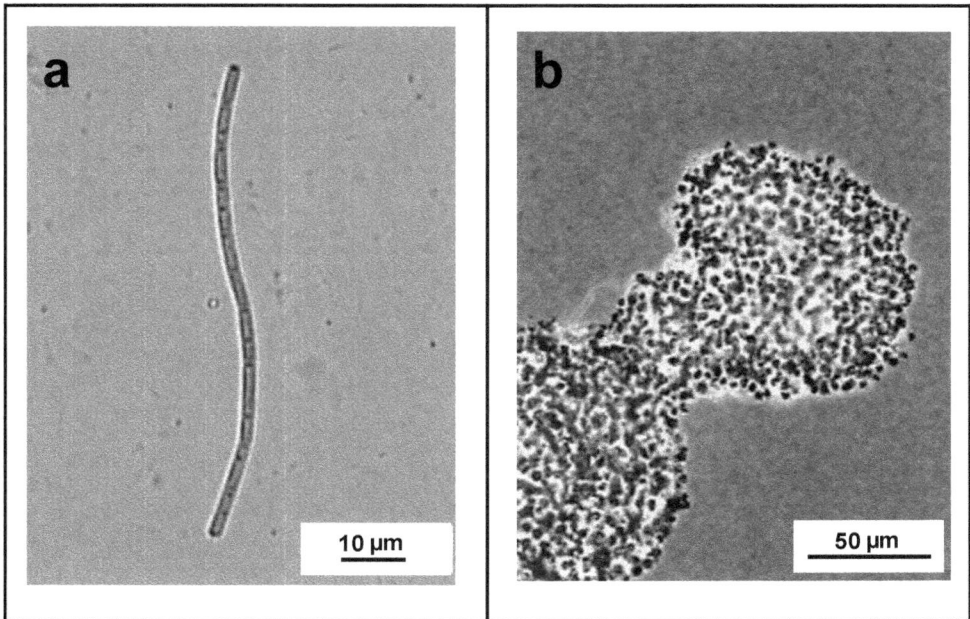

Fig. 1. The morphospecies a) *G. amphibium* BCCUSP85 and b) *M. panniformis* from Jucazinho reservoir, Northeast Brazil.

Morphospecies	Code	Origin site	cpcBA-IGS (bp)			GenBank Access Number
			cpcA	IGS	cpcB	
G. amphibium [a]	BCCUSP31	SP	143	293	198	FJ545649
G. amphibium [a]	BCCUSP79	BR	143	293	198	FJ545644
G. amphibium [b]	BCCUSP80	BR	143	86	198	JN578762
G. amphibium [b]	BCCUSP84	BR	143	296	198	JN578760
G. amphibium [a]	BCCUSP85	BR	143	291	198	FJ545646
G. amphibium [b]	BCCUSP86	BR	143	294	198	JN578759
G. amphibium [a]	BCCUSP87	BR	143	293	198	FJ392653
G. amphibium [a]	BCCUSP91	BR	143	292	198	FJ545650
G. unigranulatum [a]	BCCUSP94	BR	143	83	198	FJ545642
G. amphibium [b]	BCCUSP338	BR	143	295	198	JN578761
G. amphibium [a]	BCCUSP54	BR	143	293	198	FJ545645
G. amphibium [c]	SC2	VT	143	296	169	EF680778
G. unigranulatum [a]	BCCUSP350	BR	143	292	198	FJ545647
G. unigranulatum [a]	BCCUSP352	BR	143	83	198	FJ545643
G. unigranulatum [a]	BCCUSP96	BR	143	292	198	FJ545648
"*Oscilatoria* sp." [c]	CYA479	UG	143	296	169	AM048624
"*Oscilatoria* sp." [c]	CYA469	UG	143	296	169	AM048623
"*Planktothrix* sp." [d]	FP1	IT	143	298	169	AF212923
M. aeruginosa [e]	BCCUSP09	BR	200	69	277	AF385375
M. aeruginosa [e]	BCCUSP30	BR	200	69	277	AF385376
M. panniformis [e]	BCCUSP200	BR	200	69	277	AF385374
M. aeruginosa [e]	BCCUSP232	BR	200	69	277	AF385372
M. aeruginosa [e]	BCCUSP262	BR	200	69	277	AF385369
M. aeruginosa [e]	BCCUSP298	BR	200	69	277	AF385370
M. aeruginosa [e]	BCCUSP299	BR	200	69	277	AF385371
M. aeruginosa [e]	BCCUSP03	BR	200	69	277	AF385380
M. aeruginosa [e]	BCCUSP158	BR	200	69	277	AF385381
M. panniformis [b]	BCCUSP100	BR	200	69	277	JN578763
M. aeruginosa [e]	BCCUSP310	BR	200	69	277	AF385378
M. panniformis [b]	BCCUSP28	BR	200	69	277	JN578764
M. panniformis [b]	BCCUSP22	BR	200	69	277	JN578765
M. panniformis [b]	BCCUSP25	BR	200	69	277	JN578766
M. panniformis [b]	BCCUSP26	BR	200	69	277	JN578767
M. panniformis [b]	BCCUSP23	BR	200	69	277	JN578768
M. panniformis [b]	BCCUSP29	BR	200	69	277	JN578769
M. aeruginosa [e]	NIES-98	JP	200	69	277	AF385384
M. aeruginosa [e]	PCC7820	SC	200	69	277	AF385387
M. aeruginosa [f]	FACHB 978	CH	200	69	277	AY568682
M. aeruginosa [f]	UTEX 1939	UN	200	69	277	AY568698
M. aeruginosa [e]	PCC 7806	NE	200	69	277	AF385386
M. aeruginosa [g]	UWOCC 017	AS	200	69	277	AF195170

Tab. 1. Morphospecies, code and origin site of the used cyanobacteria strains and the GenBank access number of the sequences present in this study. cpcBA-IGS. c-phycocyanin genes cpcB, cpcA and the intervening intergenic spacer (IGS) (bp). BCCUSP - Brazilian Cyanobacteria Collection of University of São Paulo. SP Spain; BR Brazil; UG Uganda; IT Italy; JP Japan; SC Scotland; CH China; UN Unknown; NE Netherlands; SA South Africa. VT Vietnam; [a] Bittencourt-Oliveira et al (2009b); [b] This work; [c] Unpublished; [d] Pomati et al. (2000); [e] Bittencourt-Oliveira et al. (2001); [f] Wu et al. (2007); [g] Tillett et al. (2001).

Another ten sequences selected form the GenBank were also included in the analysis.

2.2 DNA extraction

DNA was extracted from fresh cells harvested at the exponential phase. Total genomic DNA was extracted using the commercial kit Gnome DNA (BIO 101, Vista, CA, USA) according to the manufacturer's instructions or according to the procedures described in Bittencourt-Oliveira et al. (2010).

2.3 PCR amplification

Amplifications were carried out in the thermocycler GeneAmp 2400 or GeneAmp PCR System 9700 (Applied Biosystems, Foster City, CA). The reaction were performed using 2.5 to 10 ng of DNA, 20 μM of each oligonucleotide primer in a total volume of 25 μL and 2.5 U of Taq DNA polymerase (Amersham Pharmacia Biotech, Piscataway, NJ) with buffer containing 1.5 mM $MgCl_2$ and 200 μM of each dNTP (Boehringer-Mannheim, Mannheim, Germany).

Amplifications for the intergenic spacer and flanking regions from cpcBA-phycocyanin operon were accomplished with the primers PCβ-F and PCα-R described by Neilan et al. (1995) using the same cycling parameters conditions according to Bolch et al. (1996). Control reactions were carried out by using the same reaction conditions and primer without DNA, and no PCR products were detected on agarose electrophoresis. All PCR reactions were repeated at least five times.

Amplification products were visualized by electrophoresis on 0,7% agarose gels stained with ethidium bromide. PCR products were purified using the Purelink Kit (Invitrogen, Carlsbad, CA, USA) according to the manufacturer's instructions. When necessary, bands were extracted and purified from the gel using the QIAquick kit (Qiagen, Hilden, Germany) as recommended by the manufacturer.

2.4 Sequencing and phylogenetic analyses

The amplified fragments were directly sequenced using the forward and reverse primers with ABI Prism® Big Dye® Terminator Cycle Sequencing Ready Kit (Applied Biosystems, Foster City, CA, USA) and 3100 ABI sequencer (Applied Biosystems, Foster City, CA, USA) according to the manufacturer's instructions. To avoid errors from PCR at least four separated amplification reactions were pooled for sequencing. The sequencing was repeated on independent PCR products. The PCRs products were sequenced on both strands at least three times.

Automated base calls for both strands were checked by manual inspection and ambiguous calls and conflicts resolved by alignment and comparison using BioEdit program (Hall, 1999) to establish a consensus sequence for each strain. Consensus sequences were aligned using ClustalW in BioEdit program (Hall, 1999) and were manually inspected. *Spirulina subsalsa* PD2002/gca (accession number AY575949) and *Cyanothece* sp. ATCC51142 (accession number CP000806) were used as outgroups for *Geitlerinema* and *Microcystis* analysis respectively.

Evolution distances between the sequences were calculated by *P distance* in the program MEGA 4.0.2 (Tamura et al., 2007).

Phylogenetic analyses were performed with MrBayes v3.1.2 (Ronquist & Huelsenbeck, 2003). An appropriate evolution model was selected using MrModeltest 2.2 (Nylander, 2004) under the Akaike Information Criterion. For the Bayesian analysis two runs of four Markov chains over 5,000,000 generations sampling every 100 generations was employed. The initial 2,500 generations were discarded as burn-in. For all analyses, posterior probability values were considered low up to 70%, moderate from 71% to 90%, and high above 90%.

3. Results and discussion

3.1 *Geitlerinema amphibium*

The phylogenetic tree including *Geitlerinema* species showed that *Geitlerinema amphibium* and *G. unigranulatum* are not genetically separated from each other (Figure 2).

The phylogenetic tree showed two Clades (I and II) (Figure 2). The Clade I encompasses three strains, one of *G. amphibium* separated from the other two of *G. unigranulatum* . These strains showed the region corresponding to the IGS with 83bp, shorter than the remaining sequences which contained up to 298bp. The evolutionary distance calculated by *P distance* between the strains from Clade I and II was up to 0.098, while among strains included on Clade II it did not exceed 0.056. This is a demonstration that the strains included in Clade I were genetically diverse from the others.

Bittencourt-Oliveira et al. (2009b) had shown that strains BCCUSP352 and BCCUSP94, belonging to Clade I, did not exhibit differences regarding cellular morphology and ultrastructure which would allow distinguish them from the other strains of *Geitlerinema*. The inclusion of *G. amphibium* BCCUSP80 in the present study reinforced the argument that these taxons are distinct species, maybe even of distinct genera because of the IGS size difference taxa.

The Clade II encompassed all the remaining strains and could be subdivided into two smaller groups (A and B). The group A included only strains of the morphospecies *G. amphibuim* , while the group B was equally constituted by strains from *G. amphibuim* or from *G. unigranulatum*, besides two strains of *"Oscillatoria* sp." (AM048623 and AM048624) identified as *Geitlerinema* sp. in Bittencourt-Oliveira et al. (2009b). The sequence *"Planktothrix* sp." (AF212923), also identified as *Geitlerinema* sp. (we refer the reader to Bittencourt-Oliveira et al. 2009b for details), appears basal to the other sequences included in Clade II. By the inclusion of four new sequences from strains of the morphospecies *G. amphibium* alterations in the tree topology did not show up, in agreement therefore with the findings of Bittencourt-Oliveira et al. (2009b).

G. amphibium and *G. unigranulatum* show overlapping morphological characteristics which make difficult the taxonomic discrimination (in the identifying sense). According to Komárek & Anagnostidis (2005), the variation interval of the *G. amphibium* cellular width ranges from (1) 1.8 to 3 (3.5-4) µm, whereas for *G. unigranulatum*, it ranges from 0.8 to 2.4 µm. The means of the studied populations by Bittencourt-Oliveira et al. (2009b) formed two distinct clusters, although the maximum values for the smaller species overlapped with the minimum values for the larger one, thus leaving no hiatus in cell width between the two taxa.

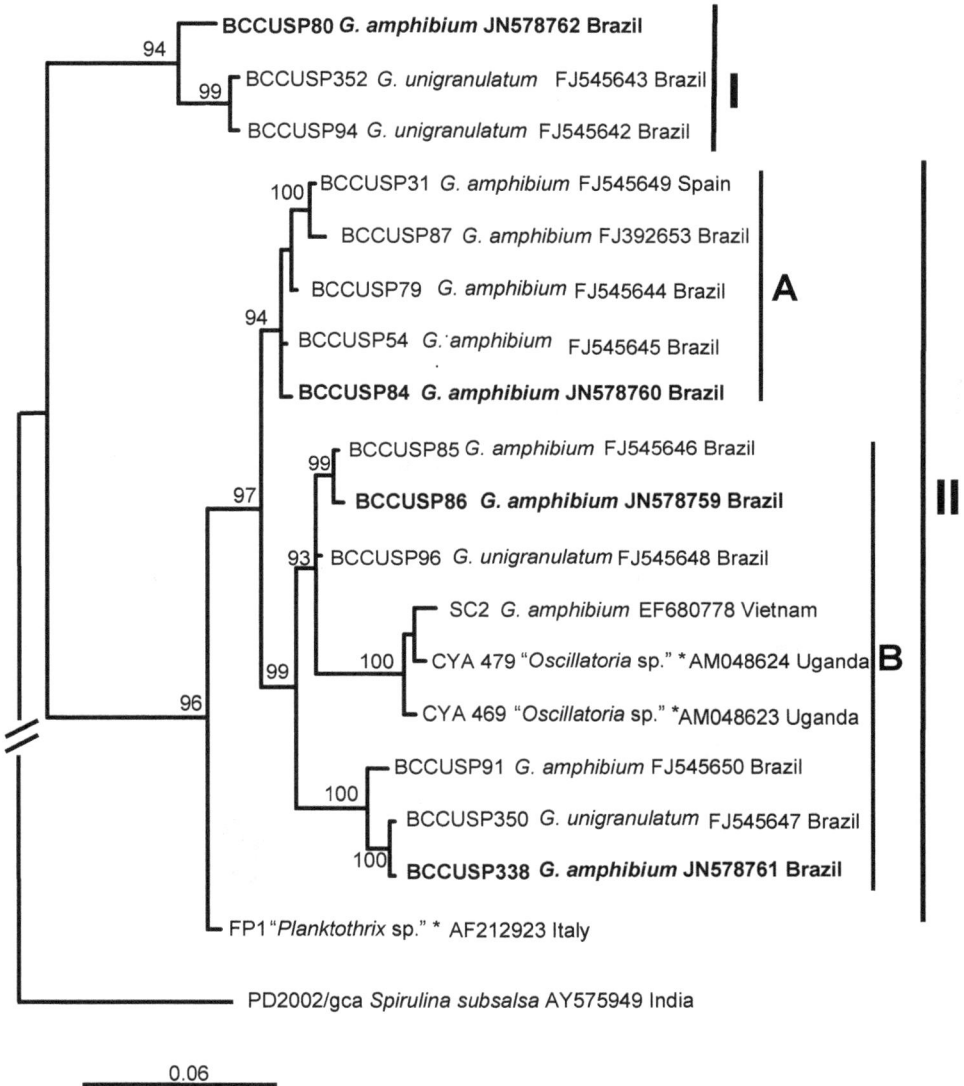

Fig. 2. Bayesian phylogetic tree with *Geitlerinema* strains. The tree was generated using intergenic spacer and flanking regions from *cpc*BA-phycocyanin operon. For the Bayesian analysis two runs of four Markov chains over 5,000,000 generations sampling every 100 generations was employed. The initial 2,500 generations were discarded as burn-in. Posterior probability (x100) is shown on each branch when higher them 70. *Spirulina subsalsa* PD2002/gca (AY575949) used as outgroup. The bar represents 0.06 substitutions. Strains in boldface were sequenced in this work. * Strains identified as *Geitlerinema* sp. in Bittencourt-Oliveira et al. (2009b).

According to the data taken by Bittencourt-Oliveira et al. (2009b), the measurements of cell lengths in strains attributable to *G. amphibium* and *G. unigranulatum* show complete overlap of maximum, minimum, and mean values. Accordingly, only when the length by width ratio (L:W) is taken into account the distinction is more accentuated.

One or two cyanophycin granules per cell (less frequently three) were positioned near the cross walls in strains attributed to both species (Figure 3). Therefore, based on cellular morphology, only through the length to width ratio was it possible to differentiate *G. amphibium* from *G. unigranulatum*. Given their quite uniform morphology and the occurrence of these taxa in the same habitat, it would be nearly impossible to distinguish them in nature. The localization and number of granules, as well as ultrastructural data, did not aid in species discrimination either. Therefore, those characteristics could not be used as diacritical features. The obtained results by Bittencourt-Oliveira et al. (2009b) did not show ultrastructural differences between *G. amphibium* and *G. unigranulatum* strains, except for the BCCUSP96 which in some trichomes exhibited slightly thickening of the apical cell, slightly folded cellular wall, thylakoids with invaginations and unidentified granules. BCCUSP96 strain was unique in having four granules per cell in some trichomes and the highest cell length-to-width ratio (more details, see Bittencourt-Oliveira et al. 2009b).

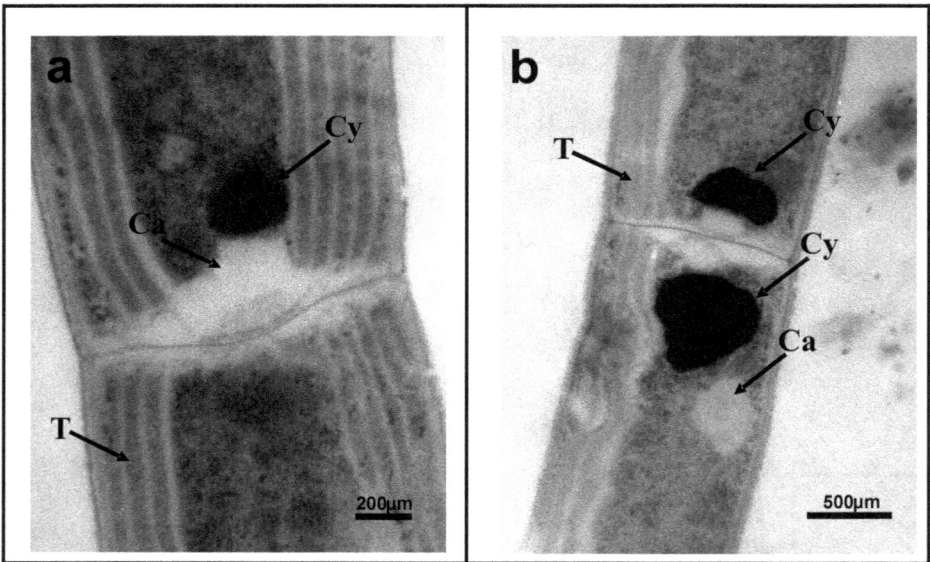

Fig. 3. Brazilian strains of *Geitlerinema* in longitudinal section of trichomes. a) *G. amphibium* BCCUSP91 and b) *G. unigranulatum* BCCUSP350, showing one and two cyanophycin granules (Cy) near the cross walls. Ca carotenoid granules. T thylacoids.

We have observed in the present study that *G. amphibium* and *G. unigranulatum* are not genetically separated from each other. The morphospecies are mixed in the phylogenetic tree and they could not be distinguished as monophyletic entities. Our findings reinforce that they should be considered synonyms as previously stated in Bittencourt-Oliveira et al. (2009b).

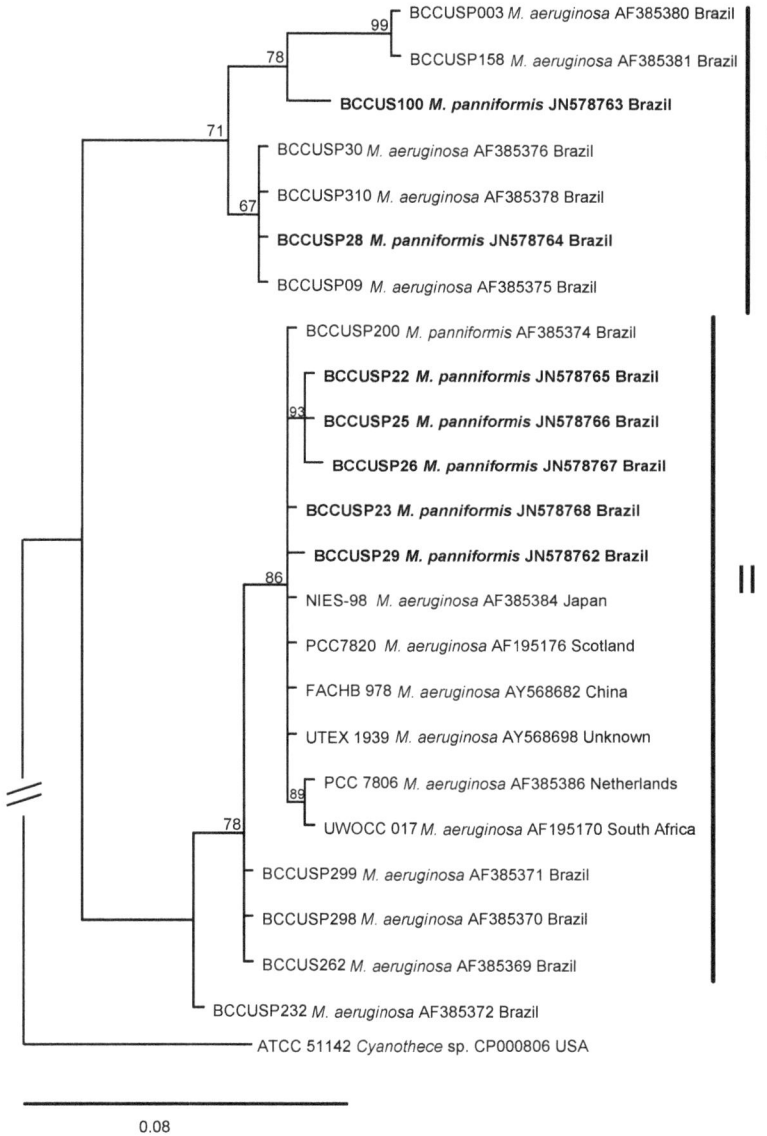

Fig. 4. Bayesian phylogenetic tree with *Microcystis aeruginosa* and *M. panniformis* strains. The tree was generated using intergenic spacer and flanking regions from *cpc*BA-phycocyanin operon. For the Bayesian analysis two runs of four Markov chains over 5,000,000 generations sampling every 100 generations was employed. The initial 2,500 generations were discarded as burn-in. Posterior probability (x100) is shown on each branch when higher them 70. *Cyanothece* sp. ATCC 51142 (CP000806) used as outgroup. The bar represents 0.08 substitutions. Strains in boldface were sequenced in this work.

3.2 *Microcystis panniformis*

Similarly as in the strains of *Geitlerinema*, *Microcystis panniformis* strains did not form a Clade isolated from *M. aeruginosa* in the phylogenetic tree generated by means of the c-phycocyanin genes *cpc*B, *cpc*A sequences and the intervening intergenic spacer (*cpc*BA-IGS) (Figure 4).

Two major groups in the *Microcystis* phylogenetic tree were formed (Clade I and II), constituted by strains of *M. panniformis* and *M aeruginosa*. We observed from the topology tree that some strains of *M. panniformis* are genetically more close to those identified as *M. aeruginosa* than others from their same morphospecies.

The posterior probability was moderate for both Clades and only the *M. aeruginosa* BCCUSP232 strain, situated as sister group of Clade II, did not shown relevant Bayesian support value for its position.

The Brazilian strains BCCUSP23 and BCCUSP200, both belonging to the morphospecies *M. panniformis*, revealed 100% identity with *M. aeruginosa* strains isolated in Japan (NIES98), China (FACHB978) or Scotland (PCC7820) as well.

Traditional taxonomy of *Microcystis* based on morphologic criteria has been questioned by several authors because of the gap lack in the variations observed in the nature and in laboratory conditions. Earlier infraspecific studies of *Microcystis* concluded that morphology does not correlate with molecular data (Bittencourt-Oliveira et al., 2001; Kurmayer et al., 2003; Otsuka et al., 1999a, 1999b, 2000, 2001; Wu et al., 2007).

Our results indicated that *M. panniformis* could also be considered synonym of *M. aeruginosa*. The similarity found between sequences of *M. panniformis* and *M. aeruginosa* coming from diverse regions all over the world, led us to stress the cosmopolitan character of the species, with strains showing ample geographical distribution in both South and North hemispheres. These results corroborate similar previous findings (Bittencourt-Oliveira et al., 2001, 2007b; Otsuka et al., 1999a, 1999b, 2000, 2001).

4. Conclusion

We conclude that *G. unigranulatum* and *M. panniformis* should be considered as synonyms of *G. amphibium* and *M. aeruginosa*, respectively, since they do not represent genetically isolated Clades inside those genera.

5. Acknowledgements

This research was supported by grants from FAPESP (State of São Paulo Research) Foundation Proc. 2004/01749-6 and 2008/56534-5, CNPq (Brazilian Council for Research and Development Proc. 300112/2008-4

6. References

Anagnostidis, K. & Komárek, J. (1985) Modern approach to the classification of the Cyanophytes 1-Introduction. *Algological Studies*, Vol. 38/39, (January, 1985) pp.291-302, ISSN 1864-131.

Anagnostidis, K. & Komárek, J. (1988) Modern approach to the classification system of Cyanophytes. 3. Oscillatoriales. *Algological Studies*, Vol.50 / 53, (January, 1988) pp.327–472, ISSN 1864-131.

Anagnostidis, K. & Komárek, J. (1990). Modern approach to the classification system of Cyanophytes, 5 - Stigonematales. *Algological Studies*, vol. 59, (January, 1990), pp.1-73, ISSN 1864-131.

Barker, G.L.A.; Konopka, A.; Handley, B.A. & Hayes, P.K. (2000). Genetic variation in *Aphanizomenon* (cyanobacteria) colonies from the Baltic Sea and North America. *Journal of Phycology*, Vol.36, No. 5, (October, 2000), pp .947-950, ISSN 1440-1835

Bittencourt-Oliveira, M. C. (2000) Development of *Microcystis aeruginosa* (Kützing) Kützing (Cyanophyceae/Cyanobacteria) under cultivation and its taxonomic implications. *Algological Studies*, Vol.99, (January, 2000), pp.29-37, ISSN 1864-131.

Bittencourt-Oliveira, M.C.; Oliveira, M.C. & Bolch, C.J.S. (2001). Genetic variability of brazilian strains of the *Microcystis aeruginosa* complex (Cyanobacteria/cyanophyceae) using the phycocianin intergenic spacer and flanking regions (*cpc*BA). *Journal of Phycology*, Vol.37, No. 5, (October, 2001), pp.810-818, ISSN 1529-8817.

Bittencourt-Oliveira, M. C. ; Kujbida, P. ; Cardozo, K.H.M.; Carvalho, V.M.; Moura, A.N.; Colepicolo, P. & Pinto, E. (2005). A novel rhythm of microcystin biosynthesis is described in the cyanobacterium *Microcystis panniformis* Komárek et al. *Biochemical and Biophysical Research Communications*, Vol. 326, No. 3, (January, 2005), pp. 687-694, ISSN 1090-2104

Bittencourt-Oliveira, M.C.; Massola Jr, N.S.; Hernandez-Marine, M.; Romo, S. & Moura, A.N. (2007a). Taxonomic investigation using DNA fingerprinting in *Geitlerinema* species (Oscillatoriales, Cyanobacteria). *Phycological Research*, Vol.55, No. 3, (September, 2007), pp.214–221, ISSN 1440-1835

Bittencourt-Oliveira, M.C.; Moura, A.N.; Gouvea-Barros, S. & Pinto, E. (2007b). HIP1 DNA fingerprinting in *Microcystis panniformis* (Chroococcales, Cyanobacteria). *Phycologia*, Vol.46, No.1, (January 2007), pp.3–9, ISSN 0031-8884.

Bittencourt-Oliveira, M.C.; Cunha, M.C.C. & Moura, A.N. (2009a). Genetic polymorphism in brazilian *Microcystis* spp. (Cyanobacteria) toxic and non-toxic through RFLP-PCR of the *cpc*BA-IGS. *Brazilian Archives of Biology and Technology*, Vol.52, No. 4, (August, 2009), pp. 901-909, ISSN 1516-8913

Bittencourt-Oliveira, M.C; Moura, A.N.; Oliveira, M.C. & Massola Jr, S.N. (2009b). *Geitlerinema* species (Oscillatoriales, Cyanobacteria) revealed by cellular morphology, ultrastrucuture and DNA sequencing. *Journal of Phycology*, Vol.45, No.3, (June, 2009), pp. 716-725, ISSN 1529-8817.

Bittencourt-Oliveira, M.C.; Piccin-Santos, P. & Gouvêa-Barros, S. (2010) Microcystin-producing genotypes from cyanobacteria in Brazilian reservoirs. *Environmental Toxicology*, (November, 2010), DOI 10.1002/tox 2010, ISSN 1522-7278

Bolch, C. J.; Blackburn S. I.; Neilan, B. A. & Grewe, P. M. (1996). Genetic characterization of strains of cyanobacteria using PCR-RFLP of the *cpc*BA intergenic spacer and flanking regions. *Journal of Phycology*, Vol.32, No.3, (June, 1996), p.445–451, ISSN 1529-8817.

Bolch, C.J.S.; Orr, P.T.; Jones, G.J. & Blackburn, S.I. (1999). Genetic, morphological and toxicological variation among globally distributed strains of *Nodularia* (Cyanobacteria). *Journal of Phycology*, Vol.35, No. 2, (April, 1999), pp.339-355, ISSN 1529-8817.

Carmichael, W. (1996). Toxic *Microcystis* and the environment. In *Toxic* Microcystis, Watanabe, M.F.; Harada, K.I.; Carmichael, W.W. & Fujiki, H. (Eds.), pp. 1-11, CRC Press, ISBN 9780849376931, Japan

Casamatta, D. A.; Vis, M. L. & Sheath, R. G. (2003). Cryptic species in cyanobacterial systematics: a case study of *Phormidium retzii* (Oscillatoriales) using RAPD molecular markers and 16S rDNA sequence data. *Aquatic Botany*, Vol.77, No. 4, (December, 2003), pp.295–309, ISSN 03043770.

Castenholz, R. W. (2001). Phylum BX. Cyanobacteria. Oxigenic Photosynthetic Bacteria. In. *Bergey's Manual of Systematic Bacteriology 2nd ed.*, Boone, D.R.; Castenholz, R.W. & Garrity, G.M., pp. 473-599, Springer, ISBN 978-0-387-98771-2, New York.

Chonudomkul, D.; Yongmanitchai, W.; Theeragool, G.; Kawachi, M.; Kasai, F.; Kaya, K. & Watanabe, M.M. (2004). Morphology, genetic diversity, temperature tolerance and toxicity of *Cylindrospermopsis raciborskii* (Nostocales, Cyanobacteria) strains from Thailand and Japan. *FEMS Microbiology Ecology*, Vol.48, No.3, (June, 2004), pp.345–355, ISSN 1574-6941

Codd G. A., Chorus I. & Burch M., (1999). Design of monitoring programmes. In *Toxic Cyanobacteria in water: a guide to their public health consequences, monitoring and management*, Chorus, I. & Bartram, J. (Eds.), pp. 313-328, E & FN Spon, ISBN 0-419-23930-8, New York.

Dyble, J.; Paerl, H.W. & Neilan, B.A. (2002). Genetic characterization of *Cylindrospermopsis raciborskii* (Cyanobacteria) isolates from diverse geographic origins based on *nif*H and *cpc*BA-IGS nucleotide sequence analysis *Applied and Environmental Microbiology*, Vol.68, No.5, pp.2567–2571, ISSN 1098-5336.

Hall, T.A. (1999). BioEdit: a user-friendly biological sequence alignment editor and analysis program for Windows 95/98/NT. *Nucleic Acids Symposium Series*, Vol.41, No. 41, (January, 1999), pp. 95-98, ISSN 1746-8272

Hannde, S.; Ballot, A.; Rohrlack, T.; Fastner, J.; Wiedner, C. & Edvardsen, B. (2007). Diversity of *Microcystis aeruginosa* isolates (Chroococcales, Cyanobacteria) form East-African water bodies. *Archives of Microbiology*, Vol.188, No. 1, (February, 2007), p.15-25, ISSN: 1432-072X

Haverkamp, T.; Acinas, S.G.; Doeleman, M.; Stomp, M.; Huisman, J. & Stal, L.J. (2008). Diversity and phylogeny of Baltic Sea picocyanobacteria inferred from their ITS and phycobiliprotein operons. *Environmental Microbiology*, Vol.10, No.1, (January, 2008), pp. 174–188, ISSN 1462-2920.

Haverkamp,T.; Schouten, D.; Doeleman, M.; Wollenzien, U.; Huisman, J. & Stal, L.J. (2009). Colorful microdiversity of *Synechococcus* strains (picocyanobacteria) isolated from the Baltic Sea. *The International Society for Microbial Ecology Journal*, Vol.3, No.4, (December, 2008), pp.397–408, ISSN 1751-7370

Iteman, I.; Rippka, R.; Marsac, N.T. & Herdman, M. (2002) rDNA analyses of planktonic heterocystous cyanobacteria, including members of the genera *Anabaenopsis* and *Cyanospira. Microbiology*, Vol.148, No.2, (February, 2002), pp.481–496, ISSN 1465-2080.

Kirkwood, A. E. & Henley, W. J. (2006). Algal community dynamics and halotolerance in a terrestrial, hypersaline environment. *Journal of Phycology*, Vol42, No.3, (June, 2006), pp.537–47, ISSN 1529-8817.

Komárek, J. & Anagnostidis, K. (1986) Modern approach to the classification system of Cyanophytes, 2 - Chroococcales. *Algological Studies*, Vol.43, No.2, (January, 1986) pp.157-226, ISSN 1864-131

Komárek, J. & Anagnostidis, K. (1989) Modern approach to the classification system of Cyanophytes, 4 - Nostocales. *Algological Studies*, Vol.56, No. 3, (January, 1989) p.247-345, ISSN 1864-131

Komárek, J. & Anagnostidis, K. (1999). Cyanoprokaryota 1. Teil: Chroococcales. In *Süßwasserflora von Mitteleuropa, Vol 19* Ettl H.; Gärtner G.; Heynig H. & Mollenhauer D. (Eds.), pp.1-545, Gustav Fischer, ISBN 978-3-8274-2051-0, Stuttgart.

Komárek, J. & Anagnostidis, K. (2005). Cyanoprokaryota II. Teil Oscillatoriales. In *Süßwasserflora von Mitteleuropa*, Büdel, B.; Krienitz, L.; Gärtner, G.; Schagerl, M. (Eds.). pp. 759, Elsevier Gmbh, ISBN 978-3-8274-2051-0, Munchen.

Kurmayer R., Christiansen G. & Chorus I., (2003). The abundance of microcystin producing genotypes correlates positively with colony size in *Microcystis* sp. and determines its microcystin net production in Lake Wannsee. *Applied and Environmental Microbiology*, Vol. 69, No. 2, (February, 2003), pp.787-795, ISSN 1098-5336

Lu,W.; Evans, E.H.; Mccoll, S.M. & Saunders, V.A. (1997) Identification of cyanobacteria by polymorphisms of PCR-amplified ribosomal DNA spacer region. *FEMS Microbiology Letters*, Vol.153, No.1, (August, 1997), pp.141–149, ISSN 1574-6968.

Lyra, C.; Suomalainen, S.; Gugger, M.; Vezie, C.; Sundman, P.; Paulin, L. & Sivonen, K. (2001). Molecular characterization of planktic cyanobacteria of *Anabaena, Aphanizomenon, Microcystis* and *Planktothrix* genera. *International Journal of Systematic and Evolutionary Microbiology*, Vol.51, No2, (March, 2001), pp.513–526, ISSN 1466-5034

Margheri, M. C.; Piccard, R.; Ventura, S.; Viti, C. & Giovannetti, L. (2003) Genotypic diversity of Oscillatoriacean strains belonging to the genera *Geitlerinema* and *Spirulina* determined by 16S rDNA restriction analysis. *Current Microbiology*, Vol.46, No. 5, (May, 2003), pp.359–64, ISSN 1432-0991

Neilan B.A.; Jacobs D. & Goodman A.E. (1995). Genetic diversity and phylogeny of toxic cyanobacteria determined by DNA polymorphisms within the phycocyanin *locus*. *Applied and Environmental Microbiology*, v.61, No. 11, (November, 1995), pp.3875–3883, ISSN 1098-5336

Neilan, B.A.; Saker,M.L. Fastner, J.; Törökés A. & Burns, B.P. (2003). Phylogeography of the invasive cyanobacterium *Cylindrospermopsis raciborskii*. *Molecular Ecology*, Vol.12, No. 1, (January, 2003), pp.133-140, ISSN 1365-294X

Nishihara, H.; Miwa, H.; Watanabe, M.; Nagashima, M.; Yagi, O & Takamura, Y. (1997). Random amplified polymorphic DNA (RAPD) analyses for discriminating genotypes of *Microcystis* cyanobacteria. *Bioscience Biotechnology and Biochemistry*, Vol.61, No.7, (July, 1997), pp.1067–1072, ISSN 1347-6947

Nylander, J.A.A. (2004). MrModeltest 2.2. Program distributed by the author. *Evolutionary Biology Centre*, Uppsala University, Upsalla

Orcutt, K.M.; Rasmussen, U.; Webb, E.A.; Waterbury, J.B.; Gundersen, K. & Bergman, B. (2002). Characterization of *Trichodesmium* spp. by genetic techniques. *Applied*

Environmental Microbiology, Vol.68, No.5, (May, 2002), pp.2236–2245, ISSN 1098-5336

Otsuka, S.; Suda, S.; Li, R.; Watanabe, M.; Oyaizu, H.; Matsumoto, S. & Watanabe, M.M. (1999a). Characterization of morphospecies and strains of the genus *Microcystis* (Cyanobacteria) for a reconsideration of species classification. Phycological Research, Vol. 47, No.3, (September, 1999), pp.189–197, ISSN 1440-1835

Otsuka, S.; Suda, S.; Li, R.; Watanabe, M.; Oyaizu, H.; Matsumoto, S. & Watanabe M.M. (1999b). Phylogenetic relationships between toxic and non-toxic strains of the genus *Microcystis* based on 16S to 23S internal transcribed spacer sequence. *FEMS Microbiological Letters,* Vol.172, No.1, (March, 1999), pp.5-21, ISSN 1574-6968

Otsuka, S.; Suda, S.; Li, R.; Matsumoto, S. & Watanabe, M.M. (2000). Morphological variability of colonies of *Microcystis* morphospecies in culture. *Journal of General and Applied Microbiology,* Vol.46, No.1, (February, 2000), pp.39-50, ISSN 1349-8037

Otsuka, S.; Suda, S.; Shibata, S.; Oyaizu, H.; Matsumoto, S. & Watanabe, M.M. (2001). A proposal for the unification of five species of the cyanobacterial genus *Microcystis* Kützing *ex* Lemmermann 1907 under the rules of the Bacteriological Code. *International Journal of Systematic and Evolutionary Microbiology,* Vol.51, No.3, (May, 2001), pp. 873-879, ISSN 1466-5034

Pomati, F.; Sacchi, S.; Rossetti, C.; Giovannardi, S.; Onodera, H.; Oshima, Y & Neilan, B.A. (2000). The freshwater cyanobacterium *Planktothrix* sp. FP1: molecular identification and detection of paralytic shellfish poisoning toxins. *Journal of Phycology,* Vol.36, No. 3, (June, 2000), pp. 553–562, ISSN 1529-8817.

Pomati, F.; Burns, B.P. & Neilan, B.A. (2004). Identification of an Na+-dependent transporter associated with saxitoxin-producing strains of the cyanobacterium *Anabaena circinalis. Applied and Environmental Microbiology,* Vol.70, No.8, (August, 2004), pp 4711–19, ISSN 1098-5336

Prabina, B. J., Kumar, K. & Kannaiyan, S. (2005). DNA amplification fingerprinting as a tool for checking genetic purity of strains in the cyanobacterial inoculum. *World Journal of Microbiology and Biotechnology,* Vol.21, No.5, (July, 2005) p.629–634, ISSN 1573-0972

Rippka, R.; Deruelles, J.; Waterbury, J.B.; Herdman, M. & Stanier, R. Y. (1979). Generic assigments, strain histories and properties of pure cultures of cyanobacteria. *Journal of General Microbiology,* Vol.111, No.1, (March, 1979), pp.1-61, ISSN 0022-1287

Romo, S., Miracle, R. & Hernandez-Marine, M. (1993). *Geitlerinema amphibium* (Ag. ex Gom.) Anagnostidis (Cyanophyceae): morphology, ultrastructure and ecology. *Algological Studies,* Vol.69, (January, 1993), pp.11–27, ISSN 1864-131

Ronquist, F. & Huelsenbeck, J.P. (2003). Mr Bayes 3: Bayesian phylogenetic inference under mixed models. *Bioinformatics,* Vol.19, No. 12, (February, 2003), pp.1572-1574, ISSN 1460-2059

Shalini; Tiwari, S.; Gupta, R. K.; Pabbi, S. & Dhar, D. W. (2007). Protocol optimization for RAPD in Cyanobacteria. *Indian Journal of Biotechnology,* Vol. 6, No. 6, (October, 2007), pp.549-552, ISSN 0972-5849

Silva, P.C.; Basson, P.I. & Moe, R.L. (1996). *Catalogue of the benthic marine algae of the Indian Ocean*, Vol 79, University of California Publications Series – Botany, ISBN 9780520098107, California

Tamura, K.; Dudley, J; Nei, M. & Kumar, S. (2007). MEGA4: Molecular Evolutionary Genetics Analysis (MEGA) software version 4.0. *Molecular Biology and Evolution*, Vol. 24, No. 8, (May, 2007), pp.1596-1599, ISSN 1537-1719

Tan, W.; Lui, Y.; Wu, Z.; Lin, S.; Yu, G.; Yu, B. & Li, R. (2010).*cpc*BA-IGS as an effective marker to characterize *Microcystis wesenbergii* (Komárek) Komárek in Kondrateva (cyanobacteria). *Harmful Algae*, Vol.9, No.6., (September, 2010), pp. 607-612, ISSN 1020-2706.

Tillett, D.; Parker, D.L. & Neilan, B.A. (2001) Detection of toxigenicity by a probe for the microcystin synthetase A gene (*mcy*A) of the cyanobacterial genus *Microcystis*: comparison of toxicities with 16S rRNA and phycocyanin operon (phycocyanin intergenic spacer) phylogenies. *Applied And Environmental Microbiology*, Vol. 67, No. 6, (June 2001), pp. 2810–2818, ISSN 1098-5336

Tomitani, A.; Knoll, A.H.; Cavanaugh, C.M. & Ohno, T. (2006). The evolutionary diversification of cyanobacteria: molecular–phylogenetic and paleontological perspectives. *Proceedings of the National Academy of Sciences of the USA*, Vol. 103, No.14, (April, 2006), pp. 5442-5447, ISSN 0027-8424

Torgan, L. C. & Paula, M. D. C. F. D. (1994). *Geitlerinema amphibium* (Ag. Ex. Gom.) Anag. (Cyanophyta-Pseudanabaenaceae) em um lago no Município de Pântano Grande, Rio Grande do Sul, Brasil. *Iheringia Serie Botânica*, Vol. 45, (January, 1994), pp.75–87, ISSN 0073-4705.

Valério, E.; Chambel, L.; Paulino, S.; Faria, N.; Pereira, P. & Tenreiro, R. (2009). Molecular identification, typing and traceability of cyanobacteria from freshwater reservoirs. *Microbiology*, Vol.155, No. 2, (February, 2009), pp. 642-656, ISSN 1465-2080

Wilson, A. E.; Sarnelle, O.; Neilan, B. A.; Salmon, T. P.; Gehringer, M. M. & Hay, M. E. (2005) Genetic variation of the bloom-forming cyanobacterium *Microcystis aeruginosa* within and among lakes: implications for harmful algal blooms. *Applied Environmental Microbiology*, Vol.71, No. 10, (October, 2005), pp.6126–6133, ISSN 1098-5336

Wilson, K.M.; Schembri, M.A.; Baker, P.D. & Saint, C.P. (2000). Molecular characterization of the toxic cyanobacterium *Cylindrospermopsis raciborskii* and design of a species-specific PCR. *Applied and Environmental Microbiology*, Vol.66, No.1, (January, 2000), pp. 332–338, ISSN 1098-5336

Wood, A.M.; Miller,S.R.; Li, W.K.W. & Castenholz, R.W. (2002) Preliminary studies of cyanobacteria, picoplankton, and virioplankton in the Salton Sea with special attention to phylogenetic diversity among eight strains of filamentous cyanobacteria. *Hydrobiologia*, Vol.473, No. 1-3, (January, 2002) p.77–92, ISSN 1573-5117

Wu, Z.; Shi, J.; Lin, S. & Li, L. (2010). Unraveling molecular diversity and phylogeny of *Aphanizomenon* (Nostocales, Cyanobacteria) strains isolated from China. *Journal of Phycology*, Vol.46, No. 5, (October, 2010), pp.1048–1058, ISSN 1529-8817

Wu, Z.X.; Gan, N.Q. & Song, L.R. (2007). Genetic diversity: geographical distribution and toxin profiles of *Microcystis* strains (cyanobacteria) in China. *Journal of Integrative Plant Biology*, Vol.49, No. 3, (March, 2007), pp.262–269, ISSN 1744-7909

Zheng, W.; Song, T.; Bao, X.; Bergman, B. & Rasmussen, U. (2002). High cyanobacterial diversity in coralloid roots of cycads revealed by PCR fingerprinting. *FEMS Microbiology Ecology*, Vol.40, No.3, (June, 2002), pp. 215–22, ISSN 1432-184X

11

Genetically Related Listeria Monocytogenes Strains Isolated from Lethal Human Cases and Wild Animals

Ruslan Adgamov[1], Elena Zaytseva[1,2],
Jean-Michel Thiberge[3], Sylvain Brisse[3] and Svetlana Ermolaeva[1*]
[1]*Gamaleya Research Institute of Epidemiology and Microbiology, Moscow,*
[2]*Research Institute of Epidemiology and Microbiology, Vladivostok,*
[3]*Institut Pasteur, Genotyping of Pathogens and Public Health, Paris,*
[1,2]*Russian Federation,*
[3]*France*

1. Introduction

The Gram-positive pathogenic bacterium *Listeria monocytogenes* is widely spread in nature [1; 12; 16; 21; 22; 55]. *L. monocytogenes* belongs to the group of pathogens that can be classified as infectious agents of sapronoses (or saprozoonoses) because their infectious reservoir is the environment [34; 37]. Characteristic features of this and similar pathogenic bacteria are abilities to grow effectively outside of the host and to infect a relatively wide range of hosts (polyhostality) [29; 32]. *L. monocytogenes* presence in natural ecosystems is supported by its isolation from a variety of animal species including mammals, birds, fish and mollusks [12; 16; 17; 22; 55]. Virulence factors, which are prerequisite to infection in humans, are involved in interactions with members of natural ecosystems including protozoa [38].

L. monocytogenes is a common contaminant of food products and causes food-borne infection in humans [10; 23; 45]. *L. monocytogenes* isolates divide into at least three distinct phylogenetic lineages. Serovar 4b strains together with strains of serovars 1/2b, 4e and 4d form phylogenetic lineage I, which has been responsible for the majority of outbreaks of food-borne listeriosis in humans [4; 23; 35; 40; 52]. Lineage II comprises strains of serovars 1/2a, 1/2c, 3a, 3c, which are associated with some recent large epidemics and many sporadic cases of foodborne listeriosis. Lineage III includes serovars 4a, 4c and a few 4b strains and is more rarely isolated from human clinical cases [52]. Epidemiologically important *L. monocytogenes* clones were identifed as being associated with listeriosis outbreaks in different countries and continents [9; 23; 35; 39; 47; 53]. The epidemic clones designated ECI, ECII and ECIV belong to lineage I, whereas ECIII belongs to lineage II [23].

Different molecular typing methods such as multilocus enzyme electrophoresis, Pulsed-Field Gel Electrophoresis (PFGE), ribotyping, random amplified polymorphic DNA (RAPD),

---

or amplified fragment length polymorphism have been used to demonstrate that food-associated and clinical *L. monocytogenes* isolates formed two distinctive although overlapping groups [2; 14; 15; 47]. In other words, although food is a major source of human infection, not all strains that contaminate food possess a similar virulence potential to cause invasive infection in humans. Analysis of domestic animal and farm environmental isolates revealed that some strains associated with human infection circulate in the agricultural complexes [20]. Although the role of the wild nature as an original source of listerial infection was suggested, the information about clone distribution among *L. monocytogenes* disseminated in natural ecosystems and their phylogenetic relationships with epidemiologically important clones is scarce [13; 19; 54].

Slowly evolving markers are necessary to analyze phylogenetic relationships and long-term epidemiology of pathogenic bacteria. Multilocus sequence typing (MLST) has been developed for the study of clonal relationships and has been successfully used for population genetics and global epidemiological analysis of different bacterial species including *Neisseria meningitidis, Streptococcus pneumoniae, Staphylococcus aureus, Streptococcus pyogenes, Campylobacter jejuni, Salmonella enterica, Clostridium difficile* [7; 8; 11; 28; 30; 39; 43; 50]. In *L. monocytogenes,* MLST currently distinguishes more than 430 STs, many of which are grouped into large clonal groups that were shown to be globally distributed (Ragon et al 2008, Chenal-Francisque et al 2011). MLST offers several advantages over other molecular typing methods. First, the DNA sequences are unambiguous and readily comparable between different laboratories, and can be stored in a shared central database to provide a broader resource for epidemiological and global population studies. Second, evolutionary genetics studies can be performed, since MLST describes variation believed to be neutral [7; 28; 30].

In this study, we applied a previously described MLST scheme [39; 43] to characterize clonal diversity of *L. monocytogenes* isolates from maternal-fetal cases of human listeriosis and from internal organs of wild animals captured in the forest territory of Russian Far East [55]. The study was aimed to reveal genetic relationships among invasive *L. monocytogenes* strains isolated from anthropogenic and wild environments and to compare isolates from Russia with those from other regions of the world, which are shared in a web database hosted at Institute Pasteur (www.pasteur.fr/mlst). Virulence of isolates belonging to the same MLST sequence type was checked by using the model of intravenous infection of laboratory mice. To evaluate diversity of virulence associated factors among isolates obtained from humans and wild animal hosts, we characterized internal sequences of the *inlA, inlB, inlC* and *inlE* genes encoding *L. monocytogenes* proteins of the internalin family, which comprises a number of surface and secreted proteins including established *L. monocytogenes* invasion factors InlA and InlB [3; 31].

2. Materials and methods

L. monocytogenes isolates and culture maintenance. In total, 40 *L. monocytogenes* isolates were included in the study (Table 1S). Some isolates of 1993-2005 were described previously [55], while others were isolated in a similar way as described [55]. The isolates were associated with infection in humans (15 isolates from maternal-fetal cases), wild small rodents (11 isolates obtained from the liver and spleen of northern red-backed vole (*Cletrionomus rutilus),* grey red-backed vole *(Cletrionomus rufocanus),* striped field mouse *(Apodemus*

No	strain	CC	ST	source	year	serovar	lineage	internalin gene alleles			
								inlA	*inlB*	*inlC*	*inlE*
1	VIMHA 009	**1**	1	stillborn	2005	4b	I	1	9	1	4
2	VIMHA 012	**1**	1	stillborn	2005	4b	I	1	9	1	4
3	VIMHA 015	**1**	1	stillborn	2005	4b	I	1	9	1	4
4	VIMHA 017	**1**	1	stillborn	2005	4b	I	1	9	1	4
5	VIMHA 010	**1**	1	stillborn	2005	4b	I	1	10	1	3
6	VIMHA 011	**1**	1	stillborn	2005	4b	I	1	9	1	3
7	VIMPA k23	**1**	64	stillborn	1998	n.d.	I	1	16	2	3
8	VIMHA 004	**2**	2	stillborn	2005	4b	I	1	1	1	1
9	VIMHA 005	**2**	2	stillborn	2005	4b	I	1	1	1	1
10	VIMHA 006	**2**	2	stillborn	2005	4b	I	1	1	1	1
11	VIMHA 007	**2**	2	stillborn	2005	4b	I	1	3	1	1
12	VIMVR 081	**2**	145	rodent	2005	4b	I	1	1	1	1
13	VIMVR 082	**2**	145	rodent	2005	4b	I	1	1	1	1
14	VIMVR 084	**2**	145	rodent	2005	4b	I	1	1	1	1
15	VIMVR 090	**2**	145	rodent	2005	4b	I	1	1	1	1
16	VIMVR 092	**2**	145	rodent	2005	4b	I	1	1	1	1
17	VIMVW 037	**2**	145	sludge	1998	4b	I	1	4	1	1
18	VIMVW 039	**2**	145	sludge	1998	4b	I	1	1	1	1
19	VIMVG 047	**19**	19	scallop	1993	1/2a	II	9	14	9	6
20	VIMPR750	**19**	314	rodent	1965	n.d.	II	9	14	6	6
21	VIMPA 064	**7**	7	stillborn	1997	n.d.	II	5	14	6	8
22	VIMPR 134	**7**	7	rodent	1952	n.d.	II	4	14	6	8

23	VIMPR 422	7	7	rodent	1952	n.d.	II	4	14	6	8
24	VIMCR 474	155	155	rodent	2006	1/2a	II	10	14	8	7
25	VIMUR 211	155	155	rodent	2006	1/2a	II	10	14	8	7
26	VIMHA 034	313	313	stillborn	2004	1/2a	II	9	14	9	6
27	VIMHA 036	313	313	stillborn	2004	1/2a	II	9	14	9	6
28	VIMHA 038	313	313	stillborn	2004	1/2a	II	9	14	9	6
29	VIMVR 062	315	315	rodent	2004	4b	I	2	1	4	1
30	VIMVG 061	315	315	scallop	1993	4b	I	2	1	4	1
31	VIMVG 062	315	315	scallop	1993	4b	I	2	1	4	1
32	VIMVG 064	315	315	scallop	1993	4b	I	2	1	4	1
33	VIMVG 065	315	315	scallop	1993	4b	I	2	5	4	1
34	VIMVG 067	315	315	scallop	1993	4b	I	2	1	4	1
35	VIMVG 077	315	315	scallop	1993	4b	I	2	1	4	1
36	VIMVG 100	315	315	fish	1993	4b	I	2	6	4	1
37	VIMVG 102	315	315	sea urchin	1993	4b	I	2	2	4	1
38	VIMVG 104	315	315	starfish	1993	4b	I	2	1	4	1
39	VIMVG 106	315	315	starfish	1993	4b	I	2	1	4	1
40	VIMVG 108	315	315	starfish	1993	4b	I	2	1	4	1

Table 1S. Strains used in the study

agrarius), korean field mouse *(Apodemus peninsulae)* and marine animals (12 isolates obtained from scallops, sea urchins and starfish). Two sludge isolates obtained on the territory where rodents were captured were also included [55]. Three historical rodent strains isolated in 1952 and 1965 were obtained from internal organs of wild rodents with the same methods as other rodent isolates were obtained in latter years [1]. The strains were kept frozen at -70°C. The cultures were plated on Brain-Heart Infusion (BHI, BD) agar and grown overnight immediately before an experiment.

PCR analysis and sequencing Overnight *L. monocytogenes* colonies were subjected to lysis by lysozyme/Proteinase K treatment as described [55]. The PCR was run with 1 μl of lysate in the "Tertsik" thermocycler (DNA Technology, Russia). PCR amplification and sequencing for MLST typing was performed as described [39] with following changes in the PCR running conditions: 20 s at 94°C, 20 s at 55 °C, 20 s at 72°C for the first 5 rounds followed by 30 cycles with timing reduced up to 5 s. For PCR amplification and sequencing of internal fragments of *inlA, inlB, inlC* and *inlE* gene, the primers were designed on the basis of the *L. monocytogenes* strain EGDe genome sequence (Glaser et al., 2001) as following: inlAF: 5′ – TAACGGGACAAATGCTCAGGC; inlAR: 5′ – TGTTAAACTCGCCAATGTGCC; inlBF: 5′ - TTTTCAGATGATGCTTTTGC; inlBR: 5′ – ATAGCGGGTTAAGTTGACTGC; inlCF: 5′ – TTTCCAGATCCCGGCCTAGC; inlCR: 5′ – ATAGCCTCAGTCTCCCCAACG; inlEF: 5′ – TCGGAAAAGCGGATGTAACAG; inlER – 5′ TGAAGCTGTTTAAATCCCACG. The PCR was performed as described above. The final elongation for 10 min was carried out for the samples used for sequencing. PCR products were purified with the Wizard® PCR Preps DNA purification kit (Promega). Sequencing was performed at the Center "Genome" (http://www.genome-center.narod.ru).

Sequence analysis Sequences were proofread and assembled in Chromas version 1.45 (Copyright© 1996-1998 Conor McCarthy, http://www.technelysium.com.au/chromas.html). DNA alignment was done with the ClustalW1.83.XP [46]. Descriptive analysis of the sequence polymorphism was performed using DnaSP version 4.10 [42]. Dendrograms were constructed with Mega version 3.1 [25]. Results of MLST typing are available from the *Listeria monocytogenes* Institut Pasteur MLST database at http://www.pasteur.fr/mlst. Sequence data of internalin gene fragments are available in the GenBank/EMBL/DDBJ databases under the accession numbers EU408789- EU408802, EU40880, EU408813, EU408816- EU408819, EU408822- EU408830, EU408830, EU408833- EU408837, EU408839- EU408844, EU408847, EU408848, EU408855- EU408864, EU408866, EU408871, EU408874- EU408878, EU408880, EU408887- EU408900, EU408902, EU408914- EU408928, EU408928, EU408931- EU408942, EU408945, EU408946, EU408953- EU408967, EU408969, EU408981- EU408984, EU408987- EU408993, EU408995, EU408998- EU409002, EU409004- EU409009, EU409012, EU409013, EU409020, EF056170- EF056174, EF056188- EF056191.

Statistics The mean values and standard errors were calculated with the use of Excel software, a part of Miscrosoft Office 2003 package. The t-test included in the same software was used for assessment of statistical significance.

3. Results

3.1 MLST typing of L. monocytogenes isolates

To define the genetic relationships of the 40 isolates included in this study, we applied the multilocus sequence typing (MLST) scheme based on sequences of the internal fragments of seven housekeeping genes that was initially developed by Salcedo and co-workers [43] and further modified [39] (Table 1). Phylogenetic analysis of concatenated sequences revealed two major branches, which corresponded to lineages I and II, as deduced based on serotype and correspondence with data for reference strains (Ragon et al., 2008). There were no lineage III strains among studied isolates. The majority of substitutions were fixed differences between lineage I and II, i.e. nucleotide sites at which all sequences in lineage I were different from all sequences in lineage II [18]. Intralineage variability was low with the exception of the *ldh* gene.

Gene	Number of substitutions (nonsynonymous)			Between lineages		Number of alleles		
	total collection	lineage I	lineage II	fixed differences	shared mutations	total colle-ction	lineage I	lineage II
abcZ	21 (2)	3 (0)	3 (1)	15	0	5	2	3
bglA	14 (0)	4 (0)	7 (0)	3	0	6	2	4
cat	27 (4)	2 (1)	8 (2)	17	0	7	3	4
dapE	34 (7)	4 (3)	2 (0)	28	0	5	3	2
dat	57 (8)	2 (0)	2 (0)	56	0	6	3	3
ldh	30 (4)	3 (3)	20 (2)	8	1	8	4	4
lhkA	16 (2)	1 (0)	0	15	0	3	2	1

Table 1. Sequence polymorphism of the seven housekeeping gene fragments

ST	lineage	CC	Serovar	Source			
				Stillborns Number of isolates (year of isolation)	Rodents Number of isolates (year of isolation)	Sea animals Number of isolates (year of isolation)	Sludge Number of isolates (year of isolation)
1	I	1	4b	6 (2005)	-	-	-
64	I	1	n.d.	1 (1998)	-	-	-
2	I	2	4b	4 (2005)	-	-	-
145	I	2	4b	-	5 (2005)	-	2 (1998)
315	I	315	4b	-	1 (2004)	11 (1993)	-
19	II	19	n.d.	-	-	1 (1993)	-
314	II	19	1/2a	-	1 (1965)	-	-
7	II	7	n.d.	1 (1997)	2 (1952)	-	-
155	II	155	1/2a	-	2 (2006)	-	-
313	II	313	1/2a	3 (2004)	-	-	-

Table 2. Source distribution of sequence types (STs) and clonal complexes (CCs).

The isolates belonged to 10 sequence types (STs) (Table 2). Seven STs were previously described, and three novel STs (ST313, ST314, ST315) are described here for the first time. Three pairs of closely related STs were revealed, as single nucleotide polymorphisms in the *ldh* gene distinguished ST1 and ST64, ST2 and ST145, and ST19 and ST314, respectively. Based on the definition of a clonal complex (CC) as "a group of profiles differing by one gene from at least one other profile of the group" [39], ST1 and ST64 belonged to the same CC, which was designated CC1 [39], ST2 and ST145 belonged to CC2, and ST314 and ST19 formed a novel clonal complex, which was designated CC19 (Table 2). Screening of ST313 and ST315 against the *L. monocytogenes* MLST database revealed that ST313 is closely related to ST20, and ST315 forms a clonal complex with ST95, ST96, ST102 and ST194. Single nucleotide substitutions in the *ldh* gene distinguished sequence types within each of these clonal complexes (data not shown).

3.2 Phylogenetic characterization of L. monocytogenes clinical isolates obtained in Russia

The *L. monocytogenes* strains isolated from the cases of human maternal-fetal infection belonged to five distinct clones (Table 2). Six and four 4b serovar isolates belonged to ST1 and ST2, respectively. The isolates were obtained in the city of Khabarovsk in the year 2005 [55]. Notably, strain F2365, the reference strain for epidemic clone I (ECI), belongs to ST1 [33; 39]. One more strain isolated from a maternal-fetal listeriosis case in 1998 belonged to ST64, a member of CC1 (Tables 1S and 2). Three serovar 1/2a clinical isolates belonged to ST313, which was so far uniquely described in Russia. The last clinical strain belonged to ST7, which was shown to be associated with multiple cases of human listeriosis in Europe as well as in Australia [39]. Thus, a majority of human isolates from Russia belonged to globally distributed clonal complexes.

3.3 Phylogenetic characterization of isolates obtained from small wild rodents and marine animals

Four distinct MLST profiles were revealed among eleven wild rodent isolates (Table 2). Only two profiles were found among twelve isolates obtained from marine animals. ST7 was revealed in strains isolated from wild small rodents (Table 2). The isolates were not related neither temporally nor territorially (Table 1 S). The profile ST145 belonging to CC2 was found in strains isolated from wild small rodents. There were no established epidemiological links between isolates from humans belonging to ST2 and these animals. ST145 strains were also isolated from sludge of the river that was territorially linked with the area of rodent capturing [51]. Still, isolates from rodents and from sludge were obtained in different years (Table 2).

ST315, which prevailed among marine isolates, has also been isolated from a rodent (Table 2). The rodent and marine animal isolates were obtained in different years, although from samples collected at closely located territories (Table 1 and [55]). In contrast, closely related rodent and marine animal isolates, which belonged to CC19, were separated by long distance and about 30 years (Table 1S).

3.4 Internalin gene diversity in L. monocytogenes isolates

To get a deeper insight into *L. monocytogenes* diversity and particularly into features that might distinguish the closely related strains isolated from different hosts, we performed partial sequencing of genes, which encode proteins of the internalin family [6]. The internal fragments of the internalin genes *inlA, inlB, inlC, inlE*, which encode functionally important LRR domains, were sequenced (Table 3). Sequences of whole LRR-coding fragments were determined for *inlB, inlC* and *inlE* genes. The sequenced fragment covered 42 % of the LRR-domain coding region of the *inlA* gene, which carries the longest LRR domain.

Gene	Fragment length	Encoded amino acids	Number of substitutions (nonsynonymous)			Between lineages (nonsynonymous)		Number of alleles		
			total collection	lineage I	lineage II	fixed differences	shared mutations	total collection	lineage I	lineage II
inlA	648	54-269	18 (6)	8 (2)	12 (5)	0	5 (2)	6	2	4
inlB	618	64-269	52 (18)	25 (10)	0	30 (9)	0	10	9	1
inlC	586	70-264	21 (7)	6 (3)	2 (0)	14 (4)	1 (0)	7	4	3
inlE	558	82-267	99 (45)	6 (3)	4 (2)	80 (30)	0	7	4	3

Table 3. Sequence polymorphism of the internalin gene fragments

The total amount of substitutions was comparable for the internalin genes and the MLST genes (Table 1 and Table 3). However, the amount of non-synonymous substitutions differed significantly: non-synonymous substitutions accounted for 30 – 50 % and 0 – 14 % of substitutions for internalins and housekeeping genes, respectively (p<0,05). The fixed nucleotide differences between two phylogenetic lineages prevailed among housekeeping genes and *inlB, inlC* and *inlE*. In contrast, there was no fixed difference for *inlA*. Shared mutations, carried by a few strains of both lineages, were found for *inlA* and *inlB* (Table 3). All lineage I human isolates carried the same *inlA* allele even though they belonged to different clonal complexes (Table 4 and Table 1S). Similarly, rodent isolates belonging to different clonal complexes were characterized by the same *inlB* allele. Analysis of the amino acid substitutions in the sequenced LRR-domains revealed that the majority of substitutions in InlC and InlE have a lineage-specific character (Table 4). In contrast, LRR-domains of InlA and InlB varied within lineages. There were no lineage-specific amino acid substitutions in InlA (Table 4).

Several alleles were revealed for at least one internalin gene within most STs with more than two isolates (Table 5). When the isolates were compared at all 11 markers (7 housekeeping and 4 internalin genes), the 40 isolates were distinguished into 17 genotypes (Fig. 1).

Human isolates belonging to ST1 differed in *inlB* and *inlE* gene sequences (Table 5 and Fig. 1). To get more evidence on relationships of these isolates with the epidemic clone I (ECI) strains, the *inlC* gene sequences were compared with the previously described *inlC* gene fragment, which distinguishes the ECI strains from other *L. monocytogenes* strains [5]. Six SNPs, which are characteristic for the ECI strains, were found to be conserved in all studied ST1 isolates . These results support the suggestion that ECI strains belong to CC1 and indicate that strains of ECI occur on the territory of Russia. The ST64 isolate, which belong to CC1, carried substitutions in the *inlB, inlC* and *inlE* genes that distinguished it from ST1 strains. The substitution in *inlC* was outside of the described *inlC* gene fragment, which is conserved among ECI strains (Chen *et al.*, 2007); hence, ST64 isolate could be considered as belonging to ECI as well.

ST	Strain[1]	InlA	InlB	InlC	InlE
		1111	111111222222	111122	111111111111111222222222222222222
		591458	6779136789045556	8124425	124444566678899011112222344455566
		448277	9231784617561272	8666743	541456724682813204615696895...
		ALDSIS	LQSIALLLIESSMNKI	KKVFVRN	AGSMLHSTESATLYIIECIASSRNIIEDII
1	VIMHA009	......	....	....	....
1	VIMHA011	......	....	....	....
1	VIMHA012	......	....	....	....
1	VIMHA015	......	....	....	....
1	VIMHA017	......	....	....	....
1	VIMHA010	......	.P..............	....	....
64	VIMPHk23	......	.H..............	....	....
2	VIMHA004	......	..NV.IP.....T...	....	....
2	VIMHA005	......	..NV.IP.....T...	....	....
2	VIMHA006	......	..NV.IP.....T...	....	....
2	VIMHA007	......	.PNV.IP.....T...	....	....
2	VIMHA008	......	..NV.IP.....T...	....	....
145	VIMVR081	......	..NV.IP.....T...	....	....
145	VIMVR082	......	..NV.IP.....T...	....	....
145	VIMVR084	......	..NV.IP.....T...	....	....
145	VIMVR090	......	..NV.IP.....T...	....	....
145	VIMVR092	......	..NV.IP.....T...	....	....
145	VIMVW039	......	..NV.IP.....T...	....	....
145	VIMVW037	......	..NV.IP.....T.N.	....	....
315	VIMVR062	...T.N	..NV.IP.....T...	.NMC...	....
315	VIMVG098	...T.N	..NV.IP.....T...	.NM....	....
315	VIMVG108	...T.N	..NV.IP.....T...	.NM....	....
315	VIMVG106	...T.N	..NV.IP.....T...	.NM....	....
315	VIMVG104	...T.N	..NV.IP.....T...	.NM....	....
315	VIMVG077	...T.N	..NV.IP.....T...	.NM....	....
315	VIMVG067	...T.N	..NV.IP.....T...	.NM....	....
315	VIMVG064	...T.N	..NV.IP.....T...	.NM....	....
315	VIMVG061	...T.N	..NV.IP.....T...	.NM....	....
315	VIMVG062	...T.N	..NV.IP.....T...	.NM....	....
315	VIMVG102	...T.N	..NV.IP.....TT..	.NM....	....
315	VIMVG065	...T.N	..NV.IP.....T.Q.	.NM....	....
315	VIMVG100	...T.N	..NV.IP.....T.Q.	.NM....	....
313	VIMHA034	.VNTL.	A.NVT.PIVQAPS..T	Q...LKS	TRNTMDPSVTGFFSLTMSVTGNEKLVTNT.
313	VIMHA036	.VNTL.	A.NVT.PIVQAPS..T	Q...LKS	TRNTMDPSVTGFFSLTMSVTGNEKLVTNT.
313	VIMHA038	.VNTL.	A.NVT.PIVQAPS..T	Q...LKS	TRNTMDPSVTGFFSLTMSVTGNEKLVTNT.
7	VIMPA064	P..T.N	A.NVT.PIVQAPS..T	Q...LKS	KRNTMDPSVTGFFSLTMSVTGNEKLVTNT.
19	VIMVG047	.VNTL.	A.NVT.PIVQAPS..T	Q...LKS	TRNTMDPSVTGFFSLTMSVTGNEKLVTNT.
314	VIMPR750	.VNTL.	A.NVT.PIVQAPS..T	Q...LKS	TRNTMDPSVTGFFSLTMSVTGNEKLVTNT.
155	VIMUR211	.VNTL.	A.NVT.PIVQAPS..T	Q...LKS	TRNTMDPSVTGFFSLTMSVTGNEKLVTNTF
155	VIMCR474	.VNTL.	A.NVT.PIVQAPS..T	Q...LKS	TRNTMDPSVTGFFSLTMSVTGNEKLVTNT.
7	VIMPR134	...T.N	A.NVT.PIVQAPS..T	Q...LKS	KRNTMDPSVTGFFSLTMSVTGNEKLVTNT.
7	VIMPR422	...T.N	A.NVT.PIVQAPS..T	Q...LKS	KRNTMDPSVTGFFSLTMSVTGNEKLVTNT.

[1]strain sources are designated with a color as follows: human isolates are red, rodent isolates are black, marine animal isolates are blue and sludge isolates are green.

Table 4. Amino acid substitutions in internalins.

ST/CC	Number of alleles				Number of
	inlA	*inlB*	*inlC*	*inlE*	*inl* profiles
ST1/CC1 n=6	1	2	1	2	3
ST2/CC2 n=5	1	2	1	1	3
ST145/CC2 n=7	1	2	1	1	2
ST7 n=3	1	2	1	1	2
ST155 n=2	1	1	1	2	2
ST313 n=3	1	1	1	1	1
ST315 n=11	1	4	1	1	4

[1]only those STs are shown that are represented by 2 and more isolates

Table 5. Variability of internalin genes within STs[1]

One of the human ST2 isolates was distinguished by its *inlB* sequence from other ST2 isolates (Table 4 and Fig. 1). A nonsynonymous substitution was found, which brought about a Q72P substitution (Table 4). Three other ST2 isolates carried an identical *inlB* allele, and the same allele was found in six of seven closely related ST145 isolates (which belonged to CC2; Tables 5 and 1S). *inlA, inlC* and *inlE* were identical in ST2 and ST145 isolates. Therefore, both housekeeping and virulence gene markers were highly similar in human ST2 isolates and murine ST145 isolates, confirming that these human and wild mouse isolates are genetically closely related.

4. Discussion

Here we applied a previously described MLST scheme [39; 43] to determine the genetic diversity of isolates acquired from clinical human cases and internal organs of wild animals, and to compare the genotypes of Russian isolates with international genotypic data. Obtained results confirmed the worldwide distribution of large clonal complexes corresponding to so-called epidemic clones. Particularly, we demonstrate for the first time that strains closely related to epidemic clone ECI, which is responsible for a number of temporally and geographically distinct outbreaks of listeriosis in North America and Europe [23], were associated with fetal-maternal cases of listeriosis in Russia.

The wide distribution of certain clones among the human population might be supported by high rates of international food trade turnover. Alternatively, human activity might play a secondary role in spreading of *L. monocytogenes* clones, which might be widely distributed in natural ecosystems, where their distribution might be supported by parasitism in wild animals. To study *L. monocytogenes* distribution in natural ecosystems, periodical surveys were performed in Russia in different years [1; 55; 56]. Surveys included animals capturing and analysis of bacterial loads in the internal organs, therefore only invasive strains that are

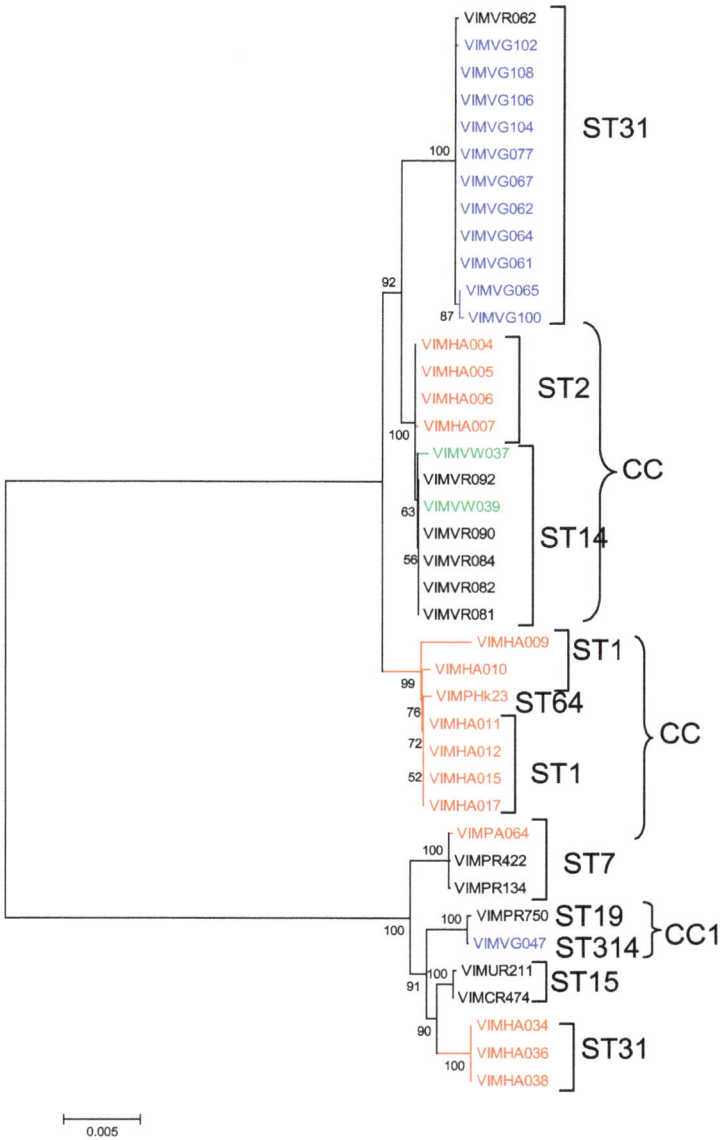

Fig. 1. The dendrogram was constructed based on the Neighbor-Joining method using concatenated sequences of the seven housekeeping and four internalin genes. The Kimura-2 parameters model was used for distance estimation. Square and curly brackets show MLST sequence types (ST) and clonal complexes (CC), respectively. The color of strain names designates the source of isolation: red for human isolates, green for rodent isolates, blue for isolates from marine animals, and black for sludge isolates. Bootstrap values obtained after 1,000 replicates are indicated at the nodes.

able to get to the internal organs were collected [52]. *L. monocytogenes* isolates studied in this work represent a few independent surveys among wild animals that were performed at different locations and in different years [1; 55]. All but one isolates, which were obtained in each survey, belonged to the same ST (see Table 2). The only exception was a survey among marine animals in 1993, when two STs were revealed: the majority of the isolates belonged to ST315, and a single isolate belonged to ST19 (Table 1S). The low diversity of animal isolates isolated each collection year might be due to the fact that the *L. monocytogenes* isolation from internal organs was successful only if outbreaks of listeriosis took place among wild animals, implying that successful surveys included epidemiologically related isolates. This suggestion is supported by the observation that secondary surveys sometimes failed to reveal *L. monocytogenes* at the same territory [55]. Still, the alternative hypothesis can not be excluded that STs found in rodents and marine animals might be prevalent in the wild environment.

Many epidemiologic studies demonstrated that lineage I is generally overrepresented among human clinical listeriosis cases, while lineage II strains are major contaminants of food products [2; 14; 15; 47]. These results led to the suggestion that lineage I strains have a higher infection potential due to intrinsic virulence features [23; 45]. Specific traits of epidemic clones include differences in both gene content and gene expression/regulation among strains [33; 44]. Similarly, particular genomic features might increase *L. monocytogenes* tropism for specific animal hosts.

Interestingly, that all lineage I *L. monocytogenes* isolates from human maternofetal listeriosis cases carried the identical *inlA* allele (Tables 1S and 4), and it is InlA that is required to cross the human maternofetal barrier and to cause fetal infection [27]. The lineage I isolates obtained from mice carried different *inlA* alleles but the same allele of the *inlB* gene. InlB was shown to be a major invasion factor for *L. monocytogenes* infection in mice [24; 26]. Mutations in InlB affect invasion efficiency and might influence *L. monocytogenes* virulence for mice [31; 49]. The fine tuning of internalin sequences might be one of the factors that influence *L. monocytogenes* tropism for specific animal hosts. The further study of naturally occurring internalin variants and their role in infection of different hosts is in progress at our laboratories.

5. Acknowledgements

The work was supported by Federal Agency on Science and Innovations (grant N 02.740.11.0310) and Russian Fund for Basic Research (grant N 09-04-00403). Platform Genotyping of Pathogens and Public Health received financial support from Institut Pasteur.

6. References

[1] Bakulov IA ,Kotliarov VM (1965) [epizootiology of listeriosis in the ussr. (the spreading of listeriosis among agricultural animals)] Veterinariia 42:28-31.
[2] Bibb WF, Gellin BG, Weaver R, Schwartz B, Plikaytis BD, Reeves MW, Pinner RW ,Broome CV (1990) Analysis of clinical and food-borne isolates of listeria monocytogenes in the united states by multilocus enzyme electrophoresis and

application of the method to epidemiologic investigations. Appl Environ Microbiol 56:2133-2141.

[3] Bierne H, Sabet C, Personnic N ,Cossart P (2007) Internalins: a complex family of leucine-rich repeat-containing proteins in listeria monocytogenes. Microbes Infect 9:1156-1166.

[4] Boerlin P ,Piffaretti JC (1991) Typing of human, animal, food, and environmental isolates of listeria monocytogenes by multilocus enzyme electrophoresis. Appl Environ Microbiol 57:1624-1629.

[5] Chen Y, Zhang W ,Knabel SJ (2007) Multi-virulence-locus sequence typing identifies single nucleotide polymorphisms which differentiate epidemic clones and outbreak strains of listeria monocytogenes. J Clin Microbiol 45:835-846.

[6] Cossart P, Pizarro-Cerdá J ,Lecuit M (2003) Invasion of mammalian cells by listeria monocytogenes: functional mimicry to subvert cellular functions. Trends Cell Biol 13:23-31.

[7] Enright MC ,Spratt BG (1999) Multilocus sequence typing. Trends Microbiol 7:482-487.

[8] Enright MC, Day NP, Davies CE, Peacock SJ ,Spratt BG (2000) Multilocus sequence typing for characterization of methicillin-resistant and methicillin-susceptible clones of staphylococcus aureus. J Clin Microbiol 38:1008-1015.

[9] Evans MR, Swaminathan B, Graves LM, Altermann E, Klaenhammer TR, Fink RC, Kernodle S ,Kathariou S (2004) Genetic markers unique to listeria monocytogenes serotype 4b differentiate epidemic clone ii (hot dog outbreak strains) from other lineages. Appl Environ Microbiol 70:2383-2390.

[10] Farber JM ,Peterkin PI (1991) Listeria monocytogenes, a food-borne pathogen. Microbiol Rev 55:476-511.

[11] Feil EJ, Maiden MC, Achtman M ,Spratt BG (1999) The relative contributions of recombination and mutation to the divergence of clones of neisseria meningitidis. Mol Biol Evol 16:1496-1502.

[12] Fenlon DR (1985) Wild birds and silage as reservoirs of listeria in the agricultural environment. J Appl Bacteriol 59:537-543.

[13] Fugett EB, Schoonmaker-Bopp D, Dumas NB, Corby J ,Wiedmann M (2007) Pulsed-field gel electrophoresis (pfge) analysis of temporally matched listeria monocytogenes isolates from human clinical cases, foods, ruminant farms, and urban and natural environments reveals source-associated as well as widely distributed pfge types. J Clin Microbiol 45:865-873.

[14] Gray MJ, Zadoks RN, Fortes ED, Dogan B, Cai S, Chen Y, Scott VN, Gombas DE, Boor KJ ,Wiedmann M (2004) Listeria monocytogenes isolates from foods and humans form distinct but overlapping populations. Appl Environ Microbiol 70:5833-5841.

[15] Harvey J ,Gilmour A (1994) Application of multilocus enzyme electrophoresis and restriction fragment length polymorphism analysis to the typing of listeria monocytogenes strains isolated from raw milk, nondairy foods, and clinical and veterinary sources. Appl Environ Microbiol 60:1547-1553.

[16] Hatkin JM, Phillips WEJ ,Hurst GA (1986) Isolation of listeria monocytogenes from an eastern wild turkey. J Wildl Dis 22:110-112.

[17] Hellström S, Kiviniemi K, Autio T ,Korkeala H (2008) Listeria monocytogenes is common in wild birds in helsinki region and genotypes are frequently similar with those found along the food chain. J Appl Microbiol 104:883-888.

[18] Hey J (1991) The structure of genealogies and the distribution of fixed differences between dna sequence samples from natural populations. Genetics 128:831-840.

[19] Ivanek R, Gröhn YT ,Wiedmann M (2006) Listeria monocytogenes in multiple habitats and host populations: review of available data for mathematical modeling. Foodborne Pathog Dis 3:319-336.

[20] Jeffers GT, Bruce JL, McDonough PL, Scarlett J, Boor KJ ,Wiedmann M (2001) Comparative genetic characterization of listeria monocytogenes isolates from human and animal listeriosis cases. Microbiology 147:1095-1104.

[21] Kalorey DR, Kurkure NV, Warke SR, Rawool DB, Malik SVS ,Barbuddhe SB (2006) Isolation of pathogenic listeria monocytogenes in faeces of wild animals in captivity. Comp Immunol Microbiol Infect Dis 29:295-300.

[22] Karunasagar I ,Karunasagar I (2000) Listeria in tropical fish and fishery products. Int J Food Microbiol 62:177-181.

[23] Kathariou S (2002) Listeria monocytogenes virulence and pathogenicity, a food safety perspective. J Food Prot 65:1811-1829.

[24] Khelef N, Lecuit M, Bierne H ,Cossart P (2006) Species specificity of the listeria monocytogenes inlb protein. Cell Microbiol 8:457-470.

[25] Kumar S, Tamura K ,Nei M (2004) Mega3: integrated software for molecular evolutionary genetics analysis and sequence alignment. Brief Bioinform 5:150-163.

[26] Lecuit M, Dramsi S, Gottardi C, Fedor-Chaiken M, Gumbiner B ,Cossart P (1999) A single amino acid in e-cadherin responsible for host specificity towards the human pathogen listeria monocytogenes. EMBO J 18:3956-3963.

[27] Lecuit M, Nelson DM, Smith SD, Khun H, Huerre M, Vacher-Lavenu M, Gordon JI ,Cossart P (2004) Targeting and crossing of the human maternofetal barrier by listeria monocytogenes: role of internalin interaction with trophoblast e-cadherin. Proc Natl Acad Sci U S A 101:6152-6157.

[28] Lemee L, Dhalluin A, Pestel-Caron M, Lemeland J ,Pons J (2004) Multilocus sequence typing analysis of human and animal clostridium difficile isolates of various toxigenic types. J Clin Microbiol 42:2609-2617.

[29] Litvin VI, Pushkareva VI ,Emel'ianenko EN (2004) [biocenosis of the natural foci of sapronotic infections (the results of 15-year observations)] Zh Mikrobiol Epidemiol Immunobiol :102-108.

[30] Maiden MC, Bygraves JA, Feil E, Morelli G, Russell JE, Urwin R, Zhang Q, Zhou J, Zurth K, Caugant DA, Feavers IM, Achtman M ,Spratt BG (1998) Multilocus sequence typing: a portable approach to the identification of clones within populations of pathogenic microorganisms. Proc Natl Acad Sci U S A 95:3140-3145.

[31] Marino M, Braun L, Cossart P ,Ghosh P (2000) A framework for interpreting the leucine-rich repeats of the listeria internalins. Proc Natl Acad Sci U S A 97:8784-8788.

[32] McLauchlin J (1997) Animal and human listeriosis: a shared problem? Vet J 153:3-5.

[33] Nelson KE, Fouts DE, Mongodin EF, Ravel J, DeBoy RT, Kolonay JF, Rasko DA, Angiuoli SV, Gill SR, Paulsen IT, Peterson J, White O, Nelson WC, Nierman W, Beanan MJ, Brinkac LM, Daugherty SC, Dodson RJ, Durkin AS, Madupu R, Haft DH, Selengut J, Van Aken S, Khouri H, Fedorova N, Forberger H, Tran B, Kathariou S, Wonderling LD, Uhlich GA, Bayles DO, Luchansky JB ,Fraser CM (2004) Whole genome comparisons of serotype 4b and 1/2a strains of the food-

borne pathogen listeria monocytogenes reveal new insights into the core genome components of this species. Nucleic Acids Res 32:2386-2395.

[34] Ortel S (1983) [significance of new results in the research on human listeriosis] Zentralbl Gynakol 105:1295-1306.

[35] Piffaretti JC, Kressebuch H, Aeschbacher M, Bille J, Bannerman E, Musser JM, Selander RK ,Rocourt J (1989) Genetic characterization of clones of the bacterium listeria monocytogenes causing epidemic disease. Proc Natl Acad Sci U S A 86:3818-3822.

[36] Pizarro-Cerdá J, Sousa S ,Cossart P (2004) Exploitation of host cell cytoskeleton and signalling during listeria monocytogenes entry into mammalian cells. C R Biol 327:115-123.

[37] Pushkareva V.I., Ermolaeva S.A., Litvin V.Yu. (2010) Hydrobionts as reservoir hosts for infectious agents of sapronoses Biol Bull 37:1-10.

[38] Pushkareva VI ,Ermolaeva SA (2010) Listeria monocytogenes virulence factor listeriolysin o favors bacterial growth in co-culture with the ciliate tetrahymena pyriformis, causes protozoan encystment and promotes bacterial survival inside cysts. BMC Microbiol 10:26.

[39] Ragon M, Wirth T, Hollandt F, Lavenir R, Lecuit M, Le Monnier A ,Brisse S (2008) A new perspective on listeria monocytogenes evolution. PLoS Pathog 4:e1000146.

[40] Rasmussen OF, Skouboe P, Dons L, Rossen L ,Olsen JE (1995) Listeria monocytogenes exists in at least three evolutionary lines: evidence from flagellin, invasive associated protein and listeriolysin o genes. Microbiology 141 (Pt 9):2053-2061.

[41] Roberts AJ, Williams SK, Wiedmann M ,Nightingale KK (2009) Some listeria monocytogenes outbreak strains demonstrate significantly reduced invasion, inla transcript levels, and swarming motility in vitro. Appl Environ Microbiol 75:5647-5658.

[42] Rozas J, Sánchez-DelBarrio JC, Messeguer X ,Rozas R (2003) Dnasp, dna polymorphism analyses by the coalescent and other methods. Bioinformatics 19:2496-2497.

[43] Salcedo C, Arreaza L, Alcalá B, de la Fuente L ,Vázquez JA (2003) Development of a multilocus sequence typing method for analysis of listeria monocytogenes clones. J Clin Microbiol 41:757-762.

[44] Severino P, Dussurget O, Vêncio RZN, Dumas E, Garrido P, Padilla G, Piveteau P, Lemaître J, Kunst F, Glaser P ,Buchrieser C (2007) Comparative transcriptome analysis of listeria monocytogenes strains of the two major lineages reveals differences in virulence, cell wall, and stress response. Appl Environ Microbiol 73:6078-6088.

[45] Swaminathan B ,Gerner-Smidt P (2007) The epidemiology of human listeriosis. Microbes Infect 9:1236-1243.

[46] Thompson JD, Higgins DG ,Gibson TJ (1994) Clustal w: improving the sensitivity of progressive multiple sequence alignment through sequence weighting, position-specific gap penalties and weight matrix choice. Nucleic Acids Res 22:4673-4680.

[47] Trott DJ, Robertson ID ,Hampson DJ (1993) Genetic characterisation of isolates of listeria monocytogenes from man, animals and food. J Med Microbiol 38:122-128.

[48] Tsai YL, Orsi RH, Nightingale KK ,Wiedmann M (2006) Listeria monocytogenes internalins are highly diverse and evolved by recombination and positive selection. Infect Genet Evol 6:378-389.

[49] Témoin S, Roche SM, Grépinet O, Fardini Y ,Velge P (2008) Multiple point mutations in virulence genes explain the low virulence of listeria monocytogenes field strains. Microbiology 154:939-948.

[50] Urwin R ,Maiden MCJ (2003) Multi-locus sequence typing: a tool for global epidemiology. Trends Microbiol 11:479-487.

[51] Vázquez-Boland JA, Kuhn M, Berche P, Chakraborty T, Domínguez-Bernal G, Goebel W, González-Zorn B, Wehland J ,Kreft J (2001) Listeria pathogenesis and molecular virulence determinants. Clin Microbiol Rev 14:584-640.

[52] Wiedmann M, Bruce JL, Keating C, Johnson AE, McDonough PL ,Batt CA (1997) Ribotypes and virulence gene polymorphisms suggest three distinct listeria monocytogenes lineages with differences in pathogenic potential. Infect Immun 65:2707-2716.

[53] Yildirim S, Lin W, Hitchins AD, Jaykus L, Altermann E, Klaenhammer TR ,Kathariou S (2004) Epidemic clone i-specific genetic markers in strains of listeria monocytogenes serotype 4b from foods. Appl Environ Microbiol 70:4158-4164.

[54] Yoshida T, Takeuchi M, Sato M ,Hirai K (1999) Typing listeria monocytogenes by random amplified polymorphic dna (rapd) fingerprinting. J Vet Med Sci 61:857-860.

[55] Zaytseva E, Ermolaeva S ,Somov GP (2007) Low genetic diversity and epidemiological significance of listeria monocytogenes isolated from wild animals in the far east of russia. Infect Genet Evol 7:736-742.

[56] Zaĭtseva EA (2005) [antibiotic susceptibility of listeria monocytogenes strains isolated in the primor'e region] Antibiot Khimioter 50:27-31.

12

Pre-Columbian Male Ancestors
for the American Continent,
Molecular Y-Chromosome Insight

Graciela Bailliet[1], Marina Muzzio[1,2], Virginia Ramallo[1,3], Laura S. Jurado
Medina[1], Emma L. Alfaro[4], José E. Dipierri[4] and Claudio M. Bravi[1,2]

[1]*Instituto Multidisciplinario de Biología Celular (IMBICE), La Plata,*
[2]*Facultad de Ciencias Naturales y Museo,*
Universidad Nacional de La Plata, La Plata,
[3]*Universidade Federal do Rio Grande do Sul, Porto Alegre,*
[4]*Instituto de Biología de La Altura (INBIAL),*
Universidad Nacional de Jujuy, Jujuy,
[1,2,4]*Argentina*
[3]*Brazil*

1. Introduction

It all began back in the 1990's when the need to compare the mitochondrial DNA for matrilineages arose. It was then when the search for polymorphic markers of the Y-specific region started, in order to test whether the histories of female and male lineages were the same.

The human Y chromosome has an intermediate mutational rate between the one from autosomes and the X chromosome. Its mode of inheritance is exclusively patrilinear and its lower effective number (1/4th of the autosomal and 1/3rd regarding the X chromosome) makes it highly susceptible to genetic drift (Jobling and Smith, 2003).

Given its specific characteristics, the Y chromosomes have shown some degree of continental differentiation, providing a way of distinguishing among European, American, African, and Asian lineages. A tetranucleotidic microsatellite (DYS19, Roewer *et al.* 1992), an Alu insert (YAP+ or M1, Hammer *et al.* 1994), a single nucleotide polymorphism (SNP) (M2) associated to YAP+ (Seielstad *et al.* 1994), and variants of the alphoid system (Santos *et al.* 1995) were the first polymorphic systems to be studied. Later on, the fruitful search for polymorphisms led Underhill and collaborators to publish the first phylogeny compiling 166 SNPs geographically correlated (Underhill *et al.* 2000). Since then the nomenclature was normalized so that different research groups could compare their work (Y Chromosome Consortium, 2002), and an improved phylogeny was published in 2008 by Karafet and collaborators. The accepted phylogeny spans in a tree of 20 major clades that represent haplogroups (Figure 2), where the accumulation of polymorphisms along the lineages determine their diversification and the configuration of sub-branches.

Haplogroups are defined by SNPs, which have a mutational rate of 13.5 10^{-10} (Anagnostopoulos et al. 1999), so that the probability of homoplasy results extremely low. Thus, the number of SNPs determines the phylogenetic status of the analyzed chromosome.

For the American continent, early studies that characterized the alphoid system and the DYS19 microsatellite showed that a large amount of Native American Y chromosomes carried the combination II-A for those loci (Pena et al. 1995, Santos et al. 1996a,b). In 1996, Underhill and collaborators described a mutation at the DYS199 locus (now named M3), that resembled the haplotype II-A described by Pena et al. (1995) in the sense that it was widely distributed in Native Americans but absent in populations of other continental origins. In a previous work of our team we correlated the three systems and described the frequency of six polymorphic sites, among which there was a new SNP, currently named SRY2627 (Bianchi et al. 1997), specific for European populations (Hurles et al. 1999).

Most males of Native American ancestry in South America carry a Y-chromosome lineage named sub-haplogroup Q1a3a (Karafet et al. 2008), which is characterized by the aforementioned SNP M3, and the derived states for M242 and M346. It has been found in all Native American populations from Alaska to the Magellan Strait, with an average frequency of 60% (Bianchi et al. 1998, Bortolini et al. 2003).

Present evidence makes researchers suppose that Q1a3a diverged from its predecessor during or shortly before the crossing of the Bering Strait (Lell et al. 2002). There is no doubt that Siberia was the main source of colonization of America, the last continent to be populated. Multidisciplinary evidence supports the hypothesis that strong bottlenecks happened during this process. Yet, when it happened and how many migratory waves occurred is still debated.

All these previous studies showed that the peopling of the Americas bore a strong founder effect regarding the paternal lineages, as a result of which most Y chromosomes described belong to a single lineage. It was only when Y-specific microsatellites began to be studied that its intrinsic diversity could be described through defining haplotypes. One of them was identified as ancestral to the diversification of Y chromosomes bearing M3. Several authors estimated the time of divergence of these lineages through maximum parsimony methods, obtaining different results: 22,770 years (13,500-58,700) (Bianchi et al. 1998); 11,456 (9,423-13,797) (Ruiz-Linares et al. 1999); 7,570 (SE 681) (Bortolini et al. 2003). Temporal differences among those three reports depended mostly on the adjustments of the estimation of mutational rates and the generation time considered. While the first two considered 27 years, Bortolini et al. used a shorter generation time of 25 years.

In 2003 Seielstad and collaborators described a new mutation at locus M242 that proved to be immediately ancestral to M3, and that now defines haplogroup Q. Haplogroup Q shows a wide distribution in the Old World, with an average frequency of 19% in Siberia that drops to 5% in Central Asia (Seielstad et al. 2003, Karafet et al. 2002). Lineages with membership in Q but lacking M3 have also been described in Native Americans. Formally designated as paragroup Q* or Q(xM3), these lineages attain high frequencies of up to 47% in some North American populations (Zegura et al. 2004, Bolnick et al. 2006), yet they are poorly represented in most Central and South American natives, with frequencies rarely surpassing 6% (Bortolini et al. 2003, Bailliet et al. 2009, Bisso-Machado et al. 2010).

Although 17 polymorphisms have been described so far inside haplogroup Q, most Native American Q lineages seem to be phylogenetically close to Q1a3a, sharing three of these SNPs (Karafet *et al.* 2008, Bailliet *et al.* 2009). It has been recently demonstrated that Q(xM3) lineages present in Native Americans, with the sole exception of some Eskimos, share the presence of M346 mutation and thus belong to clade Q1a3* (Bailliet *et al.* 2009, Bisso-Machado *et al.* 2011). Deviant from this pattern are two Q1a* lineages recently reported for one ancient Paleo-Eskimo (~4000 years BP) and one extant Eskimo (Rasmussen *et al.* 2008, Bisso-Machado *et al.* 2011).

Even though the chromosomes bearing M3 are the most frequent clade in Native populations, many research groups have tried to find out whether other haplogroups entered the American continent from Asia. In 1999, Bergen and collaborators described one SNP in the RPS4Y gene that defines what is now known as haplogroup C, present in populations of Asia, Australo-Melanesia and also the Americas, where they are can be found in populations belonging to the three major linguistic stocks, namely Eskimo-Aleuts, Na-Dene and Amerindians (Bergen *et al.* 1999, Karafet *et al.* 1999, Capelli *et al.* 2001, Hammer *et al.* 2001, Bosch *et al.* 2003, Malhi *et al.* 2008). C lineages with one further mutation at M217 belong to the C3b clade, which spread through central and eastern Asia and America. While Native North American C3 lineages share the P39 mutation that defines the monophyletic C3b group, the few C lineages known from South American populations belong to its presumed sister group C3(xP39). Although not systematically searched for in Central and South America, C lineages have so far been found in three Waorani and one Kichwa individuals from Ecuador, and two Wayuú from Venezuela, out of a total of 767 individuals from 48 populations (Lell *et al.* 2002, Bortolini *et al.* 2003, Zegura *et al.* 2004, Mazieres *et al.* 2008, Geppert *et al.* 2010).

1.1 Going narrower than haplogroups

When researchers need to work on a finer scale, they define haplotypes. Haplotypes are built from microsatellites or Short Tandem Repeats (STR), which are sequences of 2-6 nucleotides repeated in tandem. They have an average mutation rate of 3.35 10^{-3}, which translates into a possible differentiation at the father-son transmission if a high number of STRs is studied (Ballantyne *et al.* 2010). When STRs are analyzed within the lineages of a haplogroup, there are mathematical methods to estimate the time of divergence considering the mutational rate for each STR (Ballantyne *et al.* 2010).

So far the mutation rate has been estimated through the changes observed in certified father-son transmissions, and 186 STR mutational rates have been published (Ballantyne *et al.* 2010), so it is possible to chose the most adequate STRs depending on the problem to solve, and to use commercial kits for amplifying jointly 12 and 17 STRs which are resolved by capillary electrophoresis.

Lineages can be built by combining haplogroups and haplotypes. They represent monophyletic lineages that share a common ancestor and among which the diversification occurred by mutation acting at the STR level (Jobling *et al.* 2004).

There is relevant information regarding samples analyzed for C and Q haplogroups from many locations with a vast geographic distribution, which allows us to analyze the peculiarities of the geographic distribution of male lineages and their diversity in America. In the present work we analyze the distribution of major Y-chromosome haplogroups in 279 individuals from ten Native American populations from Chile, Paraguay, and Argentina, and 638 individuals from 12 cosmopolitan, urban populations of Argentina.

2. Materials and methods

We analyzed 917 samples of randomly selected male voluntary donors from 10 self-acknowledged Native American populations: Ayoreo (9) and Lengua (25) from Paraguay; Huilliche (26) and Pehuenche (18) from Chile; Wichi (118), Toba (5), Chorote (9), Mocoví (27), Mapuche (26), and Tehuelche (20) from Argentina. Twelve urban populations from Argentina were also included in the sample: Susques (19), Cochinoca (6), Rinconada (6), San Salvador de Jujuy (85), Catamarca (106), Salta (77), Tucumán (16), La Rioja (87), Mendoza (75), Tartagal (105), Aguaray (12), and Azampay (44) (Fig. 1)

Fig. 1. Map of South American sampling localities

We amplified by PCR different fragments carrying the markers of interest and then identified the polymorphism through enzymatic digestion, since restriction enzymes recognize specific sequences and cut the DNA that bears that sequence. In case there is a mutation on the sequence, the enzyme does not cut the DNA because the recognition site is different. This way, the alteration in one base is transformed into a difference in size, and it is for this reason that they are called Restriction Fragment Length Polymorphisms (RFLP).

SNP / indel	Reference SNP ID	Forward primer Reverse primer (5'-3')	PCR Size (pb)	Enzymes	Reference
YAP		caggggaagataaagaaata actgctaaaaggggatggat	150-450		Hammer, 1994
M3	rs3894	taatcagtctcctccccagca taggtaccagctcttcccaatt	201	*Mfe*I	Present paper
M9	rs3900	ccagcattaaaaactttcagg aaaacctaactttgctcaagc	340	*Hinf*I	Underhill *et al.* 1997
P27	rs9786896	ccttcaggaacttgtcgg aggacccaattattgcacac	288	*Ban*I	Present paper
M89	rs2032652	acagaaggatgctgctcagctt ccaactcaggcaaagtgagacat	87	*Nla*II	Present paper
M168	rs2032595	agtttgaggtagaatactgtttgct aatctcataggtctctgactgttc	473	*Hinf*I	Present paper
M207	rs2032658	ggggcaaatgtaagtcaagc tttctaggctgttcgctgct	164	*Dra*I	Present paper
M242	rs8179021	tcagatggcaagatttttaagtaca aaaaacacgttaagaccaatgtc	151	*Hpy*188III	Present paper
M346		agatgggaaaggcagccaa tctgtccacatgtgtccgtg	419	-1	Sengupta *et al.* 2006

Table 1. SNP and Indel. -1: Sequencing

We used 9 markers to define membership in 9 clades (Fig. 2): YAP, M168, M89, M9, P27, M207, M242, M346, and M3 (Table 1, Fig 2). Primers were designed in order to reduce the fragment length for optimal amplification with degraded DNA and, where necessary, creating a mismatch in order to generate a recognition site for RFLP (present work)(Table 1, Fig.2). The basic set of seven STRs: DYS19, DYS389 I and II, DYS390, DYS391, DYS392, DYS393 was employed to establish haplotypes (Kayser *et al.* 1997, Pascali *et al.* 1999).

In the case of the lineages belonging to the Q1a3* paragroup, we built a Median Joining Network   (Bandelt *et al.* 1999). Median Joining Networks allow researchers to estimate phylogenies at the intra-species level, without choosing between trees and employing non-recombining population data. The most popular software to compute them is NETWORK (http://www.fluxus-engineering.com/sharenet.htm), which allows the assignment of a definite weight to each character, so that less likely events could be given a higher weight (considered as more "decisive" since they are rare) than events with a high probability (since they are likely to have occurred many times). It is essential to employ a good criterion to weigh each character, with differential importance for each mutation depending on their rate, since the final network depends on it. We used a formula designed by our team (Muzzio *et al.* 2010), which establishes a precise mathematical scale that concurs with the probability of change in each marker.

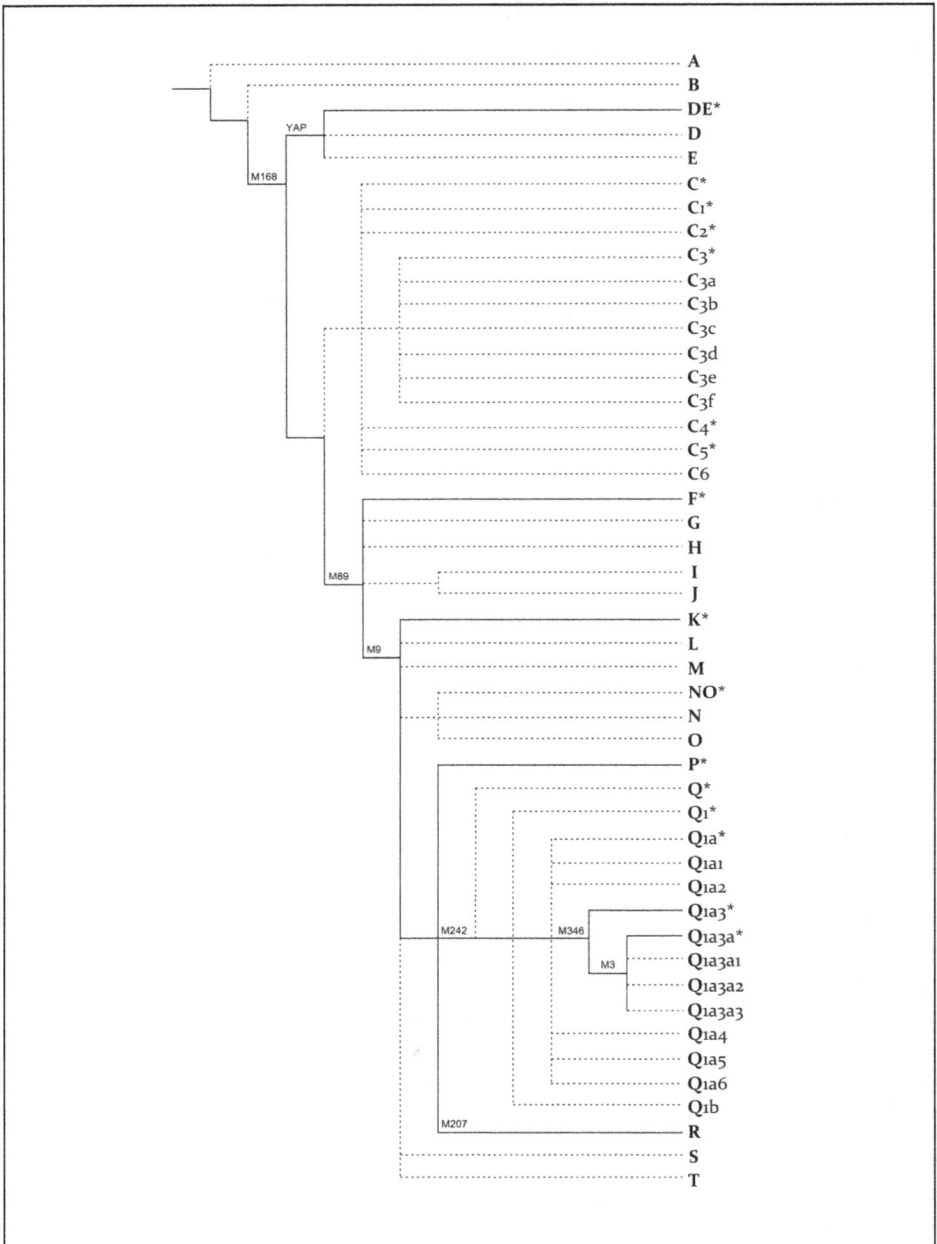

Fig. 2. Phylogenetic tree according to Karafet *et al*. [2008]. Solid lines indicate haplogroups, which can be typed by the Y-SNP RFLP assays. Markers that have been typed are indicated upon the lines. Dotted lines indicate haplogroups, which are not included in the study.

The Differentiation Index Fst (Excoffier *et al.* 1992) was applied together with the Arlequin software (Schneider, Roessli and Excoffier 2000). For the MNS analysis we used the NTSYS 2.11S (Exeter Software) from the Da Distance (Nei 1972). MNS analysis represents the information in space, where each axis represents a component that involves the variability found. In our case, the use of two axes or dimensions was enough and simplified the interpretation.

3. Results

3.1 Autochthonous lineages

Haplogroup Q1a3a attained frequencies between 46 and 96% for our Native American populations, with values equal to or higher than 80% among Wichi, Toba, Chorote, and Pehuenche (Table 2). This clade accounted for only 7-17% of the lineages in Mendoza, La Rioja, Catamarca, Tucumán, Azampay, and Aguaray but gathered together one to two thirds of patrilinages in Tartagal, SS Jujuy, Salta, Cochinoca, and Rinconada. The small urban sample of Susques stands out as having a share of 95% for Q1a3a lineages.

Populations	Province	N	Haplogroups						
			AB	DE	F (xK)	K (xQ,R)	Q1a3*	Q1a3a	R
Ayoreo	Paraguay	9	0	0	0	0.111	0.222	0.556	0.111
Lengua	Paraguay	25	0	0	0	0.040	0.280	0.680	0
Wichi	Salta	25	0	0.040	0	0	0	0.960	0
Wichi	Formosa(LY)	30	0	0	0.133	0.033	0	0.800	0.033
Wichi	Formosa(IJ)	59	0	0	0.102	0.017	0	0.831	0.051
Chorote	Salta	9	0	0	0.111	0	0	0.889	0
Mocoví	Santa Fé	27	0	0	0.074	0.037	0.074	0.481	0.333
Mapuche	Chubut	26	0	0.077	0.115	0	0.038	0.538	0.231
Tehuelche	Neuquén	20	0	0	0.150	0.050	0	0.650	0.150
Huilliche	Chile	26	0	0	0.308	0,038	0.038	0.462	0.154
Pehuenche	Chile	18	0	0.056	0.056	0	0	0.833	0.056
Susques	Jujuy	19	0	0	0.053	0	0	0.947	0
Cochinoca	Jujuy	6	0	0	0.167	0	0	0.667	0.167
Rinconada	Jujuy	6	0	0	0.167	0	0	0.667	0.167
SS Jujuy	Jujuy	85	0	0.012	0.094	0.071	0	0.459	0.365
Catamarca	Catamarca	106	0	0.094	0.208	0.009	0	0.104	0.585
Azampay	Catamarca	44	0	0.023	0.068	0	0	0.159	0.750
Salta	Salta	77	0	0.078	0.078	0.013	0.013	0.494	0325
Tartagal	Salta	105	0.010	0.105	0.152	0	0	0.362	0.371
Aguaray	Salta	12	0	0	0	0	0	0.167	0.833
Tucumán	Tucumán	16	0.063	0.125	0.250	0	0	0.125	0.438
La Rioja	La Rioja	87	0	0.149	0.241	0.057	0	0.103	0.448
Mendoza	Mendoza	75	0	0.147	0.293	0.040	0	0.067	0.453
N total		917							
Fst		0.172							

Table 2. Haplogroup frequency in populations studied. LY: Laguna Yema, IJ: Ingeniero Juarez.

A total of 374 lineages were assigned to Q1a3a, and complete STR profiles were obtained for 137 of them, resulting in 97 different haplotypes. Fixation Index (Fst) for these lineages was 0.112, and the mean gene diversity was 0.501. We observed a great allele frequency differentiation of Q1a3a haplotypes. In the MDS plot (Fig. 3), the first axis separates populations by their geographic location: Northwestern Andean populations as Rinconada, Cochinoca, and Humahuaca on the upper left side, while the Northeastern Gran Chaco populations, as Wichi, Toba, Chorote, Ayoreo, and Lengua, occupied the upper right side.

Only 13 individuals were assigned to paragroup Q1a3*, all of them derived from indigenous populations except for one donor from Salta. All cases were singletons, except for Paraguayan Ayoreo and Lengua, populations for which high frequencies of 22-30% were found. Although reduced in total number, they showed a considerably high allele frequency differentiation with a mean gene diversity of 0.478.

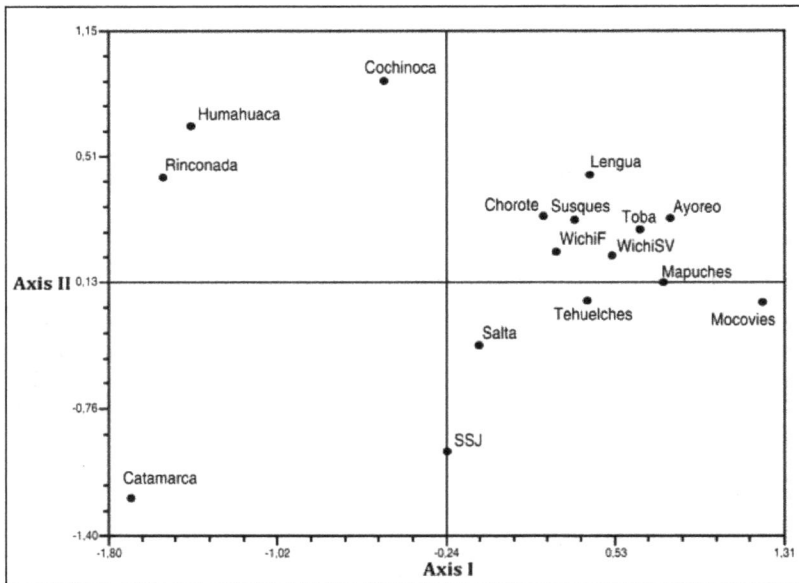

Fig. 3. Two-dimensional scaling of genetic distances (Nei, 1972) based on seven Y-chromosome STRs haplotypes belonging to Q1a3a haplogroup from 17populations from Argentina and Paraguay. SSJ: San Salvador de Jujuy. STRESS1 = 0.11876.

Network analysis of the Q1a3* haplotypes showed three Lengua at the central position, while the only haplotypes that differed in one or two allelic changes from these were from Lengua or Ayoreo populations. The other haplotypes diverged in more allele changes, while three median vectors (indicating absent haplotypes in the sample) were interposed between the central and derived haplotypes. This is concordant with the hypothesis of severe drift acting over these less frequent haplotypes. Lengua 2 and Lengua 3 carried 2 identical haplotypes each (Fig.4).

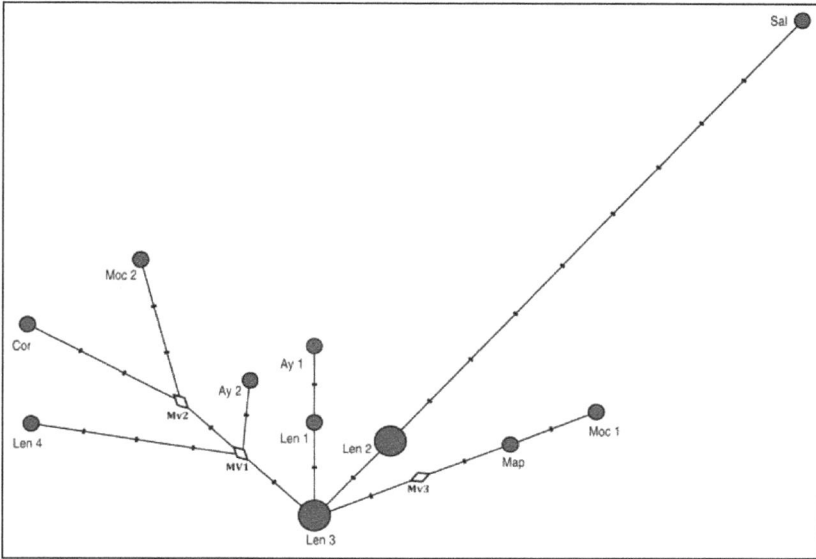

Fig. 4. Microsatellite network for South American Q1a3* haplotypes. Circle size is proportional to frequency, and dashes equal one-step STR mutation. L: Lengua, A: Ayoreo, Mo: Mocoví, Ma: Mapuche, Cor: Córdoba. Mv 1, 2, and 3 are median vectors not found in the sample.

3.2 Allochthonous haplogoups

The most frequent allochtonous haplogroups were R and F(xK). High frequencies of 32-83% for R were found in all the urban populations except for the three highland samples of Rinconada, Susques, and Cochinoca. Conversely, R never surpassed the maximum of 34% in the Native Americans ones, and was even absent in four of them. Paragroup F(xK) showed overall frequencies lower than 30% in the urban populations, and values under 15% in the Native ones, with the remarkable exception of 31% in Chilean Huilliches. Surprisingly, frequencies for F(xK) were equal or higher than those for R in Amerindian populations of Formosa: Wichi, Chorote, Huilliche, Tehuelche, and Pehuenche (Table 1).

Haplogroup DE had a maximum frequency of around 15% in Mendoza and La Rioja, and paragroup K(xQ,R) did not exceed values of ~7% when only reasonably sized samples (i.e. N≥10) were considered. We only found two Y chromosomes that belonged to the AB haplogroups.

3.3 Genetic differentiation among total haplogroups

Population variation was 17.23% and the differentiation coefficient observed among populations was quite high (Fst= 0.17).

Figure 5 represents the two-dimensional Da distance matrix (Nei, 1972). The stress value was 0.0297, which suggests good adjustment. A highlight in the figure represents the first axis (R1), which explains 100% variation and separates all Native American populations from the urban ones, with the exception of those from Jujuy: SS Jujuy, Rinconada, Cochinoca, and Susques. The high frequencies of Q1a3a, Q1a3*, and K(xQ,R) determine that portion of the plot. The three populations of Wichi plus Susques, Chorote, Pehuenche and Rinconada and Cochinoca (these last two undistinguishable) represent one group, while the other one comprises Mapuche, Tehuelche, Huilliche, Mocoví, and SS Jujuy samples. The samples from Lengua and Ayoreo were situated far in relation to the aforementioned, probably because of their highest proportion of Q1a3*.

The other half of the plot is influenced by the foreign haplogroups AB, DE, F(xK), and R, whose frequencies determine the position of the urban and semi-urban populations positions, excluding those from Jujuy. Salta, Aguaray and Azampay constitute one group; La Rioja and Mendoza form another group, and finally Tartagal, Catamarca, and Tucumán do not configure or take part in any group (Fig. 5).

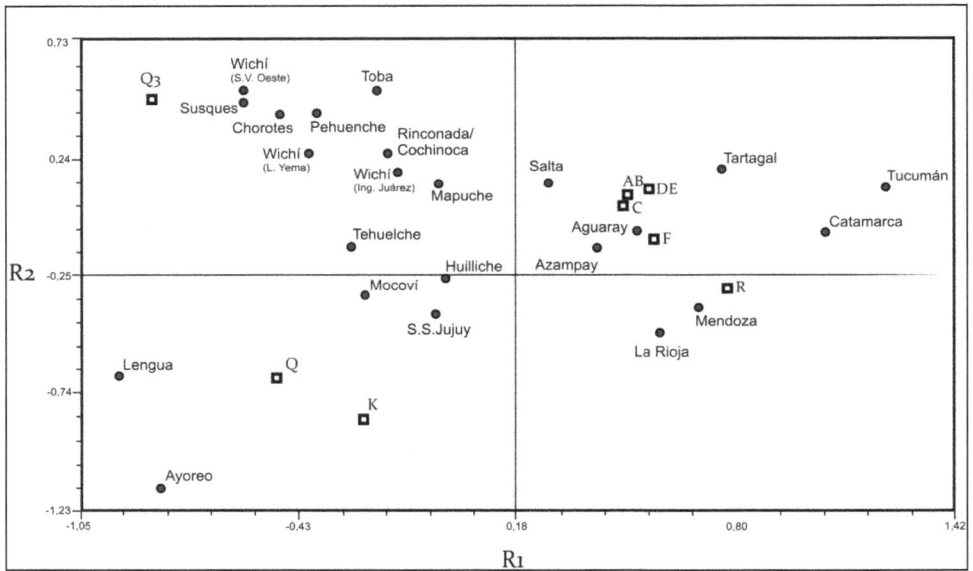

Fig. 5. Two-dimensional scaling of Nei's (1972) distance matrix; ● Populations. ◘ Haplogroups; Stress=0.02970

4. Conclusion

We summarize below a few important facts about the autochthonous and allochthonous haplogroups that were studied in this research work:

4.1 Autochthonous haplogroups to America

Q1a3a (bearer of the derived state for M242, M346, and M3) is the most frequent and widely distributed clade in the Americas (Underhill *et al.* 1996; Bianchi *et al.* 1998; Bortolini *et al.* 2003; Bisso Machado *et al.* 2010, Geppert *et al.* 2010; Toscanini *et al.* 2011) and is considered autochthonous to that Continent. In Q1a3a, 5 mutations have been identified: a) M19 T-A which defines Q1a3a1 (Underhill *et al.* 1996) has been found in 22 of 33 Ticuna and 2 of 19 Wayuu (Bortolini *et al.* 2003), and in 2 Toba from Argentina (Toscanini *et al.* 2011); b) M194 T-C which defines Q1a3a2 (Underhill *et al.* 2001) was described in one Maya (Shen *et al.* 2000); c) M199 is an insertion defining Q1a3a3 (Underhill *et al.* 2001), which was found in 1 Suruí (Shen *et al.* 2000); d) SA01 C-T which defines the new sublineage Q1a3a4, which has been identified in the Andean populations of South America (Jota *et al.* 2011). However, none of these variants have been found in the series of samples analyzed in the present work.

Members of Q1a3* are the sister group to major sub-haplogroup Q1a3a (Seielstad et al. 2003, Bortolini et al. 2003, Bailliet et al. 2009). This clade has a very low frequency and patchy distribution in South America (Bortolini et al. 2003, Bailliet et al. 2009) but attains higher frequencies in North America (Bolnick et al. 2006, Malhi et al. 2008). Out of the 17 polymorphisms described for Q by Karafet et al. (2008), all Q* haplogroups so far analyzed for America present the derived states for M242, P36.2, MEH2, and M346 (Karafet et al. 2008; Bailliet et al. 2009; Bisso-Machado et al. 2010; Geppert et al. 2011; Toscanini et al. 2011)), except for one ancient Paleo-Eskimo and one extant Eskimo individuals assignable to Q1a* (Rasmussen et al. 2008, Bisso-Machado et al. 2011).

It should be highlighted that Q1a3* was found in the 9.2% of 885 males from 16 ethnic groups of Siberia and East Asia. The age for this subhaplogroup was estimated in South Siberia at about 4.5±1.5 thousand years ago (Ka), while the divergence time between clade Q1a3* and American-specific haplogroup Q1a3a was equal to 13.8±3.9 Ka, pointing to a relatively recent entry date to America (Malyarchuk *et al.* 2011).

Haplogroup C is present in Asia in variable frequencies. In Mongolia, C3-M217 and C3a-M48) are the benchmark haplogroups, with frequencies of 13% and 46% respectively (Chen *et al.* 2011), while among the Kazakhs those are present in 9% and 57% of the cases (Nasidze *et al.* 2005).

In North America the subhaplogroups C3* and C3b were described with frequencies averaging 5.8% (Bergen *et al.* 1999; Zegura *et al.* 2004; Bolnick *et al.* 2003). Only C3* was found in South America, among 2 Wayuu described by Zegura *et al.* (2004) and 3 Waorani and 1 Kichwa from the Ecuador Amazon region (Geppert *et al.* 2010).

4.2 Allochthonous haplogroups

A-B are almost exclusive to sub-Saharan Africa. While clade A chromosomes occurs with high frequencies of 30-66% in Southern and Eastern African populations and are also present at lower values in North and Central Africa (Hassan *et al.* 2008, Cruciani *et al.* 2002), haplogroup B chromosomes are specially frequent in Central and Western Africa (Hammer *et al.* 2001; Underhill *et al.* 2001; Jobling and Tyler-Smith, 2003).

Our YAP+ chromosomes are most probably members of clade E, widely distributed in Africa and West Eurasia.

Under paragroup F(xK) we have probably detected an assortment of lineages of both European and Middle Eastern/North African origin belonging to haplogroups G, H, I and J, whose presence in Native and cosmopolitan populations of Argentina has already been reported (Corach *et al.* 2010, Blanco-Verea *et al.* 2010).

R is the most frequent haplogroup in Europe (Jobling and Tyler-Smith 2003), and is also the most common haplogroup in Argentinean urban populations (Ramallo *et al.* 2009a, Corach *et al.* 2010).

Contemporary self-acknowledged Native American populations keep bearing an important number of paternal Native lineages. Although this also happens in admixed urban contexts, the populations from Jujuy differ from the rest because of their high Native American contribution. Such component comes from the ethnographic and historical characteristics of Jujuy, since it was one of the most highly populated regions during pre-Columbian times and offered a strong resistance to the Spanish colonization (Hernández, 1992; Pucci, 1998). On the other hand, there is evidence of a lower proportion of admixture probably due to altitude, which may have acted as a barrier or dissuasive effect for the inhabitants of a European origin (Dipierri *et al.* 1997, 1998, 2000).

Our results also show a connection among Mapuche, Huiliche, and Tehuelche, which is possible to be interpreted within a historical context: Mapuche from Argentina and Huiliche from Chile have the same origin, and contact between them is quite well documented (Martínez Sarasola, 1992).

In reference to the foreign haplogroups, it is the first time that AB is described for Argentina; so far the African presence had only been found through YAP+ chromosomes (Bravi *et al.* 2000n).

The third most frequent haplogroup in the Argentine populations studied thus far in our laboratory is F(xK), while K(xQ,R) is a minor haplogroup among South American samples and involves subhaplogroups of Asian origin (Su *et al.* 2000; Hammer *et al.* 2001; Su *et al.* 1999; Underhill *et al.* 2001).

The high frequency of R can be explained by the strong European migration that took place during the late XIX and early decades of the XX centuries, specifically with the arrival of Italian and Spaniard migrants. Something similar was observed for other American countries such as Brazil (Bortolini *et al.* 2003), Mexico (Rangel-Villalobos *et al.* 2008), and the United States (Zegura *et al.* 2004; Bolnick *et al.* 2006). Even the R1b subhaplogroup was described in 11% urban samples from the city of La Plata, Argentina (Bianchi *et al.* 2007).

4.3 Concluding remarks

Native American male lineages found in self-acknowledged Native American populations can also be found in urban contexts, although at lower frequencies. Likewise, even among self-acknowledged Native American populations foreign haplogroups are present, depending on the recent history of human migrations.

The distribution of autochthonous lineages is the result of a complex admixture process that occurred in many Latin American populations. We are currently employing those Native traces to explain other historical events such as the peopling of the Americas, by describing the possible bottlenecks and founder effects that the lineage distribution shows.

5. Acknowledgment

We thank all DNA donors for making this work possible. We are very grateful to Néstor O. Bianchi and Susana Salceda for their help during different stages of our work. Grant sponsors: CONICET, CICPBA, ANPCyT, UNJu, Antorchas Foundation of Argentina. G Bailliet and CM Bravi are members to the "Consejo Nacional de Investigaciones Científicas y Técnicas de la República Argentina" (CONICET).

6. References

Anagnostopoulos, T.; Green, P.M.; Rowley, G; Lewis, C.M. & Giannelli, F. (1999). DNA variation in a 5-Mb region of the X chromosome & estimates of sex-specific/type-specific mutation rates. *American Journal Human Genetics*, Vol.64, No.2, pp.508-517.

Bailliet, G.; Ramallo, V.; Muzzio, M.; García, A.; Santos, M.R.;. Alfaro, E.L.; Dipierri, J.E.; Salceda, S.; Carnese, F.R.; Bravi, C.M.; Bianchi, N.O. & Demarchi, D.A. (2009). Brief Communication: Restricted Geographic Distribution for Y-Q* Paragroup in South America. *American Journal of Physical Anthropology*, Vol.140, pp.578–582.

Ballantyne, K.N.; Goedbloed, M.; Fang, R.; Schaap, O.; Lao, O.; Wollstein, A.; Choi, Y.; van Duijn, K.; Vermeulen, M.; Brauer, S.; Decorte, R.; Poetsch, M.; von Wurmb-Schwark, N.; de Knijff, P.; Labuda, D.; Vézina, H.; Knoblauch, H.; Lessig, R.; Roewer, L.; Ploski, R.; Dobosz, T.; Henke, L.; Henke, J.; Furtado, M.R. & Kayser, M. (2010). Mutability of Y-Chromosomal Microsatellites: Rates, Characteristics, Molecular Bases, and Forensic Implications. *The American Journal of Human Genetics*, Vol.87, pp.341–353.

Bandelt, H.-J.; Forster, P. & Röhl, A. (1999). Median-joining networks for inferring intraspecific phylogenies. *Molecular Biology Evolution*, Vol.16, pp.37-48.

Bergen, A.W.; Wang, C.Y.; Tsai, J.; Jefferson, K.; Dey, C.; Smith, K.D.; Park, S.C.; Tsai, S.J. & Goldman, D. (1999). An Asian-Native American paternal lineage identified by RPS4Y resequencing and by microsatellite haplotyping. *Annual of Human Genetics*, Vol. 63, pp.63–80

Bianchi, N.O.; Bailliet, G.; Bravi, C.M.; Pena, S.D. & Rothhammer, F. (1997). Origin of Amerindian Y-chromosome as inferred by the analysis of six polymorphic markers. *American Journal of Physical Anthropology*, Vol.102, pp.79–89.

Bianchi, N.O.; Catanesi, C.I.; Bailliet, G.; Martinez-Marignac, V.L.; Bravi, C.M.; Vidal-Rioja, L.B.; Herrera, R.J.& Lopez-Camelo, J.S. (1998). Characterization of ancestral and derived Y-chromosome haplotypes of New World native populations. *American Journal of Human Genetics*, Vol.63, pp.862–1871.

Bisso-Machado, R.; Jota, M.S.; Ramallo, V.; Peixão-Côrtes, V.R.; Lacerda, D.R.; Salzano, F.M.; Bonatto, S.L.; Santos, F.R. & Bortolini, M.C. (2011). Distribution of Y-Chromosome Q Lineages in Native Americans. *American Journal of Human Biology*, Vol.23, No.4, pp.563-566.

Blanco-Verea, A.; Jaime, J.C.; Brión, M. & Carracedo, A. (2010). Y-chromosome lineages in native South American population. *Forensic Science International Genetics*, Vol.4, No. 3, pp. 187-93.

Bolnick, D.A.; Bolnick, D.I. & Smith, D.G. (2006). Asymmetric male and female genetic histories among Native Americans from Eastern North America. *Molecular Biology Evolution*,Vol.23, No.11, pp.2161-2174

Bosch, E.; Calafell, F.; Rosser, Z.H.; Nørby, S.; Lynnerup, N.; Hurles, M.E. & Jobling M.A. (2003). High level of male-biased Scandinavian admixture in Greenlandic Inuit shown by Y-chromosomal analysis. *Human Genetics*, Vol.112, No4, pp.353-63.

Bortolini, M.C.; Salzano, F.M.; Thomas, M.G.; Stuart, S.; Nasanen, S.P.K.; Bau, C.H.D.; Hutz, M.H.; Layrisse, Z.; Petzl-Erler, M.L.; Tsuneto, L.T.; Hill,.K.; Hurtado, A.M.; Castro de Guerra, D.; Torres, M.M.; Groot, H.; Michalski, R.; Nymadawa, P.; Bedoya, G.; Bradman, N.; Labuda, D. & Ruiz-Linares, A.(2003). Y-chromosome evidence for differing ancient demographic histories in the Americas. *American Journal Human Genetics*, Vol.73, No.3:524-539.

Bravi, C.M.; Bailliet, G.; Martìnez-Marignac,V.L. & Bianchi, N.O. (2000). Origin of YAP+ lineages of the human Y-chromosome. *American Journal Physical Anthropology*, Vol. 112, No.2.pp.149-158.

Capelli, C.; Wilson, J.F.; Richards, M.; Stumpf, M.P.; Gratrix, F.; Oppenheimer, S.; Underhill, P.; Pascali, V.L.; Ko, T.M. & Goldstein, D.B. (2001). A predominantly indigenous paternal heritage for the Austronesian-speaking peoples of insular Southeast Asia and Oceania. *American Journal of Human Genetics*, Vol.68, pp.432–443.

Chen, Z.; Zhang, Y.; Fan, A.; Zhang, Y.; Wu, Y.; Zhao, Q.; Zhou, Y.; Zhou, C.; Bawudong, M.; Mao, X.; Ma, Y.; Yang, L.; Ding, Y.; Wang, X. & Rao, S. (2011) Brief Communication: Y-chromosome haplogroup Analysis Indicates That Chinese Tuvans Share Distinctive Affinity With Siberian Tuvans. *American Journal of Physical Anthropology*, Vol.144, pp.492–497.

Corach, D.; Lao, O.; Bobillo, C.; van Der Gaag, K.; Zuniga, S.; Vermeulen, M.; van Duijn, K.; Goedbloed, M.; Vallone, P.M.; Parson, W.; de Knijff, P. & Kayser, M. (2010). Inferring continental ancestry of argentineans from Autosomal, Y-chromosomal and mitochondrial DNA. *Annals Human Genetics*, Vol.74, No.1, pp.65-76.

Cruciani, F.; Santolamazza, P.; Shen, P.; Macaulay, V.; Moral, P.; Olckers, A.; Modiano, D.; Holmes, S.; Destro-Bisol, G.; Coia, V.; Wallace, D.C.; Oefner, P.J.; Torroni, A.; Cavalli-Sforza, L.L.; Scozzari, R. & Underhill, P.A. (2002). A Back Migration from Asia to Sub-Saharan Africa Is Supported by High-Resolution Analysis of Human Y-Chromosome Haplotypes". *American Journal of Human Genetics*, Vol. 70, pp.1197–1214.

Dipierri, J.E.; Alfaro, E.; Martinez-Marignac,V.L.; Bailliet, G.; Bravi, C.M.; Cejas, S. & Bianchi, N.O. (1998). Paternal directional mating in two amerindian subpopulations from the northwest of Argentina. *Human Biology*, Vol.70, No.6, pp.1001-1010.

Dipierri, J.E.; Alfaro, E.L.; Peña, J.A.; Jacques, C. & Dogoujon, J.M. (2000). GM, KM inmunoglobulin allotypes and other serum genetic markers (HP, GC, PI and TF) among south American populations living at different altitude (Jujuy province, Argentina) Admixture Estimates. *Human Biology*, Vol.72, No.2, pp.305-319.

Dipierri, J.E.; Alfaro, E.L.; Vullo, C. & Gutiérrez, N. (1997). El polimorfismo HLA en el noroeste argentino. III Jornadas Nacionales de Antropología Biológica, Rosario Argentina. Rosario, Santa Fe, Argentina

Excoffier, L. Smouse, P.E. & Quattro, J.M. (1992). Analysis of molecular variance inferred from metric distances among DNA haplotypes: Application to human mitochondrial DNA restriction data. *Genetics* 131:479-491.

Geppert, M.; Baeta, M.; Nuñez, C.; Martínez-Jarreta, B.; Zweynert, S.; Vacas-Cruz, O.W.; González-Andrade, F.; González-Solorzano, J.; Nagy, M. & Roewer, L. (2011) Hierarchical Y-SNP assay to study the hidden diversity and phylogenetic relationship of native populations in South America. *Forensic Science International: Genetics*, Vol.5, pp.100–104

Hammer, M.F. (1994). Arecent insertion of an Alu element on the Y chromosome is a useful marker for human population studies. *Molecular Biology Evolution*, Vol.11, pp.749–761.

Hammer, M.F.; Karafet, T.M.; Redd, A.J.; Jarjanazi, H.; Santachiara-Benerecetti, S.; Soodyall, H. & Zegura, S.L. (2001). Hierarchical patterns of global human Y-chromosome diversity. *Molecular Biology Evolution*, Vol.18, pp.1189–1203

Hammer, M.F.; Redd, A.J.; Wood, E.T.; Bonner, M.R.; Jarjanazi, H.; Karafet, T.; Santachiara-Benerecetti, S.; Oppenheim, A.; Jobling, M.A.; Jenkins, T.; Ostrer, H. & Bonne-Tamir, B. (2000). Jewish and Middle Eastern non-Jewish populations share a common pool of Y-chromosome biallelic haplotypes. *Proceeding of National Academic of Science*, Vol.97, pp. 6769–6774.

Hassan, H.Y.; Underhill, P.A.; Cavalli-Sforza, L.L. & Ibrahim, M.E. (2008). Y-chromosome variation among Sudanese: restricted gene flow, concordance with language, geography, and history. *American Physical Anthropology*, Vol.137, No.3, pp.316–23

Hernandez, I. (1992). Los indios de la Argentina. Colecciones Mapfre 1492, Madrid.

Hurles, M.E.; Veitia, R.; Arroyo, E.; Armenteros, M.; Bertranpetit, J.; Perez-Lezaun, A.; Bosch, E.; Shlumukova, M.; Cambon-Thomsen, A.; McElreavey, K.; Lopez De Munain, A.; Rohl, A.; Wilson, I.J.; Singh, L.; Pandya, A.; Santos, F.R.; Tyler-Smith, C. & Jobling, M.A. (1999). Recent male-mediated gene flow over a linguistic barrier in Iberia, suggested by analysis of a Y-chromosomal DNA polymorphism. *American Journal of Human Genetic*, Vol.65, No.5, pp.1437-48.

Jobling, M.A. & Gill, P. (2004). Encoded evidence: DNA in forensic analysis. *Natural Review Genetics*, Vol.5, No.10, pp.739-51.

Jobling, M.A. & Tyler-Smith, C. (2003). The human Y chromosome: An evolutionary marker comes of age. *National Review of Genetics*, Vol.4, pp.598–612.

Jota, M.S.; Lacerda, D.R.; Sandoval, JR.; Vieira, P.P.R.; Santos-Lopes, S.S.; Bisso-Machado, R.; Paixão-Cortes, V.R.; Revollo, S.; Paz-Y-Miño, C.; Fujita, R.; Salzano, F.M.; Bonatto, S.L.; Bortolini, M.C. and Santos, F.R. & The Genographic Consortium. (2011). A new subhaplogroup of native American Y-Chromosomes from the Andes. *American Journal of Physical Anthropology*, Article first published online: 13 SEP| DOI: 10.1002/ajpa.21519

Karafet, T.M.; Mendez, F.L.; Meilerman, M.B.; Underhill, P.A.; Zegura, S.L. & Hammer, M.F. (2008). New binary polymorphisms reshape and increase resolution of the human Y chromosomal haplogroup tree. *Genome Research*, Vol.18, No5, pp.:830-8.

Karafet, T.M.; Zegura, S.L.; Posukh, O.; Osipova, L.; Bergen, A.; Long, J.; Goldman, D.; Klitz, W.; Harihara, S.; de Knijff, P.; Wiebe, V. & Griffiths, R.C.,Templeton, A.R. y Hammer, M.F. (1999). Ancestral Asian Source(s) of New World Y-Chromosome Founder Haplotypes. *American Journal of Human Genetics*, Vol.64, pp.817–831,

Karafet, T.M.; Osipova, L.P.; Gubina, M.A.; Posukh, O.L.; Zegura, S.L. & Hammer, M.F. (2002). High levels of Y-chromosome differentiation among native Siberian populations and the genetic signature of a boreal hunter-gatherer way of life. *Human Biology*, Vol.74, pp.761–789.

Kayser, M.; Caglia, A.; Corach, D.; Fretwell, N.; Gehrig, C.; Graziosi, G.; Heidorn, F.; Herrmann S.; Herzog B.; Hidding M.; Honda K.; Jobling M.; Krawczak M.; Leim K.; Meuser S.; Meyer E.; Oesterreich W.; Pandya A.; Parson W.; Penacino G.; Perez-

Lezaun A.; Piccinini A.; Prinz M.; Schmitt C.; Schneider P. M.; Szibor R.; Teifel-Greding J.; Weichhold G.; de Knijff P & Roewer L. (1997). Evaluation of Y-chromosomal STRs: a multicenter study. *International Journal of Legal Medicine*, Vol.110, pp.125-133

Lell, J.T.; Sukernik, R.I.; Starikovskaya, Y.B.; Su, B.; Jin, L.; Schurr, T.G.; Underhill, P.A. & Wallace, D.C. (2002). The dual origin and Siberian affinities of Native American Y chromosomes. *American Journal Human Genetics*, Vol.70, pp.192-206.

Malyarchuk, B.; Derenko, M.; Denisova, G.; Maksimov, A.; Wozniak, M.; Grzybowski, T.; Dambueva, I. & Zakharov, I. (2011). Ancient links between Siberians and Native Americans revealed by subtyping the Y chromosome haplogroup Q1a. *Journal of Human Genetics*, Vol.1, pp6

Malhi, R.S.; Gonzalez-Oliver, A.; Schroeder, K.B.; Kemp, B.M. ; Greenberg, J.A.; Dobrowski, S.Z.; Smith, D.G.; Resendez, A.; Karafet, T; Hammer, M. Zegura, S. & Brovko, T. (2008). Distribution of Y chromosomes among Native North Americans: A study of Athapaskan population history. *American Journal of Physical Anthropology*, Vol 137, No 4, pp. 412-424.

Martinez Sarasola, C.(1992). In: EMECE (ed.): Nuestros paisanos los indios. Vida, historia y destino de las comunidades indígenas de la Argentina.

Mazières, S.; Guitard, E.; Crubézy, E.; Dugoujon, J.M.; Bortolini, M.C.; Bonatto, S.L.; Hutz, M.H.; Bois, E.; Tiouka, F.; Larrouy, G.& Salzano FM. (2008) Uniparental (mtDNA, Y-chromosome) polymorphisms in French Guiana and two related populations--implications for the region's colonization. Annual of Human Genetics, Vol 72, No1, pp.145-56.

Muzzio, M.; Muzzio, J.C.; Bravi, C.M. & Bailliet, G. (2010). Technical note: A method for assignment of the weight of characters. *American Journal Physical Anthropology*, Vol.143, No.3, pp.488-92.

Nasidze, I.; Quinque, D.; Dupanloup, I.; Cordaux, R.; Kokshunova, L. & Stoneking, M. (2005). Genetic evidence for the Mongolian ancestry of Kalmyks. *American Journal of Physical Anthropology*, Vol.128, pp.846-854.

Nei, M. (1972). Genetic distance between populations. American Naturalist, Vol.106, pp.283-292.

Pascali,V.L.; Dobosz, M. & Brinkmann, B. 1999. Coordinating Y-chromosomal STR research for the Courts. *International Journal of Legal Medicine*, Vol.112, No.1, pp.1.

Pena, S.D.J.; Santos, F.R.; Bianchi, N.O.; Bravi, C.M.; Carnese, F.R.; Rothhammer, F.; Gerelsaikhan, T.; Munkhtuja, B. & Oyunsuren, T. (1995). A major founder Y-chromosome haplotype in Amerindians. *Nature Genetics*, Vol.11, pp.15.

Pucci, R.(1998). El tamaño de la población aborigen del Tucumán en la época de la conquista: balance de un problema y propuesta de una estimación. *Población y Sociedad*, Vol.5, pp.239-270.

Ramallo,V.; Mucci, J.M.; García, A.; Muzzio, M.; Motti, J.M.B.; Santos, M.R.; Perez, M.E.; Alfaro, E.L.; Dipierri, J.E.; Demarchi, D.A.; Bravi, C.M. & Bailliet,G. (2009a). Comparison of Y chromosome haplogroup frequencies in 8 Provinces of Argentina. Forensic Science International: Genetics Supplement Series, Vol.2, pp.431–432

Ramallo, V.; Muzzio, M.; Motti, J.B.M.; Salceda, S. & Bailliet, G. (2009b). Linajes masculinos y su diversidad en comunidades wichi de Formosa. *Revista del Museo de Antropología*, Vol.2, pp.67-74

Rangel-Villalobos, H.; Muñoz-Valle, J.F.; González-Martín, A.; Gorostiza,A.; Magaña, M.T. & Páez-Riberos, L.A. (2008). Genetic admixture, relatedness, and structure patterns among Mexican populations revealed by the Y-chromosome. American *Journal Physical Anthropology*, Vol.135, No.4, pp.448-461.

Roewer, L.; Arnemann, J.; Spurr, N.K.; Grzeschik, K.-H. & Epplen, J.T. (1992). Simple repeat sequences on the human Y chromosome are equally polymorphic as their autosomal counterparts. *Human Genetics*, Vol.89, pp.389–394.

Ruiz-Linares, A.; Ortiz-Barrientos, D.; Figueroa, M.; Mesa, N.; Munera, J.G.; Bedoya, G.; Velez, I.D.; Garcia, L.F.; Perez-Lezaun, A.; Bertranpetit, J.; Feldman, M.W. & Goldstein, D.B. (1999). Microsatellites provide evidence for Y chromosome diversity among the founders of the New World. *Proceedings National Academy Science USA*, Vol.96, pp.6312–6317.

Santos, F.R.; Pena, S.D.J.; & Tyler-Smith, C. (1995b). PCR haplotypes for the human Y chromosome based on alphoid satellite variants and heteroduplex analysis. *Gene*, Vol.165, pp191–198.

Santos, F.R.; Bianchi, N.O.; & Pena, S.D.J. (1996 a). Worldwide distribution of human Y chromosome haplotypes. *Genome Research*, Vol.6, pp.601–611.

Santos, F.R.; Rodríguez-Delfin, L.; Pena, S.D.J.; Moore, J.; & Weiss, K.M . (1996b). North and South Amerindians may have the same major founder Y chromosome haplotype. *American Journal of Human Genetics*, Vol.58, pp.1369–1370.

Schneider, S.; Roessli, D. & Excoffier, L. (2000). Arlequin ver.2.000: A software for population genetics data analysis. Genetics and Biometry Laboratory. Unversity of Geneva. Switzerland.

Seielstad, M.T.; Hebert, J.M.; Lin, A.A.; Underhill, P.A.; Ibrahim, M.; Vollrath, D.; Cavalli-Sforza, L.L.. (1994). Construction of human Y-chromosomal haplotypes using a new polymorphic A to G transition. *Human Molecular Genetics*, Vol.3,No.12, pp.2159-2161.

Seielstad, M.; Yuldasheva, N.; Singh, N.; Underhill, P.; Oefner, P.; Shen, P. &; Wells, R.S. (2003). A novel Y-chromosome variant puts an upper limit on the timing of first entry into the Americas. *American Journal Human Genetics*, Vol.73, No.3, pp.700-705.

Sengupta, S.; Zhivotovsky, L.A.; King, R.; Mehdi, S.Q.; Edmonds, C.A.; Chow, C.E.; Lin, A.A.; Mitra, M.; Sil, S.K.; Ramesh, A.; Usha Rani, M.V.; Thakur, C.M.; Cavalli-Sforza, L.L.; Majumder, P.P. & Underhill, P.A. (2006). Polarity and temporality of high-resolution Y- chromosome distributions in India identify both indigenous and exogenous expansions and reveal minor genetic influence of Central Asian pastoralists. *American Journal Human Genetics*, Vol.78, pp.202–221.

Shen, P.; Wang, F.; Underhill, P.A.; Franco, C.; Yang, W.H.; Roxas, A.; Sung, R.; Lin, A.A.; Hyman, R.W.; Vollrath, D.; Davis, R.W.; Cavalli-Sforza, L.L. & Oefner, P.J. (2000). Population genetic implications from sequence variation in four Y chromosome genes. *Proceeding of National Academy of Science USA*, Vol.97, pp.7354–7359.

Su, B.; Xiao, J.; Underhill, P.; Deka, R.; Zhang, W.; Akey, J.; Huang, W.; Shen, D.; Lu,.D.; Luo, J.; Chu, J.; Tan, J.; Shen,.P.; Davis,.R.; Cavalli-Sforza, L.; Chakraborty, R.; Xiong, M.; Du, R.; Oefner, P.; Chen, Z. & Jin, L. (1999). Y-chromosome evidence for a northward migration of modern humans into Eastern Asia during the last Ice Age. *American Journal Human Genetics*, Vol.65, pp.1718–1724.

Su, B.; Jin, L.; Underhill, P.; Martinson, J.; Saha, N.; McGarvey, S.T.; Shriver, M.D.; Chu, J.; Oefner, P.; Chakraborty, R.; Deka, R.. (2000). Polynesian origins: Insights from the Y chromosome. *Proceeding of National Academy of Science USA*, Vol.97, pp.8225–8228.

Toscanini, U.; Gusmão, L.; Berardi, G.; Gomes, V.; Amorim, A.; Salas, A. & Raimondi, E. (2011). Male lineages in South American native groups: Evidence of M19 traveling south. *American Journal Physical Anthropology.* Aug 8. doi: 10.1002/ajpa.21562. [Epub ahead of print]

Underhill, P.A.; Jin, L.; Lin, A.A.; Mehdi, S.Q.; Jenkins, T.; Vollrath, D.; Davis, R.W.; Cavalli-Sforza, L.L. & Oefner, P.J. (1997). Detection of numerous Y chromosome biallelic polymorphisms by denaturing high-performance liquid chromatography. *Genome Research*, Vol.7, No10, pp.996-1005.

Underhill, P. A.; Jin, L.; Zemans, R.; Oefner, P. J. & Cavalli-Sforza, L. L. (1996). A pre-Columbian Y chromosome-specific transition and its implications for human evolutionary history. *Proceeding of National Academy of Science USA*, Vol.93, pp.196-200.

Underhill, .PA.; Shen, P.; Lin, A.A.; Jin, L.; Passarino, G.; Yang, W.H.; Kauffman, E.; Bonne-Tamir, B.; Bertranpetit, J.; Francalacci, P.; Ibrahim, M; Jenkins, T.; Kidd, J.R.; Mehdi, S.Q.; Seielstad, M.T.; Wells, R.S.; Piazza, A.; Davis, R.W.; Feldman, M.W.; Cavalli-Sforza, L.L. & Oefner, P.J. (2000). Y chromosome sequence variation and the history of human populations. *Nature Genetics*, Vol.26, pp.358-361.

Underhill, P.A.; Passarino, G.; Lin, A.A.; Shen, P.; Mirazon, L.M.; Foley, R.A.; Oefner, P.J.; Cavalli-Sforza, L.L. (2001). The phylogeography of Y chromosome binary haplotypes and the origins of modern human populations. *Annual Human Genetics*, Vol.65, pp.43–62.

Y Chromosome Consortium. (2002). A nomenclature system for the tree of human Y-chromosomal binary haplogroups. *Genome Research*, Vol.12, pp.339–348.

Zegura, S.L.; Karafet, T.M.; Zhivotovsky, L.A. & Hammer, M.F. (2004). High-resolution SNPs and microsatellite haplotypes point to a single, recent entry of Native American Y chromosomes into the Americas. *Molecular Biology Evolution*, Vol.21, pp. 164–175.

13

Genetic Diversity and the Human Immunodeficiency Virus Type-1: Implications and Impact

Orville Heslop
University of the West Indies,
Jamaica

1. Introduction

1.1 Overview

Genetic diversity is the tendency of individual genetic characteristics in a population to vary from one another or the potential of a genotype to change or deviate when exposed to environmental or genetic factors (Dale & Park, 2010). The extent to which the trait or the genotype tends to vary describes genetic variability and at the molecular level variability is measured by determining the rate of mutation. Genetic diversity however measures the number of the actual variation of species in a population (Dale & Park, 2010).

The scope of references to the numerous microbes which demonstrate both genetic diversity and variability within their genomes are beyond the limits of this chapter. Our focus is therefore confined generally to those species which would best explain the nature, mode and factors which influence mutational changes which can be measured.

1.2 Factors influencing diversity

The genetic response of micro-organisms to host and various environmental selective pressures may vary differentially among species. The outcome characterized by the influence of these factors often result in minor or major mutational changes and host adaptability.

Organisms of one genotype such as mumps, rubella and the measles virus do not demonstrate the type variability or diversity as the influenza, human immunodeficiency virus (HIV) and hepatitis viruses. Consequently the predictability of the epidemiology of highly genetically diversified agents is often more challenging. The etiology of an influenza outbreak is invariably the result of genetic re-assortment of different influenza strains. Metzulan et al., (2004) report that genetically distinct multiple viruses can combine to cause an influenza B epidemic in a community and that the frequent re-assortment among these viruses play a role in generating the genetic diversity of influenza B viruses (Metzulan et al., (2004). There are several factors which influence diversity and are dependent on the species involved, the host with various immunological selective pressures, environmental pressures as well as treatment intervention resulting in development of resistance genes to antibacterial, antiviral, antifungal and antiprotozoal agents. Resistance to several antimicrobials used as therapeutic agents relative to the various and specific categories of microorganisms have been well documented (Murray et al., (2007). The human immunodeficiency virus is demonstrably the most genetically diverse microorganism and is an important reference point in understanding and appreciating the impact of genetic diversity. Whatever the resulting impact of mutational changes, it is important to determine how these changes impact the definitive host, their influence on transmission of the infecting agent and whether these changes are measurable and can be used as an epidemiological tool in clinical and laboratory diagnosis.

1.3 Measuring genetic diversity

The etiological agent in an outbreak can be determined based on phenotypic characteristics which are the observable properties of the organism measured under controlled laboratory settings. Genetic characteristics are, however, more difficult to measure employing various molecular methods and highly trained individuals. The major difference between phenotypic (physiological) and genetic characteristics, is observed in altered physiological changes as opposed to properties of micro-organisms which are genetically inherited. The importance of genotyping as an epidemiological tool in determining mutant genes in drug resistance cannot be overemphasized. This involves the genome of micro-organisms and is the focal point for single or multiple mutational changes.

Behavioural changes of micro-organisms are the results of mutational changes in genes and consequently are responsible for the various features exhibited by these organisms. Included in the changing behavioral patterns of microorganisms and in particular bacteria, may range from differences in biochemical reactions, growth requirements, tolerance to various temperatures and resistance to therapeutic agents. These changes may be demonstrated under controlled laboratory conditions by assays and specific methods designed to determine phenotypic and genotypic characteristics.

2. Background of the problem

The first cases of acquired immunodeficiency syndrome (AIDS) were recognized in the USA during 1981 (Morbidity Mortality Weekly., 1981). By 1983 the human immunodeficiency virus - type 1 (HIV-1) had been identified and was later confirmed to be the aetiologic agent of AIDS (Barré-Sinoussi et al., 1983). At the end of 2000 an estimated 60 million cases of human immunodeficiency virus –type 1 (HIV-1) infection were reported globally (Barré-Sinoussi, 1983; Piot, 2001). The extensive genetic variation which characterizes HIV-1 is an important feature of the HIV/AIDS epidemic (Robertson et al., 2000). The genetic variability of HIV-1 is generated by the lack of proof reading ability of the reverse transcriptase, the rapid turn-over of HIV-1 *in vivo*, host selective immune pressures and recombination events during viral replication (Heyndrickx, 2000; Ho, 1995; Leitner, 2005; Michael, 1995; Tatt, 2001). This has several public health implications although the biological significance is not known (Loussert-Ajaka, 1998; Tatt, 2001). The HIV-1 subtypes vary in the geographical distribution of the virus (Heyndrickx, 2000; Tatt, 2001). The differences in HIV-1 subtypes and geographic diversity impact the development of efficacious vaccines, diagnostic testing and antiviral therapy (Borrow, 1997; Loussert-Ajaka, 1998; Piot, 2001). On the other hand, the genetic hetero-geneity in HIV may aid the surveillance of transmission patterns within the epidemic (Robertson et al., 2000).

The World Health Organization reports that the economies of Caribbean countries, which rely heavily on tourism components are substantially affected by the HIV epidemic (UNAIDS/ WHO, 2004). In addition, most countries in the Caribbean are limited in their capacity to control and track the evolution of the epidemic. It is therefore necessary to strategize surveillance programmes which include the determination of existing HIV-1 subtypes to understand the changing nature of the epidemic and measures to control spread of the virus in the Caribbean (d'Cruz-Grote, 1996; Delwart, 1995).

In Jamaica, the first case of AIDS was reported in 1983 and the number of reported cases increased to over 12,000 by 2007 (Ministry of Health, Jamaica, 2007). Being a tourist destination increases Jamaica's potential for introduction of different HIV-1 subtypes and circulating recombinant forms (CRFs) of the virus into the Jamaican population. The aim of this study is to determine the HIV-1 subtypes in HIV-1 infected persons and the molecular epidemiology of the virus in Jamaica. The main objectives are to:

i.   determine current subtypes and recombinant forms of HIV-1
ii.   investigate the relationships between HIV-1 subtype strains in terms of transmission and disease progression in Jamaica
iii. evaluate the heteroduplex mobility assay (HMA) as a sub typing method in the HIV/AIDS epidemic in Jamaica

3. Literature review

3.1 Human Immunodeficiency Virus - type 1 (HIV-1)

Human immunodeficiency virus - type 1 (HIV-1), a member of the genus *Lentivirinae*, and family Retroviridae is closely related to its primate retrovirus HIV-2 and is the cause of acquired immunodeficiency syndrome (AIDS) (Chiu, 1985; Peeters, 2000). While HIV-1 is responsible for the vast majority of cases in the AIDS pandemic, HIV-2 is almost exclusively confined to Africa and, to a large extent, is responsible for the epidemics seen in that region (Peeters, 2000).

HIV-1 is divided into three genetic groups: group M (major or main), group O (outlier) and N (new or non-M, non-O) group M being responsible for the AIDS pandemic accounting for over 90% of HIV infections worldwide (Hemelaar et al., 2006). Nine pure subtypes of HIV-1 group M are currently known (A-D, F- H, J and K). These subtypes are further into sub-subtypes such as F (F1 and F2 and A (A1, A2 and A3). Subtypes and sub-subtypes may form additional mosaic forms when individuals are dually or multiply- infected by different strains of HIV-1 giving rise to known recombinant forms (Burke, 1997) - CRFs, - drug resistance differences). To date, over 40 CRFs are recognized (Taylor et al., 2008).

3.2 Morphology and genomic organization of human immunodeficiency virus -type 1 (HIV-1)

Structure of HIV-1 A schematic presentation of the structure of the human immunodeficiency virus - type 1 (HIV-1) as described by Hahn, (1994 - 42) is shown in

Figure 1. Like the other human retroviruses the mature virion of HIV comprises icosahedral cores containing the RNA genome, the RT enzyme and gag proteins surrounded by an envelope which is acquired as the virion buds through the host cell membrane (Folks & Khabbaz, 1998). Morphologically HIV-1 differs from HTLV-1 with respect to the nucleoid, the dense inner part of the core, and the pronounced glycoprotein spikes of the envelope. However the overall diameter of the mature particles of both viruses is 100-120 nm in diameter (Folks, 1998; Hahn, 1994). The viral core contains the viral genome, which comprise two identical copies of single stranded positive sense RNA, each of which encodes the complete structural, enzymatic and regulatory proteins (Hahn, 1994).

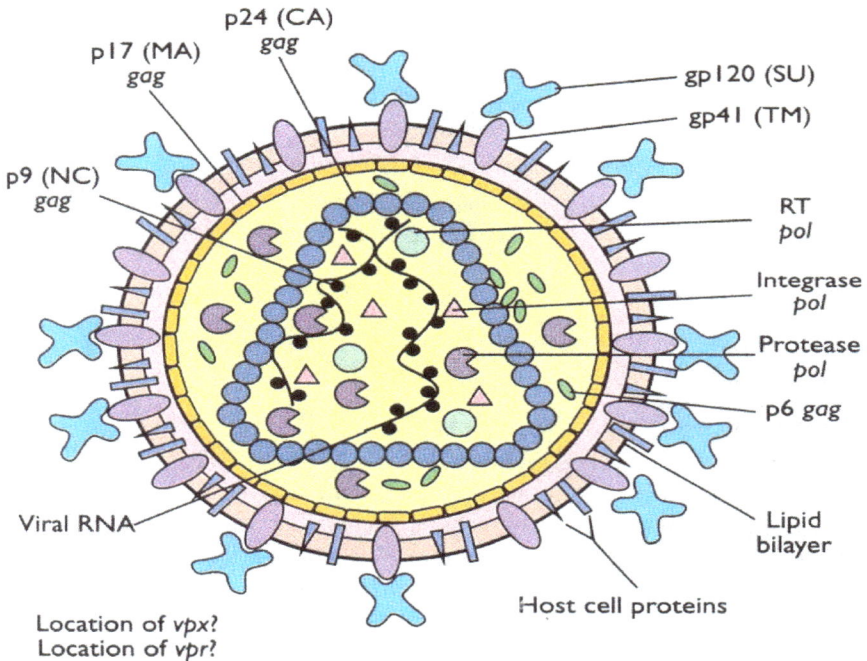

Fig. 1. Human immunodeficiency virus - type 1 (HIV-1) Adapted: Hahn et al (Hahn, 1994)

The structural proteins that form the nucleocapsid and matrix shell are induced by the *gag* gene. The nonstructural gene products of the HIV-1 *pol* gene region include the reverse transcriptase (RT), integrase and protease (Folks, 1994; Guatelli, 2002; Hahn, 1994). The lipid envelope encoded by the *env* gene contains structural proteins with a surface domain (SU, gp120) and transmembrane domain (TM, gp41) presented as trimeric spikes on the surface of the virion (Folks, 1994; Guatelli, 2002; Hahn, 1994; Kwong, 1998). The viral envelope contains conserved (C) and variable (V) regions, is poorly immunogenic, and contains binding sites for CD4 T cells on gp120 and chemokine co-receptors (Kwong, 1998; Wyatt, 1998a; 1998b). These, so-called, co-receptors are masked by variable loops V2 and V3 of gp120 until the CD4 molecule is engaged (Rizzuto, 1998).

3.3 Genome organization of HIV-1

The complete sequence of HIV-1 contains a 9.2 kb genome (Folks, 1994). The genomic map of the HIV-1 viral genome is well documented and is shown in Figure 2. (Hahn, 1994). In keeping with the general structure of primate lentiviruses, the HIV-1 genome comprises 9 genes, including 3 structural, *gag*, polymerase (*pol*), envelope (*env*), and 6 nonstructural, *tat, rev, vip, nef, vpu* and *vpr* genes which are flanked by long terminal repeat (LTR) sequences. The HIV-1 LTR is approximately 640 bp long and is segmented into the U3, R, and U5 regions. The genes of HIV-1, corresponding proteins and functions have been reviewed, as shown in Table 1. Guatelli et al (2002).

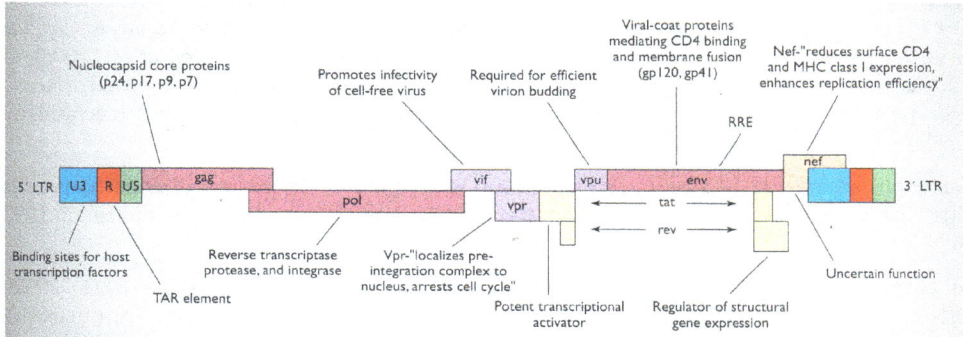

Fig. 2. Genome arrangement/map of the HIV-1 genome - adapted: Hahn et al (1994)

3.4 HIV-1 diversity and molecular epidemiology

3.4.1 HIV-1 diversity

The genetic diversity of HIV-1 has resulted in differences among the subtypes in the LTR sequences, transcriptional promoters, accessory and regulatory genes such as *nef, tat, rev* and *vpu* (Geretti, 2002). The basis of genetic diversity of HIV-1 subtypes is, to a large extent, the consequence of a high mismatch error rate in the reverse transcriptase (RT). In addition, there is an absence of exonuclease proof reading activity (Spira et al., 2003). Other factors which contribute to genetic diversity include the replicative rate of each viral subtype, mutational changes arising in each replicative cycle, genomic recombination and viral fitness (Robertson, 1995; Spira, 2003). Evolutionary changes in the genome may result from host, environmental and other selective pressures (Hu, 2005; Piot, 2002; Quinones-Mateu, 1999; Renjifo, 2002).

In comparison to the *env* gene the *pol* and the *gag* regions are much more intolerant to mutation primarily because they encode core protein sequences which are relatively inflexible. Genetic changes in *pol* and *gag* may give rise to drug resistance as some antiretroviral drugs are directed against the RT and protease proteins encoded by these genes (Spira et al., 2003). For example, 6.8%–10% variation has been shown in the RT of clade B isolates from Ethiopia whereas, intra-clade differences of 3.5–5% have been reported in isolates from Africa, India and South America (Loemba, 2002; Loemba, 2002a; 2002b). Differences in the degree of variation among subtypes also have been observed in LTR sequences and in the transcriptional promoters involved in HIV replication (Montano, 1997).

Other variations observed among HIV-1 subtypes include 14.4%–23.8% differences in the sequences of viral regulatory protein, *nef* (Jubier-Maurin et al., 1999). Subtype C contains a unique *rev* protein and an enlarged *vpu* product while subtype D expresses a *tat* protein with a C-terminus deletion (Gao & Robertson, 1998). Also a significantly shorter envelope V1-V2 loop sequences and fewer potential N-linked glycosylation sites have been shown in subtypes C and A compared to subtype B (Chohan et al., 2005).

Differences in co-receptor usage and syncytia inducing capacity between subtypes have also been reported (Cocchi, 1995; Deng, 1996; Fenyö, 1998; Tersmette, 1988; Worgall, 1999). For example some subtypes may exhibit enhanced susceptibility to CCR5 (R5) inhibition (Marozsan et al., 2005). Subtype B viruses for example appear as R5 viruses in early infection and emerge as X4 viruses in advanced disease while recombinant subtype BG strains appear to be X4 viruses regardless of the clinical stage of disease (Pérez-Alvarez et al., 2006). In contrast subtypes A and C appear infrequently as X4 viruses, even in advanced disease. Subtype C viruses however, are able to switch from CCR5 to CXCR4 using mechanisms similar to those used by subtype B viruses (Pollakis et al., 2004).

Variations among HIV-1 strains also have been seen in the ability to infect Langerhans cells. This tropism has been linked to vaginal/cervical transmission of the virus (Dittmar, 1997; Hu, 1999).

Subtype diversity in response to antiretroviral drugs has been reported (Loemba 2002a, 2002b; Miller, 2007; Quinones-Matue, 1998; Velazquez-Campoy, 2002). Changes in nucleotide sequences play an important role in conferring resistance to each antiretroviral drug or even entire classes of nucleoside and non-nucleoside reverse transcriptase inhibitors (NRTI and NNRTI) and protease inhibitors (PI) (Velazquez-Campoy, 2002; Quinones-Matue, 1998; Tantillo, 1994). While the majority of the reports on resistance to antiretroviral drugs and subtype diversity have focused on subtype B it has been noted that Y181C and Y1811 mutations, respectively can render HIV-1 group O and all HIV-2 strains resistant to the entire NNRTI class of drugs (Leitner, 2005; McCutchan, 2004; Papuashvili, 2005;Tantillo, 1994).

3.4.2 Global distribution of HIV-1 subtypes, sub-subtypes, circulating recombinant forms (CRFs), and unique recombinant forms (URFs)

Majority of the subtypes, subtypes and several CRFs are found in Africa (Geretti, 2002; Leitner, 2005; McCutchan, 2004; Yang, 2005). The phylogenetic classification of HIV-1 subtypes including sub-subtypes, CRFs and URFs are well documented (Tables 1.) (Taylor, 2008) The profile of HIV-1 subtypes fluctuates in some countries. For example, in 1996 the dominant subtypes in Eastern Europe were subtypes B in homosexuals, C in heterosexuals and G in nosocomially infected patients but by 2003 subtype A1 accounted for more than 90% of infections (Papuashvili et al., 2005). In Western Europe subtype B initially dominated the epidemic but the prevalence of non-subtype B has been noted to be steadily increasing (deMendoza, 2005; Monno, 2005; Taylor, 2005). Non-B subtypes were also noted to be increasing in Canada (Akouamba et al., 2005). Emerging recombinant forms are presently seen in several regions including sub-subtypes B/CRF_O1AE, A/CRFO2_AG and A/CRFO6-cpx seen in South East Asia, Uzbekistan and Estonia, respectively (Carr, 2005; Tovanabutra, 2004; Zetterberg, 2004;). A new CRF designated CRF18_cpx was recently isolated in Cuba (Thomson et al., 2005). With the exception of Cuba and Jamaica, Caribbean islands have reported subtype B as virtually the only subtype present in the other regions of the Caribbean (Cleghorn, 2000; Gittens, 2003, Heslop, 2009; Thomson, 2005; Vaughan, 2003). Five non-B subtypes including A, C, D, E and J have been reported in Jamaica (Heslop et al., 2009). It is important to note that the predominant subtype B in the United States and other countries is associated with a homosexual epidemic, while its association with a heterosexual epidemic has been reported only in the Caribbean (Avila et al., 2002).

Classification	Definition	Examples
Subtypes or clades	Genetically related HIV-1 strains that are essentially phylogenetically equidistant, generating a starlike, rather than a treelike, phylogeny	Subtypes A, B, C, D, F, G, H, J, and K are currently known; A through D are highly prevalent, others have low prevalence and limited geographic distributions
Sub-subtypes	Distinct lineages within a subtype; genetic distance between sub-subtypes is smaller than that between subtypes	Subtypes A and F are subdivided into sub-subtypes A1 through A4 and F1 and F2, respectively; mostly these circulate in Central and West Africa
Intersubtype recombinant forms	Mosaic strains with segments from two or more subtypes alternating across the genome	Common in mixed-subtype epidemics; thought to result from infection of a person with more than one HIV-1 subtype
Circulating recombinant forms	Specific recombinant forms that are spreading in a population; new forms are defined when three people without direct epidemiologic linkage are found to be infected; the assigned name reflects sequence of discovery and subtype composition, with "cpx" indicating forms containing three or more subtypes	Currently, 43 forms are described; CRF01_AE and CRF02_AG are found principally in Southeast Asia and West Africa, respectively; others have more limited distributions
Unique recombinant forms	Intersubtype recombinant forms recovered from only a single person	Hundreds of forms have been described on the basis of partial or complete genome sequences; their potential for epidemic spread is unknown
Geographically distinct lineages	Lineages, often country-specific, that are distinguishable phylogenetically; unlike sub-subtypes, they are not phylogenetically equidistant within subtypes	Thai B, Indian C, West vs. East African D, and Former Soviet Union A (FSU-A)

Adapted from Taylor et al (2008).

Table 1. Phylogenetic Classification and global distribution of HIV-1

3.5 Molecular detection of HIV-1 infection and determination of diversification among the virus subtypes

3.5.1 Molecular detection of HIV-1 infection

The HIV-1 DNA polymerase chain reaction (PCR) assay which detects HIV proviral DNA in PBMC and HIV-1 reverse transcriptase polymerase chain reaction (RT-PCR) assay for detection of HIV-1 RNA virions in plasma are highly sensitive and independent of the host responses (Diagnostic Tests, 1998; Eisenstein, 1990). The nucleic acid amplification methods involve exponential amplification of viral nucleic acid sequences allowing detection of low numbers of HIV infected lymphocytes in patients (Schochetman, 1989).

The PCR assays enable the resolution of early HIV infection before seroconversion and have been particularly useful in the early identification of HIV infected infants born to HIV seropositive women where antibody tests may be ineffective due to passive transfer of maternal antibodies (Sheppard, 1993).

3.5.2 Human Immunodeficiency Virus -type 1 subtype determination

Several techniques for HIV-1 subtype determination have been described (Myers, 1995; Ou Cy, 1992). These comprise molecular based methods including genetic sequencing, probe hybridization assay, restriction length polymorphism (RFLP) analysis, subtype-specific PCR, combinatorial melting assay, heteroduplex mobility assay (HMA) and a phenotypic method of serotyping (Cornelissen, 2007; Delwart, 1995; Luo,1998; Murphy, 1999; Peeters, 1998; Robbins, 1995; van Harmelen, 1999). Of these methods HMA has been shown to be reliable, rapid and inexpensive (Delwart, 1995; Loussert-Ajaka, 1998; van der Auwera, 2000). Currently the two most widely used methods for HIV-1 subtype determination are genetic sequencing and phylogenetic analysis and HMA. Molecular analysis has been used extensively in the HIV/AIDS pandemic to determine transmission patterns of HIV-1 (Leitner et al, 1999). Recently routine HIV-1 genotyping has been proposed as a tool to identify dual HIV-1 infections (Cornelissen et al, 2007).

3.5.3 DNA sequencing and phylogenetic analysis

Sequencing and phylogenetic analysis are still the most accurate approach for characterizing viral genomes in identifying new subtypes, sub-subtypes, recombinant, complex and unique forms of HIV-1 and are well documented (Leitner, 2005; Loussert-Ajaka, 1998; Tatt, 2001). These methodologies have been important tools in determining sequence relatedness, mutational changes relating to viral ancestry, the relative time of viral introduction, mutations associated with antiretroviral drug resistance, viral transmission patterns and other epidemiological information (Leitner et al., 1998). The definitive method of subtyping HIV-1 is the genetic sequencing of the envelope (*env*), group antigen (*gag*) or polymerase (*pol*) The main advantages of this method are that it allows the determination by phylogenetic analysis of the relatedness among HIV sequences, the study of the evolution of HIV quasispecies in different hosts linked to a common source of infection, and the study of virus transmission patterns among different risk groups (Diaz, 1997; Holmes, 1995; Leitner, 1999). However genetic sequencing is expensive, time consuming and requires highly qualified personnel (Hu, 2005; Leitner, 2005; Loussert-Ajaka, 1998; Tatt, 2001). In addition phylogenetic methods have been widely used to investigate alleged person to person transmission of the virus, for example, from health care professionals to their patients (Holmes, 1993; Ou, 1992). These molecular techniques have been useful in confirming transmissions in household settings, in forensic investigations including sexual offences, and cases of intentional HIV infection (Albert, 1994; APRI newsletter, 2001; Morb Mortal Wkly Report, 1994).

3.5.4 Heteroduplex mobility assay (HMA)

The heteroduplex mobility assay (HMA) has been shown to be a reliable and standardizable method of HIV-1 subtyping (Holmes, 1995; Ou, 1992). Its usefulness as a tool to study HIV diversity in different populations has been demonstrated (APRI Newsletter, 2001; Holmes, 1993; Hu, 2005). The HMA may be used to distinguish strains of HIV-1 and to trace viral quasispecies within individuals and within populations (Albert, 1994; Hu, 2005). It is cost effective and correlates well with genetic sequencing and characterization by phylogenetic analysis (Delwart, 1993; Heyndrickx, 1998). The HMA has been introduced by UNAIDS in several developing countries as a tool for monitoring subtype distribution. The combination of results from HMA *gag* and *env* genes allows the detection of intersubtype recombinant strains. This is not possible when only one genomic region is typed (Heyndrickx, 2000; Loussert-Ajaka, 1998; Tatt, 2001).

The hypervariable V3 region of *env* is most frequently used for genetic subtyping by the HMA. This is done particularly when there is a need to investigate epidemiologically linked infections (Mulder-Kampinga, 1993; Scarlatti, 1993; Weiser, 1993). Other hypervariable regions of *env* such as V1/V2 and V4/V5 are used less in the HMA to trace epidemiological linkage (Lamers, 1993; Simmonds, 1990). As opposed to the variable region of V3, subtyping of the more conserved *tat* gene region by HMA was particularly successful in identifying epidemiological relationships in cases of person to person transmission of HIV-1 (Simmonds et al., 1990).

4. Molecular characterization and epidemiology of Human Immunodeficiency Deficiency Virus - type 1 (HIV-1) in Jamaica

4.1 Introduction

This chapter describes the study which was undertaken to determine the HIV-1 subtypes in Jamaica and assess the heteroduplex mobility assay (HMA) as a method for tracking the HIV-1 epidemic. The HMA was introduced by the Joint United Nations Programme on HIV/AIDS (UNAIDS) in several developing countries as a tool for monitoring HIV-1 subtype distribution (Heyndrickx, 2000; Loussert-Ajaka, 1998). An abbreviated account of the clinical, immunological and laboratory characteristics of a representative sample of the cohort of HIV-1 infected patients which was studied is also presented. In this chapter the terms HIV-1 isolates and HIV-1 proviral DNA are used inter- changeably.

4.2 Materials and methods

4.2.1 Study population

The study population included EDTA blood samples collected from 1341 consecutive HIV-1 infected patients presenting at designated health care centres situated in several parishes throughout Jamaica were received by the Microbiology Department, University Hospital of the West Indies (UHWI), a tertiary care centre, for leucocyte immunophenotyping. Jamaica has 14 parishes from which samples were obtained from 12 parishes. Hanover and Trelawny were the only parishes that were not included. A data abstraction form was used to collect socio-demographic and clinical information from the hospital records of 94 patients who were seen at UHWI. The medical records of the patients seen at other healthcare facilities were not available for the study.

The peripheral blood mononuclear cells (PBMC) were separated from remnant EDTA blood samples of each patient using density gradient centrifugation with Histopaque® 1077 (Sigma–Aldrich Inc., St. Louis, Mo, USA) and the cells were kept frozen at − 20⁰C until required.

4.2.2 Immunophenotyping and T lymphocyte enumeration

Absolute CD3, CD4 and CD8 T lymphocyte counts were determined on EDTA blood samples by flowcytometry using the FACSCOUNT system and reagents (Biosciences

Immunocytometry Systems, San Jose, CA). The manufacturers' instructions were followed. The established reference ranges for T lymphocyte subsets in the Jamaican adult and paediatric populations (unpublished data) were used to evaluate the immune status of the patients. At least 1 EDTA blood sample from a healthy blood donor was included, as a control, with each test run.

4.2.3 Polymerase Chain Reaction (PCR) amplification of HIV-1 and *gag* genes

Polymerase chain reaction (PCR) was performed on PBMC from 318 patients. Two microlitre aliquots of PBMC was used as the DNA template for two-step nested polymerase chain reactions (PCR) carried out in a Perkin Elmer 9600 Thermal Cycler (Perkin-Elmer Corp. Norwalk, Conn.) using the PCR mixtures and amplification programs described in the HIV-1 *env/gag* HMA subtyping kit manuals, with modifications (Delwart version 5; Heslop, 2005; van der Auwera, 2000). The HIV-1 HMA subtyping kits which include plasmid clones of the complete genome of HIV-1 subtypes A–J of the major M group of HIV-1 sequences and primer pairs for PCR amplification of the HIV-1 *env* and *gag* genes were obtained from the National Institutes of Health (NIH) AIDS Research and Reference Program (Heslop, 2005; van der Auwera, 2000). The PCR core reagents used were commercially prepared (Invitrogen, Life Technologies, Grand Island, NY).

Due to the high rate of failed amplifications with primers supplied in the HMA kit, the *gag* gene primer pairs were replaced with DT1/DT7 and DT3/DT6 as first and second round primers (Delwart, 1995; Tatt, 2000; van der Auwera, 2000). The plasmids containing cloned *env* and *gag* genes, respectively, were amplified using second round *env* and gag primers. A negative control in which the DNA template was replaced by reagent grade water was included with each PCR assay.

The second round PCR amplification of the *env* gene using the ED5/ ED12 primers yielded a 1.2 Kb fragment spanning the V1-V5 coding region of HIV-1 gp120, while the ES7/ES8 primer pair yielded a 666 bp fragment spanning the V3-V5 coding region of gp120. Second round amplification of the *gag* gene resulted in a 748 bp fragment of the HIV-1 p17/ p 24 gene.

The PCR products were resolved by agarose gel electrophoresis carried out in 1.5% agarose gel containing ethidium bromide (0.5 µg/ml) at 80 volts for 45 minutes and visualized by ultraviolet (UV) illumination. The appropriate kilobase markers and controls were included in each run.

4.2.4 Genotyping of HIV-1 *env* and *gag* genes by Heteroduplex Mobility Assay (HMA)

The heteroduplex mobility assays (HMA) were performed on HIV-1 isolates from 180 patients. The HMA was performed as previously described with modifications (Loussert-Ajaka, 1998).

Both the *gag*-HMA and *env*-HMA were performed on each HIV-1 isolate. The subtype reference strains were selected following a pilot run of the HMA on 28 samples. Eleven *env* and 12 *gag* reference subtypes were selected from a panel of reference subtypes to include subtypes A, B, C, D, E, F, G , H and J in the HMA procedure. All subtype B reference strains (4 from the *gag* and 3 from the *env*) included in the kit were used in the HMA due to the dominance of this clade in the Caribbean and North America. In this method equal volumes of the patient HIV strain DNA and reference strain DNA are mixed under certain conditions to form heteroduplexes. This is followed by polyacrylamide gel electrophoresis. The HIV subtype is identified by the heteroduplex with the highest electrophoretic mobility which is the most closely matched heteroduplex formed from the most closely related strains. For the formation of heteroduplexes 5μl aliquots of a mixture containing equal volumes (5μl each and 1μl annealing fluid) of the second round PCR products of the HIV-1 isolates and the reference subtype strains were separated by polyacrylamide gel electrophoresis using Criterion™ precast gels which contained 5% polyacrylamide and 6M urea (Bio-Rad Laboratories, Hercules, CA). The heteroduplexes were visualized under UV light after ethidium bromide (0.5μg/ml) staining. The HIV-1 subtype assigned was that of the heteroduplex with the highest electrophoretic mobility.

4.2.5 Genotyping of HIV-1 *env* and *gag* genes by DNA sequencing and phylogenetic analysis

A total 63 HIV-1 isolates including 5 which were untypable by HMA were subjected to DNA sequencing and phylogenetic analysis of the purified PCR product or cloned PCR product. The sequencing primers for direct DNA sequencing were the second round PCR primers of the *gag* and *env* gene regions, respectively as described above.

4.2.6 Purification and cloning of amplicons of HIV-1 *env* and *gag* genes

The PCR products of the *gag* and *env* genes were purified using a commercially prepared kit (QIAGEN Inc. Valencia, CA) and cloned using the pGEM-T Easy Vector System Kit (Promega Corporation, Madison, WI). The manufacturers' instructions were followed.

4.2.7 Confirmation of *env* and *gag* clones by Polymerase Chain Reaction (PCR)

The recombinant clones were subjected to *env*-DNA PCR and *gag*-DNA PCR, respectively followed by agar gel electrophoresis to confirm the presence of the cloned DNA inserts, as described above.

4.2.8 Purification of HIV-1 *env* and *gag* gene clones

The plasmid vectors carrying the HIV-1 *env* and *gag* gene clones were purified using the PlasmidPURE™ DNA Miniprep kit (Sigma, St Louis, MO). The procedure followed was in accordance with the manufacturer's instructions. The final product of this procedure was highly purified plasmid DNA. The yield of cloned DNA was estimated by agar gel electrophoresis of 1-5 μl of the DNA using 1.5% agarose gel and read against an appropriate kilobase marker with ethidium bromide staining as described above. The *env* and *gag* gene clones were stored at -20°C until required.

4.2.9 Cycle sequencing of HIV-1 *env*, *gag* amplicons and clones

Commercially prepared reagents (BigDye® Terminator v3.1 Cycle Sequencing Kit, Applied Biosystems) were used for sequencing following the manufacturer's instructions. Briefly, sequence reactions were performed on the different HIV-1 DNA templates, including

purified PCR products, purified cloned inserts, double stranded control DNA (pGEM®
3Zf(+), 0.2μg/μl or 200ng/μl) and negative control DNA templates supplied with the kit.
Sequence reactions were carried out in a Perkins Elmer 9600 thermocycler using the
following parameters:

96°C for 10 seconds |
50°C for 5 seconds | 25 cycles
60°C for 4 minutes   ↓

Following purification of the sequence reactions using the DyeX 2 (QIAGEN, Valencia, CA)
DNA sequencing was performed using the ABI 3100 analyser (Applied Biosystems, Foster
City, CA).

4.2.10 Sequence editing and phylogenetic analysis of HIV-1 *env* and *gag* sequences

Following the cycle sequencing procedure single strand sequences generated by forward and
reverse sequencing primers were assembled into contiqs (a contiguous alignment of
overlapping sequences) using the Sequencher™ 4.0 software program (Gene Codes
Corporation, Ann Arbor, MI). The formed contiqs were edited for ambiguous base pairs
introduced during sequencing. The chromatograms generated were subsequently compared
for relationships between the HIV-1 isolates and GenBank reference strains. All sequences
were imported in the FASTA format for alignment by the ClustalX software programme
(http://inn-prot.weizmann.ac.il/software/ClustalX.html)   (Thompson   et   al.,   1997).
Alignments were carried out firstly by pairwise matching of the sequences of interest,
followed by multiple alignments with referenced HIV-1 sequences from GenBank (http://hiv-
web.lanl.gov/ALIGN_CURRENT/ALIGN-INDEX.html).   The multiple aligned sequences
were edited prior to the construction of phylogenetic trees. The PAUP 4.0 software programme
(Beta version 8 for Windows 95/98/ME/NT/2000/XP, Sinauer Associates, Inc, Sunderland,
MA) was used to construct rooted and unrooted neighbour-joining phylogenetic trees using
the SIV$_{CPZ}$ sequence (simian derived virus) of CPZ.GA.CPZGAB (GenBank accession number
X52154, *htpp://hiv-web.lanl.gov/ALIGN_CURRENT/ALIGN-INDEX.html*) as the out-group
(Swofford, 2003). The phylogenetic trees were subsequently bootstrapped (using 1000
replications) to establish confidence and statistical reliability (Hall, 2001).

4.2.11 Statistical analyses

Univariate and multivariate analyses of the patients' sociodemographic, clinical and
laboratory data were performed using the Statistical Packages for Social Sciences (SPSS)
Version 8 software.

4.3 Results

4.3.1 Demographic, clinical and immunological characteristics of HIV-1 infected patients

As shown in Table 2, the study population was predominantly heterosexual (90/94, 96%)
with a preponderance of females (68/94, 72%) and did not contain any injecting drug users
(0/94, 0%) while the only non-injecting drug used was marijuana (1/94, 1%). Other risk
factors for HIV infection included inconsistent condom use (21/94, 20%), history of

ulcerative (12/94, 13%) and non-ulcerative (15/94, 16%) STI. The percentages of antiretroviral drug therapy naïve and antiretroviral drug treated patients were 56% (53/94) and 44% (41/94), respectively. The clinical manifestations in the HIV-infected persons are summarized in Table 3 and a variety was seen. Constitutional symptoms including weight loss occurred in more than three quarters of the patient population and skin rashes in almost a half while opportunistic and respiratory infections were present in over one third. Other STI were common, over 20% and weight loss was found in just under 20%. There was 1 case of Kaposi's sarcoma while 13% of the population were asymptomatic.

The CD4 T cell counts were available for 73 patients. No trend was noted in the CD4 T cell counts by age and duration of diagnosis of HIV-1 infection for the cohort. However, when the patient population was separated into antiretroviral therapy naïve and treated categories, CD4 counts were statistically significantly higher in untreated patients ($p<0.005$, Table 4). A significant decreasing trend in CD4 counts with age in untreated patients ($p<0.05$) was observed (Table 5). In contrast no significant correlation was observed with age and CD4 counts in antiretroviral drug treated patients (Table 14). In multivariate analyses both increasing age in antiretroviral therapy naïve patients and absence of antiretroviral drug therapy were identified as independent risk factors for low CD4 counts ($p= 0. 022$; 95% confidence interval (CI), 1.014, 1.197 and $p=0.003$; 95% CI, 2.096, 35.982, respectively).

Characteristics	94
Mean age (years) ± SD	37 ±10
Gender	
Male	26(28)
Female	68(72)
Asymptomatic/Pre-AIDS	3(14)
Symptomatic/AIDS	81(86)
Risk Factors	
Sexual practices	
Heterosexual	90(96)
MSM	4(15)
Inconsistent condom use	21(20)
History of STI	
Ulcerative	12(13)
Non-ulcerative	15(16)
Drug use	
Marijuana	1(1)
Cocaine	0(0)
IV drug	0(0)
Antiretroviral drug therapy	
Yes	41(44)
No	53(56)

*MSM = Men who have sex with men. The cohort of HIV-1 infected patients was predominantly heterosexual.

Table 2. Characteristics of 94 HIV-1 infected patients presenting at the University Hospital of the West Indies (UHWI)*

Manifestation	Frequency (%)
Constitutional symptoms	73(77)
Skin rashes	46(49)
Opportunistic infection	35(37)
Respiratory infection	34(36)
Other STI	21(22)
Vomiting/diarrhoea	13(14)
Lymphadenopathy	11(13)
Neurological	6(6)
Kaposi sarcoma	1(1)
Asymptomatic	13(14)

* Respiratory infections include cough, dyspnea, bacterial pneneumonia, pneumocystis carinii pneumonia, otitis media. Opportunistic infections include candidiasis, toxoplasmosis, cytomegalo virus infection, scabies, molluscum contagiosum, ulcers, hairy leukoplakia of tongue. Neurological manifestations include meningitis and neuralgia.

Table 3. Clinical manifestations in 94 HIV- infected patients*

CD4 cells/μl		
Category (N)	Mean (± SD)	Range
Untreated (32)	378 (341)	5- 1305
Treated(41)	192 (163)	1- 691
Total (73)	291(140)	1- 1305

* N = Number of patients in each category. CD4 T Lymphocyte counts were available for 73/94 patients. The CD4 cells counts were statistically significantly higher in the untreated patients (p< 0.005)

Table 4. CD4+ T Lymphocyte counts in antiretroviral drug treated and untreated HIV- infected patients*

CD4+ cells/μl		
Age group/years(N)	Mean (± SD)	Range
≤ 20 (2)	879 (315)	656 - 1102
21 - 30 (13)	486 (345)	19 - 1305
31 - 40 (12)	284 (270)	7 - 684
41 - 50 (2)	294 (392)	16 - 571
51 - 60 (3)	8 (2)	5 - 9

* N = Number of patients in each category. A decreasing trend in CD4 counts with age in untreated patients (p<0.05) was observed.

Table 5. CD4+ T Lymphocyte counts in 32 antiretroviral drug therapy naive patients by age*

4.3.2 HIV-1 DNA-PCR assay

The PCR amplicons obtained included *env* gene, 141/318 (44.3%), *gag* gene, 170/318 (53.5%). Of the total 318 samples tested, positive results in one or both genes were obtained in 251 (79%), 113 (113/318, 35.5%) were amplified in both genes while neither the *env* or *gag* gene was amplified in 67 (21.0%) samples. The results of the HIV-1 *env* and *gag* DNA – PCR assays are summarized in Table 6. As shown in Figure 3, all *gag* (35/35,100%) and *env* (22/22, 100%) reference subtype plasmids were amplified successfully by all second round primers.

Category	Positive(%)	Negative(%)
Env-PCR	141(44.3)	177(55.7)
Gag-PCR	170(53.5)	148(46.5)
Env + gag-PCR	113(35.5)	205(64.5)

Table 6. HIV-1 DNA – PCR assays in 318 HIV- infected patients

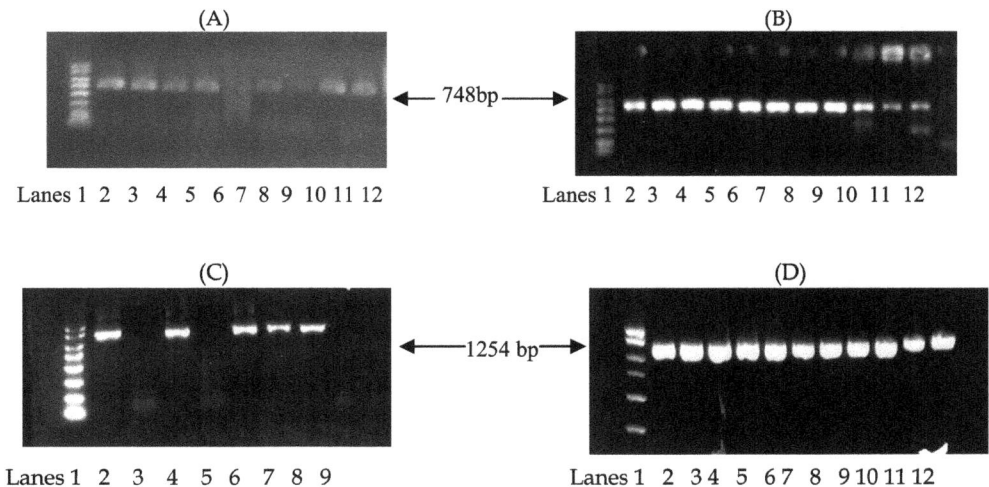

Fig. 3. DNA polymerase chain reaction (PCR) assay: Photographs (A-D) of agarose gel electrophoresis of PCR products. Lane 1 in each gel contains the base pair (bp) marker. Gels A and B show the 748 bp PCR product of the *gag* gene region of HIV-1 isolates and reference strains. Gels C and D show the 1254 bp PCR product of the *env* gene region of HIV-1 isolates and reference strains.

4.3.3 Heteroduplex mobility assay (HMA) analysis of HIV-1 proviral DNA/isolates

The *env* and *gag* HMA were performed on the PCR amplicons of 180 samples. Of these 174 (96.7%) were unambiguously genotyped while 6 samples (3.3%) were indeterminate. The majority of HIV-1 isolates were subtype B (170/180, 94.4%) while 1 (0.6%) isolate each of subtypes A, C, D and E was found. The representative HMA polyacrylamide gel photographs in Figures. 4-5 illustrate the endpoint of the HMA procedure and comparative electrophoretic mobilities of different HIV subtypes which were found.

1
m, A1 B1 B2 B3 B4 C1 D1 E1 F1 G1 H2 J Blk

11
m A1 B1 B2 B3 B4 C1 D1 E1 F1 G1 H2 J Blk

111
m, A1 B1 B2 B3 B4 C1 D1 E1 F1 G1 H2 J Blk

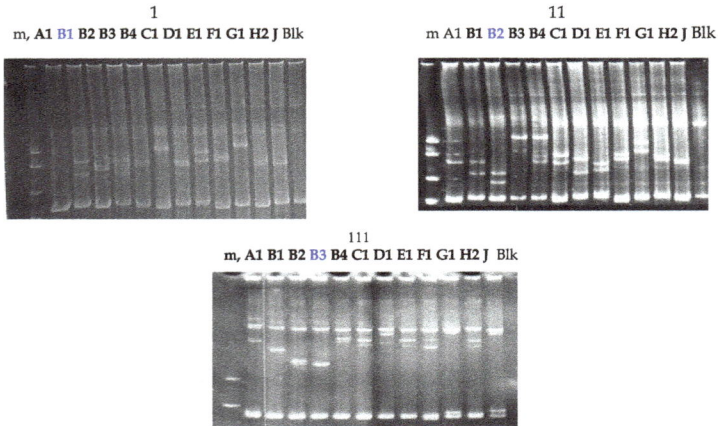

Fig. 4. Heteroduplex mobility assay (HMA) of HIV-1 strains. Photographs (1-III) of polyacrylamide gels showing subtype B strains which formed heteroduplexes with 3 subtype B reference strains B1 (I), B2 (II) and B3 (III). Heteroduplexes of 11 subtype reference strains including A-J strains were tested. The subtypes assigned are those of the heteroduplexes with the highest mobility formed between amplicons of the patient's isolate and subtype reference strains. The first lane (m) of each gel contains the Kb marker and the last lane (BLK) the negative control.

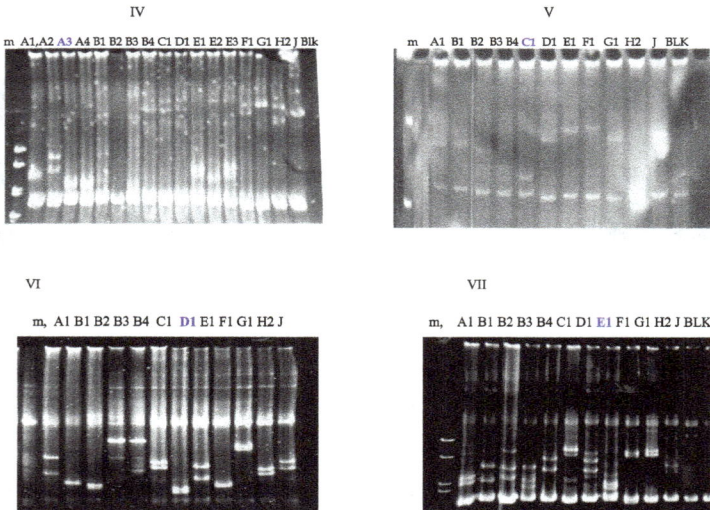

IV
m A1,A2 A3 A4 B1 B2 B3 B4 C1 D1 E1 E2 E3 F1 G1 H2 J Blk

V
m A1 B1 B2 B3 B4 C1 D1 E1 F1 G1 H2 J BLK

VI
m, A1 B1 B2 B3 B4 C1 D1 E1 F1 G1 H2 J

VII
m, A1 B1 B2 B3 B4 C1 D1 E1 F1 G1 H2 J BLK

Fig. 5. Heteroduplex mobility assay (HMA) of HIV-1 non-B subtype strains. Photographs (IV-VII) of polyacrylamide gels showing subtypes A (IV), C (V), D (VI), E (VII). Heteroduplexes of 11 subtype reference strains A-J were tested. The subtypes assigned are those of the heroduplexes with the highest mobility formed between amplicons of the patient's isolate and reference subtype strains. The first lane (m) of each gel contains the Kb marker and the last lane (Blk) the negative control.

4.3.4 DNA sequencing and phylogenetic analysis of HIV-1 proviral DNA/isolates

A total 63 HIV-1 isolates were subjected to automated DNA sequencing which yielded 54 DNA sequences (54/63, 86%) including 26 *gag* and 28 *env* gene region sequences. Representative neighbor-joining phylogenetic trees constructed from the *gag* and *env* gene sequences, respectively are shown in Figures 6-7. Of the 54 sequences 50 (50/54, 93%) were assigned subtype B, 2 (2/54, 4%) subtype D and 1 each (1/54, 2%) assigned subtypes A and J, respectively. In the *gag* gene 87% (22/26) clustered with HIV-1 B subtype reference strains (BUS98, BFR, BTH90), 2 (8%) with D subtype reference strain (DUG94), 1 (4%) each with A subtype and CRF reference strains (A1 UG92, 18CPXCM) and J subtype reference strain (JSE93). The subtype assignments of the gag gene sequences were supported by bootstrap values ≥ 70% in 96% (25/26) of isolates. On the other hand all (28/28, 100%) *env* gene sequences clustered with B subtype reference strains (BUS98, BFR, BTH90). The subtype assignments of the majority of *env* gene sequences (25/28) were supported by bootstrap values > 90%. The relationships between the Jamaican strains were assessed by the unrooted phylogenetic trees constructed from the HIV-1 *gag* and *env* sequences.

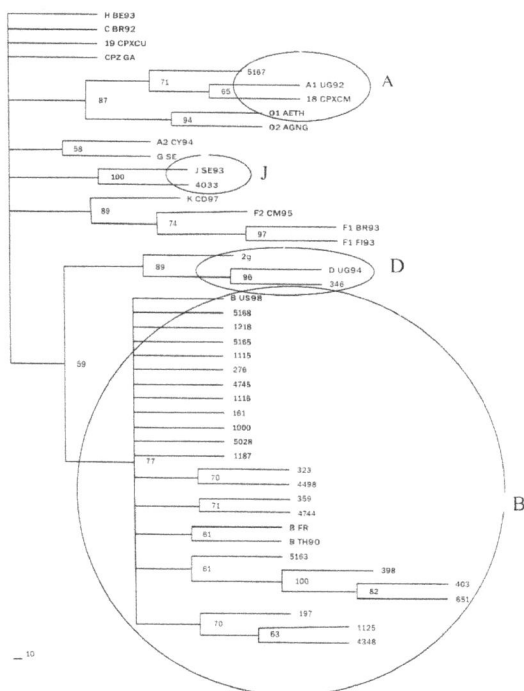

Fig. 6. Phylogenetic analysis of HIV-1 *gag* gene sequences of Jamaican isolates: Neighbour-joining phylogenetic tree constructed from 26 HIV-1 *gag* sequences, reference subtype strains (B US98, B FR, B TH90, D UG94, J SE93, A1 UG92) and circulating recombinant forms (18CPXCM). Numbers at the nodes of the tree indicate bootstrap values expressed as the percentage of 1000 replicates supporting each subtype in the tree.

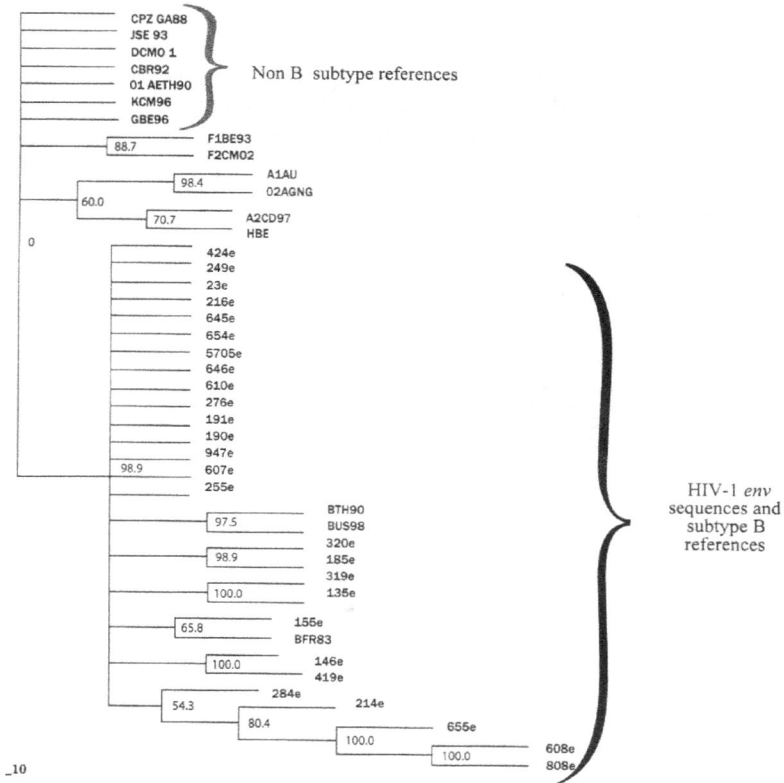

Fig. 7. Phylogenetic analysis of HIV-1 *env* gene sequences of Jamaican isolates: Neighbour-joining phylogenetic tree constructed from 27 Jamaican HIV-1 *env* sequences and subtype reference strains. The Jamaican isolates clustered with subtype B reference strains (BUS98, BTH90, BFR83). Supporting bootstrap values, indicated by numbers at internal nodes, are expressed as the percentage of 1000 replicates supporting each subtype in the tree.

4.3.5 Comparison of HIV-1 subtype assignment by HMA and phylogenetic analysis

For the subtype B viruses which were characterized by both methods there was an overall 86% (38/44) concordance between phylogenetic analysis and HMA. Of the 6 discordant subtype B isolates, 5 were indeterminate and 1 identified as subtype D in the HMA. For non-B subtypes the rate of concordance was lower (67%, 2/3) as the 2 subtype D viruses were identified as subtype E and subtype B, respectively by HMA. The HMA was concordant with phylogenetic analysis in discriminating the only subtype A virus in the sample.

4.3.6 Geographical distribution of HIV-1 subtypes in Jamaica

The geographical distribution of HIV-1 subtype strains among the parishes in Jamaica is shown in Figure 8. Multiple subtypes were found in 3 of the 14 parishes sampled.

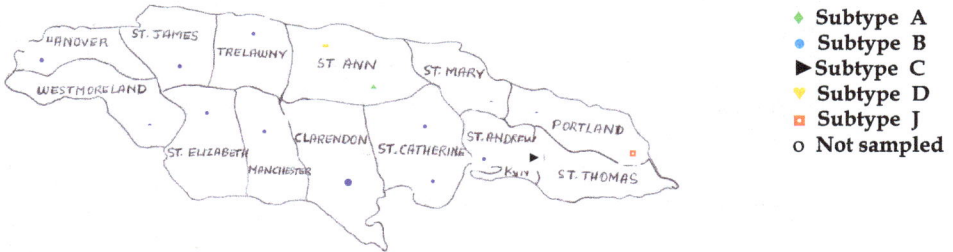

Fig. 8. Map of Jamaica showing the geographical distribution of the different HIV-1 subtype strains in 14 parishes: Multiple subtypes were found in the parishes of St. Ann (population: 52,600), Portland (population: 33,000) and the capital city Kingston (population: 157,000).

4.4 Discussion

This study of the molecular epidemiology of HIV-1 is the largest and most recent of its kind from the English- speaking Caribbean (Cleghorn, 2000; Gittens, 2003; Vaughan, 2003). Another important difference between the present study and similar studies conducted in the other English-speaking Caribbean countries is the fact that the *gag* gene of the HIV-1 isolates was genotyped (Gittens, 2003; Vaughan, 2003).

The patients whose records were reviewed are representative of the cohort of adult HIV-1 infected patients from whom peripheral blood mononuclear cells (PBMC) were obtained for HIV-1 isolation and molecular typing. The exact time of contracting HIV infection could not be ascertained in these patients. However at the time of diagnosis a spectrum of AIDS related conditions which defined both primary and late stage HIV-1 infection were seen in the majority of patients whereas a few were asymptomatic (JAMA, 2006).

The explanation for the late presentation at hospital of some of the patients is beyond the scope of this study. What is notable, however, is the relatively long period of time between diagnosis and having the required CD4 cell counts done to assess the immune status and implement antiretroviral regimens in some patients according to current guidelines (JAMA, 2006; Office of AIDS Research Advisory Council, 2008). The late presentation, delay in immune monitoring and antiretroviral therapy are reflected in the proportion of patients whose CD4+ T lymphocyte counts were already below 350 and 200 cells/μl, respectively, at the time of testing. This may be due to the fact that CD4 cell determinations and antiretroviral drugs only became widely accessible to Jamaican patients under the Global HIV/AIDS Programme in mid- 2004.

The absence of intravenous drug users from the cohort is also worthy of note. Cleghorn et al (Cleghorn et al., 2000) also reported an absence of intravenous drug users from a cohort of HIV infected patients in Trinidad and Tobago. However other known risk factors for HIV-1 infection such as history of other STI, inconsistent condom use and multiple sexual partners were comparable to that seen in other cohorts including that reported from Trinidad and

Tobago and other Caribbean /West Indian islands (Cleghorn, 2000; Fleming, 1999; Gibney, 1999; Murphy, 1999).

In this Jamaican cohort increasing age in antiretroviral drug naïve patients and absence of antiretroviral drug therapy in all age categories were the only independent risk factors identified for decreasing CD4 T cell counts. This is in keeping with what is known of the natural progression of HIV infection (Fox et al., 2008). Although disease progression in HIV-1 infected patients has been linked with HIV-1 subtypes the impact of different subtypes was not assessed in this study as the overwhelming majority of patients were infected with subtype B (Baeten , 2007; Kaleebu, 2002; Kanki, 1999). The impact of antiretroviral treatment on disease progression was evident as the CD4 T cell counts were significantly lower in the group of antiretroviral drug treated patients and similar in the group of patients on antiretroviral therapy irrespective of age (Harari, 2004; Hogg, 1999).

A panel of HIV-1 primer sets from international sources was used in the PCR amplifications. Nonetheless failures in amplifications of either *gag* or *env* gene were observed in about half of the isolates and failures in both genes in about a fifth of isolates. The low sensitivity of the HIV PCR assays and the frequent non-concordance between the PCR results for the *gag* and *env* gene regions might be due to the marked genetic heterogeneity of the virus in both genes (Delwart, version 5; Swofford, 2003). This is supported by the fact that all reference plasmids were successfully amplified using the primer sets from the National Institutes of Health (NIH) prepared kits which failed to amplify a number of the HIV-1 isolates. Other authors have encountered the problem of unamplifiable HIV-1 strains with the *env* primers from the HMA kits due to the broad heterogeneity within the gp120 region of the *env* gene (Agwale et al., 2001). For the above mentioned reasons antiretroviral drug use in HIV-infected patients in Jamaica might not have contributed substantially to genetic variation in the *env* gene as reported by others (Agwale et al., 2001). Six HIV-1 subtypes were identified in this study including subtypes A, B, C, D, E and J. It was not entirely surprising to find that the majority of HIV-1 isolates in Jamaica were subtype B which accounted for over 90% of infections. This is in keeping with reports, from other English speaking Caribbean Islands, which did not include Jamaican samples (Gittens, 2003; Vaughan, 2003). Subtype B is also the predominant subtype in North America, Western Europe, Australia and South America (Distler, 1995; Laukkanen, 2000; Ramos, 1999).

The percentages of subtypes A, C, D, E and J found in this study were low. However the mere presence of these non-B HIV-1 subtypes in the HIV/AIDS population in Jamaica is of great importance as it signals high genetic diversity of the virus and the HIV/AIDS epidemic. It might be essential that these subtypes be considered in vaccine modeling. Outside of Central Africa a limited number of HIV-1 variants are usually circulating in each country, rarely more than 2 or 3, representing multiple introductions or in some cases locally generated CRFs (Pérez et al., 2006). Subtype A is found in East and Central Africa, Central Asia and Eastern Europe with a global prevalence of 12.3% (Hemelaar et al., 2006). Subtype C, the dominant subtype in India and China, is the most prevalent subtype worldwide accounting for approximately 50% of infections (388, 389). Subtype D is generally limited to East and Central Africa with sporadic cases observed in Southern and Western Africa (Vogt et al., 1986). Subtype E is a recombinant strain (CRF01_AE), and not a distinct subtype as initially thought, which co-circulates with subtype B within the intravenous drug user (IDU) population and fishermen in Thailand and also occurs in

Vietnam and South East Asia (Entz, 2000; Heyndrickx, 2000; Wasi, 1995). Subtype J occurs in various locations with a less than 1% prevalence (Archer, 2007; Hemelaar, 2006).

The DNA sequencing and phylogenetic analyses performed in the study did not allow identification of inter-subtype recombinants as the subtype of the HIV-1 isolates were assigned on the DNA sequence of a single gene region, either *env* or *gag* (Louwagie, 1993; Tatt, 2000). On the other hand genotyping by HMA analysis was performed on the *gag* and *env* gene sequences of each isolate. This method also failed to identify any recombinant strain in Jamaica except for the subtype E strain which was re-assigned subtype D by phylogenetic analysis (Heslop et al., 2005). The subtype A virus identified in this Jamaican sample appears to be related, in the *gag* gene, to CRF18- cpx which originated in Central Africa and, currently, is circulating in Cuba (Thomson et al., 2005). It has been reported that the genome of CRF18-cpx contains multiple segments clustering with subtypes A1, F, G, H and K as well as segments failing to cluster with any subtypes (Thomson et al., 2005). The multiple subtypes of HIV-1 in circulation in Jamaica should be monitored for the likely emergence of inter-subtype recombinant strains as the epidemic progresses. Recombinant strains have been observed mostly in areas where multiple subtypes co-circulate (Heyndrickx et al., 2000). The emergence of recombinant strains in other Caribbean and Latin American countries, including Cuba and Brazil, following sudden shifts in HIV-1 subtype distribution from a predominance of subtype B is well documented (Pérez, 2006; Santos, 2006; Tatt, 2000; Thomson, 2005).

The limited divergence of the Jamaican HIV-1 strains, from the subtype reference sequences, in the *gag* and *env* gene is another important observation. This was evident from the high bootstrap values supporting the subtype classification of this sample especially subtype B viruses. This might at least be partly explained by the fact that the cohort was largely antiretroviral therapy naïve. Secondly the Jamaican HIV-1 isolates which were not amplified by the NIH primer sets might include the more divergent strains and these were not genotyped (Vaughan et al., 2003). Gittens et al (Gittens et al., 2003) reported a broad genetic diversity of *env* gene sequences in Barbados suggesting multiple introductions of subtype B viruses to the island (10). Similarly Cleghorn et al reported a significant subcluster within the B subtype in Trinidad (Vaughan et al., 2003).

The discordance between HMA and phylogenetic analysis in the assignment of subtypes is not unique to the present study (Loussert-Ajaka, 1998; Novitski, 1996; Swofford, 2003; Thompson, 1994). The HMA has been shown to give excellent results for the detection of subtypes B and F, the prevalent subtypes in Caucasian patients originating from Western countries and Romania, respectively. Conversely extensive viral variation might create problems in countries like Africa where different HIV subtypes have been circulating longer (Loussert-Ajaka et al., 1998). This raises the question of whether further problems with the HMA might occur in Jamaica with its multiple subtypes of the virus. For example, in previous studies of international cohorts, the HIV-1 strains which were deemed un-typeable/indeterminate by HMA and were subsequently assigned by DNA sequencing turned out to be highly divergent subtypes A-D or G related strains (Loussert-Ajaka, 1998; Novitski, 1996; Swofford, 2003; Thompson, 1994). In one study almost two thirds of the subtype D isolates were incorrectly genotyped by HMA (10). In the present Jamaican sample the 2 subtype D viruses were incorrectly genotyped, by HMA, as subtypes B and E, respectively (UNAIDS, 2004 - 12). The viruses belonging to subtypes B and D have been shown to be closely related with respect to *gag*, *env* and *pol* gene sequences and probably diverged relatively recently. Therefore separation of subtype B and subtype D is not as well defined as between other subtypes (Tatt, 2000; UNAIDS/WHO, 2006). Nonetheless the HMA remains the recommended genotyping method second only to DNA sequencing. Consequently it has been recommended that the plasmid selection in the HMA kit be constantly revised to cover viral diversification (Loussert-Ajaka, 1998; Thompson, 1994).

Limitations of the study include the fact that a substantial proportion of samples were not amplifiable by the primer sets which were used. The failure to sequence both the *env* and *gag* genes of each isolate to indentify inter-subtype recombinants is another limitation. A more comprehensive study of the genetic diversity of the Jamaican HIV-1 isolates should include customized or more conservative primers, such as those recently described, and sequencing *pol* gene regions (Agwale, 2001; Gittens, 2003; Tatt, 2000). More extensive genomic sequence analyses which were not performed in this study might also reveal more genetic diversity and linkages among Jamaican isolates.

4.4.1 Conclusions and recommendations

Transmission of the multiple subtypes of HIV-1 in the Jamaican population is predominantly heterosexual and is not linked to intravenous drug use. HIV-1 subtype B is the dominant subtype contributing to the HIV/AIDS epidemic amidst high genetic regions of the virus in Jamaica. The results also emphasize the need for genotyping of multiple genetic regions, to be attempted, to ensure positive results against the background of frequent failures of the necessary PCR assays.

The HMA is well established as a genotyping method in the Jamaican setting. Our results confirm the value of this cost- effective and reliable method in tracking the HIV/AIDS epidemic. This does not abrogate the need for DNA sequencing and phylogenetic analysis which should be used as an adjunct methodology to resolve indeterminate HMA results and ensure the accurate assignment of non-B subtype strains of HIV-1.

The implications of HIV-1 genetic diversity and its impact on diagnosis, treatment, progression and prevention of disease continue to be better understood. Therefore it is important that the molecular epidemiology of HIV-1 in Jamaica continue to be monitored especially in light of pending vaccines and multiple subtypes of the virus found in Jamaica. We recommend the use of the HMA as a major method supported by DNA sequencing and phylogenetic analysis for tracking the transmission of the virus. A more comprehensive use of DNA sequencing which involves other genetic regions of the virus such as the *pol* gene and more extensive characterization of the genome are required to better elucidate the molecular epidemiology of this virus in Jamaica. Further studies also should include optimizing the HMA through the development of PCR primers and reference clones of HIV-1 which are customized for Jamaican strains of HIV-1.

5. References

Agwale, SM., Robbins, KE., Odama, L., Saekhou, A., Zeh, C., Edubio, A., Njoku, OM., Sani-Gwarzo, N., Gboun, MF., Gao, F., Reitz, M., Hone, D., Folks, TM., Pieniazek, D., Wambebe, C., & Kalish ML. (2001). Development of an *env* gp41–based heteroduplex mobility assay for rapid human immunodeficiency virus type 1 subtyping. *Journal Clinical Microbiolology*, Vol. 39, pp. 2110-4.

Akouamba, BS., Vie,l J., Charest, H., Merindol, N., Samson, J., Lapointe, N., Brenner, BG., Lalonde, R., Harrigan, PR., Boucher, M., & Soudeyns, H. (2005). *HIV*-1 genetic diversity in antenatal Cohort. Canada. *Emerging Infectious Diseases*, Vol. 11, pp. 1230-4.

Albert, J., Wahlberg, J., Leitner, T., Escanilla, D., & Uhlen, M. (1994). Analysis of a rape case by direct sequencing of the HIV-1 *pol* and *gag* genes. *Journal of Virology*, Vol. 68, pp. 5918-24.

APRI Newsletter. (2001). Day A. Nonhuman DNA testing increases DNA's power to identify and convict criminals. The silent witness. Vol. 6, pp.1-2.

Archer, J., & Robertson, DL. (2007). Understanding the diversification of HIV-1   groups M and O. AIDS Vol. 21, pp. 1693-1700.

Avila, MM., Pando, MA., Carrion, G., Peralta, LM., Solomon, H., et al. (2002). Two HIV-1 epidemics in Argentina: different genetic subtypes associated with different risk groups. *Journal Acquired Immune Deficiency Syndrome*, Vol. 29 (4), pp. 422-6.

Baeten JM, Chohan B, Lavreys L, Chohan V, McClelland RS, Certain L, Mandaliya K, Jaoko W, Overbaugh J. HIV-1 subtype D infection is associated with faster disease progression than subtype A inspite of similar plasma HIV-1 loads. J infect Dis 2007; 195: 1177-80.

Barré-Sinoussi, F., Chermann, JC., Rey, F., Nugeyre, MT., Chamaret, S., Gruest, J., Dauguet, C., Axler-Blin, C., Vézinet-Brun, F., Rouzioux, C., Rozenbaum, W., Montagnier, L.. (1983). Isolation of a T-lymphotrophic retrovirus from a patient at risk for acquired immune deficiency syndrome. *AIDS*, Vol. 220, pp. 868-71.

Borrow, P., Lewicki, H., Wei, X., Horwitz, MS., Peffer, N., Meyers, H., Nelson, JA., Gairin, JE., Hahn, BH., Oldstone. MB., & Shaw, GM (1997). Antiviral pressure exerted by HIV–1 specific cytotoxic T lymphocytes (CTLs) during primary infection demonstrated by rapid selection of CTL escape virus. *Nat Med*, Vol. 3, pp. 205- 11.

Burke, DS. (1997).Recombination in HIV: an important evolutionary strategy. *Emerging Infectious Diseases*, Vol. 3, pp. 253- 9.

Carr, JK., Nadai, Y., Eyzaguirre, L., Saad, MD., Khakimov, MM., Yakubov, SK., Birx, DL., Graham, RR., Wolfe, ND., Earhart, KC., & Sanchez, JL. (2005). Outbreak of a West African recombinant of HIV-1 in Tashkent, Uzbekistan. *Journal of Acquired Immune Deficiency Syndrome*, Vol. 39, pp. 570-5.

Chiu, IM., Yaniv, A., Dahlberg, JE., Gazit, A., Skuntz, SF., Tronick, SR., & Aaronson, SA (1984). Nucleotide sequence evidence for relationship of AIDS retrovirus to lentivirus. Nature, Vol. 317, pp. 366-8.

Chohan, B., Lang, D., Sagar, M., Korber, B., Lavreys, L., Richardson, B., & Overbaugh, J. (2005). Selection for human immunodeficiency virus type 1 envelope glycosylation variants with shorter V1-V2 loop sequences occurs during transmission of certain genetic subtypes and may impact viral RNA levels. *Journal of Virology*, Vol. 79, pp. 6528-31 clades in the in the Caribbean using *pol* gene sequences. AIDS *Res Hum Retroviruses,*

Cleghorn, FR., Jack, N., Carr, JK., Edwards, J., Mahabir, B., Sill, A., & McDanal, CB. (2000). Connolly SM, Goodman D, Bennetts RQ, O'Brien TR, Weinhold KJ, Bartholomew C, Blattner WA, Greenberg ML. A distinctive clade B HIV type 1 is heterosexually transmitted in Trinidad and Tobago. *Proc Natl Acad Sci* USA; Vol. 97, pp. 10532-7.

Cocchi, F., DeVico, AL., Garzino-Demo, A., Arya, SK., Gallo, RC., Lusso, P. (1995). Identification of RANTES, MIP-1 alpha, and MIP-1 beta as the major HIV-suppressive factors produced by CD8+ T cells. *Science,* Vol. 270, pp. 1811- 5.

Cornelissen M, Jurriaans S, Kozaczynska K, Prins JM, Hamidjaja RA, Zorgdrager F, Bakker M, Back N, van der Kuyl AC. Routine HIV-1 genotyping as a tool to identify dual infections. AIDS 2007; 21: 807-11

Murphy G, Belda FJ, Pau CP, Clewley JP, Parry JV. Discrimination of subtype B and non-subtype B strains of human immunodeficiency virus-type 1 by stereotyping: correlation with genotyping. J Clin Microbiol 1999; 37: 1356-60.

Dale, J W & Park, SF. (2010). Mutation and variation, In: *Molecular genetics of Bacteria,* A John Wiley and Sons, Ltd, Publication, (5th Ed.), 37-72.

d'Cruz-Grote D. Prevention of HIV infection in developing countries. Lancet. 1996; 348: 1071-4.

de Mendoza, C., Rodriguez, C., Colomina, J., Tuset, C., Garcia, F., Eiros, JM., Corral, A., Leiva, P., Aguero, J., Torre-Cisneros ,J., Pedreira, J., Viciana, I., del Romero, J., Saez, A., Ortiz de Lejarazu, R., & Soriano, V. (2005). Spanish HIV Seroconverter Study Group. Resistance to nonnucleoside reverse-transcriptase inhibitors and prevalence of HIV type 1 non-B subtypes are increasing among persons with recent infection in Spain. *Clinical Infectious Diseases,* Vol. 41, pp. 1350-4.

Delwart, EL, Herring, B., Rodrigo, AG., & Mullins, JI. (1995). Genetic subtyping of human immunodeficiency virus using a heteroduplex mobility assay. *PCR Methods Application,* Vol. 4, pp.202-16.

Delwart, EL., Busch, MP., Kalish, ML., Mosley, JW., & Mullins, JI. (1195). Rapid molecular epidemiology of human immunodeficiency virus transmission. *AIDS Res Human Retroviruse,* Vol. 11, pp. 1081-93.

Delwart, EL., Herring, B., Learn, Jr. GH., Rodrigo, AG., & Mullings, JI. Heteroduplex mobility analysis HIV-1 *env* subtyping kit protocol version 5.

Delwart, EL., Shpaer, EG., Louwagie, J., McCutchan, FE., Grez, M., Rubsamen- Waigmann, H., & Mullins, JI. (1993). Genetic relationship determined by a DNA heteroduplex mobility assay: analysis of HIV-1 *env* genes. *Science,* Vol. 262, pp. 1257-61.

Deng, H., Liu R., Ellmeie,r W., Choe, S., Unutmaz, D., Burkhart, M., Di Marzio, P., Marmon, S., Sutton, RE., Hill, CM., Davis, CB., Peiper, SC., Schall, TJ., Littman, DR., & Landau NR. (1996). Identification of a major co-receptor for primary isolates of HIV-1. *Nature,* Vol. 381, pp. 661-6.

Diagnostic Tests for AIDS. (1988). *The Medical Letter on Drugs and Therapeutics* Vol. 2, pp. 299- 303.

Diaz, RS., Zhang, L., Busch, Mp., Mosley, JW., & Mayer, A.(1997). Divergence of HIV-1 quasispecies in an epidemiologic cluster. *AIDS,* Vol. 11: pp. 514-22

Ding, J., Jacobo-Molina, A., Nanni, RG., Boye,r PL., Hughes, SH., Pauwels, R., Andries, K., Janssen, PA., & Arnold, E. (1994). Locations of anti-AIDS drug binding sites and resistance mutations in three dimensional structure of HIV-1 reverse transcriptase. Implications for mechanisms of drug inhibition and resistance. Journal of Molecular Biology Vol. 243, pp. 369-87.

Distler, O., McQueen, PW., Tsang, ML., Byrne, C., Neilan, BA., Evans, L., Penny, R., Cooper, DA., & Delaney, SF. (1995). Characterization of the V3 region of HIV-1 isolates from Sydney, Australia. *AIDS Res Hum Retroviruses,* Vol. 11, pp. 423-5.

Dittmar, MT., Simmons, G., Hibbitts, S., O'Hare,. M, Louisirirotchanakul, S., Beddows, S., Weber, J., Clapham, PR., & Weiss, RA. (1997). Langerhans cell tropism of human immunodeficiency virus type 1 subtype A through F isolates derived from different transmission groups. *Journal of Virology*, Vol. 71, pp. 8008-13.

Eisenstein, B. (1990). A new method of using molecular genetics for medical diagnosis. *New England Journal of Medicine*, Vol. 322, pp. 178-183.

Entz, AT., Ruffolo, VP., Chinveschakitvanich, V., Soskoline, V., & van Griensven, GJ. (2000). HIV-1 prevalence, HIV-1 subtypes and risk factors among fishermen in the Gulf of Thailand and the Adaman sea. *AIDS*, Vol. 14, pp.1027-34.

Fenyö EM, Morfeldt-Månson L, Chiodi F, Lind B, von Gegerfelt A, Albert J, Olausson E, Asjö B. Distinct replicative and cytopathic characteristics of human immunodeficiency virus isolates. J Virol 1998; 62: 4414-9

Fleming DT, Wasserheit JN. From epidemiological synergy to public health policy and practice: the contribution of other sexually transmitted diseases to sexual transmission of HIV infection. Sex Transm Infect 1999; 75:3-17.

Folks, TM., & Khabbaz, RF. (1998). Retroviruses and associated diseases in humans. In: Topley and Wilson's Microbiology and Microbial Infections. 9th Edn. Collier, L., Ballows, A., Sussman, M., & Arnold. New York, PP. 781-802.

Fox J, Scriba TJ, Robinson N, Weber JN, Phillips RE, Fidler S. Human immunodeficiency virus (HIV)-specific T helper responses fail to predict CD4+ T cell decline following short course treatment at primary HIV-1 infection. Clin Exp Immunol 2008; 152: 532-7.

Gao, F., Robertson, DL., Carruthers, CD., Morrison, SG., Jian, B., Chen, Y., Barré-Sinoussi, F., Girard, M., Srinivasan, A., Abimiku, AG., Shaw, GM., Sharp, PM., & Hahn BH. A comprehensive panel of near-full-length clones and reference sequences for non-subtype B isolates of human immunodeficiency virus -type 1. *Journal of Virology*, Vol. 72, pp. 5680-98.

Geretti, AM. (2006). HIV-1 subtypes: epidemiology and significance for HIV management. *Current Opinion Infect Diseases*, Vol. 19, pp. 1-7.

Gibney L, Choudhury P, Khawja Z, Sarker M, Islam N, Vermund S. HIV/AIDS in Bangladesh: an assessment of biomedical risk factors for transmission. Int J STD AIDS 1999; 10:338-46.

Gittens, MV., Roth, WW., Roach, T., Stringer HG. Jr., Pieniazek, D., Bond, C., & Levett PN. (2003). The molecular epidemiology and drug resistance determination of HIV type 1 subtype B infection in Barbados. *AIDS Res Hum Retroviruses*, Vol. 19, pp.313- 9.

Guatelli, JC., Siliciano, RF., Kuritzkes, DR., & Rickman, DD. (2002). Human Immuno-deficiency Virus. In: *Clinical Virology*. 2nd Ed. Rickman, DD., Whitley, RJ & Hayden, FG (Eds). American Society for Microbiology, Washington, pp 685- 729

Hahn, BH (1994). Viral genes and their products. In: *Textbook of AIDS Medicine*. S Broders, TC Merigan, D Bolognesi (Eds) .Williams and Wilkins; Baltimore. Vol. 21, p. 208.

Hall, BG. (2001). A how-to manual for molecular biologists. *Sinauer Associates, Inc.* Sunderland.

Harari A, Petitpierre S, Vallelian F, Pantaleo G. Skewed representation of functionally distinct populations of virus-specific CD4 T Cells in HIV-1- infected subjects with progressive disease: changes after antiretroviral therapy. Blood 2004; 103: 966-72.

Hemelaar, J., Gouws, E., Ghys, PD., & Osmanov, S. (2004). Global and regional distribution of HIV-1 genetic subtypes and recombinants in. AIDS 2006; 20: W13-W23.

Heslop, OD., Smikle, MF., & Vickers, IE. (2009). Christian NA, Harvey KM, Figueroa JP, Brown SE, Christie CDC, Bain B, Barton EN. High genetic diversity in human immunodeficiency virus type-1 (HIV-1) subtypes in Jamaica. *West Indian Medical Journal*, Vol. 58 (3), pp. 195- 200.

Heslop, OD., Smikle, MF., Deer, D., Christian, NA., Vickers, IE., Harvey, KM., Figueroa, JP., Christie, CD., Bain, B., & Barton, EN. (2005) Human immunodeficiency virus type-1 (HIV-1) subtypes in Jamaica. *West Indian Medical Journal*, Vol. 54, 279- 82.

Heyndrickx, L., Janssens W., Zekeng, L., Musonda, R., Anagonou, S., Van der Auwera, G., Coppens S, Vereecken, K., De Witte, K., Van Rampelbergh, R., Kahindo, M., Morison, L., McCutchan, FE., Carr, JK., Albert, J., Essex, M., Goudsmit, J., Asjö, B., Salminen, M., Buvé, A., & van Der Groen, G. (2000). Simplified strategy for detection of recombinant human immunodeficiency virus type 1 Group M Isolates by *gag/env* heteroduplex mobility assay. *Journal of Virology*, Vol. 74, pp. 363-70.

Heyndrickx, L., Janssens, W., Coppens, S., Vereecken, K., Willems, B., Fransen, K., Colebunders, R., Vandenbruaene, M., & van der Groen, G.(1998). HIV-1 C2V3 *env* diversity among Belgian individuals. *AIDS Res Hum Retroviruses*, Vol. 14, pp. 1291-6.

Ho, DD., Neuman, AU., Perelson, AS., Chen, W., Leonard, JM., & Markowitz, M. (1995). Rapid turnover of plasma virions and CD4 lymphocytes in HIV-1 infection. *Nature*, Vol. 373, pp. 123-6.

Hogg RS, Yip B, Kully C, Craib KJ, O'Shaughnessy MV, Schechter MT, Montaner JS. Improved survival among HIV infected patients after initiation of triple-drug antiretroviral regimens. Can Med Assoc J 1999; 160: 659-65.

Holmes, EC., Zhang, LQ., Robertson, P., Cleland, A., Harvey, E., Simmonds, P., & Leigh Brown, AJ. (1995). The molecular epidemiology of human immunodeficiency virus type 1 in Edinburgh. *Journal of Infectious Diseases*, Vol. 171, pp. 45-53.

Holmes, EC., Zhang, LQ., Simmonds, P., Rogers, AS., & Brown, AJ.(1993). Molecular investigation of human immunodeficiency virus (HIV) infection in a patient of an HIV-infected surgeon. *Journal of Infectious Diseases*, Vol. 167: pp. 1411-14.

Hu, DJ., Buve,, A, Baggs J., van der Groen, G., & Dondero, T. (1999). What role does HIV-subtype play in transmission and pathogenesis? An epidemiological perspective. *AIDS*, Vol. 13, 873-81.

Hu, DJ., Subbarao, S., Vanichseni, S., Mock, PA., Ramos, A., Nguyen, L., Chaowanachan, T., Griensven ,F., Choopanya, K., Mastro, TD., & Tappero, JW (2005). Frequency of HIV-1 dual subtype infections, including intersubtype superinfections, among injection drug users in Bangkok, Thailand. *AIDS*, Vol. 19, pp. 303-8.

JAMA (2006). Treatment for adult HIV infection: 2006 recommendations of the international AIDS Society. Vol. 296, pp. 827-843.

Jeeninga, RE., Hoogenkamp, M., Armand-Ugon, M., de Baar M, Verhoef, K., & Berkhout, B. (2000). Functional differences between the long terminal repeat transcriptional promoters of human immunodeficiency virus -type 1 subtypes A through G. *Journal Virology*, Vol. 74, pp. 3740-51.

Jubier-Maurin, V., Saragosti, S., Perret, JL., Mpoudi, E., Esu-Williams, E., Mulanga, C., Liegeois, F., Ekwalanga, M., Delaporte, E., & Peeters M. Genetic characterization of the *nef* gene from human immunodeficiency virus type 1 group M strains representing genetic subtypes A, B, C, E, F, G and H. *AIDS Res Hum Retroviruses*, Vol. 15, pp. 23-32.

Kaleebu P, French N, Mahe C, Yirrell D, Watera C, Lyagoba F, Nakiyingi J, Rutebemberwa A, Morgan D, Weber J, Gilks C, Whitworth J. Effect of human immunodeficiency virus (HIV) type 1 envelope subtype A and D on disease progression in a large cohort of HIV positive persons in Uganda. J infect Dis 2002; 185: 1244-50.

Kanki PJ, Hamel DJ, Sankalé JL, Hsieh C, Thior I, Barin F, Woodcock SA, Guèye-Ndiaye A, Zhang E, Montano M, Siby T, Marlink R, NDoye I, Essex ME, MBoup S. Human immunodeficiency virus type 1 subtypes differ in disease progression. J infect Dis 1999; 179: 68-73.

Kwong, PD., Wyatt, R., Robinson, J., Sweet, RW., Sodroski, J., & Hendrickson, WA (1998). Structure of an HIV gp120 envelope glycoprotein in complex with the CD4 receptor and a neutralizing human antibody. *Nature*, VOl. 393, pp. 648-59.

Lamers, SL., Sleaman, JW., She, JX., Barrie, KA., Pomeroy, SM., Barrett, DJ., & Goodenow MM. (1993). Independent variation and positive selection in *env* V1 and V2 domains with maternal-infant strains of human immunodeficiency virus type 1 *in vivo*, *Journal of Virology*, Vol. 67, pp. 3951-60.

Laukkanen, T., Carr, JK., Janssens, W., Liitsola, K., Gotte, D.,, McCutchan, FE., Op de Coul, E., Cornelissen, M., Heyndrickx, L., van der Groen, G., & Salminen, MO. (2000). Virtually full-length subtype F and F/D recombinant HIV-1 from Africa and South America. *Virology*, Vol. 269, pp.95-104.

Leitner, T & Fitch, WM. (1999).The phylogenetics of known transmission histories. In: The Evolution of HIV. Crandall KA, (ed). *Johns Hopkins University Press*, Baltimore pp. 315.

Leitner, T., Foley, B., Hahn B et al. (2005). HIV sequence compendium 2005. *Los Alamos National Laboratory*, New Mexico.

Leitner, T., Korber, B., Daniels, M., Calef, C., & Foley, B (Eds). (2005). HIV-1 subtypes and circulating recombinant form (CRF) reference sequences. In: *HIV Sequence Compendium*, pp. 41-8. Los Alamos National Laboratory, NM 87545, USA.

Loemba, H., Brenner, B., Parniak, MA., Ma'ayan, S., Spira, B., Moisi, D,. Oliveira, M., Detorio, M, & Wainberg, MA. (2002). Genetic divergence of human immunodeficiency virus type 1 (HIV-1) Ethiopian clade C reverse transcriptase (RT) and rapid development of resistance against non- nucleoside inhibitors of RT. *Antimicrobial Agents and Chemotherapy*, Vol. 46, 2087-94.

Loemba, H., Brenner, B., Parniak, MA., Ma'ayan, S., Spira, B., Moisi, D., Oliveira, M., Detorio, M., Essex, M., & Wainberg, MA.(2002) Co-receptor usage and HIV-1 inter-clade C polymorphisms in the protease and reverse transcriptase (RT) of HIV-1 isolates from Ethiopia and Botswana. *Antiviral Therapy*, Vol. 7pp. 141-8.

Loemba, H., Brenner, B., Parniak, MA., Ma'ayan, S., Spira, B., Moisi, D., Oliveira, M., Detorio, M., Essex, M., & Wainberg, MA. (2002). Polymorphisms of cytotoxic T-lymphocyte (CTL) and T-helper epitopes within reverse transcriptase RT) of HIV-1 subtype C from Ethiopia and Botswana following selection of antiretroviral drug resistance. *Antiviral Res Vol*. 56, pp. 129-42.

Loussert-Ajaka, I., Menu, E., Apetrei, C., Peters, M., Damond, F., & Mauclere, P. (1998). HIV-1 diversity and the reliability of the heteroduplex mobility assay. *AIDS* Res and Human Retroviruses, Vol. 14, pp. 877-83.

Louwagie, J., McCutchan, FE., Peeters, M., Brennan, TP., Sanders-Buell, E., Eddy, GA., van der Groen, G., Fransen, K., Gershy-Damet, GM., & Deleys, R. (1993). Phylogenetic analysis of the *gag* genes from 70 international HIV-1 isolates provides evidence for the multiple genotypes. *AIDS*, Vol. 7, pp.769- 80.

Luo, CC., Downing, RG., Dela, Torre N., Baggs, J., Hu, DJ., Respess, RA., Candal, D., Carr, L., George, JR., Dondero, TJ., Biryahwaho, B., & Rayfield MA. (1998). The development and evaluation of a probe hybridisation method for subtyping HIV type 1 infection in Uganda. *AIDS Res Hum Retroviruses*, Vol. 14, pp. 691-4.

Marozsan, AJ., Moore, DM., Lobritz, MA., Abraha, A., Reeves, JD., & Arts EJ.(2005). Differences in the fitness of two diverse wild-type human mmuno- deficiency virus type 1 isolates are related to the efficiency of cell binding and entry. *Journal of Virology*, Vol. 79, pp. 7121-34.

Matsuzaki, Y., Sugawara, K., Takashita, E., Muraki, Y., Hongo, S., Katsushima, N., Mizuta, K., & Nishimura H. (2004). Genetic diversity of influenza B virus: The frequent reassortment and co-circulation the genetically distinct reassortant viruses in a community. *Medical Virology*, Vol. 74, pp. 132-140.

Michael, NL. (1999). Host genetic influences on HIV-1 pathogenesis. *Current Opinion Immunology*, Vol. 11, pp. 466-4.

Miller, MD., Margo,t N, McColl, D., & Cheng, AK.(2007). K65R development among subtype C HIV-1 infected patients in tenofovir DF clinical trials. *AIDS*, Vol. 21, pp. 265-6.

Ministry of Health, Jamaica. National HIV/STI prevention and control program facts and figures. (January-June 2007). HIV/AIDS epidemic update.

Monno, L., Brindicci, G., Lo Caputo, S., Punzi, G., Scarabaggio, T., Riva, C., Di Bari, C., Pierotti, P., Saracino, A., Lagioia, A., Mazzotta, F., Balotta, C., & Angarano, G. (2005). HIV-1 subtypes and circulating recombinant forms (CRFs) from HIV-infected patients residing in two regions of Central and Southern Italy. *Journal of Medical Virology*, Vol. 75, pp. 483-90.

Montano, MA., Novitsky, VA., Blackard, JT., Cho, NL., Katzenstein, DA., & Essex M. (1997). Divergent transcriptional regulation among expanding human immunodeficiency virus type 1 subtypes. *Journal of Virology* Vol. 71, 8657- 65.

Morb Mortal Wkly (MMW) Report. (1994). Centers for Disease Control and Prevention (CDC). Human immunodeficiency virus transmission in household settings - United States. Vol. 43, pp. 353-6.

Morb Mortal Wkly (MMW, 1981) Report. Pneumocystis pneumonia – Los Angeles.; ASM press, Vol. 30, pp. 250-5.

Mulder-Kampinga, GA., Kuiken, C., Dekker, J., Scherpbier, HJ., Boer K., & Goudsmit, J. (1993). Genomic human immunodeficiency virus type 1 RNA variation in mother and child following intra-uterine virus transmission. *Journal of General Virology*, Vol. 74, pp. 1747-56.

Murphy G, Belda FJ, Pau CP, Clewley JP, Parry JV. Discrimination of subtype B and non-subtype B strains of human immunodeficiency virus-type 1 by serotyping: correlation with genotyping. J Clin Microbiol 1999; 37: 1356-60.

Murray, PR. Baron, EJ. Jorgensen, JH. Landry, ML. Pfaller, MA (Eds). (2007) *Manual of Clinical Microbiology* (9th Ed), ASM Press, Washington D.C.

Myers G, Hahn BH, Mellors JW et al. Human Retroviruses and AIDS 1995: A Compilation and Analysis of Nucleic Acid and Amino Acid Sequences. Los Alamos National Laboratory, Los Alamos 1995.

Novitski, V., Arnold, C., & Clewley, JP. (1996). Heteroduplex mobility assay for subtypingHIV-1: Improved methodology and comparison with phylo- genetic analysis of sequence data. *Journal Virology Methods*, Vol. 59, pp. 61-72.

Office of AIDS Research Advisory Council (OARAC) January 2008; Guidelines for the use of antiretroviral

Ou, CY., Ciesielski, CA., Myers, G., Bandea, CI., Luo, CC., Korber, BT., Mullins, JI., Schochetman, G., Berkelman, RL., & Economou, AN.(1992). Molecular epidemiology of HIV transmissions in a dental practice. *Science*, Vol. 256, pp. 1165-71.

Papuashvili, MN., Novokhatsky, AS., & Shcherbakova, TI. (2005). Characteristics of HIV-1 *env* V3 loop sequences for subtype A1 variant spread in Eastern Europe. *Infect Genet Evol*, Vol. 5, pp. 45-53.

Peeters, M., & Sharp, PM. (2000). Genetic diversity of HIV-1: the moving target. *AIDS*, Vol. 14, pp. S129-S140.

Peeters, M., Liegeois, F., Bibollet-Ruche, F., Patrel, D., Vidal, N., Esu-Wiliams, E., Mboup, S., Mpoudi Ngole, E., Koumare, B., Nzila, N., Perret, JL., & Delaporte E. (1998). Subtype-specific polymerase chain reaction for the identification of HIV-1 genetic subtypes circulating in Africa. AIDS, Vol. 12, pp.671-3.

Pérez, L., Thomson, MM., Bleda, MJ., Aragonés, C., González, Z., Pérez, J., Sierra, M., Casado, G., Delgado, E., & Nájera, R. (2006). HIV Type 1 molecular epidemiology in Cuba: high genetic diversity, frequent mosaicism and recent expansion of BG intersubtype recombinant forms. *AIDS Res Hum Retroviruses*, Vol. 22, pp. 724-33.

Pérez-Alvarez, L., Muñoz, M., Delgado, E., Miralles, C., Ocampo, A.,& García, V. (2206). Thomson M, Contreras G, Nájera R; Spanish Group for Antiretroviral Resistance Studies in Galicia. Isolation and biological characterization of HIV-1 BG intersubtype recombinants and other genetic forms circulating in Galicia, Spain. Journal of Medical Virology, Vol. 78, pp. 1520-8.

Piot, P., & Bartos, M. (2002). The epidemiology of HIV and AIDS. In: *AIDS in Africa,* (2nd ed). Essex, M., Mboup, S., Kanki, PJ., Marlink, PJ., & Tlou, SD. (Eds). Kluwer Academic/Plenum, New York, p. 200.

Piot, P., Bartos, M., Ghys, PD., Walker, N., Schwartländer, B. (2001). The global impact of HIV/AIDS. *Nature*, VOL. 410, pp. 968 –73.

Pollakis, G., Abebe, A., Kliphuis, A., Chalaby, MI., Bakker, M., Mengistu, Y., Brouwer, M., Goudsmit, J., Schuitemaker, H., Paxton, WA. (2004). Phenotypic and genotypic comparisons of CCR5-and CXCR4-tropic human immuno-deficiency virus type 1 biological clones isolated from subtype C-infected individuals. Journal of Virology Vol. 78, pp. 2841-52.

Quinones-Mateu, ME., & Arts, EJ. (1999). Recombination in HIV-1: Update and implications. *AIDS Reviews*, Vol. 1, pp. 89-100.

Quinones-Matue, ME., Albright, JL., Mas, A., Soriano, V., & Arts, EJ. (1998). Analysis of pol gene heterogeneity, viral quasi-species, and drug resistance in individuals infected with group O strains of human immunodeficiency virus type 1. Journal of Virology, Vol. 72, pp. 9002-15.

Ramos, A., Tanuri, A., Schechter, M., Rayfield, MA., Hu, DJ., Cabral, MC., Bandea, CI., Baggs, J., & Pieniazek, D. (1999). Dual and recombinant infection: an integral part of the HIV-1 epidemic in Brazil. Emerging Infect Diseases, Vol. 5, pp.65-74.

Renjifo, B & Essex, M. HIV-1 subtypes and recombinants.(2002). In: *AIDS in Africa*, (2nd edn). Essex, M., Mboup, S., Kank,i PJ., Marlink, PJ., & Tlou, SD. (Eds). Kluwer Academic/Plenum. New York, p. 138.

Rizzuto, CD., Wyatt, R. (1998). Hernandez-Ramos N, Sun Y, Kwong PD, Hendrickson WA, Sodroski JA. Conserved HIV gp120 glycoprotein structure involved in chemokine receptor binding. Science, Vol. 280, pp. 1949-53.

Robbins KE, Kostrikis LG, Brown TM, Anzala O, Shin S, Plummer FA, Kalish ML. Genetic analysis of human immunodeficiency virus -type 1 strains in Kenya: a comparison using phylogenetic analysis and a combinatorial melting assay. AIDS Res Hum Retroviruses 1999; 15: 329-35.

Robertson, DL., Anderson, JP., Bradac, JA., Carr, JK., Foley, B., Funkhouser, RK., Gao, F., Hahn ,BH., Kalish, ML., Kuiken, C., Learn ,GH., Leitner, T., McCutchan, F., Osmanov, S., Peeters, M., Pieniazek, D., Salminen, M., Sharp, PM., Wolinsky, S., & Korber, B. (2000) HIV-1 nomenclature proposal. *Science*, Vol. 288, pp. 55-6.

Robertson, DL., Hahn, BH., & Sharp, PM. (1995). Recombination in AIDS viruses. *Journal Molecular Evolution*, Vol. 40, pp. 249-59.

Sankale, JL., M'Boup, S., Kim, B., Tovanabutra, S., Hamel, DJ., Brodine, SK., Kanki, PJ., & Birx, DL.(2004). HIV-type 1 circulating recombinant form CRFO9_cpx from West Africa combines subtypes A, F, G and may share with CRFO2_AG and Z321. *AIDS Res Hum Retroviruses*, Vol. 20, pp. 819-26.

Santos, AF., Sousa, TM., Soares, EA., Sanabani, S., Martinez, AM., Sprinz, E., Silveira, J., Sabino, EC., Tanuri, A., & Soares, MA. (2006). Characterization of a new circulating recombinant form comprising HIV-1 subtypes C and B in southern Brazil. *AIDS*, Vol. 20, pp. 2011-19.

Scarlatti, G., Leitner, T., Halapi, E., Wahlberg, J., Marchisio, P., Clerici-Schoeller, MA., Wigzell, H., Fenyö, EM., Albert, J., & Uhlén M. (1993). Comparison of variable region 3 sequences of human immunodeficiency virus type 1 from infected children with RNA and DNA sequences of the virus populations of their mothers. *Proc Natl Acad Sci USA*, Vol. 90, pp. 1721-5.

Schochetman, G., Epstein, JS., & Zuck ,TF. (1989). Serodiagnosis of infection with   the AIDS virus and other human retroviruses. *Ann Rev Microbiol* , Vol. 43, pp. 629-59.

Sheppard, HW., Busch, MP., Louie, PH., Madej, R., & Rodgers, GC. (1993). HIV-1 PCR and isolation seroconverting and seronegative homosexual men: absence of long-term immunosilent infection. *Journal Acquir Immune Defic Syndr*, Vol. 6, pp. 1339-46.

Simmonds, P., Balfe, P., Ludlam, CA., Bishop, JO., & Brown, AJ. (1990)Analysis of sequence diversity in hypervariable regions of the external glycoprotein of human immunodeficiency virus-type 1. *Journal of Virology*, Vol. 64 pp. 5840-50.

Spira, S., Wainberg, MA., Loemba, H., Turner, D., & Brenner, BG.(2003). Impact of clade diversity on HIV-1 virulence, antiretroviral drug sensitivity and drug resistance. *Journal of Antimicrobial Chemotherapy*, Vol. 51, pp. 229-40.

Swofford, DL. (2003). PAUP. Phylogenetic analysis using Parsimony (and other methods). Version 4. *Sinauer   Associates, Sunderland* agents in HIV-1 infected adults and adolescents.

Tatt, ID., Barlow, KL., & Clewley, JP. (2000). A *gag* gene heteroduplex mobility assay   for subtyping HIV-1. *Journal of Virology Methods*, Vol. 87, pp.41-51.

Tatt, ID., Barlow, KL., Nicol, A., & Clewley, JP. (2001). The public health significance of HIV-1 subtypes. *AIDS*, Vol. 15 (suppl 5): pp. s59-s71.

Taylor, BS., Sobieszczyk, ME., McCutchan, FE., & Hammer, SM. (2008). The challenge of HIV-1 subtype diversity. *N England Journal of Medicine*, Vol. 358, pp. 1590-602.

Tersmette, M., de Goede, RE., Al, B., Winkel, IN., Gruters, RA., Cuypers, HT., Huisman, HG., & Miedema, F. Differential syncytium-inducing capacity of human immunodeficiency virus isolates: frequency detection of synctium-inducing isolates in patients with acquired immunodeficiency syndrome (AIDS) and AIDS-related complex. Journal of Virology Vol. 62, pp. 2026-32. Thompson, JD., Gibson, TJ., Plewniak, F., Jeanmougin, F., & Higgins, DG. (1997). The Clustal_X windows interface: Flexible strategies for multiple sequence alignment aided by quality analysis tools. *Nucleic Acids Res*, Vol. 25, pp. 4876-82.

Thompson, JD., Higgins, DG., & Gibson, TJ. (1994). Clustal W: improving the sensitivity of progressive multiple sequence alignment through sequence weighting, position-specific gap penalties and weight matrix choice. *Nucleic Acids Res*, pp. 22, pp. 4673-80.

Thomson, MM., Casado, G., Posada, D., Sierra, M., & Najera, R.(2005). Identification of a novel HIV-1 complex circulating recombinant form (CRF18_cpx) of African origin in Cuba. *AIAS, Vol,* 19, pp. 1155-63.

Tovanabutra, S., Beyrer, C., Sakkhachornphop, S., Razak, MH., Ramos, GL., Vongchak, T., Rungruengthanakit, K., Saokhieo, P., Tejafong, K., Kim, B., De Souza, M., Robb, ML., Birx, DL,, Jittiwutikarn, J., Suriyanon, V., Celentano, DD., & McCutchan, FE. (20004). The changing molecular epidemiology of HIV type 1 among Northern Thai drug users, 1999 to 2002. *AIDS Res Hum Retroviruses*, Vol. 20, pp. 465- 75.

UNAIDS (December 2004). USAIDS epidemic update. Joint United Nations Programme on HIV/AIDS (UNAIDS) World Health Organization (WHO)

UNAIDS/WHO (December 2006). Joint United Nations Programme on HIV/AIDS.

van der Auwera, G., & Heyndrickx, L. (December 2000). HIV-1 group M *gag* heteroduplex mobility analysis (HMA) subtyping kit Protocol Version 3, pp. 1- 15.

van Harmelen, J., van der Ryst, E., Wood, R., Lyons, SF., & Williamson, C. (1999).Restriction fragment length polymorphism analysis for rapid *gag* subtype determination of human immunodeficiency virus type 1 in South Africa. Journal Virology Methods, Vol. 78, pp.51-9.

Vaughan, HE., Cane, P., Pillay, D., & Tedder RS. (2003). Characterization of HIV -type 1 clades in the in the Caribbean using *pol* gene sequences. *AIDS Res   Hum Retroviruses*, Vol. 19, pp. 929-32.

Vaughan, HE., Cane, P., Pillay, D., & Tedder, RS. (2003). Characterization of HIV -type 1.

Velazquez-Campoy, A., Vega ,S.,& Freire, E. (2002). Amplification of the effects of drug resistance mutations by background polymorphisms of HIV-1 protease from African subtypes. *Biochemistry*, Vol. 41, pp. 8613-9.

Vogt MW, Witt DJ, Craven DE, Byington R, Crawford DF, Schooley RT, Hirsch MS. Isolation of HTLV-III/LAV from cervical secretions of women at risk for AIDS. Lancet 1986; i: 525-7, Vol. 19, pp. 929-32.

Wasi, C., Herring, B., Raktham, S., Vanichseni, S., Mastro, TD., Young, NL., Rübsamen-Waigmann, H., von Briesen, H., Kalish, ML., & Luo CC. (1995). Determination of HIV-1 subtypes in injecting drug-users in Bangkok, Thailand using peptide binding enzyme immunoassay and heteroduplex mobility assay: evidence of increasing infection with HIV-1 subtype E. *AIDS*, Vol. 9, pp.843-9.

Weiser, B., Burger, H., Hachman, S., Hsu, YJ., & Gibbs, R. Use of serial HIV-1 sequences from a pregnant woman and her twins to study timing of vertical transmission. *Vaccines* Vol. 1, pp. 31-6.

Worgall S, Connor R, Kaner RJ, Fenamore E, Sheridan K, Singh R, Crystal RG. Expression and use of human immunodeficiency virus type 1 co-receptors by human alveolar macrophages. J Virol 1999; 73: 5865-74.

Wyatt, R., & Sodroski, J. (1998). The HIV-1 envelope glycoproteins: fusogens, antigens, and immunogens. *Science*, Vol. 280, pp. 1884-8.

Wyatt, R., Kwong, PD., Desjardins, E.,, Sweet, RW., Robinson, J., Hendrickson, W A., Sodroski ,JG. (1998) The antigenic structure of the HIV gp120 envelope glycoprotein. *Nature*, Vol. 393, pp. 705-11.

Yang, C., Li, M., Mokili, JL., Winter, J., Lubaki, NM., Mwandagalirwa, KM., Kasal, MJ., Losoma, AJ., Quinn, TC., Bollinger, RC., Lal, & RB. (2005). Genetic diversification and recombination of HIV type 1 group M in Kinshasa, Democratic Republic of Congo. *AIDS Res Hum Retroviruses*, Vol. 21, pp. 661-6.

Zetterberg, V., Ustina, V., Liitsola, K., Zilmer, K., Kalikova, N., Sevastianova, K., Brummer-Korvenkontio, H., Leinikki, P., & Salminen, MO. (2004). Two viral strains and a possible novel recombinant are responsible for the explosive injecting drug use-associated HIV type 1 epidemic in Estonia. *AIDS Res Hum Retroviruses*, Vol. 20, pp. 1148-56.

14

Approaches for Dissection of the Genetic Basis of Complex Disease Development in Humans

Nicole J. Lake, Kiymet Bozaoglu,
Abdul W. Khan and Jeremy B. M. Jowett
Baker IDI Heart and Diabetes Institute, Melbourne,
Australia

1. Introduction

When genome-wide association (GWA) studies emerged as an approach for dissecting the genetic architecture of complex disease, there were great hopes for its potential to unlock the mysteries of complex human disease, a full dissection of which had largely evaded numerous genetic linkage analyses and candidate gene studies. Based on the hypothesis that common variants influenced much of common disease development, GWA focus was initially on common allelic variants with a minor allele frequency of at least 5%. Six years and thousands of these associated loci down the track, GWA studies have seemingly reached a fork in the road. It has become apparent that the influence of common variants on complex disease development can only explain a modest proportion of phenotypic variation. Furthermore difficulty in finding the causal variant for an associated locus has slowed anticipated translation towards the clinic. Consequently there has been wide discussion surrounding the limitations of GWA studies and value of a continued focus on common variants. A key debate therefore has arisen at this nexus that compares the Common Disease-Common Variant against the emerging Common Disease-Rare Variant hypothesis that is now gathering support.

In this chapter we discuss whether further progress can be made with the GWA approach and contrast its potential against a future with increasingly affordable whole-genome sequencing. We also explore the various sequencing strategies that seek to make progress within tight research budgets. Success in elucidating the genetic architecture of complex human disease has the potential to affect many by laying the foundations for understanding the complex mechanisms of disease facilitating translation into new drug treatments and the development of biomarkers for assessment of onset of disease.

2. Background

Complex human diseases represent a serious global health concern. Development of complex common diseases including cardiovascular disease (CVD) and diabetes relies on an intricate network of interactions between genes and environment. The continuum over which a complex disease can be defined is itself influenced by continuous traits such as CVD being influenced by weight and cholesterol levels. It is thus not unexpected that approaches to decode such complex genetic puzzles have been slow in their progress.

GWA studies survey the association of single nucleotide polymorphisms (SNPs) with a disease or trait to identify genetic loci. Commercial chips use a 'tag' SNP for each region of linkage disequilibrium to capture variation across genomes by high-throughput genotyping technologies (Manolio et al. 2009; Witte 2010). The large numbers of samples that can be processed relatively inexpensively provides GWA study designs with excellent statistical power to detect modest effect sizes that have been integral to its success. The utility of GWA produced data has enabled efficient replication studies and consequent reduction of false-positive findings. The advent of GWA was an important advance on the existing approaches used for dissecting human disease and the first approach tailored for complex disease research. Prior to this, linkage analysis used comparatively widely spaced polymorphic markers and determined their co-segregation with the disease or trait in affected families. While this proved effective for isolating rare, highly penetrant variants associated with Mendelian diseases, it struggled to provide disease gene identification due to broad localisation intervals. Additionally the variable heritability of many complex phenotypes and challenges in accurate trait measurement may have contributed to the shortcomings of those linkage studies (Hirschhorn et al. 2005). On the other hand, candidate gene studies offered higher resolution but ignored most of the genome. The reliance on a correct hypothesis for the candidate gene's involvement in disease risk can also result in false-positive associations (Witte 2010) or true associations that were difficult to replicate (Manolio et al. 2009).

The Common Disease-Common Variant (CDCV) hypothesis provided inspiration and optimism that GWA studies were a feasible approach for characterising the genetic architecture of complex common diseases. Largely brought to attention by Lander and Reich (Reich et al. 2001), the CDCV had some attractive qualities. The well established 'Out of Africa' model illustrated proliferation of the global population from a small group of founders, a dynamic which could have allowed mutations to become common (Iyengar et al. 2007). CDCV supporters (Lander 1996; Collins et al. 1998) explained that these common variants that were once neutral, or even beneficial, had become harmful in the context of dramatic environmental changes (Kryukov et al. 2007). As the HapMap project progressed, efficiently characterising patterns of linkage disequilibrium, testing of the CDCV hypothesis became a possibility using the emerging platforms and GWA study design. Free from the constriction of a functional hypothesis and candidate gene focus, GWA studies have revealed numerous loci previously not implicated in complex disease and further allowed investigation of the genetic basis of commonality between multiple complex phenotypes (Frazer et al. 2009). However despite these successes, the validity of the CDCV hypothesis is under challenge (Frazer et al. 2009) and the translation of the findings to useful mechanistic knowledge has been problematic.

3. Genome-Wide Association study designs

A GWA study can be undertaken using several different designs. The traditional GWA study approach has been a case-control design, but cohort studies and family-based approaches also have merit. Each offers unique advantages and drawbacks. Cost, time and intentions for data application are key influences over the optimal choice of design.

3.1 Case-control

Traditionally, disease associated loci were illustrated by comparison of population-based affected 'case' and unaffected 'control' groups. Control groups can be shared between studies without introducing bias (2007; Witte 2010). Case-control designs can thus be organised easily and at minimum costs for large group sizes. The ability to survey unrelated individuals is a strength of GWA studies for ease of ascertainment of study participants. However it also raises issues such as confounding effects of population stratification (Thomas et al. 2002; Wacholder et al. 2002) where the genotyped variants may have different frequencies among ethnically defined strata of a population. The differences in allele frequencies between these strata can create false-positive associations (Dickson et al. 2010). Initially there were significant concerns for the impact of population stratification on the GWA approach (Thomas et al. 2002), however recent findings (Wacholder et al. 2002; 2007; Goldstein 2009; Hindorff et al. 2009) have largely quelled this discussion by showing the confounding effects of stratification were overestimated. This is not to say however that it does not exist, but rather that careful consideration of population samples and genotyping of specific sets of variants can reduce bias and provide the ability to detect and adjust for it if present.

3.2 Cohort

Cohort design is distinct from case-control GWA study design where a sample of the population, the cohort, is randomly ascertained and traits relevant for the disease of interest measured (Manolio 2009). A primary advantage of cohort design is that it allows avoidance of dichotomising phenotypes that reduces the statistical power of GWA (McCarthy et al. 2008). It is thus particularly useful for analysis of quantitative disease-related traits such as body mass index and fasting blood glucose levels. Additionally, environmental exposures and other covariates can easily be investigated in this design (Manolio 2009), allowing dissection of the relative influences of environmental and genetic factors.

3.3 Family-based

Although they require more effort to ascertain, phenotype and sample, family-based studies offer many advantages. Using related groups reduces the likelihood of population stratification (Hirschhorn et al. 2005; Lasky-Su et al. 2010; Witte 2010), while the increased commonality of alleles among families reduces complexity associated with genetic heterogeneity. The ability to perform joint linkage and association analysis is another asset of a family-based approach (Visscher et al. 2008). Relative to unrelated populations, a family study can offer increased efficiency and statistical power per individual genotyped by enabling additional pairings such as sibling and parent-child. General belief that family studies offer substantially reduced power compared to unrelated samples (Hirschhorn et al. 2005; McCarthy et al. 2008; Lasky-Su et al. 2010; Witte 2010) stems from misinterpretation of an earlier publication by Risch and Merikangas (Risch et al. 1996) and as explained by Blangero (Blangero 2004). Their findings were based on comparison of an ill-advised family design with an optimal association design (Blangero 2004). For common complex diseases, the statistical power of a more efficient family design can be shown to be similar to GWA of unrelated individuals (Figure 1) (Laird et al. 2006). Indeed it has been argued that the advantages offered by a related group outweigh the small statistical power lost (Visscher et

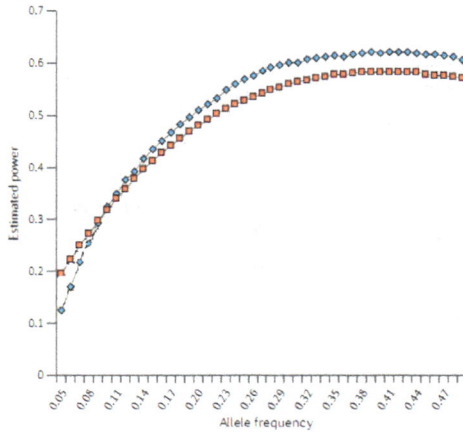

Fig. 1. The statistical power of GWA for blue case-control compared to red family trio (affected child and parents). This is modeled for a disease with 14% prevalence, similar to CVD in Australia. It should also be noted that a family design is more powerful for low frequency variants. Adapted from(Laird et al. 2006)

al. 2008). Additionally family designs allow investigation of the influence of parent of origin effects of an allele which is being increasingly recognised as an important factor in assessing disease risk (Kong et al. 2009; Manolio et al. 2009; Agopian et al. 2011). The finding that variants from parental origin are likely to confer increased risk for complex diseases (Kong et al. 2009) supports this approach. Despite favour for one or the other (Bell et al. 1997; Visscher et al. 2008), complementation of both related and unrelated groups will be important for a robust approach.

4. Limitations of Genome-Wide Association studies

GWA studies have been an unrefuted success in finding trait/disease-associated SNPs (TASs) (Hindorff et al. 2009) and candidate genes, however they have produced disappointment in establishing causal variants and disease mediating genes, that has arisen in part due to the unexpected complexity of gene transcriptional regulation and variable linkage disequilibrium structure. Furthermore, for the overwhelming majority of complex diseases studied, the collective effects of associated variants can only explain a small proportion of the trait heritability (Manolio et al. 2009). There are several possible explanations for this so called "missing heritability" that are further explored in this section.

4.1 GWA only detects associated loci and not the causal variant directly

In most cases GWA studies do not directly identify the causal variant. Once a region has been characterised as harbouring a causal locus, detailed study of all genetic variants in linkage disequilibrium with the TAS is needed to discover the causal variant. This is a difficult and time consuming process, and so a candidate gene approach relying on previously known functions of neighboring genes and their relationship to the associated trait is often used to provide direction. If the TAS lies within a gene coding sequence, causative gene selection can be straightforward, however examples of associated SNPs within gene introns regulating another more distant gene show that selection based on proximity can by myopic (Jowett et al. 2010; Ragvin et al. 2010). Furthermore it is difficult to assign TASs that lie within an intergenic gene desert. Given that approximately 88% of TASs are located in intronic and intergenic regions (Hindorff et al. 2009) the process of identifying the causal gene is problematic. Table 1 lists how SNP location in relation to gene structure may affect a trait influencing gene.

Once the causal variant is identified, understanding how it contributes to the disease phenotype also remains challenging, particularly for non-coding polymorphisms. On account of SNP selection on high density SNP chips, GWA studies have driven investigation into intronic and intergenic regulation of gene expression. Increased understanding about the role of non-coding RNA has also illustrated a means for intergenic variants to influence gene expression. Characterisation of intergenic SNPs within long non-coding RNA has begun (Pasmant et al. 2011), and may be expected to continue.

SNP location		Potential mechanisms
Coding region	Promoter region	Affect transcription factor binding[1]
	UTR	Affect enhancer activity, binding of translational regulators[2]
	Exon	Missense mutation[3]
	Non-coding RNA target site	Affect miRNA regulation[4]
	Non-coding RNA sequence	Influences miRNA production[5]
Intron	Splice junction	Influence alternative splicing[6]
	Outside splice junction	Influence transcriptional activity[6]
Intergenic	Intergenic non-coding DNA	Influences gene regulation[7]
	Non-coding RNA sequence	Influence long non-coding RNA target regulation and heterochromatin formation[8]

Table 1. Potential biological consequences of a genetic variant in relation to its location. Examples of these have been described in the following; [1](Hindorff et al. 2009), [2](He et al. 2009), [3](Rutter 2010), [4](Zhang et al. 2011), [5](Ryan et al. 2010), [6](Cooper 2010), [7](Jia et al. 2009), [8](Pasmant et al. 2011).

4.2 GWA cannot discriminate against true and 'indirect' associations

Recently Dickson et al (Dickson et al. 2010) found that the strength of an association signal can be strongly affected when there is a large difference in allele frequency between the genotyped common tagging variant and causal variant. This concept of a TAS representing a diluted signal from a neighboring causal variant was named a 'synthetic association' (Dickson et al. 2010). The hypothesis proposes that when one or more rare causal variants are enriched in a haplotype with a particular allele of the common variant, the association may be falsely assigned. This raises the possibility that a common variant association may instead represent several rare variants of potentially larger effect, and consequently the true strength of the association may have been underestimated. Furthermore these rare variants may act over large distances up to megabases to create synthetic associations (Dickson et al. 2010). This suggests that the region typically selected for re-sequencing (~500kb) may have been underestimated, and thus causal variants missed. The synthetic association hypothesis is supported by the Crohn's disease associated loci NOD2 which features three rare coding variants that drive a strong association signal at nearby common variants (Wang et al. 2010).

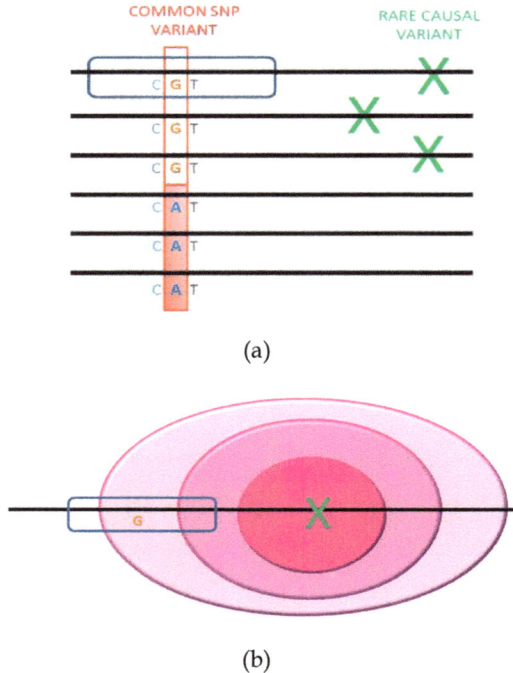

(a)

(b)

Fig. 2. (a) Synthetic Associations occur when causal rare variants are inherited more commonly with one tag SNP allele (e.g. "G") than the other ("A") allele (black lines represent different individuals). This rare variant may be outside the region of linkage disequilibrium defined by the common SNP (shown by the blue box). Adapted from (Cirulli et al. 2010). (b) Because the rare variant is too infrequent to be detected by GWAS, the true strength of the association signal will be missed. A diluted version of this signal will instead be represented by the common variant (Dickson et al. 2010).

However, debate has arisen over the validity of the synthetic association paradigm and its applicability to all common TAS (Orozco et al. 2010; Anderson et al. 2011; Wray et al. 2011). Since rare variants are likely to have arisen relatively recently, they are unlikely to be shared by divergent ethnic groups. Consequently, common variants that reflect a signal driven by rare alleles, or synthetic associations, are also expected to be largely population specific (Orozco et al. 2010; Anderson et al. 2011). A locus that harbours a cluster of low frequency variants that confer high risk of disease should be amenable to linkage mapping (Orozco et al. 2010; Anderson et al. 2011). Therefore, associations that are sensitive to linkage analysis but not ethnically replicated potentially represent synthetic associations. In light of widespread trans-ethnic associations and discrepancies between linkage and GWA loci, it has been argued that synthetic associations are unlikely to explain the majority of GWA signals (Orozco et al. 2010; Waters et al. 2010; Anderson et al. 2011). The hypothesis remains to be further evaluated with the aid of high-throughput sequencing efforts (Orozco et al. 2010). If synthetic associations prove common, commentary about the large role of regulatory variants (Hardy et al. 2009; Hindorff et al. 2009) in complex phenotypes may be a little premature as non-coding common variants may instead represent a causal variant within a coding region megabases away.

4.3 GWA is limited to common variants

The limitation of GWA to detecting only common variants is central to literary debate over missing heritability sources. In the context of the CDCV hypothesis this was initially a logical approach, however large missing heritability for almost all complex diseases (Table 2) suggest that the CDCV may not be sustainable and that the influence of common variants on complex phenotypes has been overestimated. Despite the impressive number of variant sites genotyped by high density SNP chips, a full genome wide survey remains incomplete, with up to 20% of common SNPs not sufficiently tagged (Frazer et al. 2009). This has arisen mainly due to the remarkable and perhaps originally underestimated genetic diversity of the human species. Another confounder has been the observed complex patterns of linkage disequilibrium (LD), where so called "LD blocks" tend to fragment and fail when lower frequency variants are considered. The result is that for the most part, lower frequency variants are missed by GWA studies. Inability to detect these low frequency variants has been suggested widely in the literature as contributing significantly to missing heritability (Hirschhorn et al. 2005; Hardy et al. 2009; Manolio et al. 2009; Manolio et al. 2009; Schork et al. 2009; Cirulli et al. 2010; Dickson et al. 2010; Park et al. 2010). In this context the potential for rare variants with moderate effect sizes to influence common disease has been recognised (Manolio et al. 2009; Schork et al. 2009; Carvajal-Carmona 2010). Furthermore this paradigm has been suggested as more consistent with evolutionary genetics where less harmful variants with smaller effect sizes remain under less pressure from the action of negative selection and are therefore more likely to be common (Schork et al. 2009).

	Associated loci	Heritability	Heritability explained
Type 1 diabetes (T1D)	45 [1]	0.8 [2]	31% [3]
Type 2 diabetes	46 [4,5]	0.42 [2]	10% [6]
Crohn's disease	71 [7]	0.55 [2]	23% [7]
Height	51 [8]	0.8 [9]	5% [9]

Table 2. Despite large numbers of associated loci identified, many complex diseases and traits have only a small proportion of heritability explained. Heritability is the proportion of phenotype explained by genetics.[1] (Burren et al. 2011), [2](So et al. 2011), [3](Clayton 2009), [4](Kooner et al. 2011), [5](Parra et al. 2011), [6](Imamura et al. 2011), [7](Fransen et al. 2011), [8](Zhao et al. 2010), [9](Yang et al. 2010).

4.4 Missing heritability

Despite numerous GWA studies and the identification of dozens of trait associated variants for each complex disease, a large proportion of heritability remains unexplained for almost all complex phenotypes examined (Table 2). The study of height illustrates the gap between discovered and estimated heritability. Extensive GWA has revealed 51 loci significantly associated height, a complex trait with high heritability of 0.80, yet these variants combined only explained 5% of the heritability (Yang et al. 2010; Zhao et al. 2010). The small contribution of these common variants to complex phenotypic variance has also limited utility of these markers for disease predication (Hirschhorn 2009). If the proportion of heritability explained is regarded as a measure of the success of the GWA approach, then its performance in elucidating complex disease genetics may appear underwhelming. However against the backdrop of many additional contributing factors such as rare variants, copy number variants, epigenetic modifications and epistasis, the achievement of the GWA approach can be considered solid progress.

4.4.1 Copy number variants

Copy number variants (CNVs) including insertions and deletions ranging from a few bases up to megabases in size have been highlighted as having potential to contribute to complex phenotypes (Frazer et al. 2009; Manolio et al. 2009; Cirulli et al. 2010). A broad study across 8 complex diseases by the Welcome Trust Case Control Consortium using existing datasets to tag CNVs found that for those CNVs typable on the employed platforms, they were unlikely to play a major role in the determination of the genetic basis of common diseases (Craddock et al. 2010). A key caveat as highlighted in the study was what could be measured by the SNP chip platforms. As noted by Alkan et al (Alkan et al. 2011), these platforms fail to detect CNV's less than 500 bp in size. These small CNVs thought to arise during recombination events and consequently are both difficult to detect and difficult to genotype. However, new sequencing data is emerging showing that this class of small CNV is not only pervasive but also likely to contribute significantly to disease risk. Therefore it is too early to dismiss the potential contribution of CNVs to complex disease development.

4.4.2 Epigenetics

Heritable epigenetic modifications to DNA and histones may be another potential contributor to complex disease processes (Bell et al. 2011). Epigenetic marks include post-translational modification of histones, non-coding RNA and DNA methylation (Rakyan et al. 2011). Epigenetic reprogramming of zygotes is used in arguing against meiotic transmission of epigenetic states, however several observations suggest that retention is possible (Rideout et al. 2001; Rakyan et al. 2003; Blewitt et al. 2006; Bell et al. 2011). Results from the study of monozygote twins show that they have more similar epigenetic profiles relative to dizygotic twins (Fraga et al. 2005). A complicating factor for investigation of epigenetic changes is that they are often tissue specific. Unlike GWA studies where most tissues are suitable for identifying germ line genetic variation, easily accessible samples such as blood may not efficiently reflect robust epigenetic alterations in other tissues that carry an impact on disease development. Since many tissue types that are important to disease development are effectively inaccessible due to unacceptable risks to the patient, such as brain, the development of comprehensive epigenetic profiles may be limited. Furthermore, intra tissue heterogeneity presents an additional complication where a tissue, such as adipose, may be infiltrated with other cell types such as macrophages as observed in states of obesity. Reliable solutions that resolve these confounding issues are yet to be developed.

4.4.3 Environment

Phenotypic variation is traditionally considered as a result of two sources of variation; genetic and environmental (Richards 2006). While the magnitude of environmental effects on complex disease is disputed (Hardy et al. 2009; Rappaport et al. 2010), their influence on common disease development is generally accepted (Bell et al. 1997; Manolio 2009; Manolio et al. 2009; Cirulli et al. 2010). It has been proposed that the smaller the odds ratio, the more likely that environmental factors predominate (Bodmer et al. 2008) and hence a need for a more comprehensive measure of environmental exposure and its interaction with genetic variants is required (Hemminki et al. 2006; Murcray et al. 2009; Eichler et al. 2010; Rappaport et al. 2010). However if a study does record environmental factors they are usually by participant self report and can be biased by inaccurate and variable recall, leading some researchers to dismiss their inclusion as they generate too much noise. It is therefore not surprising that gene-environment interactions are currently poorly understood.

The novel characterisation of the 'exposome' by Rappaport and Smith (Rappaport et al. 2010) provides a basis to explore gene-environment interactions. The exposome represents all exposures that impact the internal chemical environment of humans (Rappaport et al. 2010). GWA studies have the potential to identify SNPs that demonstrate heterogeneity between subgroups defined by an environmental exposure (Murcray et al. 2009) and could potentially reveal loci that might otherwise have been dismissed. A good starting point for maximising GWA information about gene-environment interactions is to perform environmental-wide association studies as undertaken by Patel et al (Patel et al. 2010). This approach using exposome biomarkers can be used to propose novel environmental factors for further study, such as γ-tocopherol which was positively associated with type 2 diabetes risk (Patel et al. 2010). Notably γ-tocopherol is the most abundant component of Vitamin E in the US diet, and represents up to 50% of total Vitamin E in insulin-target tissues (Patel et al. 2010) making it a plausible candidate. A GWA study with γ-tocopherol may reveal candidates for gene-environment interaction points.

Evidence has shown that the environment can trigger heritable epigenetic changes (Jirtle et al. 2007; Ng et al. 2010) however this requires more investigation to determine the extent of influence on gene expression. Indeed it has been argued that GWA studies might be misleading without consideration of environmental factors (Bell et al. 1997). Additionally the observation that some environmental risk factors can have a 'heritable' component, such as a heritability estimate of 0.6 for regular tobacco use (Kendler et al. 2000), suggests that the environmental component of complex disease can be "contaminated" with heritability arising from behavioral and risk taking psychological traits (Petronis 2010).

4.4.4 Is the metric wrong?

Realisation that epigenetic variation can contribute to complex phenotypic variation (Richards 2006; Petronis 2010; Bell et al. 2011) has important implications for heritability estimates. Discussion around the potential for human transgenerational epigenetic inheritance (Richards 2006; Petronis 2010; Bell et al. 2011; Rakyan et al. 2011) has suggested that the genetic contribution to complex phenotypes may have overestimated. Therefore attributing heritability estimates solely to genetic variants may be short sighted if epigenetic marks are significantly heritable, as Slatkin suggests (Slatkin 2009). Heritability can differ temporally and between environments (Heath et al. 1998; Turkheimer et al. 2003; Visscher et al. 2008) implying that dissecting phenotypic variation is not as simple as currently treated. Daily variance in some traits such as blood glucose and difficulty in standardising measurement have also contributed to variable estimates of trait heritability. However on balance, the estimates today are not likely to be sufficiently far from the actual so as to remove all of the missing heritability for complex diseases.

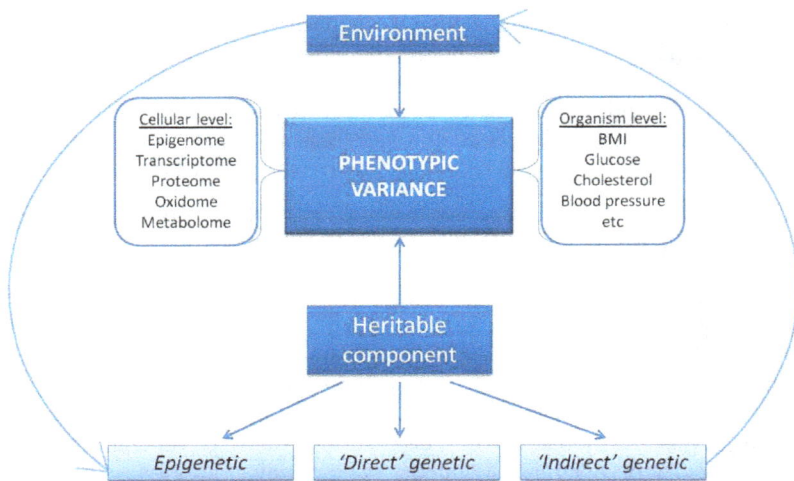

Fig. 3. Heritability estimates represent the genetic variation contributing to phenotypic variation and may be considered as divisible into 3 components; epigenetic, 'direct' genetic and 'indirect' genetic. Epigenetics and 'direct' genetics influence the phenotype directly, whereas 'indirect' genetics represent the heritable portion of behaviours that contribute to environmental influences, such as heritability of regular tobacco use. The environment may induce epigenetic changes suggesting that the two traditional components of phenotypic variance, namely the environment and the heritable portion, may not be wholly distinct.

4.4.5 Epistasis

The extent of missing heritability encouraged revision on the factors that influence genetic effects on disease development. Epistasis in genetics is defined as the interaction between genes; a phenomenon where one gene influences the expression of another. For example the E4 allele of the apolipoprotein E (ApoE) gene is associated with increased serum cholesterol levels, but only when the individual possesses the A2A2 genotype of the low-density lipoprotein receptor (LDLR) gene (Tyler et al. 2009). ApoE's phenotypic effect thus depends on the LDLR locus. Epistatic effects amplify the complexity between genotype and phenotype, suggesting that the combination of genotypes inherited may be instrumental to disease risk (Moore et al. 2009), and it has been argued that epistasis is likely to be a ubiquitous component of the genetic architecture of common complex disease (Moore 2003; Moore et al. 2009). Humans have evolved robust biological systems that use network redundancy to confer resistance to genetic or environmental fluctuations (Moore et al. 2009). Disease could be perceived as an accumulation of parallel pathway 'breakdowns', whereby an important pathway and all of its 'backups' become dysfunctional. In this context, the particular combination of alleles inherited will determine disease phenotype. This concept could explain failure to replicate significant SNP associations and the low effect sizes of each. When considered in a multilocus model, a variant may exert a larger effect than estimated with a single locus approach (Wang et al. 2005). Indeed a "genometype" as proposed by Moore (Moore 2009) would ideally be considered as a basis for calculating clinical risk.

The computational burden associated with statistical analysis of epistasis on a genome-wide scale is unfortunately a limitation in dissecting complex disease genetics. To consider all possible pair-wise interactions alone, approximately 500 billion SNP-pairs would have to be examined (Zhang et al. 2011). Development of efficient two-locus epistasis tests has been challenging as reflected in a recent study by Bell et al (Bell et al. 2011). A two-locus analysis on type 2 diabetes GWA data identified 79 significant pair-wise results, but all included a TCF7L2 SNP which was the most significant in a single-locus analysis. It was realised that the strong TCF7L2 signal was driving the two-locus significance results, ultimately obscuring multilocus findings and signaling a hurdle for such analysis. The complex computation for investigation of genome wide epistasis therefore remains a significant challenge.

5. Approaches to addressing GWA limitations

Despite its shortcomings, GWA studies have produced data that is rich in information. Further optimisation and re-analysis may yield additional insights from this approach while identification of disease genes within the loci is expected to improve current understanding of complex disease development.

5.1 Transcriptome analysis enables prioritisation of candidate variants

Considering that most trait associated loci feature variants that do not affect protein sequence, it is anticipated that many of these loci have a regulatory role on gene expression (Heap et al. 2010), although as discussed above this assumption may be somewhat premature. Genomic regions that contribute to phenotypic variation by influencing transcription, mRNA stability and splicing are termed expression quantitative trait loci or eQTLs (Heap et al. 2010). Integration of expression profiles with GWA data can reveal eQTLs by identifying transcripts that correlate with the TASs (Figure 4). This approach allows empirical ranking of juxtaposed candidate disease-mediating genes that can be taken forward into *in vitro* and *in vivo* functional biological assays to determine the role and mechanism in the disease process. Translation of these eQTLs into enhanced biological knowledge of disease has been seriously impeded by the dearth of large scale human datasets that carry both genetic and transcript data from the same individual that facilitate identification of the disease mediating gene(s) (McCarthy et al. 2008). However some datasets do exist and a demonstration of this approach has been reported where variants in an intronic region of FTO were found to be strongly correlated with gene expression of a more distant gene RBL2 (Jowett et al. 2010). Similarly, approaches relying on conservation of synteny and nucleotide sequence among vertebrate

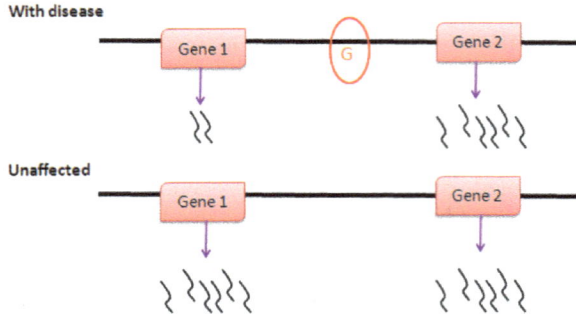

Fig. 4. GWA has found the circled SNP to be associated with a disease. It can be difficult though to assign this SNP to a candidate gene. Transcriptome analysis can make this easier by comparing the expression levels of flanking genes to each other, and between affected and unaffected individual. The decreased expression of gene 1 is associated with the disease, suggesting that gene 1 is the better candidate gene.

species have highlighted a second gene IRX3 flanking FTO that may contribute to disease processes (Ragvin et al. 2010). Further function validation is required to resolve the relative contribution of these genes at this locus.

The study of eQTLs has shown that they generally explain a greater proportion of the phenotype than risk alleles (Freedman et al. 2011), suggesting that consideration of expression data may uncover missing heritability. It has also reinforced that eQTL associations can be tissue-specific (Freedman et al. 2011), differ between *in vivo* and *in vitro* systems and also between cell types (Goring et al. 2007; Farber et al. 2009; Hardy et al. 2009). Evaluation of expression profiles using a single platform for liver, adipose tissue and blood showed a 30% overlap in the three tissues (Nica et al. 2008). This observation suggests that there is value in using any available transcriptome data set for analysis, but also highlights that the majority of expression patterns might be tissue-specific (Nica et al. 2008). The present lack of a comprehensive data set detailing expression profiles of normal tissues calls for caution in interpreting significant associations derived from tissue–specific expression data. This issue has encouraged the Genotype-Tissue Expression project, an idea which was pre-empted in the literature (Goring et al. 2007; Nica et al. 2008). Allelic-specific expression can be used to characterise an eQTL in an individual heterozygous for the risk allele and offers a means of avoiding sources of error that can confound assessment of expression profiles across multiple individuals (Heap et al. 2010). Gene expression can also differ temporally, especially for developmentally important genes. Model organisms such as the zebrafish have utility in investigating the effects of non-coding elements on gene expression in a particular context (Ragvin et al. 2010). Defining the appropriate tissue and time point to glean human expression data from is not clear, and is an area earmarked for further development (Freedman et al. 2011).

5.2 A wider region should be considered for fine mapping

Identification of synthetic associations discussed above has shown that a wider region encompassing the TAS should be considered during fine mapping. If rare variants can act over megabase distances to drive an association signal (Dickson et al. 2010), then the average region size selected for investigation (~500kb) may be too narrow to capture causal variants. This approach may increase the explained heritability for complex phenotypes as the combined influence of multiple rare variants could exhibit a larger effect collectively than the synthetically associated common variant (Dickson et al. 2010) where the effect size has eroded due to distance and LD. Following up on non-reproducible common associations and loci that have known linkage peaks within a few megabases might increase the likelihood of finding a causal variant.

5.3 Integrate with Epigenetic-Wide Association studies

Recent technology advances have allowed contemplation of epigenome-wide association (EWA) studies (Rakyan et al. 2011), where up to 450,000 DNA methylation sites can be measured on a single chip. Although the primary focus of the analysis of these results will be on DNA methylation, there is evidence to suggest that information might be obtained on other epigenetic marks such as histone modifications (Bernstein et al. 2007) and can thus have a broader application. Development of epigenetic maps by the International Human Epigenome Consortium (Rakyan et al. 2011) will empower EWA synonymous to the utility of the HapMap in GWA studies. Additional insights might arise where loci harbouring genetic variation influence the methylation state of the region (methQTLs) (Rakyan et al. 2011). Therefore there is potential utility in combining data from GWA and EWA studies as illustrated by a recent study where a haplotype-specific DNA methylation locus for a type 2 diabetes risk variant was identified (Bell et al. 2010).

While EWA holds great promise for revealing genetic architecture of complex disease, it also presents unique challenges. A significant difference from GWA is that epigenetic association with a phenotype can be causal as well as consequential (Rakyan et al. 2011). Although this adds another layer of complexity to EWA analysis, it can be dissected by appropriate study design. Randomly ascertained longitudinal cohort studies will be integral to determining the temporal origins of the reported epigenetic association and elucidating whether its origin is pre or post onset of the phenotype (Rakyan et al. 2011). Furthermore with thorough detailing of environmental changes, this study design may reveal an environmental cue for epigenetic change. However this presents a double-edged sword, with environmental factors such as smoking able to confound EWA (unlike GWA) and inflate effect size estimates (Rakyan et al. 2011). Additionally ascertainment, phenotyping and sample collection of a large cohort as would be required for such a study is not a trivial undertaking. Looking at this issue from another perspective, it presents a means to investigate environmental backgrounds which increase epigenetic risk variants. Such information has scope for application in a clinical setting, and could be useful in encouraging behavioral changes in patients with a 'vulnerable' epigenetic locus.

5.4 Reduce statistical rigour

The statistical rigour applied to GWA study data analysis has been suggested as potential explanation for the observed missing heritability (Yang et al. 2010). Variants with weak but relevant effect sizes which fall under stringent thresholds resulting from adjustment for multiple testing may potentially be excluded from analysis to reduce the false positive rate. While commonly each TAS is evaluated individually, Yang et al considered a model that evaluates the variance explained by a group of TASs collectively (Yang et al. 2010) using a software program that estimated the variance explained by the group on a chromosome or on the whole genome (Yang et al. 2010). Application of this approach to human height GWA studies revealed that as much as 40% of missing heritability may be found by such accumulation of the variance explained (Yang et al. 2010). The biological plausibility of considering ~300,000 loci to influence this trait is debatable. However the principal has been demonstrated and further functional studies are required to empirically determine appropriate statistical rigour that may differ from one phenotype to the next. Application to predict disease onset found improvement with the whole genome method (Makowsky et al. 2011), however genotyping thousands of variants is unlikely to be sufficiently cost effective and carry a substantially large clinical benefit to warrant its widespread use.

5.5 Increase study size

Even a relatively large GWA study (5000 cases and 5000 controls) has relatively low power to detect an association; 0.9% at a P-value of 10^{-7} (Kraft et al. 2009). Therefore increasingly large cohorts have been studied in addition to the combining existing sets into the meta-analysis studies (Zeggini et al. 2008; Barrett et al. 2009; Park et al. 2010). While this has been productive to an extent, increasing size also increases the potential for increased genetic heterogeneity and thereby raises the background noise level. Additionally since such large cohorts are impractical to collect from a single clinic or site, they must be undertaken at multiple participating centers that increase the chance for differences in phenotyping accuracy. The point at which additional loci might be discovered against increasing amount of background noise might have been reached as demonstrated by a recent GWA study for central obesity and fat distribution using waist and waist-hip measurements. The cohort consisted of ~100,000 participants, but the results yielded only three associated loci, with each explaining a mere fraction (0.05%, 0.04% and 0.02%; combined total 0.11%) of the total trait variance (Lindgren et al. 2009).

An alternate to increasing study size is to increase diversity, that is to undertake similarly sized GWA studies in a range of diverse ethnic groups. It has been established that some SNPs are population specific (2010) and that any given variant may show a different effect size between differing ethnic groups (Dickson et al. 2010). Since many GWA studies had an initial focus on European populations, examining non-European groups may reveal additional loci. Such studies are now starting to appear and have successfully identified new sets of loci (Kooner et al. 2011; Parra et al. 2011). While it is apparent that a causal variant may not be consistently associated among different populations due to unique genetic and environmental factors, a population-specific variant should also not be discounted as having limited clinical application. This is because of the commonality of intra- and inter-cellular molecular pathways in humans. That is any loci leading to identification of a disease gene will reveal a pathway that will be present across all members of the species, and as such is expected to have broad application.

The observed missing heritability suggests that many inherited factors remain to be discovered. For example, if further pursuing common variants, a crude statistical estimate suggested that there may be up to 800 additional variants yet to be discovered for type 1 diabetes (So et al. 2011). However Goldstein argues that analytical approaches based on discovery of common variants have been exhausted and that if any loci are left to discover, it will not be worth the time and cost of ever larger studies to detect them (Goldstein 2009).

This stance is supported by Park et al who proposed that additional loci discovered in larger studies will generally have smaller effects (Park et al. 2010) and therefore may be limiting for detecting large effect variants. Such estimates have provided additional momentum to a paradigm shift that analysis of complex disease should be redirected toward rare variants with potentially large effects.

6. Case study: Type 2 diabetes

The pursuit for genetic factors that underlie Type 2 Diabetes (T2D) encapsulates the challenges and rewards of GWA studies. Almost 350 million people suffer from diabetes mellitus, with approximately 90% of those having developed T2D specifically (Danaei et al. 2011). Characterised by insulin resistance and abnormal beta-cell function (Imamura et al. 2011), the genetic contribution of T2D was well established by family and twin studies (Groop et al. 1996; Poulsen et al. 1999). Current heritability estimates for T2D predict ~40% of the phenotype is explained by genetics (So et al. 2011).

With the incidence and consequent economic impact of this common complex disease projected to increase considerably (Colagiuri 2010), there has been a strong focus on identification of the genetic basis of T2D development. Although GWA has facilitated the identification of over 45 genetic susceptibility variants (Kooner et al. 2011), the promise of clinical application remains unfulfilled.

6.1 The GWA study boom

Prior to the advent of GWA studies, linkage analysis and candidate gene studies were used to discover loci associated with T2D. Despite extensive efforts spanning a decade (Frayling 2007) progress was slow. Notable achievements included identification of PPARG and KCNJ11 as candidate genes (Altshuler et al. 2000; Gloyn et al. 2003), and TCF7L2 as the common variant driving linkage at a T2D risk gene region (Grant et al. 2006). Additionally linkage analysis identified several hundred loci, of which over 50 were replicated by 5 or more independent studies (Lillioja et al. 2009). As GWA studies emerged from 2007, a sudden increase of T2D associated loci followed, reflecting the potential of GWA to provide genetic clues of disease etiology (Table 3).

6.2 The role of epigenetic modification

As has been observed with almost all complex diseases, only ~10% of the heritability can be explained despite over 45 associated loci reported (Imamura et al. 2011). As discussed above, there are several possibilities that may explain the missing heritability including rare variants, copy number polymorphisms and inherited epigenetic polymorphisms. Epigenetic modification as a result of environmental influences and in particular the fetal environment has been highlighted as having potential to predispose to T2D (Ling et al. 2009; Liguori et al. 2010). The Dutch Hunger Winter Study revealed an association between prenatal exposure to famine and decreased glucose tolerance (Ravelli et al. 1998). Effects on methylation were also investigated with all but one famine exposed individual showing significant hypomethylation of the known T2D candidate gene insulin-like growth factor 2 (IGF2) relative to their unaffected sibling (Heijmans et al. 2008). This example highlights the vulnerability of the fetal genome to the prenatal environment, and the ability of epigenetic changes to persist between generations.

Candidate genes	Gene product	Effect size odds ratio
KCNQ1	Potassium voltage-gated channel	1.43
TCF7L2	Transcription factor	1.37
DUSP9	Dual specificity phosphatase	1.27
UBE2E2	Ubiquitin-conjugating enzyme	1.19
IRS1	Insulin receptor substrate	1.19
IGF2BP2	Insulin-like growth factor mRNA binding protein	1.17
FTO	Dioxygenase that repairs alkylated DNA and RNA by oxidative demethylation	1.15
KCNJ11	Potassium inwardly-rectifying channel	1.15
THADA	Thyroid adenoma associated protein	1.15
PPARG	Peroxisome proliferator-activated receptor	1.14
HHEX-KIF11-IDE	HHEX: homeobox family transcription factor, KIF11: kinesin-like protein, IDE: insulin-degrading zinc metallopeptidase	1.13

Table 3. Top ranking candidate genes for type 2 diabetes as per estimated effect size. Odds ratio's as listed in (Imamura et al. 2011).

6.3 The slow walk from association to mechanism

While T2D GWA studies have been successful for revealing associated loci, it has proved difficult to understand the mechanisms through which these loci mediate their biological consequences. Here we examine this aspect of translation of GWA identified loci to useful biological knowledge using several examples to illustrate.

6.3.1 IGF2BP2

An intronic locus within translation regulator IGF2 mRNA binding protein IGF2BP2 was significantly associated with T2D, implicating IGFBP2 and supporting the candidacy of IGF2 and its regulation in T2D. Previously unpublished data using transcriptome analysis combined with genetic variation data in a large human cohort showed a nominally significant association between IGFBP2 and the intronic TAS supporting the hypothesis that this locus mediates its effects through regulating the expression of IGFBP2 (Figure 5). The methodology for this analysis was as reported for the FTO loci by Jowett and colleagues (Jowett et al. 2010). Investigation of top T2D GWA candidates in the Dutch Hunger Winter cohort showed that IGF2BP2 variants were most strongly associated with increased risk of T2D, impaired glucose tolerance and area under the curves (AUC) for oral glucose tolerance testing (OGTT) (van Hoek et al. 2009). The presence of the IGF2BP2 risk allele in those exposed to famine was nominally significantly associated with lower AUC for glucose. That two factors which confer risk to T2D are together associated with increased glucose uptake may seem contradictory, but in light of a similar observation for another risk allele (van Hoek et al. 2009) this can be explained. IGF2BP2 variant may instead represent a protective allele, acting to repair the detrimental consequences of fetal malnutrition on glucose tolerance in later life (van Hoek et al. 2009). IGF2BP2's mode of action in T2D development is still unclear, with recent research revealing its capacity to both initiate and repress

(a)

Chromosome coordinates (Mb)

(b)

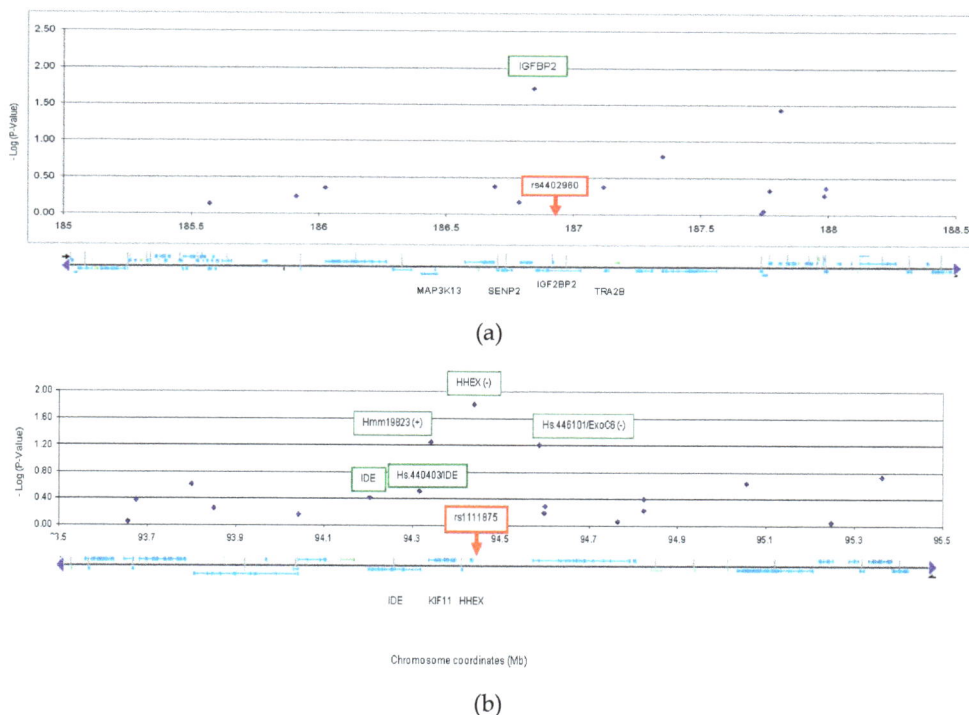

Fig. 5-a. Association of the rs4402960 SNP with the expression levels of genes within 5Mb of the SNP coordinate. Strong association of the IGFBP2-residing SNP with IGFBP2 is observed, suggesting that this locus mediates its effects through IGFBP2 regulation of expression. 5-b. Association of the rs1111875 SNP with the expression levels of genes within 5Mb of the SNP coordinate. Strong association of the SNP with HHEX is observed, suggesting that this loci mediates its effects through HHEX regulation of expression. The intron/exon structure of the genes in this region is illustrated by the lines and blocks beneath the x-axis (thin line, intron; block, exon). The negative log of the P-value is plotted against the coordinates in megabase pairs (Mb). The methodology for this analysis was as reported for the FTO loci by Jowett and colleagues (Jowett et al., 2010).

translation of targets (Boudoukha et al. 2010; Dai et al. 2011). This reported association between IGF2 hypomethylation and IGF2BP2 risk variant suggests scope for an increased risk in an individual that possesses both marks. Investigation of IGF2 expression levels in the Dutch Winter Cohort may provide insight into IGF2BP2's role by reporting a condition that exacerbates the variants risk effects.

6.3.2 HHEX-KIF11-IDE

The HHEX-IDE-KIF11 locus encompasses an extended region of high linkage disequilibrium spanning three genes, two of which were considered plausible candidates; the homeobox HHEX transcription factor involved in pancreatic development (Prokopenko et al. 2008) and the insulin degrading enzyme IDE. Two different approaches suggested that the locus might regulate HHEX expression as the mediator of its biological consequences (Ragvin et al. 2010) (Jowett unpublished). As described above, relying on a large human cohort with transcriptomic and genetic variation data, we observed that HHEX was the most strongly associated transcript with the associated variant. IDE was not associated, while KIF11 was not within detectable limits of expression and could not be commented on further. In the second approach it was observed that the TAS for this locus was within a highly conserved noncoding element (HCNE) exhibiting synteny between a number of vertebrate species including mouse, chicken, frogs and zebrafish (Ragvin et al. 2010). Subsequent reporter gene studies of the enhancer region found that expression was specific to the pancreatic islet cells in a zebrafish model (Ragvin et al. 2010). However there is also evidence for a role for IDE in T2D development (Farris et al. 2003; Pivovarova et al. 2009), suggesting potential for an associated locus to modulate expression of more than one gene in the region.

6.3.3 CDKN2A-CDKN2B

With the intergenic SNP lying 125kb upstream of the closest protein-coding gene, the CDKN2A-CDKN2B locus also demonstrates a difficult scenario for assigning a candidate gene. These genes encode cyclin-dependent kinase inhibitors known to have an important role in β-cell function and regeneration (Bao et al. 2011), and thus are plausible biological candidates for T2D. This SNP is not lost in the 'gene desert', but rather it maps to a long non-coding RNA ANRIL, which lies anti-sense to CDKN2B and has been shown to have a pivotal role in regulating CDKN2A/B expression in mice (Pasmant et al. 2011). ANRIL achieves this by interacting with polycomb repressive complexes to form heterochromatin at the locus, ultimately leading to its repression (Aguilo et al. 2011). Given that over expression of the CDKN2A orthologue in a rodent model induces a T2D phenotype (Krishnamurthy et al. 2006), the regulatory action of ANRIL is likely important in T2D development. As more is learnt about the abundance and function of non-coding RNAs in general, it may not be unexpected that this mode of regulation might apply in more GWA loci where there are no obvious candidate genes (Pasmant et al. 2011).

6.3.4 TCF7L2

The TCF7L2 intronic SNP association is highly replicated and carries one of the largest effect sizes of all GWA loci identified to date. Possessing two risk alleles increases T2D risk by 80% (Florez et al. 2006). The loss of TCF7L2 has been observed to induce impaired β-cell function and apoptosis (Shu et al. 2008; da Silva Xavier et al. 2009), providing an explanation of its effect on T2D development. Paradoxically, decreased protein levels in diabetic islets were correlated with increased TCF7L2 transcript levels (Le Bacquer et al. 2011). Correlation of the risk allele with a 'more open' chromatin state and enhanced access for transcriptional machinery (Gaulton et al. 2010) may provide a possible explanation for observed increase in mRNA (Billings et al. 2010), but how reduced protein levels result from increased mRNA expression remains to be explained. Reports that different isoforms may exert opposing effects on β-cell survival (Le Bacquer et al. 2011) may contribute to an explanation of this paradox. Future experiments to measure the relative expression levels of the pro and anti-apoptotic splice variants in islet cells from people with diabetes may be informative. TCF7L2 illustrates that understanding a strong candidate gene's role in disease pathogenesis can be difficult, even despite extensive effort.

As further mechanistic work progresses providing biological information of which genes mediate the consequences of genetic variation and how this happens, links between these genes may become apparent thus revealing potential insight into networks and pathways that control disease mechanism. For example, IGF2 has been shown to compete with insulin as an IDE substrate (Misbin et al. 1983; Ding et al. 1992), and thus IDE is likely to act in the same pathway as IGF2BP2. Additionally, HHEX and TCF7L2 are both targets of the WNT signaling cascade (Frayling 2007) and may operate collectively on the output of this pathway. Such commonality of pathways between T2D associated loci highlight the potential for a broader understanding of the etiological basis of an association, and provide a rationale for direct testing for evidence of genetic interaction.

Limitations	Type 2 diabetes example	Approach
A SNP may be regulating a gene other than the obvious candidate	FTO intronic SNP	Use transcriptomics to determine a variant's mechanism of action
Difficulty in establishing the mechanism of action for a strong candidate gene	TCF7L2	Characterise alternative spice forms
Only detects loci associated, and not the causal variant, and as such a region may harbour more than one plausible candidate	HHEX-IDE region	Consider one TAS at a time; use model systems of highly conserved non-coding elements
Associated SNP is a large distance from closest gene	CDKN2A-CDKN2B locus	Look beyond protein-encoding genes
Large missing heritability	Only ~10% explained	Extend analysis to consider rare variants, epigenetic and epistatic effects , collect large trans-ethnic samples

Table 4. Analysis of GWA data has been challenging. Approaches to overcoming these limitations for type 2 diabetes loci may be useful for other TASs.

6.4 The rewards of GWA studies

Despite the numerous loci identified by GWA studies, translation of this data into novel therapeutics and risk prediction tools has been slow. The primary cause has been difficulty in establishing the disease mediating gene, where in many cases annotation by proximity can be misleading. Even with a clear understanding of causality, a gene product may not necessarily represent a chemically tractable drug target. Characterisation of T2D locus SLC30A8 represents a potential exception, where the associated SNP encoded a missense mutation (Arg → Trp) in the C-terminus of the protein (Rutter 2010). SLC30A8 encodes a zinc transporter involved in insulin storage and secretion (Imamura et al. 2011) and has also been implicated in type 1 diabetes (Wenzlau et al. 2007). A person with two copies of the risk allele experience a 53% increased risk of developing diabetes (Rutter 2010) most likely due to reduced transporter activity. A drug that enhances the activity of this transporter may potentially increase insulin secretion and thus lower blood glucose levels (Imamura et al. 2011), however chemical agonists are harder to develop then antagonists, thus the prospect of new drugs from this research remains confined to the future.

The use of T2D risk alleles in disease prediction to date has also found limited utility in clinical practice. Adding a genetic risk score to established prediction models based on clinical traits such as BMI and family history showed only limited improvement in predictive power (Lyssenko et al. 2008) and is not sufficient to warrant the additional cost of genotyping. Other studies (de Miguel-Yanes et al. 2011) have suggested that such clinical applications might carry relatively greater benefit for younger people since they may not have developed age-dependent clinical risk factors such as hypertension and dyslipidemia. Discussion about the potential for a genetic risk score alone to influence healthy patients to make lifestyle changes (Grant et al. 2009) suggests that using raw information may be an attractive alternative to producing numerical scores. Clearly there is some way to go before results from GWA studies can be used to their full fruition, but the progress in risk loci characterisation gives hope that novel insights will be gained with potential benefits eventually feeding into clinical practice.

7. Future direction: Common Disease-Rare Variant hypothesis

In light of the large proportion of unexplained heritability for most common complex diseases, new tactics have been employed to advance its genetic characterisation. The Common Disease-Rare Variant (CDRV) hypothesis has become the most attractive theory for explaining the failures of GWA studies (Nielsen 2010). This model proposes that rare variation is the major contributor to complex diseases, with genes or genomic regions potentially harbouring many different rare variants (Schork et al. 2009). In contrast to the CDCV, rare variants are expected to have arisen relatively recently, or have become rare due to negative selection (Schork et al. 2009). Therefore the Multiregional evolution hypothesis that describes multiple founder groups contributing to the origin of modern humans is supportive of the CDRV rather than the Out of Africa model (Iyengar et al. 2007). Candidate genes proposed by GWA for type 1 diabetes (Nejentsev et al. 2009), hypertriglyceridemia (Johansen et al. 2010) and Crohn's disease (Momozawa et al. 2011) have been found to harbour rare missense mutations, highlighting the plausibility of the CDRV theory. Particularly alluring is that less frequent variation is more likely to be functional (Schork et al. 2009). Indeed the majority of rare (<1%) missense mutations present in humans are somewhat deleterious in nature (Kryukov et al. 2007), providing hope that rare variants will carry larger effect sizes and hopefully more obvious in mechanism than the often cryptic common variation.

7.1 New approaches

Despite claims that expectations for finding missing heritability of complex disease in rare variants may be overoptimistic (Carvajal-Carmona 2010), recent studies have highlighted the promise of the study of rare variants in complex disease (Bowden et al. 2010; Holm et al. 2011). DNA sequencing is the most efficient method for the identification of rare variants and while only a few years ago this would be impractical due to costs and time for large sample numbers, the advent of next generation sequencing technologies has bought this approach into the realm of possibility (Ng et al. 2009; Cirulli et al. 2010; Nielsen 2010). To capture all genetic variation, whole-genome sequencing (WGS) would be the preferred approach and has already found utility in characterisation of rare monogenic disease (Choi et al. 2009). While WGS costs are becoming more affordable, they remain impractical for large sample sizes, prompting development of sequencing approaches that will reduce costs such as target enrichment and sample multiplexing. Cohort design for rare variants also forms an important consideration, with large extended pedigree designs offering the most efficient method for generating sufficient copies of the rare variant to undertake meaningful statistical analysis. These strategies (Figure 6) will represent the next stage of genetic dissection for complex diseases.

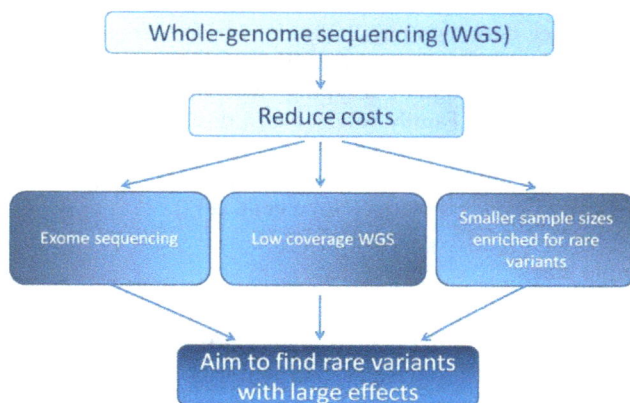

Fig. 6. Until WGS is affordable, alternatives such as exome sequencing, low-coverage WGS and smaller sample sizes reduce costs for a sequencing approach.

7.1.1 Exome sequencing

The exome (all genomic coding sequence) represents approximately 2.3% of the genome (Lehne et al. 2011) and is the genomic region most likely to harbour variants of large effect. Consistent with this has been the experience from GWA studies where stronger association signals for coding variants than non coding variants have been reported (Lehne et al. 2011). Given its smaller size it is also less expensive and faster to sequence than the whole genome allowing a large number of samples to be analysed and increasing statistical power for evaluation of effects. Successful identification of rare causal variants by this approach has been recently reported for several Mendelian diseases (Ng et al. 2010; Ng et al. 2010) and also for complex disease traits (Bowden et al. 2010) and disorders (Sanders 2011). However focusing solely on the exome ultimately ignores the majority of the genome where regulatory variation is located. Sequencing exomes in blocks that include flanking regions can increase genome coverage, with 40kb flanks equating to 34% of the genome (Lehne et al. 2011). The size of the flanking region considered will be cost-limited, but a focus on annotated regulatory elements could be used in a gene specific manner and customised to limit the amount of sequence required and reduce costs (Lehne et al. 2011). Ultimately exome sequencing is considered a temporary measure for making advances before the cost of WGS becomes affordable.

7.1.2 Low-coverage whole-genome sequencing

Another approach for reducing costs is to sequence genomes at reduced coverage, with only 250 samples required to find 99% of synonymous variants (2010). Although this technique still allows a genome to be sequenced several times, there may be incomplete genome coverage and an expected increase in error rate (Nielsen 2010). It has been noted that marginal increase in error rate of only a few percent can equate to substantially reduced statistical power (Nielsen 2010). Computational biology approaches, such as imputation, uses the data from sequenced individuals to infer the identity of missing nucleotides in a new data set, and has been shown to increase the statistical power of GWA studies (Nielsen 2010). Imputation accuracy decreases parallel to the variant's frequency, with an error rate of 35% reported for minor alleles observed only twice in a low-coverage sample (2010). However the increased risk of pursuing a typing error variant is of concern, especially for cohorts of unrelated individuals. Family cohort based designs are superior in this regard as genotyping errors will be more transparent through precedence of Mendelian inheritance. Nevertheless, low-coverage WGS has been demonstrated as being effective for detecting rare variants influencing complex diseases (Holm et al. 2011).

7.1.3 Smaller sample sizes

The advent of sequencing approaches for the identification of rare variants has placed a limit on the number of samples that can be analysed using current technologies, due to available resources. As such, study designs that required smaller sample sizes have been favoured (Cirulli et al. 2010; Zeggini 2011). These include extreme trait design, isolated populations and cohorts of families (Cirulli et al. 2010; Zeggini 2011). Isolated populations and family studies offer not only increased allele frequencies of rare or even private variants, but also the potential for reduced phenotypic, environmental and genetic heterogeneity (Cirulli et al. 2010; Zeggini 2011).

7.1.4 Successful application of sequencing approaches

The 1000 Genomes Project uses WGS, exome sequencing and low-coverage WGS to catalog 95% of variants with frequency >1% and coding variants with frequencies as low as 0.1% (Durbin et al. 2010). An exciting recent publication by Holm et al (Holm et al. 2011) shows successful application of this dataset and imputation for a project studying the complex cardiac disorder sick sinus syndrome in the isolated Icelandic cohort. They identified a rare variant (MAF 0.004) in MYH6 that was strongly associated with disease (OR 12.5). The success of Holm et al (Holm et al. 2011) can be in part attributed to their isolated population study design and its reduced genetic heterogeneity. The variant described by Holm et al may not be present in other populations (Zeggini 2011), limiting its utility in variant specific tests of prediction. This is consistent with the low-coverage 1000 Genomes pilot that highlighted the potential for rare variants to be ethnically unique with most of the ~9 million new SNPs discovered being population specific (Via et al. 2010) with non-synonymous variants more susceptible to this trend than synonymous SNPs (2010). Gene centric as opposed to variant centric measurements may be necessary to allow utility of risk assessments across diverse population groups.

7.1.5 Modern linkage analysis

A recent study of plasma levels of the adipocytokine adiponectin levels (Bowden et al. 2010) illustrated the usefulness of reliance on linkage analysis for finding families enriched with large effect variants. A key finding from this paper was that only a few families may contribute significantly to the originally observed linkage signal (Bowden et al. 2010). These families were examined with exome and direct sequencing to determine the causal variant for adiponectin levels, and a rare variant (MAF 0.011) identified that explained 17% of the phenotypic variation in the whole sample, and 63% in carrier families. Association methods enabled detection of this variant in other families. This success validates Blangero's argument that linkage analysis was unjustly discarded as a method for dissecting complex disease development (Blangero 2004), and that a combination of both sequencing and family linkage data will be an efficient method for the study of rare variants in common complex disease (Bowden et al. 2010; Cirulli et al. 2010).

8. Conclusion

The era of GWA studies has taught us a much about the influence of common variants on complex diseases, and revealed the influence of many previously unknown loci. As the disease mediating genes are in turn identified, it is anticipated that novel insights into the molecular pathways affecting disease risk will arise. However it has become clear that the contribution of common variants to common complex diseases has been overestimated, as substantial missing heritability for many complex phenotypes shows that only small proportion of overall phenotypic variation can be accounted for the vast majority of traits studied. Systems biology approaches combining large scale genetic and transcriptomic datasets appear key to the identification of the disease genes and the dearth of such well characterised large cohorts has no doubt contributed to the slow pace of translation.

The study of complex diseases field now looks towards rare variants and emerging next-generation sequencing technology for the next major steps forward, with family-based approaches carrying several advantages over competing designs. Until WGS becomes economically viable, affordable alternative approaches will provide the platform for advancement including exome targeted sequencing and imputation. In due course, it is anticipated that technological advances such as those under development by Pacific Biosciences will evolve to enable WGS for large sample sizes at high statistical power, allowing comprehensive characterisation of human genetic variation.

Although drug focused outcomes of genetic studies are in their infancy, there is excitement for future success (Hirschhorn 2009; Manolio et al. 2009; Cirulli et al. 2010). It is acceptable to have a proportion of heritability unexplained if the variants found can initiate development of an effective, low cost drug (Manolio et al. 2009), therefore a variant's usefulness might be based on a cost benefit ratio for treatment advancement rather than the proportion of heritability explained. For now the future of genetic analysis of complex disease looks bright as new technologies will allow great insights into the genome that were inconceivable just a decade ago.

9. References

(2007). "Genome-wide association study of 14,000 cases of seven common diseases and 3,000 shared controls." *Nature* 447(7145): 661-678.
(2010). "A map of human genome variation from population-scale sequencing." *Nature* 467(7319): 1061-1073.

Agopian, A. J., et al. (2011). "MI-GWAS: a SAS platform for the analysis of inherited and maternal genetic effects in genome-wide association studies using log-linear models." *BMC Bioinformatics* 12: 117.

Aguilo, F., et al. (2011). "Long Noncoding RNA, Polycomb, and the Ghosts Haunting INK4b-ARF-INK4a Expression." *Cancer Res* 71(16): 5365-5369.

Alkan, C., et al. (2011). "Genome structural variation discovery and genotyping." *Nat Rev Genet* 12(5): 363-376.

Altshuler, D., et al. (2000). "The common PPARgamma Pro12Ala polymorphism is associated with decreased risk of type 2 diabetes." *Nat Genet* 26(1): 76-80.

Anderson, C. A., et al. (2011). "Synthetic associations are unlikely to account for many common disease genome-wide association signals." *PLoS Biol* 9(1): e1000580.

Bao, X. Y., et al. (2011). "Association between Type 2 Diabetes and CDKN2A/B: a meta-analysis study." *Mol Biol Rep*.

Barrett, J. C., et al. (2009). "Genome-wide association study and meta-analysis find that over 40 loci affect risk of type 1 diabetes." *Nat Genet* 41(6): 703-707.

Bell, C. G., et al. (2010). "Integrated genetic and epigenetic analysis identifies haplotype-specific methylation in the FTO type 2 diabetes and obesity susceptibility locus." *PLoS One* 5(11): e14040.

Bell, D. A., et al. (1997). "Genetic analysis of complex disease." *Science* 275(5304): 1327-1328; author reply 1329-1330.

Bell, J. T., et al. (2011). "A twin approach to unraveling epigenetics." *Trends Genet* 27(3): 116-125.

Bell, J. T., et al. (2011). "Genome-wide association scan allowing for epistasis in type 2 diabetes." *Ann Hum Genet* 75(1): 10-19.

Bernstein, B. E., et al. (2007). "The mammalian epigenome." *Cell* 128(4): 669-681.

Billings, L. K., et al. (2010). "The genetics of type 2 diabetes: what have we learned from GWAS?" *Ann N Y Acad Sci* 1212: 59-77.

Blangero, J. (2004). "Localization and identification of human quantitative trait loci: king harvest has surely come." *Curr Opin Genet Dev* 14(3): 233-240.

Blewitt, M. E., et al. (2006). "Dynamic reprogramming of DNA methylation at an epigenetically sensitive allele in mice." *PLoS Genet* 2(4): e49.

Bodmer, W., et al. (2008). "Common and rare variants in multifactorial susceptibility to common diseases." *Nat Genet* 40(6): 695-701.

Boudoukha, S., et al. (2010). "Role of the RNA-binding protein IMP-2 in muscle cell motility." *Mol Cell Biol* 30(24): 5710-5725.

Bowden, D. W., et al. (2010). "Molecular basis of a linkage peak: exome sequencing and family-based analysis identify a rare genetic variant in the ADIPOQ gene in the IRAS Family Study." *Hum Mol Genet* 19(20): 4112-4120.

Burren, O. S., et al. (2011). "T1DBase: update 2011, organization and presentation of large-scale data sets for type 1 diabetes research." *Nucleic Acids Res* 39(Database issue): D997-1001.

Carvajal-Carmona, L. G. (2010). "Challenges in the identification and use of rare disease-associated predisposition variants." *Curr Opin Genet Dev*.

Choi, M., et al. (2009). "Genetic diagnosis by whole exome capture and massively parallel DNA sequencing." *Proc Natl Acad Sci U S A* 106(45): 19096-19101.

Cirulli, E. T., et al. (2010). "Uncovering the roles of rare variants in common disease through whole-genome sequencing." *Nat Rev Genet* 11(6): 415-425.

Clayton, D. G. (2009). "Prediction and interaction in complex disease genetics: experience in type 1 diabetes." *PLoS Genet* 5(7): e1000540.

Colagiuri, R., Brown, J., and Dain, K (2010). A call to action on diabetes. I. D. Federation. Brussels.

Collins, F. S., et al. (1998). "A DNA polymorphism discovery resource for research on human genetic variation." *Genome Res* 8(12): 1229-1231.

Cooper, D. N. (2010). "Functional intronic polymorphisms: Buried treasure awaiting discovery within our genes." *Hum Genomics* 4(5): 284-288.

Craddock, N., et al. (2010). "Genome-wide association study of CNVs in 16,000 cases of eight common diseases and 3,000 shared controls." *Nature* 464(7289): 713-720.

da Silva Xavier, G., et al. (2009). "TCF7L2 regulates late events in insulin secretion from pancreatic islet beta-cells." *Diabetes* 58(4): 894-905.

Dai, N., et al. (2011). "mTOR phosphorylates IMP2 to promote IGF2 mRNA translation by internal ribosomal entry." *Genes Dev* 25(11): 1159-1172.

Danaei, G., et al. (2011). "National, regional, and global trends in fasting plasma glucose and diabetes prevalence since 1980: systematic analysis of health examination surveys and epidemiological studies with 370 country-years and 2.7 million participants." *Lancet* 378(9785): 31-40.

de Miguel-Yanes, J. M., et al. (2011). "Genetic risk reclassification for type 2 diabetes by age below or above 50 years using 40 type 2 diabetes risk single nucleotide polymorphisms." *Diabetes Care* 34(1): 121-125.

Dickson, S. P., et al. (2010). "Rare variants create synthetic genome-wide associations." *PLoS Biol* 8(1): e1000294.

Ding, L., et al. (1992). "Comparison of the enzymatic and biochemical properties of human insulin-degrading enzyme and Escherichia coli protease III." *J Biol Chem* 267(4): 2414-2420.

Durbin, R. M., et al. (2010). "A map of human genome variation from population-scale sequencing." *Nature* 467(7319): 1061-1073.

Eichler, E. E., et al. (2010). "Missing heritability and strategies for finding the underlying causes of complex disease." *Nat Rev Genet* 11(6): 446-450.

Farber, C. R., et al. (2009). "Future of osteoporosis genetics: enhancing genome-wide association studies." *J Bone Miner Res* 24(12): 1937-1942.

Farris, W., et al. (2003). "Insulin-degrading enzyme regulates the levels of insulin, amyloid beta-protein, and the beta-amyloid precursor protein intracellular domain in vivo." *Proc Natl Acad Sci U S A* 100(7): 4162-4167.

Florez, J. C., et al. (2006). "TCF7L2 polymorphisms and progression to diabetes in the Diabetes Prevention Program." *N Engl J Med* 355(3): 241-250.

Fraga, M. F., et al. (2005). "Epigenetic differences arise during the lifetime of monozygotic twins." *Proc Natl Acad Sci U S A* 102(30): 10604-10609.

Fransen, K., et al. (2011). "The quest for genetic risk factors for Crohn's disease in the post-GWAS era." *Genome Med* 3(2): 13.

Frayling, T. M. (2007). "Genome-wide association studies provide new insights into type 2 diabetes aetiology." *Nat Rev Genet* 8(9): 657-662.

Frazer, K. A., et al. (2009). "Human genetic variation and its contribution to complex traits." *Nat Rev Genet* 10(4): 241-251.

Freedman, M. L., et al. (2011). "Principles for the post-GWAS functional characterization of cancer risk loci." *Nat Genet* 43(6): 513-518.

Gaulton, K. J., et al. (2010). "A map of open chromatin in human pancreatic islets." *Nat Genet* 42(3): 255-259.

Gloyn, A. L., et al. (2003). "Large-scale association studies of variants in genes encoding the pancreatic beta-cell KATP channel subunits Kir6.2 (KCNJ11) and SUR1 (ABCC8) confirm that the KCNJ11 E23K variant is associated with type 2 diabetes." *Diabetes* 52(2): 568-572.

Goldstein, D. B. (2009). "Common genetic variation and human traits." *N Engl J Med* 360(17): 1696-1698.

Goring, H. H., et al. (2007). "Discovery of expression QTLs using large-scale transcriptional profiling in human lymphocytes." *Nat Genet* 39(10): 1208-1216.

Grant, R. W., et al. (2009). "The clinical application of genetic testing in type 2 diabetes: a patient and physician survey." *Diabetologia* 52(11): 2299-2305.

Grant, S. F., et al. (2006). "Variant of transcription factor 7-like 2 (TCF7L2) gene confers risk of type 2 diabetes." *Nat Genet* 38(3): 320-323.

Groop, L., et al. (1996). "Metabolic consequences of a family history of NIDDM (the Botnia study): evidence for sex-specific parental effects." *Diabetes* 45(11): 1585-1593.

Hardy, J., et al. (2009). "Genomewide association studies and human disease." *N Engl J Med* 360(17): 1759-1768.

He, M., et al. (2009). "Functional SNPs in HSPA1A gene predict risk of coronary heart disease." *PLoS One* 4(3): e4851.

Heap, G. A., et al. (2010). "Genome-wide analysis of allelic expression imbalance in human primary cells by high-throughput transcriptome resequencing." *Hum Mol Genet* 19(1): 122-134.

Heath, A. C., et al. (1998). "Interaction of marital status and genetic risk for symptoms of depression." *Twin Res* 1(3): 119-122.

Heijmans, B. T., et al. (2008). "Persistent epigenetic differences associated with prenatal exposure to famine in humans." *Proc Natl Acad Sci U S A* 105(44): 17046-17049.

Hemminki, K., et al. (2006). "The balance between heritable and environmental aetiology of human disease." *Nat Rev Genet* 7(12): 958-965.

Hindorff, L. A., et al. (2009). "Potential etiologic and functional implications of genome-wide association loci for human diseases and traits." *Proc Natl Acad Sci U S A* 106(23): 9362-9367.

Hirschhorn, J. N. (2009). "Genomewide association studies--illuminating biologic pathways." *N Engl J Med* 360(17): 1699-1701.

Hirschhorn, J. N., et al. (2005). "Genome-wide association studies for common diseases and complex traits." *Nat Rev Genet* 6(2): 95-108.

Holm, H., et al. (2011). "A rare variant in MYH6 is associated with high risk of sick sinus syndrome." *Nat Genet* 43(4): 316-320.

Imamura, M., et al. (2011). "Genetics of type 2 diabetes: the GWAS era and future perspectives [Review]." *Endocr J*.

Iyengar, S. K., et al. (2007). "The genetic basis of complex traits: rare variants or "common gene, common disease"?" *Methods Mol Biol* 376: 71-84.

Jia, L., et al. (2009). "Functional enhancers at the gene-poor 8q24 cancer-linked locus." *PLoS Genet* 5(8): e1000597.

Jirtle, R. L., et al. (2007). "Environmental epigenomics and disease susceptibility." *Nat Rev Genet* 8(4): 253-262.

Johansen, C. T., et al. (2010). "Excess of rare variants in genes identified by genome-wide association study of hypertriglyceridemia." *Nat Genet* 42(8): 684-687.

Jowett, J. B., et al. (2010). "Genetic variation at the FTO locus influences RBL2 gene expression." *Diabetes* 59(3): 726-732.

Kendler, K. S., et al. (2000). "Tobacco consumption in Swedish twins reared apart and reared together." *Arch Gen Psychiatry* 57(9): 886-892.

Kong, A., et al. (2009). "Parental origin of sequence variants associated with complex diseases." *Nature* 462(7275): 868-874.

Kooner, J. S., et al. (2011). "Genome-wide association study in individuals of South Asian ancestry identifies six new type 2 diabetes susceptibility loci." *Nat Genet.*

Kraft, P., et al. (2009). "Genetic risk prediction--are we there yet?" *N Engl J Med* 360(17): 1701-1703.

Krishnamurthy, J., et al. (2006). "p16INK4a induces an age-dependent decline in islet regenerative potential." *Nature* 443(7110): 453-457.

Kryukov, G. V., et al. (2007). "Most rare missense alleles are deleterious in humans: implications for complex disease and association studies." *Am J Hum Genet* 80(4): 727-739.

Laird, N. M., et al. (2006). "Family-based designs in the age of large-scale gene-association studies." *Nat Rev Genet* 7(5): 385-394.

Lander, E. S. (1996). "The new genomics: global views of biology." *Science* 274(5287): 536-539.

Lasky-Su, J., et al. (2010). "On genome-wide association studies for family-based designs: an integrative analysis approach combining ascertained family samples with unselected controls." *Am J Hum Genet* 86(4): 573-580.

Le Bacquer, O., et al. (2011). "TCF7L2 splice variants have distinct effects on beta-cell turnover and function." *Hum Mol Genet* 20(10): 1906-1915.

Lehne, B., et al. (2011). "Exome localization of complex disease association signals." *BMC Genomics* 12: 92.

Liguori, A., et al. (2010). "Epigenetic changes predisposing to type 2 diabetes in intrauterine growth retardation." *Frontiers in Endocrinology* 1.

Lillioja, S., et al. (2009). "Agreement among type 2 diabetes linkage studies but a poor correlation with results from genome-wide association studies." *Diabetologia* 52(6): 1061-1074.

Lindgren, C. M., et al. (2009). "Genome-wide association scan meta-analysis identifies three Loci influencing adiposity and fat distribution." *PLoS Genet* 5(6): e1000508.

Ling, C., et al. (2009). "Epigenetics: a molecular link between environmental factors and type 2 diabetes." *Diabetes* 58(12): 2718-2725.

Lyssenko, V., et al. (2008). "Clinical risk factors, DNA variants, and the development of type 2 diabetes." *N Engl J Med* 359(21): 2220-2232.

Makowsky, R., et al. (2011). "Beyond missing heritability: prediction of complex traits." *PLoS Genet* 7(4): e1002051.

Manolio, T. A. (2009). "Cohort studies and the genetics of complex disease." *Nat Genet* 41(1): 5-6.

Manolio, T. A., et al. (2009). "The HapMap and genome-wide association studies in diagnosis and therapy." *Annu Rev Med* 60: 443-456.

Manolio, T. A., et al. (2009). "Finding the missing heritability of complex diseases." *Nature* 461(7265): 747-753.

McCarthy, M. I., et al. (2008). "Genome-wide association studies for complex traits: consensus, uncertainty and challenges." *Nat Rev Genet* 9(5): 356-369.

Misbin, R. I., et al. (1983). "Inhibition of insulin degradation by insulin-like growth factors." *Endocrinology* 113(4): 1525-1527.

Momozawa, Y., et al. (2011). "Resequencing of positional candidates identifies low frequency IL23R coding variants protecting against inflammatory bowel disease." *Nat Genet* 43(1): 43-47.

Moore, J. H. (2003). "The ubiquitous nature of epistasis in determining susceptibility to common human diseases." *Hum Hered* 56(1-3): 73-82.

Moore, J. H. (2009). "From genotypes to genometypes: putting the genome back in genome-wide association studies." *Eur J Hum Genet* 17(10): 1205-1206.

Moore, J. H., et al. (2009). "Epistasis and its implications for personal genetics." *Am J Hum Genet* 85(3): 309-320.

Murcray, C. E., et al. (2009). "Gene-environment interaction in genome-wide association studies." *Am J Epidemiol* 169(2): 219-226.

Nejentsev, S., et al. (2009). "Rare variants of IFIH1, a gene implicated in antiviral responses, protect against type 1 diabetes." *Science* 324(5925): 387-389.

Ng, S. B., et al. (2010). "Exome sequencing identifies MLL2 mutations as a cause of Kabuki syndrome." *Nat Genet* 42(9): 790-793.

Ng, S. B., et al. (2010). "Exome sequencing identifies the cause of a mendelian disorder." *Nat Genet* 42(1): 30-35.

Ng, S. B., et al. (2009). "Targeted capture and massively parallel sequencing of 12 human exomes." *Nature* 461(7261): 272-276.

Ng, S. F., et al. (2010). "Chronic high-fat diet in fathers programs beta-cell dysfunction in female rat offspring." *Nature* 467(7318): 963-966.

Nica, A. C., et al. (2008). "Using gene expression to investigate the genetic basis of complex disorders." *Hum Mol Genet* 17(R2): R129-134.

Nielsen, R. (2010). "Genomics: In search of rare human variants." *Nature* 467(7319): 1050-1051.

Orozco, G., et al. (2010). "Synthetic associations in the context of genome-wide association scan signals." *Hum Mol Genet* 19(R2): R137-144.

Park, J. H., et al. (2010). "Estimation of effect size distribution from genome-wide association studies and implications for future discoveries." *Nat Genet* 42(7): 570-575.

Parra, E. J., et al. (2011). "Genome-wide association study of type 2 diabetes in a sample from Mexico City and a meta-analysis of a Mexican-American sample from Starr County, Texas." *Diabetologia* 54(8): 2038-2046.

Pasmant, E., et al. (2011). "ANRIL, a long, noncoding RNA, is an unexpected major hotspot in GWAS." *FASEB J* 25(2): 444-448.

Patel, C. J., et al. (2010). "An Environment-Wide Association Study (EWAS) on type 2 diabetes mellitus." *PLoS One* 5(5): e10746.

Petronis, A. (2010). "Epigenetics as a unifying principle in the aetiology of complex traits and diseases." *Nature* 465(7299): 721-727.

Pivovarova, O., et al. (2009). "Glucose inhibits the insulin-induced activation of the insulin-degrading enzyme in HepG2 cells." *Diabetologia* 52(8): 1656-1664.

Poulsen, P., et al. (1999). "Heritability of type II (non-insulin-dependent) diabetes mellitus and abnormal glucose tolerance--a population-based twin study." *Diabetologia* 42(2): 139-145.

Prokopenko, I., et al. (2008). "Type 2 diabetes: new genes, new understanding." *Trends Genet* 24(12): 613-621.

Ragvin, A., et al. (2010). "Long-range gene regulation links genomic type 2 diabetes and obesity risk regions to HHEX, SOX4, and IRX3." *Proc Natl Acad Sci U S A* 107(2): 775-780.

Rakyan, V. K., et al. (2003). "Transgenerational inheritance of epigenetic states at the murine Axin(Fu) allele occurs after maternal and paternal transmission." *Proc Natl Acad Sci U S A* 100(5): 2538-2543.

Rakyan, V. K., et al. (2011). "Epigenome-wide association studies for common human diseases." *Nat Rev Genet* 12(8): 529-541.

Rappaport, S. M., et al. (2010). "Epidemiology. Environment and disease risks." *Science* 330(6003): 460-461.

Ravelli, A. C., et al. (1998). "Glucose tolerance in adults after prenatal exposure to famine." *Lancet* 351(9097): 173-177.

Reich, D. E., et al. (2001). "On the allelic spectrum of human disease." *Trends Genet* 17(9): 502-510.

Richards, E. J. (2006). "Inherited epigenetic variation--revisiting soft inheritance." *Nat Rev Genet* 7(5): 395-401.

Rideout, W. M., 3rd, et al. (2001). "Nuclear cloning and epigenetic reprogramming of the genome." *Science* 293(5532): 1093-1098.

Risch, N., et al. (1996). "The future of genetic studies of complex human diseases." *Science* 273(5281): 1516-1517.

Rutter, G. A. (2010). "Think zinc: New roles for zinc in the control of insulin secretion." *Islets* 2(1): 49-50.

Ryan, B. M., et al. (2010). "Genetic variation in microRNA networks: the implications for cancer research." *Nat Rev Cancer* 10(6): 389-402.

Sanders, S. S. (2011). "Whole-exome sequencing: a powerful technique for identifying novel genes of complex disorders." *Clin Genet* 79(2): 132-133.

Schork, N. J., et al. (2009). "Common vs. rare allele hypotheses for complex diseases." *Curr Opin Genet Dev* 19(3): 212-219.

Shu, L., et al. (2008). "Transcription factor 7-like 2 regulates beta-cell survival and function in human pancreatic islets." *Diabetes* 57(3): 645-653.

Slatkin, M. (2009). "Epigenetic inheritance and the missing heritability problem." *Genetics* 182(3): 845-850.

So, H. C., et al. (2011). "Evaluating the heritability explained by known susceptibility variants: a survey of ten complex diseases." *Genet Epidemiol* 35(5): 310-317.

So, H. C., et al. (2011). "Evaluating the heritability explained by known susceptibility variants: a survey of ten complex diseases." *Genet Epidemiol*.

Thomas, D. C., et al. (2002). "Point: population stratification: a problem for case-control studies of candidate-gene associations?" *Cancer Epidemiol Biomarkers Prev* 11(6): 505-512.

Turkheimer, E., et al. (2003). "Socioeconomic status modifies heritability of IQ in young children." *Psychol Sci* 14(6): 623-628.

Tyler, A. L., et al. (2009). "Shadows of complexity: what biological networks reveal about epistasis and pleiotropy." *Bioessays* 31(2): 220-227.

van Hoek, M., et al. (2009). "Genetic variant in the IGF2BP2 gene may interact with fetal malnutrition to affect glucose metabolism." *Diabetes* 58(6): 1440-1444.

Via, M., et al. (2010). "The 1000 Genomes Project: new opportunities for research and social challenges." *Genome Med* 2(1): 3.

Visscher, P. M., et al. (2008). "Genome-wide association studies of quantitative traits with related individuals: little (power) lost but much to be gained." *Eur J Hum Genet* 16(3): 387-390.

Visscher, P. M., et al. (2008). "Heritability in the genomics era--concepts and misconceptions." *Nat Rev Genet* 9(4): 255-266.

Wacholder, S., et al. (2002). "Counterpoint: bias from population stratification is not a major threat to the validity of conclusions from epidemiological studies of common polymorphisms and cancer." *Cancer Epidemiol Biomarkers Prev* 11(6): 513-520.

Wang, K., et al. (2010). "Interpretation of association signals and identification of causal variants from genome-wide association studies." *Am J Hum Genet* 86(5): 730-742.

Wang, W. Y., et al. (2005). "Genome-wide association studies: theoretical and practical concerns." *Nat Rev Genet* 6(2): 109-118.

Waters, K. M., et al. (2010). "Consistent association of type 2 diabetes risk variants found in europeans in diverse racial and ethnic groups." *PLoS Genet* 6(8).

Wenzlau, J. M., et al. (2007). "The cation efflux transporter ZnT8 (Slc30A8) is a major autoantigen in human type 1 diabetes." *Proc Natl Acad Sci U S A* 104(43): 17040-17045.

Witte, J. S. (2010). "Genome-wide association studies and beyond." *Annu Rev Public Health* 31: 9-20 24 p following 20.

Wray, N. R., et al. (2011). "Synthetic associations created by rare variants do not explain most GWAS results." *PLoS Biol* 9(1): e1000579.

Yang, J., et al. (2010). "Common SNPs explain a large proportion of the heritability for human height." *Nat Genet* 42(7): 565-569.

Zeggini, E. (2011). "Next-generation association studies for complex traits." *Nat Genet* 43(4): 287-288.

Zeggini, E., et al. (2008). "Meta-analysis of genome-wide association data and large-scale replication identifies additional susceptibility loci for type 2 diabetes." *Nat Genet* 40(5): 638-645.

Zhang, L., et al. (2011). "Functional SNP in the microRNA-367 binding site in the 3'UTR of the calcium channel ryanodine receptor gene 3 (RYR3) affects breast cancer risk and calcification." *Proc Natl Acad Sci U S A* 108(33): 13653-13658.

Zhang, X., et al. (2011). "Tools for efficient epistasis detection in genome-wide association study." *Source Code Biol Med* 6(1): 1.

Zhao, J., et al. (2010). "The role of height-associated loci identified in genome wide association studies in the determination of pediatric stature." *BMC Med Genet* 11: 96.

Permissions

The contributors of this book come from diverse backgrounds, making this book a truly international effort. This book will bring forth new frontiers with its revolutionizing research information and detailed analysis of the nascent developments around the world.

We would like to thank Mahmut Caliskan, for lending his expertise to make the book truly unique. He has played a crucial role in the development of this book. Without his invaluable contribution this book wouldn't have been possible. He has made vital efforts to compile up to date information on the varied aspects of this subject to make this book a valuable addition to the collection of many professionals and students.

This book was conceptualized with the vision of imparting up-to-date information and advanced data in this field. To ensure the same, a matchless editorial board was set up. Every individual on the board went through rigorous rounds of assessment to prove their worth. After which they invested a large part of their time researching and compiling the most relevant data for our readers. Conferences and sessions were held from time to time between the editorial board and the contributing authors to present the data in the most comprehensible form. The editorial team has worked tirelessly to provide valuable and valid information to help people across the globe.

Every chapter published in this book has been scrutinized by our experts. Their significance has been extensively debated. The topics covered herein carry significant findings which will fuel the growth of the discipline. They may even be implemented as practical applications or may be referred to as a beginning point for another development. Chapters in this book were first published by InTech; hereby published with permission under the Creative Commons Attribution License or equivalent.

The editorial board has been involved in producing this book since its inception. They have spent rigorous hours researching and exploring the diverse topics which have resulted in the successful publishing of this book. They have passed on their knowledge of decades through this book. To expedite this challenging task, the publisher supported the team at every step. A small team of assistant editors was also appointed to further simplify the editing procedure and attain best results for the readers.

Our editorial team has been hand-picked from every corner of the world. Their multi-ethnicity adds dynamic inputs to the discussions which result in innovative outcomes. These outcomes are then further discussed with the researchers and contributors who give their valuable feedback and opinion regarding the same. The feedback is then collaborated with the researches and they are edited in a comprehensive manner to aid the understanding of the subject.

Apart from the editorial board, the designing team has also invested a significant amount of their time in understanding the subject and creating the most relevant covers. They scrutinized every image to scout for the most suitable representation of the subject and create an appropriate cover for the book.

The publishing team has been involved in this book since its early stages. They were actively engaged in every process, be it collecting the data, connecting with the contributors or procuring relevant information. The team has been an ardent support to the editorial, designing and production team. Their endless efforts to recruit the best for this project, has resulted in the accomplishment of this book. They are a veteran in the field of academics and their pool of knowledge is as vast as their experience in printing. Their expertise and guidance has proved useful at every step. Their uncompromising quality standards have made this book an exceptional effort. Their encouragement from time to time has been an inspiration for everyone.

The publisher and the editorial board hope that this book will prove to be a valuable piece of knowledge for researchers, students, practitioners and scholars across the globe.

List of Contributors

Tomáš Pánek and Ivan Čepička
Charles University in Prague, Czech Republic

Birgül Özcan
Mustafa Kemal University, Faculty of Sciences and Letters, Biology Department, Turkey

Ahmed L. Abdel-Mawgood
Faculty of Agriculture, El-Minia University, Egypt

Katarzyna Wolska
University of Natural Science and Humanities n Siedlce, Poland

Piotr Szweda
Gdansk University of Technology, Poland

Pei-Chun Liao
Department of Biological Science & Technology, National Pingtung University of Science & Technology, ROC

Shong Huang
Department of Life Science, National Taiwan Normal University, Taiwan, R.O.C.

Andrea Akemi Hoshino and Paula Macedo Nobile
Universidade Estadual de Campinas, SP, Brazil

Juliana Pereira Bravo
Universidade Estadual Paulista, SP, Brazil

Karina Alessandra Morelli
Fundação Oswaldo Cruz, RJ, Brazil

Inês Bártolo and Nuno Taveira
Centro de Investigação Interdisciplinar Egas Moniz (CiiEM), Instituto Superior de Ciências da Saúde Egas Moniz, Monte de Caparica and Unidade dos Retrovírus e Infecções Associadas, Centro de Patogénese Molecular, Faculdade de Farmácia de Lisboa, Portugal

Otávio Noura Teixeira
Programa de Pós-Graduação em Engenharia Elétrica, Universidade Federal do Pará, Brazil
Laboratório de Computação Natural, Centro Universitário do Estado do Pará, Brazil
Movimento Evolucionário e Cooperativo para a Construçao do Artificial - MEC2A, Brazil

Renato Simões Moreira, Walter Avelino da Luz Lobato and Hitoshi Seki Yanaguibashi
Programa de Pós-Graduação em Engenharia Elétrica, Universidade Federal do Pará, Brazil
Movimento Evolucionário e Cooperativo para a Construçao do Artificial - MEC2A, Brazil

Roberto Célio Limão de Oliveira
Programa de Pós-Graduação em Engenharia Elétrica, Universidade Federal do Pará, Brazil

Denitsa Teofanova, Peter Hristov, Aneliya Yoveva and Georgi Radoslavov
Institute of Biodiversity and Ecosystem Research, Bulgarian Academy of Sciences, Bulgaria

Maria do Carmo Bittencourt-Oliveira and Viviane Piccin-Santos
Universidade de São Paulo, Brazil
Universidade Estadual Paulista, Brazil

Ruslan Adgamov and Svetlana Ermolaeva
Gamaleya Research Institute of Epidemiology and Microbiology, Moscow, Russian Federation

Elena Zaytseva
Research Institute of Epidemiology and Microbiology, Vladivostok, Russian Federation
Gamaleya Research Institute of Epidemiology and Microbiology, Moscow, Russian Federation

Jean-Michel Thiberge and Sylvain Brisse
Institut Pasteur, Genotyping of Pathogens and Public Health, Paris, France

Marina Muzzio and Claudio M. Bravi
Instituto Multidisciplinario de Biología Celular (IMBICE), La Plata, Argentina
Facultad de Ciencias Naturales y Museo, Universidad Nacional de La Plata, La Plata, Argentina

Virginia Ramallo
Universidade Federal do Rio Grande do Sul, Porto Alegre, Brazil
Instituto Multidisciplinario de Biología Celular (IMBICE), La Plata, Argentina

Emma L. Alfaro and José E. Dipierri
Instituto de Biología de La Altura (INBIAL), Universidad Nacional de Jujuy, Jujuy, Argentina

Graciela Bailliet and Laura S. Jurado Medina
Instituto Multidisciplinario de Biología Celular (IMBICE), La Plata, Argentina

Orville Heslop
University of the West Indies, Jamaica

Nicole J. Lake, Kiymet Bozaoglu, Abdul W. Khan and Jeremy B. M. Jowett
Baker IDI Heart and Diabetes Institute, Melbourne, Australia

www.ingramcontent.com/pod-product-compliance
Lightning Source LLC
Chambersburg PA
CBHW070715190326
41458CB00004B/983